AF828906

Natural Disaster Risk Management

Ulrich Ranke

Natural Disaster Risk Management

Geosciences and Social Responsibility

Springer

Ulrich Ranke
Burgdorf
Germany

ISBN 978-3-319-20674-5 ISBN 978-3-319-20675-2 (eBook)
DOI 10.1007/978-3-319-20675-2

Library of Congress Control Number: 2015945566

Springer Cham Heidelberg New York Dordrecht London
© Springer International Publishing Switzerland 2016
This work is subject to copyright. All rights are reserved by the Publisher, whether the whole or part of the material is concerned, specifically the rights of translation, reprinting, reuse of illustrations, recitation, broadcasting, reproduction on microfilms or in any other physical way, and transmission or information storage and retrieval, electronic adaptation, computer software, or by similar or dissimilar methodology now known or hereafter developed.
The use of general descriptive names, registered names, trademarks, service marks, etc. in this publication does not imply, even in the absence of a specific statement, that such names are exempt from the relevant protective laws and regulations and therefore free for general use.
The publisher, the authors and the editors are safe to assume that the advice and information in this book are believed to be true and accurate at the date of publication. Neither the publisher nor the authors or the editors give a warranty, express or implied, with respect to the material contained herein or for any errors or omissions that may have been made.

volcano picture credit: Jochen Scheffl/fotolia.com

Printed on acid-free paper

Springer International Publishing AG Switzerland is part of Springer Science+Business Media
(www.springer.com)

Contents

List of Figures

List of Tables

Chapter 1
Introduction

Natural disasters occur every day and can occur at almost at any place on the globe. Nevertheless the geological situation and the geomorphological implication make the occurrence at specific places more probable than at others. Even very rare disasters occur, such as the meteorite impact in Siberia in 2012, an event that according to statistics only happens less than once a year. But if it had been a meteorite of just several centimeters in diameter, it would have created an impact crater able to eradicate a mid-size town from the landscape.

On August 22nd, 2003 a renowned group of international tsunami experts gathered in the small town of Angera on the west coast of Java island to commemorate the 125th anniversary of the Krakatau eruption which caused that era's largest tsunami. Krakatau was the second biggest volcano eruption in history and its eruption caused a pyroclastic surge and triggered a tsunami that killed 36,000 people. Today there are only three minor islands to be seen as the whole caldera is drowned. But since then one larger island, the Anak Krakatau, has constantly been building up by about 7 m/year showing that the history of Krakatau is far beyond being finished. In that meeting the 150 tsunami experts discussed the present status of international tsunami knowledge, on how tsunamis are generated, what casualties have thus far been experienced worldwide, what a possible early warning system could look like, and what kind of disaster prevention should be established. Among the experts was the director of the world famous Pacific Tsunami Warning Center (PTWC), Laura Kong, who started her presentation with the words: "Tsunami is not a question if, it is a question when." One year later the biggest tsunami ever recorded in modern history hit the northern tip of the island of Sumatra killing 170,000 people in the Aceh province alone and causing a death toll of a total of more than 230,000 people all around the Indian Ocean. Another question is why it took another 7 years until Indonesia got a functioning early warning system at its disposal that already proved its reliability, successfully forecasting a tsunami in 2009 offshore Yogyakarta. The answer is, although so many experts were gathered in Angera, early warning was given no priority in the political agenda of the country. The expert meeting was just seen as a gathering of technical people. It took another earthquake cum tsunami in the year 2006—offshore

© Springer International Publishing Switzerland 2016
U. Ranke, *Natural Disaster Risk Management*,
DOI 10.1007/978-3-319-20675-2_1

Yogyakarta—until the Indonesian President Susilo Bambang Yudiyono declared the matter top of the political agenda. The political will, together with broad international support, especially from the German government, finally succeeded in realizing this ambitious target. Today the Indonesian islands are much better protected from the impacts of a tsunami; although a tsunami will neither today nor in future be hindered from occurring.

In Italy six geologists and the representative of the local emergency management bureau were recently sentenced to six years in prison on involuntary manslaughter, for not having rendered the right information regarding the earthquake of April 6th, 2009, killing 309 people and leaving 80,000 homeless in the small town of L'Aquila in central Italy. Although the city of L'Aquila is centered at the highest earthquake risk area of Italy and has experienced many earthquakes in its history, the judges justified their verdict that the local emergency management commission had gathered a week before the quake and came up with the finding that there would be no higher risk, although the area experienced increased seismic activity the days before. The commission's findings were found to be "unprecise, useless, incomplete and contradictory." Based on a disaster-preparedness report (EU-Microdis 2007) it turned out that the Italian government showed only little interest in evaluating the post-disaster situation. For several months following the disaster, the 67,500 homeless quake survivors lived in one of 170 tent camps or in public buildings. According to the post-disaster damage assessment the affected area of the city can be sectioned into three parts. In one sector, where buildings were better maintained relative to the rest of the central city, no one died. In another section of the city a few people died, mostly due to collapse of "external architectural details" or the falling of roof tiles from buildings. The greatest concentration of deaths came in the section of the city where reinforced concrete buildings collapsed partly or completely. The well-known risk expert David Alexander of the University of Firenze, said in an interview (Forbes Magazine: 22.10.2012) that instead of sentencing scientists ("you cannot predict an earthquake accurately") it would have been better had the Italian officials responded to the previous disasters and based their judgment on scientific evidence; at least with "common sense for not to replicate errors made during the earthquakes that occurred decades ago and to address compliance to national building codes." He pointed out that the "Governments have neglected to take steps to improve infrastructure" and argued that "It is not the responsibility of the scientists to control building stability or even to reinforce the structures." Furthermore he fears, that the "threat of prosecution will lead to silence scientific voices." Different from the L'Aquila sentence, the many earthquakes along the San Andreas Fault (such as the Northridge earthquake of 1994 that claimed 57 lives, or that of 1989 in San Francisco with 67 dead) never in history have the disaster-monitoring scientists and the local emergency managers been accused. More than 5000 natural scientists in Italy addressed the president of Italy to reconsider the verdict but with no reaction thus far. The scientists claim that although building codes were established for Italy, most of the houses that collapsed did not comply with the building code, not even the local hospital. The rigidity of the L'Aquila verdict is therefore unique. As a reaction many scientists are considering withdrawing from risk management and will refrain from commenting on risk situations any more. Especially in Italy, a country that is exposed to more natural hazards than most of the other European countries and that has with Naples a city with more than 3 million people living at the foot of the Vesuvius volcano, many scientists are deeply worried about ending up in court.

The catastrophe of Hurricane Katrina flooding New Orleans is another example of natural disasters affecting societies. But other than the tsunami where the impact overthrew the people around the Indian Ocean without prior warning or at L'Aquila where the city is known to be earthquake-prone, the Hurricane Katrina flood can be seen as a more or less man-made disaster. Hurricane Katrina is seen today as the worst natural disaster to ever happen in the United States. The hurricane windspeed exceeded 280 km/h and was accompanied with torrential rains and resulted in flooding the city of New Orleans up to a level of 7.6 m. Eighty percent of the city area was flooded, 1800 people lost their lives, and the economic damages were estimated to have reached US$125 billion, making this catastrophe the biggest economic disaster prior to Fukushima. The victims were not equally distributed among the social classes of the local population. Altogether the impact was primarily on the socially underprivileged, black, poor, and older Afro-Americans (Beaudoin 2007). The evacuation of the city caused the greatest displacement of American citizens and further amplified the already existing social inadequacy. Large riots and tremendous turmoil followed, showing many Americans the previously unbelievable social disparity within American society, pointed out by many people as an example of the failure of state (Pirsching 2006).

On May 12th, 2008 a highly destructive earthquake struck Central China. The Wenchuan earthquake had a magnitude of M 8.0 and its epicenter was located on the western flank of the Chengdu basin in Sichuan province (Huang and Li 2009). The earthquake originated at the Longmenshan Fault zone that follows the southern flanks of the Tibetan Massif. The rupture zone was more than 300 km long and was a result of the convergence of the Tibetan Plateau with the northward propagating Eurasian plate. The earthquake was the most destructive event in China's recent history and led to 70,000 fatalities, almost 20,000 persons missing, and 350,000 people injured. More than 20 million houses were badly damaged, and more than five million people were made homeless. Confusing numbers were given on the death toll of the schoolchildren; some releases quoted 10,000 dead children whereas others gave a number up to 20,000, making children and teachers 25 % of all victims. The high percentage of this particularly high death toll alarmed the Chinese government and launched an investigation of the situation. The quake struck in the early afternoon when most of the younger children were taking a rest. Almost 7000 public school buildings were completely destroyed, even in areas more than a 100 km away from the epicenter, whereas most of the other official buildings were able to withstand the impact. Also astonishing, most of the schools for elite pupils and communist cadre claimed a death toll of only a few children. The government investigation revealed that the normally two- to three-storied public schools were really of poor quality, mainly made up of unreinforced bricks, with thin iron rods and low-grade cement "that crumbled like charcoal." The government explained the poor construction quality with the fact that in years before, the rapidly increasing population forced the authorities to "rush to build schools." Local civil engineers blamed the authorities for not complying with the 1976 national building code outlawing unreinforced brick buildings for schools. Furthermore they pointed out that many of the school buildings were located at places that were not suitable for housing: places close to rivers and in areas of unstable soils and quicksands, prone to liquefaction. The private damaged houses were in most places located in steep-sided valleys, with soils predominantly made up of a mixture of large boulders, sand, and gravel, not at all suitable for construction sites. Most of these houses were also made of unreinforced bricks. According to a study carried out by Tang et al. (2011) the earthquake induced more than 56,000 landslides in steep mountainous areas that additionally caused more than 2000 fatalities. The earthquake-induced landslides also produced extensive damage to housing settlements, irrigation channels, and rivers. Highways and bridges were blocked or destroyed, and the city of Wenchuan and many other towns became isolated. Many aftershocks of magnitude 6.0 (Richter scale) of the main earthquake triggered a series of mass movements that clogged a nearby river, forming a series of 35 different size lakes. The water levels in the lakes rose steadily, threatening more than 700,000 people downstream. To prevent a flash flood the Chinese military services blasted the debris to allow a controlled water run-off.

The emergency response to the earthquake by local and national authorities was immediate and very decisive. The central government within hours established an Earthquake Rescue and Relief Headquarters that provided comprehensive technical, medical, and food assistance for restoring livelihoods and the physical environment for the 20 million residents who were stricken. The government established an "impressive number of temporary camp[s]" (EERI 2008) equipped with hospitals, pharmacies, and schools and provided food, water, sanitation, electricity, communication, and safety. The camps offered training programs and job searches for residents, and residents even took or created jobs within the camp itself. Also as in many emergency situations in the developing world many survivors preferred to stay in small tent camps near their former homes. These people reasoned that they have to take care of crops, cattle, and poultry. Overnight the government mobilized 130,000 soldiers for rescue and provision of livelihood of the survivors and to maintain law and order. In addition, many private individuals and social groups from all over the country came to Sichuan province to help. The Wenchuan earthquake was one of the very few cases where the Chinese government accepted foreign aid and aid personnel, a political decision seen by many observers to be related to the Olympic Games to start some months later in Beijing. As in most emergency cases worldwide the biggest workload had been carried out by the survivors themselves. With the help of local nongovernmental organizations, community groups, and volunteers, search and rescue activities were initiated before formal assistance arrived. Observers realized a very high level of nongovernmental involvement which is not usually normal within the Chinese political and cultural context (EERI ibid). The media played an important role in all the rescue operations. As relief and recovery efforts became more effective, the role of many of the civil society groups became smaller.

Just a month later the Chinese government passed the "Wenchuan Earthquake Disaster Recovery and Reconstruction Act" that provided the legal basis for disaster mitigation activities: damage assessment, temporary housing schemes, reconstruction planning, financing, recovery, and rehabilitation. A special team for reconstruction planning was established and a general reconstruction plan was developed for the three affected provinces comprising detailed plans on "rural development, urban–rural housing, infrastructure, public service facilities, productivity distribution and industrial restructuring, market service system, disaster prevention and mitigation, ecological rehabilitation, and land utilization" (EERI ibid). The central element of China's post-disaster management concept was to pair affected regions with unaffected regions across China. For example, the city of Chengdu was sponsored by the city of Shanghai. The contributions included provision of human resources, "in kind" support from planning institutes and other agencies within the sponsor region, provision of temporary housing units, and donations and financial support. As of August 25th, 2008 the government announced that almost 1.5 million disaster-affected people had been relocated; about 180,000 had been organized to work outside the disaster zone, and about 680,000 people had found jobs in their hometowns (Xinhua News Agency 2008). About 90 % of the 140,000 damaged business outlets had been reopened. Almost 663,000 temporary houses had been constructed and another 2500 were being installed. Nearly all of the 50,000 km of roads damaged by the earthquake have been restored. According to Watts (2008), the reconstruction is estimated to have cost about US$150 billion, an amount that was equivalent to one-fifth of the entire tax revenue of China for a single year. This included providing new homes for 4 million refugees, replacing schools, and creating jobs for 1 million people. Rebuilding the infrastructure of the mountain areas will be a challenge in respect to the tectonically and climatic unstable conditions. One of the most pressing challenges is to ensure higher seismic standards and construction quality in the rebuilding. The reconstruction plan calls for higher earthquake resistance levels of infrastructure construction in the quake-hit regions, especially for schools and hospitals. In rural areas, reconstruction planners recommend that technicians advise residents on safe rebuilding, but many villagers are moving ahead and reconstructing with a variety of traditional and recycled materials, particularly reclaimed bricks (EERI, ibid).

2010 was a year that struck Pakistan with a heavy flood. Although the country is subject to yearly flooding during monsoon and had already been devastated along the river Indus in the years 2007, 2009, and was later in 2011 subject to another flood, the flood of 2010 was by far the worst of the last decades. The 2010 monsoon rainfall almost doubled the amount of water compared to a "normal" monsoon season; it was the highest since 1994 and ranked second highest of the last 50 years. The United Nations has rated the floods in Pakistan as the greatest humanitarian crisis in recent history. In fact more people have been affected in Pakistan from the disaster than the 2004 Southeast Asian tsunami and the recent earthquakes in Kashmir and Haiti combined. The 2010 Pakistan flood began in late July resulting from heavy monsoon rains in mountain ranges of the northern provinces that a couple of days later affected the entire Indus River basin. The rainfall was supposed by NASA to be most probably a result of the global La Niña effect. According to meteorologists the so-called jet stream—a seasonal wind system normally flowing at an altitude between 7 and 12 km from west to east—was seen as the central cause of this weather phenomenon. At that time the jet stream was what meteorologists call "frozen-in," with the result that the weather conditions remained unchanged causing rising temperatures and extreme rainfall. Such a "frozen" jet stream was also seen as the root cause for the heat wave in Russia and the flooding in the United Kingdom in 2007. Rainfall of 200 mm/day was recorded in most parts of the country, with a maximum of almost 300 mm/day.

By mid-August in total 1800 people had been killed by the flood and about 20 million people displaced from their homes in 36 districts. A reported 3.5 million children were at risk from waterborne diseases, and 6 million people were in need of food. Twenty percent of Pakistan's total land area was under water, and 20 million people were affected by destruction of their property, their arable land, and infrastructure. Two million houses were destroyed beyond repair and 30,000 km^2 of land were washed away. In the aftermath of the catastrophe malaria, dengue fever, and cholera spread in the most affected regions. Ten million people were reported to have no access to clean drinking water. The Pakistani economy faced a considerable loss of more than US$40 billion due to heavy damage to infrastructure and crops. The wheat crop damage alone was estimated to be over US$500 million. The International Monetary Fund (IMF) estimated the gross domestic product to drop to 10 % and the inflation rate to increase to 25 %, mostly as the main export product, cotton, experienced a loss in production of 25 %. The flood destroyed almost the entire harvest of sugarcane, rice, and wheat, worth US$3 billion. According to Pakistan cotton industry information the flood destroyed two million bales of cotton, by far the biggest export sector of the country, with the result that cotton prices worldwide promptly started to rise. As a reaction to the catastrophe, there was a great deal of universal willingness to help. In total the European Union and the United States contributed about half a billion US dollars for reconstruction and recovery. But it turned out that the contribution by the country itself was only small. Former US Foreign Secretary Hillary Clinton therefore demanded that the Pakistani government significantly increase the country's aid budgets by increasing the tax base.

As a reaction of the 2004 tsunami-induced economic crisis, the former government of the Republic of Maldives made an announcement that the government intended to create a "Sovereign Wealth Fund" with money earned from tourism to be used to purchase land elsewhere for the Maldives. The those days acting-President Rasheed's statement on the "Future of the Flood Threatened Maldive Islands" was the first political leader's address to the world that there were places on Earth where the climate-induced sea-level rise had shown the first serious signs. The water level is rising steadily and beginning to threaten existence on the islands, the same as many other small states' islands such as Tuvalu, Tonga, Fiji, Samoa,

Vanuatu, Funafuti, and others. There is furthermore the well-known example of the island of South Talpati offshore Bangladesh, as the first island that vanished from the earth map; the same will happen with the Tegua Atoll of Vanuatu. The same thing happened in Papua New Guinea where the government evacuated the 980 inhabitants of Carteret Island in 2005 (Jacobeit and Methmann 2007).

The more than 1000 islands forming the State of Maldives lie less than 1 m above sea level, making the 380,000 inhabitants—and by the way the capital Male with its 36,000 inhabitants is the most densely populated area of the world—most vulnerable from flooding. But not only is the land size diminishing every day, also the ingress of sea water is destroying the island's fresh water reservoirs. So the prime minister addressed his people that in the long term, "The Maldives will not survive as an island state." He urged his country that they have to face one day leaving the islands and seeking new homes somewhere. He therefore started to negotiate to buy land in the Indian Union, in Sri Lanka, as well as in Australia. The government of the Tuvalu Islands did the same. It negotiated with Australia and New Zealand to get shelter in these countries. But the Australian government will only accept up to 90 Tuvaluans every year, as it claims that there is no real risk from sea level rise to Tuvalu and the Tuvaluans are coming as "economic refugees" and not for climate reasons.

For Maldivians, India and Sri Lanka pose the first choice for evacuation, as they share the same language, culture, and ethnic heritage, especially with the Indian Federal States of Tamil Nadu and Kerala. The money for land acquisition would be taken from the income from the more than 600,000 tourists every year that made up to about 30 % of the gross domestic product of 2008. But the Maldivian population did not accept these "visions" and were afraid of being forced to leave the country. So after heavy riots and political turmoil and following the opposition's charge that the Nasheed administration was no longer able to govern the country in the way Islam was demanding, Nasheed resigned in February 2012 and was later arrested.

The Maldive example shows impressively what can happen when a reasonably founded and seriously thought over and meaningful oriented strategy fails, most probably due to not incorporating the population at risk in the decision-making process. Instead they were just presented a government's decision. Those who were deeply affected by the political decision were not given the chance to express their views and especially in that sector overwhelmingly dealing with their everyday life. A broad and extensive discussion should have been institutionalized, giving everyone the feeling his or her fears, experiences, and "vision on livelihood" were properly taken up by the authorities. Next to the individual, representatives of all social groups (religious leaders, the political opposition, representatives from industry and science, etc.) should have been gathered and given ample time to express their views at a national roundtable. It is clear that such a discussion would have taken much effort and much time, but the time problem was then, and is still today, not threatening the island's existence overnight. The government should have given such a socially comprehensive discussion enough space although respecting the fact that the rising sea level allows no way out other than a change in policy. Thus the fears of the population were neither focused on the technical matters of the sea-level rise nor on the financial aspects, but found its expression in the field of emotions and feelings; in the last consequence it was formulated on behalf of Islamic beliefs and traditions.

But the "Maldive Vision" of resettling in other places has simultaneously an international dimension. A question that immediately arose regarded the kind of political status such persons would be attributed after resettling. Are they still Maldivians, who are now living in India or will they be given Indian nationality or will they be treated just as emigrants or ethnic minorities? The International Law on Refugees as it is laid down in the Charter of the United Nations distinguishes only between "Refugees" and "Internally Displaced Persons" (IDP), who are forced to leave their country or parts due to military, racial, or ethnic conflicts. Sometimes such refugees were forced to leave their country for decades, such as the many Afghanis settling in Pakistan in the 1980s, during which time the people have no internationally accepted political representation. In such cases the United Nations High Commissioner for Refugees (UNHCR) takes over representing these peoples. A similar representation exists for IPDs. The basic legal definition on what forms a nation requires, according to international law, a nation's territory and a nation's population. But when territory is flooded due to the rise in sea level and no territory exists anymore, the basic definition of what constitutes a nation is no longer valid. Although the United Nations Environmental Program (El-Hinnawi 1985) introduced the term "climate refugee" into the public debate, there is no internationally accepted legal authorization for the UNHCR to take over care of the climate refugees. The question for UNHCR is whether the refugees left their home deliberately or whether they were forced to do so. In the case of the Maldives the idea was to seek shelter in another country. The decision was supposed to be definitively taken voluntarily and not a subject of "forced migration." The United Nations since then has put the notion on the agenda of the United Nation Security Council several times, but the five permanent members nor the industrialized nations, as well as many advanced countries were not inclined to take up the matter, although it was stated that already in 1990 the amount of climate refugees was estimated to be about 25 million more than those refugees of wars and conflicts (Myers 2001). The IPCC stated in 1990 that next to climate-induced sea-level rise desertification, soil erosion, and heat waves will also make climate refugees a substantial problem of the future. In the Rio UNCED Agenda in Chap. 12 the notion of climate refugees was already made (see also Stern 2007). In order to object to the reluctance of the industrialized states the governments of the affected island states formed the Alliance of the Small Island States (AOSIS) to raise their voices in order to fight for a worldwide reduction of greenhouse gases. They argue that they release almost no CO_2 into the atmosphere but are the ones who suffer first.

On January 12th, 2010 at an early hour an earthquake of magnitude 7.0 struck the Caribbean nation of Haiti. The quake's epicenter hit just 15 km west of Port-au-Prince and its two million inhabitants. According to the USGS the plate movement produced the biggest earthquake since 1751. Haiti and its eastern neighbor the Dominican Republic lie on the northern edge of the Caribbean tectonic plate, where the Caribbean plate moves along the North America plate by a left-lateral strike-slip motion and compression at a speed of about 20 mm/year. There are two major faults defining this plate boundary: the septentrional fault system, which runs through northern Haiti, and in the south the Enriquillo–Plantain Garden fault system. And it is this fault system which moves at a speed of 7 mm/year thus making it responsible for nearly half of the overall movement between the Caribbean and North American plates. Here along the Enriquillo–Plaintain Garden fault system the earthquake seemed to have been triggered. According to assessments made by the Global Seismic Monitor at the German Research Centre for Geosciences (GFZ) the focal point of the earthquake was at a depth of 17 km. Thus far the Enriquillo–Plantain Garden fault system had not produced any major earthquake but is nevertheless supposed to be the source of the large earthquakes in 1860, 1770, and 1751. For some years there has been a fault monitoring carried out by the University of Texas and in 2008 it showed seismic models with slip rates of around 8 mm/year. And due to the fact that the last known major earthquake near Haiti was in 1751, Texas A&M warned that this could add up to yield of about 2 m of accumulated strain deficit, leading to an earthquake of magnitude 7.2 if all were released in a single event. The January 12th main shock did not produce observable surface displacement but apparently caused a significant uplift of the Léogâne delta. Thirty-three aftershocks have also been recorded, ranging from magnitude 5.9 to 4.2. As the earthquake took place on land, there were serious concerns that it would generate a larger tsunami. Therefore a local tsunami warning was issued for Haitian coasts within 100 km of the epicenter. The PTWC (Hawaii) recorded a 12-m high wave at Santo Domingo in the Dominican Republic and decided that there was no threat to coastal areas away from the epicenter and resulted in a subsequent cancellation of the tsunami warning. Another reason for the devastating impact of the quake is that Haiti does not have any real construction standards and that in many of the quarters there were even more squatters than livable homes. According to an estimate by the mayor of Port-au-Prince, about 60 % of buildings were shoddily built and unsafe in normal circumstances.

But the earthquake had not only that extreme salience normal for earthquakes of this magnitude but also it struck a nation that is characterized by a highly vulnerable society. Haiti has a population of nearly 10 million people of whom 80 % live under the poverty line. Most Haitians live on less than US$2 a day making the country the poorest country in the western hemisphere. More than two thirds of the labor force do not have formal jobs. Haiti's deep and widespread poverty results from a long history of state failure to establish an idea of a functioning nation. As are other fragile states, Haiti is characterized by widespread poverty and inequality, economic decline and unemployment, institutional weakness, corruption, violence, and conflicts (Verner and Heinemann 2006). The country's "conflict–poverty trap" results from two main factors: a socioeconomic factor characterized by a rapidly increasing population, by no jobs and no income which forces the people to leave rural areas and to migrate into the bigger cities, no education, and no social security; and second by failure of the states that have not been able to provide the least necessary public goods (water, electricity, transportation, etc.) but have also failed to establish law and order, security, and powerful state institutions. Still today the state provision of infrastructure and basic services is limited to the capital, Port-au-Prince, and some other urban centers.

> Haiti's income distribution is among the world's most inequitable with a Gini coefficient of 0.66. Nearly half of Haiti's households are trapped in absolute poverty and live on less than a dollar a day. Social indicators such as literacy, life expectancy, infant mortality, and child malnutrition show that poverty is extensive. About 40 % of people cannot read and write; some 20 % of children suffer from malnutrition; nearly half the population has no access to healthcare; and more than four-fifths have no clean drinking water (Verner and Heinemann ibid)

Two years after the devastating earthquake half a million people of out of the former 1.5 million are still living in tents according to information from the . But many of those who have been given shelter in the refugee camps have not found secure living conditions. Many displaced persons have been victims of forced eviction from the camps; some of them have now been displaced two or three times. The problem is that many of the refugees have never been formally registered and have neither passport nor identification cards and are thus not eligible for national and international support. UNHCR has therefore focused its efforts not only in providing emergency relief items but also in providing quake victims with civil documents. Many of the IDP were handed birth certificates to give them a legal existence. The Haitian Government is internationally under severe pressure to finally update their civil registry system and to make it accessible to people all over Haiti.

After a heavy rainfall in the summer of 2013 from thunderstorm Cleopatra large-scale flooding occurred in many parts of Italy. Streets were flooded, bridges collapsed, and trees overturned. Sixteen people died in Sardinia; in the southern Italian city of Catanzaro the losses were calculated up to €10 million. In Venice it was already the fifth flood in that year. The water level rose by about 45 cm. There was a great uproar in the population claiming that the national weather service had given the warning about the thunderstorm far too late. The government immediately allocated €20 million and reiterated that it has, "Done all what it can do." The head of the Civil Protection Authority stated, "that there was a warning 12 h before the event broadcasted to all districts and communities." He added that "[T] he population shall ask their respective communities, why the authorities haven't reacted upon this warning," and added "[E]arly warning is one thing, but early action cannot be compelled by central government."

References

Beaudoin, C.E. (2007): Media effects on public safety following a Natural Disaster: Testing Lagged dependent variable models. - Journalism & Mass Communication Quarterly Vol.84, No. 4, p. 695-714, Sage Journals (online: http://intl-jmq.sagepub.com)

EERI (2008): Learning from Earthquakes - The Wenchuan, Sichuan Province, China, Earthquake of May 12, 2008.- Special Earthquake Report, Earthquake Engineering Research Institute, Oakland CA

EU-Microdis (2007): European Commission 6th Framework Programme, Global Change and Ecosystems, GOCE-CT-2007-036877
El-Hinnawi , E. (1985): Environmental refugees. – UNEP, Nairobi
Huang, R., & Li, W. (2009): Analysis of the geo-hazards triggered by the 12 May 2008 Wenchuan Earthquake, China. Bulletin of Engineering Geology and the Environment, Vol.68, p. 363-371, Springer – online
Jacobeit, C. & Methmann, C. (2007): Klimaflüchtlinge – Die verleugnete Katastrophe. – Greenpeace, Universität Hamburg
Myers, N. (2001): Environmental refugees - A global phenomenon of the 21st Century. – Philosophical Transactions of the Royal Society: Biological Sciences, Vol. 357, p.167-182, London
Pirsching, M. (2006): Good Bye New Orleans - Der Hurrikan Katrina und die amerikanische Gesellschaft.- Leykam, Graz
Stern, N. (2007): The economy of climate change - The Stern Review.- Cambridge University Press, Cambridge MD
Tang, C., Zhu, J., Qi, X. & Ding, J. (2011): Landslides induced by the Wenchuan earthquake and the subsequent strong rainfall event: A case study in the Beichuan area of China.- Engineering Geology, Elsevier B.V. (online: doi:10.1016/j.enggeo.2011.03.013)
Verner, D. & Heinemann, A. (2006): Social Resilience and State Fragility in Haiti - Breaking the Conflict-Poverty Trap. – The World Bank, "en breve", No. 94 (online: http://www.worldbank.org/lac)
Watts, J. (2008). Sichuan quake: China's earthquake reconstruction to cost $150bn, The Guardian, August 15th 2008. (online: www.guardian.co.uk)
Xinhua News Agency (2008). "China Quake Death Toll Remains Unchanged at 69 226," . – Xinhua News Agency (online: www.chinaview)

Chapter 2
Natural Disaster and Society

2.1 Major Natural Disasters and Their Socioeconomic Impact

2.1.1 Natural Disaster Triggered Nuclear Accident

On March 11th, 2011 at 2 o'clock in the afternoon local time an earthquake of 9.0 on the Richter scale occurred in the Japan trench almost 200 km away from the coast. The earthquake that is known as Tohoku event triggered a tsunami that piled up waves up to 30 m. Automatically the four nuclear power plants along the east coast were shut down and the systems were switched over to emergency power supply. But the earthquake generated a massive tsunami that ran ashore and with its up to 14-m high waves destroyed most of the technical and emergency infrastructure of the Fukushima-Daiichi nuclear power plant. Due to this impact the emergency cooling system failed and the heat inside the reactor blocks rose continuously, triggering a series of hydrogen explosions destroying four of the five power-generating blocks, although the reactor containments remained stable. Nevertheless it came to a core melting in the aftermath of the destruction. The accident set free massive radioactive fallout and released extremely contaminated fluids into the Pacific Ocean. Only the prevailing northward and seaward winds prevented a nationwide catastrophe. The reactor blocks were destroyed although nuclear engineers worldwide claim that it is possible to construct nuclear power plants that resist every earthquake magnitude. The engineers further claim that the Fukushima disaster was not originated by the earthquake but by the tsunami. Macfarlane (2012), newly appointed head of the US Nuclear Regulatory Commission, stated the difference in understanding between geoscientists and engineers is that thus far it is not an established fact that "Nuclear engineers can and do integrate knowledge of Earth processes adequately." Undoubtedly one of the reasons for the tsunami to hit the area that strongly, was the fact that for constructing the plant, the steep cliff at Daiichi was dug out to place the reactor blocks just at sea level; thus requiring less energy for the cooling system. The natural topography about a 100 m behind the power plant area shows an increase to an

© Springer International Publishing Switzerland 2016
U. Ranke, *Natural Disaster Risk Management*,
DOI 10.1007/978-3-319-20675-2_2

average level of about 30 m whereas the tsunami waves were about 14 m high. A TEPCO manager was cited to have confessed that choosing this particular spot for the power plant was "not a good decision."

The authorities promptly evacuated an area of 20 km around the power plant, a radius that was later extended to more than 30 km. More than 90 % of the 400,000 people originally living in the area were successfully evacuated within a couple of days and sheltered in provisional camps. The radioactive fallout of locally more than 3000 Bq/m^2, Cesium 134, and Cesium 137 was measured in the topsoil covering a small corridor of 30 km length from the plant site towards the northwest in the direction of the city of Fukushima. In all, 16,000 people lost their lives; 90 % of them drowned in the cold water of the Japan Sea. Still 4500 are missing. It is anticipated that the area will be off-limits for at least the next 20–30 years to let the radiation level decay naturally (it takes 10 years to reduce the radiation by 100 mSv).

Following this, the Japanese government resigned and the new government declared a moratorium in nuclear energy production (that was already watered down in late 2012), a drastic turnaround, as all governments before Fukushima were inclined to double energy production from nuclear energy. Together with the Japanese government the Italian Government declared that it would refrain from starting nuclear energy production, and the United States, Russia, France, India, China, and Brasil reiterated they would not deviate from their original nuclear-based energy path. By law Japan limited the runtime of its 54 nuclear power plants to maximum 40 years and shut down almost all the plants in the months to come. Many critics in the country, mostly from the production sector, raised massive fears of an energy deficiency to come with enormous consequences for economic growth. In fact that did not materialize as it was possible to balance the energy demand by bringing additional coal- and gas-fired plants on the network and by importing much energy from abroad. In 2012 Japan imported 25 % more energy than before. But the direct socioeconomic consequences were even much more dramatic. One of the many outcomes of the accident in the Fukushima district was due to energy failure and, due to the fact that the area had to be evacuated completely, was no longer able to produce. The result was that, for instance, the car manufacturer Toyota was forced to reduce production by 30 %, which in contrast, for a couple of months made the German car maker Volkswagen the biggest car manufacturer in the world.

Furthermore within 48 h in Germany the government declared a radical turnaround in its energy policy, banning nuclear power generation in the country and shutting down 9 of its 27 nuclear power plants immediately. Although it had, months before, decided to extend the runtimes for more than 15 years, toppling a decision of the former German government that decided—in accordance with national and international energy producers—gradually to step out of nuclear energy. The European Commission estimates that power generation costs in Europe will double in the next years. The German government is now prioritizing the use of renewable energy especially wind and solar. But such a supply scenario requires massive investments to improve the power grid capability to transport the power from the North Sea to the consumer in the south. According to a press release by the German Ministry of Environment (BMU) of February 2013 the costs for a higher share of renewable energy in national energy production might amount up to €1 billion.

2.1.2 World Food Price Increase Due to Drought

In 2012 the United States of America was stricken by the largest drought in 25 years. There was no rain for weeks, resulting in a drying up of the maize, wheat, corn, and soybean crops and in the death of thousands of cattle due to a dramatic shortage of fodder. In 32 of the Midwestern states, especially the region called the corn belt (consisting of up to two thirds of the territory of the United States) a state of emergency had been declared. According to information from the US Weather Agency NOAA, the 2012 summer experienced temperatures on average of 25 °C, thus lying 1.2 °C higher than the average of the whole twentieth century, with the month of July showing the highest ever recorded monthly temperature in the history of the United States. The temperatures even overstepped the thus far peak temperatures of the year 1936, when that drought was called the "Dust Bowl Year." The prices for fodder immediately peaked to US$330 per ton and thus doubled the 2011 price level. But the 2012 drought was not the only one in recent US history. In 2000 and 2004 there had been severe droughts in the Midwest, with serious impacts on food and agricultural production. The heat wave that had stricken the country was seen as a reaction of ongoing climate change and was, according to the US National Climatic Data Center, overlain by the La Niña phenomenon that radically changed the wind and precipitation regime over the United States. The 2012 drought resulted in a drop in agricultural production in the United States of about 20 % for maize, and led to an increase in prices for all agricultural commodities; for instance, the price for maize increased by 45 %.

Automatically the US drought food production situation had a deep impact on world food prices. The Global Price Index for food that is yearly assessed by the FAO, showing the high price levels that were reached since 2010 remained at a very high level. And the experts of FAO, WFP, and IFAD see food price volatility and high food prices to continue as a result of the anticipated climate projections that will lead to more extreme weather events in the future. The result will be a drop in crop yield and food supply, especially in the developing countries. Stern (2007, p. 65ff.) reported that a 3 °C increase in world temperature will bring up to 500 million people more to the brink of malnutrition. The FAO (2013) furthermore reiterated that only economic growth will lead to better nutritional status. Today's tense food security scenario is also an outcome of a number of other factors. First: as a reaction of the energy price increase in the last years many industrialized nations (especially Brazil but also Germany) converted large parts of their arable land to produce energy plants for biofuel. Second: the food base of some Asian countries that thus far was dominated by a rice and cereal food basis is steadily changing towards a high energy diet (meat). But an animal-source food base requires the provision of one unit food calorie, eight units of cereals (maize, corn, soybeans, etc.) and thousands of liters of water. Third: the expected increase in world population and even more the changing living standards especially in countries such as China, India, and Brazil will lead to an increase in food demand of 50–80 % (IPCC ibid), a demand that can only be settled if the size of irrigated land is increased by 30 % to 60 million hectares and for rainfed land by 20 % (20 million hectares), provided that the precipitation distribution does not change. The WBGU (2007) instead estimates a drop in crop yield of up to 50 % in areas where the precipitation distribution will alter because of climate change. The volatility of the world food market furthermore brought many international investors to the floor who discovered this macroeconomic sector and began speculating on the rise or fall of food prices. Even Germany's largest bank lately reiterated that it will continue to provide agricultural-investment products after the German Bank concluded that speculation is not the only reason for higher prices for farm commodities (Bloomberg News Agency of 13. Jan. 2913). Fourth: hydrometeorological effects such as saltwater intrusion in coastal waters, mass movements, desertification, soil erosion, hurricanes, and floods will further lead to a deterioration of food production worldwide. The decrease of income from only desertification and salting of soil is according to the World Resources Institute (WRI 2007) estimated to top about US$50 billion.

If food prices increase further this will definitively have serious repercussions for the countries of the world. Those countries with a strong economy will most probably be able to buffer the price increase by higher consumer prices and by diversifying the food base. But even now in upcoming economies the price burden might result in a deformation of the rental economy. Fragile states that are today at the brink of poverty will be thrown back to the 1990s (FAO ibid.).

2.1.3 Sea-Level Rise and the Survival of Small Island States

The ongoing global temperature increase already resulted in a sea-level rise, although this increase is admittedly still small when seen from a worldwide perspective. But there are regions on Earth that are becoming more and more vulnerable even by this increase. The so-called small island states especially those in the Pacific Ocean or other areas in the big river deltas where most of them lie just 2–3 m above sea level. The islands of Tuvalu, Samoa, Kiribati, Carteret Island, the Maldives, and many others came into public focus when it became obvious that many of them would disappear from global maps if climate change continued unaltered. There are already today two examples of islands that have vanished and where the people have had to seek refuge at other places: the Carteret Islands of Papua New Guinea, the small island of South Talpati in the Ganges Delta, and some tiny islands on Tuvalu. But it is not the global temperature increase and with it the melting of the arctic inland ice shield that is the dominant factor of the rising level. It is instead more an effect of a thermosteric increase in water volume due to a temperature-induced decrease in density. And it is believed (IPCC 2007) that this increase will continue for the decades to come. According to IPCC (ibid) the global sea level has risen in the Pacific Islands from 2003–2008 by about 2.5 mm, in comparison to an increase of 0.5 mm in the time span 1961–2003. The total sea-level increase of the last 100 years is estimated at 20 cm and will result in a sea level of about 90 cm in the next 100 years. Together with the rising sea level, significant adverse effects (tropical storm surges, floods, salinization of fresh water ponds) will strike that part of the Pacific region in the future. Already today the number of victims due to weather-related disasters reached 20 % of the total population (1.2 million in 1990).

Life hasn't changed over the last thousand years on these islands; the people still have a subsistence-based living. The economic basis is the sea: fish and raw materials come from there. Small but nevertheless significant developments in medical services and in the economy resulted in relatively high urban growth rates; for instance, that of the Republic of Kiribati led to a current population of 100,000 people, an increase from several thousand after the Second World War. The capital of Kiribati today amounts to more than 60,000 people with a population density of 160 people/km^2, making this place one of the most densely settled places in the whole Pacific region. The lifestyle of the people on the small island states did not change much over the last 2000 years, but the high population growth rates brought along serious environmental degradation. Pollution of the water resulted in lower fish catches and poorer water quality.

It is feared that further global warming will lead to two main changes: one is the average temperature increase will lead to a continuously rising sea level. And second, the change in climate conditions will result in more extreme weather events including tropical cyclones, heavy storms, and flooding. Hotter days and less rainfall will dry up the freshwater reservoirs on the islands. The warming of the sea water will furthermore lead to acidization of the water with serious consequences for the coral reefs, to coastal erosion, and salinization of the freshwater resources. According to IPCC (2001),

> [T]he high exposure to natural hazards (tropical cyclones, storm surge, droughts, tsunamis, and volcanic eruptions), the limited physical sizes of the islands, their relative isolation and simultaneously great distance to major markets, limited natural resources and over exploitation by human activities leading to degradation of natural systems, their thin water lenses and decreasing fresh water reservoirs, a strong import dependence and high sensitivity to external markets, the generally rapid population growth and urbanization together with specific industrial activity and the generally poorly developed infrastructure by on the other hand an extensive tourism dependency.

This makes all the small island states in the Pacific Ocean economically, socially, and physically highly vulnerable.

The problem of the rising sea level is that it is definitively generated by the industrialized countries (the largest CO_2 producers in the world are the United States, China, and India) but those countries face only a minor risk from the sea-level rise, whereas the population of the small island states are threatened with the loss of their countries forever and are responsible for only 0.03 % of the world's CO_2 emissions (IPCC ibid), a fact that is internationally already accepted but that didn't really change the attitude of the international community. The risk of vanishing from international maps has been on the international agenda for many years. There has been a series of international conferences and resolutions, but none of them really brought results. Meanwhile the small island states of the Pacific joined in an organization called Alliance of Small Island States (AOSIS) and founded a regional Secretariat of the Pacific Community (SPC) to better address their matters in international discussions. Their main pleas are that the industrialized nationals should finally start complying with the Kyoto CO_2 emission targets and that the international society should define the legal status of what is called a "climate refugee." The president of Maldives Nasheed stated that "the rising sea level is the fate of the country and over the long term the archipelago would no longer exist as nation" and therefore he announced to start talks with Sri Lanka and India to negotiate to "one day" resettle his people in one of the states. But he also sees chances to resettle in Australia and New Zealand. The favorite countries, however, will be India and Sri Lanka as both share their religious and ethnic origins as well as the social and economic system. But the latter two countries already face big social and economic problems, poverty, overpopulation, and an agriculture-dominated economy. Following his announcement there was a big uproar in the country and his government was toppled soon after. Nevertheless he initiated a huge discussion within the United Nations focusing on two different aspects: why should the small islands states—that were contributing almost nothing of the worldwide CO_2 emissions—have to be the first victims of the sea-level rise, and what kind of an internationally guaranteed legal status such emigrants might be given.

The many different definitions the people were already given in the public debate shows how complex it will be to solve this problem of climate refuges, forced migrants, economic refugees, or climate refugees. Those who are affected by the rising sea level argue that they are forced by nature, whereas those identified as probable destinations have already begun to discuss the multidimensional effects of climate change, but only as it affects their own interests. It is mainly the OECD countries that still refuse to take up the notion on the international agenda of the United Nations. But resettling will most likely not be solved in a way the asylum seekers have in mind. It is highly unlikely that one recipient nation will grant asylum to an entire country. Instead they will most probably be divided up with all the negative consequences; it is strongly feared by asylum seekers that it will lead to the more or less disappearance of their social and cultural identity. To create precedents one country in the region already has clarified its political standpoint: Australia. The country fears to be overrun by refugees from all over Asia. Consequently the government declared all refugees entering the country without an official permit will be detained in special camps and returned to their countries. Recently, an article published in the Australian magazine, Security Solutions, argued that forced migration due to climate change is a security threat for the receiving nations (Soderblom 2008). The article used Tuvalu as a case study to suggest a link between forced migration and terrorism. The article claims that if migrants such as Tuvalu's 10,000 people migrate to Australia, then millions of poor and unskilled regional neighbors will come begging for a new life. The New Zealand government on the other hand will allow 75 Tuvaluans per year to settle in the country in order to support the national labor scheme. Tuvalu is one of five Pacific island states the country selected to be likely candidates for the permanent resettlement of the entire population. Therefore it has been discussed to increase the amount of people to be allowed to enter to up to 500 Tuvaluans per year. It is also expected that some residents may be relocated to the island state of Niue, which is in free association with New Zealand, and was largely abandoned after Cyclone Heta struck its shores in 2004. Such an immigration scheme will not really solve the problem, as Tuvalu will not be emptied for over 100 years.

In 2013 Ioane Teitiota, a Kiribati citizen sought political asylum in New Zealand claiming to be a climate refugee. The island of Kiribati, 4000 km north of New Zealand, is threatened by the sea-level rise. He justified his application on the Human Rights Charter of the United Nations Refugee Convention, although the status of a climate refugee is not incorporated in the convention. The New Zealand authorities have rejected his appeal at the court of first instance, but Ioane Teitiota has appealed to the High Court. If his petition is accepted, it would be the first time that the status of "climate refugee" has been jurisdicially acknowledged, a judgment that would give hundreds of thousands of climate refugees in the Pacific Ocean and Indian Ocean small island states a rationale for asylum seeking.

2.1.4 Respiratory Hazards from Ash Clouds

The vast amount of ashes emitted during the Mt. St. Helens volcanic eruption in 1980 triggered an intense debate on how the ashes are posing a respiratory risk to the population. The ashes were then covering large parts of the densely populated midwestern United States. Similar discussions came up during the Soufriere Hills eruption (Montserrat) in 1995 as well as after the Merapi eruption in 2006 in Indonesia. Since Mt. St. Helens quite a number of investigations have been carried out on this issue (Horwell and Baxter 2006). Although the volcanic event is normally short in time, the ashes may remain in the air for several years like the ashes of Krakatoa that traveled the globe for roughly a decade (Winchester 2003). After the Mt. St. Helens eruption regional hospitals had an up to fivefold increase in visits from normal.

Volcanic ashes are made up of hard, sharp-edged silica grains less than 2 mm diameter. As they are emitted by the hot volcanic air high up into the stratosphere the grains are able to disseminate over large parts of the globe, thus also affecting regions that were not subject to the eruption. The epidemiological science mostly deals with ash particles that are even smaller and occur in the range of 10 microns and less. Although those of about 10 microns preferably affect human airways (thorax, bronchi), ash particles of less than 4 microns pose a hazard to the lungs themselves and may cause silicosis and lung cancer. Freshly erupted ash particles are furthermore often covered with a not yet weathered and not yet oxidized cover of acids, hydrocarbons, and trace metals, but also often have adsorbed sulfur and other noxious elements from the gases. These elements are then breathed in together with the sharp-edged particles leading to asthma-type diseases. Although there is little toxicological and epidemiological evidence that a short duration of one or two days of exposure to ash particles of 10 microns may cause a hazard, it is nevertheless obvious that pre-existing respiratory diseases will be amplified by exposure to these particles. Particles of sizes smaller than 4 microns are believed to definitively cause chronic diseases, especially in children.

2.1.5 Failure of State (Hurricane Katrina, United States)

On August 2005 Hurricane Katrina formed from a "normal" tropical storm (Category 1; Saffir–Simpson Scale) to one of the biggest hurricanes to ever hit United States coasts. Katrina was the twelfth tropical storm of that season having its source in the central Atlantic Ocean. The storm first crossed Florida and then entered the Gulf of Mexico. There it rapidly developed to a category 5 hurricane that headed for the Mississippi Delta. Windspeeds of 280–350 km/h were recorded and accompanied with torrential rains over the coastal region. The hurricane made landfall on August 28th, just south of the city of New Orleans, and led to the flooding of the city. New Orleans is located below sea level and at the time was almost entirely surrounded by flood-protecting levees giving the city a so-called bowl or bathtub morphology. This made New Orleans and the areas surrounding the city highly vulnerable to floods from the Mississippi River and to storm surges from the Gulf of Mexico. One dam protected the town towards the north against Lake Pontchartrain and the other from the Gulf of Mexico. Several times in the past New Orleans was hit by hurricanes; the last time was in 1965 by Hurricane Betsy. That day 13,000 people had to be evacuated; 40 lost their lives. In the years before Katrina there were quite a number of hazard scenarios all indicating the high level of vulnerability. The last one was in the year 2004 and pinpointed that about 1 million inhabitants were at risk of losing their homes, 400,000 from diseases and illness, and 60,000 were thought to lose their lives. There were already computer-aided scenarios at hand that defined how to handle a potential disaster.

The hurricane devastated the shoreline about 300 km along the Gulf Coast. The water reached the deeper parts of the city two hours before (!) the hurricane itself made landfall. The enormous windspeed and the torrential rain damaged the dam north of the lake and one along the Central Industrial Canal passing through the center of the town. The water rose quickly and immediately destroyed the levees on both sides of the canal. The water flooded the central city parts of "Orleans Bowl," "Orleans East Bowl," and "St. Bernhard," where the later famous Ninth Ward was located. But the levees in the north protecting the city against Lake Ponchartrain also failed very soon. So the water entered the city mainly from the north. At the end an area of about 250 km^2 was flooded, mostly by 4 m water. At many places the water levels even reached heights of more than 7 m. After the flood was over, it took about 40 days for the city to dry up again. Eighty percent of the city area was flooded, 1800 people lost their lives, and the economic damages were estimated to have reached US$125 billion, making this catastrophe the biggest economic disaster prior to Fukushima.

Right upon the first alerts that Katrina would make landfall at the mouth of the Mississippi Delta and would strike New Orleans, the first evacuation orders were given by the city‘s mayor, at first only for the coastal regions but shortly after also for the entire city. By the time Katrina hit, almost 1 million inhabitants had left in city in about 430,000 vehicles resulting in huge traffic congestion on the major and suburban roads. It was estimated that only about 10 % (about 80–90,000) of the inhabitants were still in the city. They were directed to several official shelters, among them the famous Sports Arena (Super Dome). There about 15,000 people were sheltered, with very poor service facilities (toilets, etc.) that led to massive protests and a multitude of hardships for the evacuees. A total of 1464 deceased victims were officially reported by the Louisiana state authorities, and 350 victims lost their lives outside New Orleans (Jonkmann et al. 2009). The victims were not equally distributed among the social classes of the local population. Although the white part of the population had jobs, cars, and mobile phones at their disposal and thus were able to follow the evacuation advice by local government, most of the victims were black, poor, and unemployed. To them the warning did not come in time. But even if informed in time, they wouldn't have had the chance to act accordingly. Furthermore as there is no registration of people living in US cities, the New Orleans city government did not know that about 50,000 more people lived in the city than expected. Altogether, this social group had to bear the biggest burden. And after having declared a national state of emergency the National Guard cleared the flooded parts of the city. Large riots and tremendous turmoil followed, showing a thus far unbelievable social disparity within American society, that, as discussed before, is pointed out by many people as an example of the failure of the state, the first one in an industrialized nation for decades (Pirsching 2006). As data were not available for all victims for post-disaster analysis, the published figures on the victims concerned (only) totalled 829. Of these only 1 % were under 1 year, but 60 % were older than 60. The age ratio of New Orleans before Katrina was about 12 % older than 60. The explanation for this given by the emergency managers was that this age group would have especially required help on evacuation that they did not get. Also it is stated that many of the elders strictly rejected the evacuation orders. The male–female ratio did not reveal any gender preference on the casualties. On the race indicator it turned out that about 50 % were black African Americans, 40 % white Caucasian, and the rest were mostly Hispanics or of Asian origin. As the ratio of African Americans in New Orleans before Hurricane Katrina was even higher, this fact is taken by Louisiana state officials that the often-raised accusation that the flood mostly killed "poor blacks" is not the fact. A statistical analysis revealed that the average percentage of victims from Katrina coincides with data from other flood events worldwide (Jonkmann ibid.).

Today, more than eight years later New Orleans is not the city it was in 2005. Immediately after Katrina the city invested US$8 billion to increase the dam heights and to set up a citywide pumping system capable of pumping the water from a (normal) hurricane out of the city bowl. These improvements turned out to be very successful as the next Hurricane Isaac that hit New Orleans in September 2012 (the same day the Republican Party wanted to nominate Mitt Romney for president) "only" resulted in smaller damages, although it has to be stated that Hurricane Isaac was not as strong as Katrina (Saffir Simpson Scale 1–2). Furthermore most of the houses have been reconstructed. Some areas that were at high risk were completely evacuated and the people were resettled in lower-risk areas. The proprietors were compensated for their losses not on the reconstruction cost basis, but for the worth of their property before the hurricane, with the consequence that those who suffered most were compensated least. In fact the population of New Orleans became in the years after "more white"; even the proportion of Hispanics was reduced by about 10 %. Today large areas of the lowlands near Lake Pontchartrain that were formerly settled by many "underprivileged" are no longer living quarters and the houses are becoming ruins. Many of those Katrina evacuated did not came back in the following years as most of them did not have any property to live on, and the number of inhabitants was reduced to about 200,000 in the years after Katrina; meanwhile about 350,000 are living in New Orleans again (Pirsching ibid.).

2.2 Natural Disaster and the Society of Risk

2.2.1 Population Dynamics and Risk

Since 2011 the world's population has exceeded 7 billion people, who are estimated to comprise 6 % of the total population that has ever lived on our planet (110 billion) since about the Stone Age according to information of the US Population Reference Bureau (2011). China is the country that today hosts the largest population (1.38 billion) followed by India (1.25 billion) and the United States of America with 320 million. The population dynamic of the next 40–50 years will continue, but the increases will definitively slow down. The population dynamic will see India in 2050 with an additional 1.4 billion people as the most populated country on Earth, and China will experience no net increase. Another phenomenon in this regard that concerns population development experts is the rapid increase of elderly persons worldwide, reaching ages that are (normally) not attributed to an "effective" production capacity. Today there are already about 900 million people over the age of 60 worldwide and it is expected that by the middle of this century that number will rise to 2.4 billion, most of them in the high-income countries, whereas at the same time in some developing countries the number of people under the age of 25 will increase from 40 to 60 %. But the mere increase in population is no longer seen as the Earth's most striking issue. What population dynamics experts worry is that these people will then make use of much more of the natural resources than today, an effect that can already be seen in some Asian countries where the people changed from a cereal-based food to a more meat-based food. And for each volume of meat about eight volumes of grains must be invested (ecological footprint).

A third factor of concern in population development is the fact that the general increase in population is superimposed by an extreme trend for migration into the larger megaconurbations, like Calcutta, Tokyo, or Lagos. Although only five such megacities with more than 10 million inhabitants existed in 1975, the number will increase to 26 in the year 2015, most of them in Asia and Latin America. The population increase of the megacities will be about 60 million yearly. Today one in two people lives in a city and in only about 35 years, two out of three will. Although 80 % of US Americans today live in cities, it is anticipated that such a population development will also hold true for the densely populated nations in Africa, Asia, and Latin America. So that in the year 2050 more than 50 % (4 billion) of the entire world will be living in megacities. In Asia, Africa, and Latin America we will experience a doubling of the city population in 30 years to about 2.6 billion. Figure 2.1 shows that the majority of the megacities of the world will be mainly located in Asia.

Another striking feature for the population at risk from natural disasters is the changing climate. Scientific evidence presented by the Intergovernmental Panel on Climate Change (IPCC 2001, 2007; UNFCCC 2009) already overwhelmingly indicated that "climate change is, without doubt, occurring and the Earth is warming" (IPCC 2007). Furthermore the IPCC concluded that most probably global warming

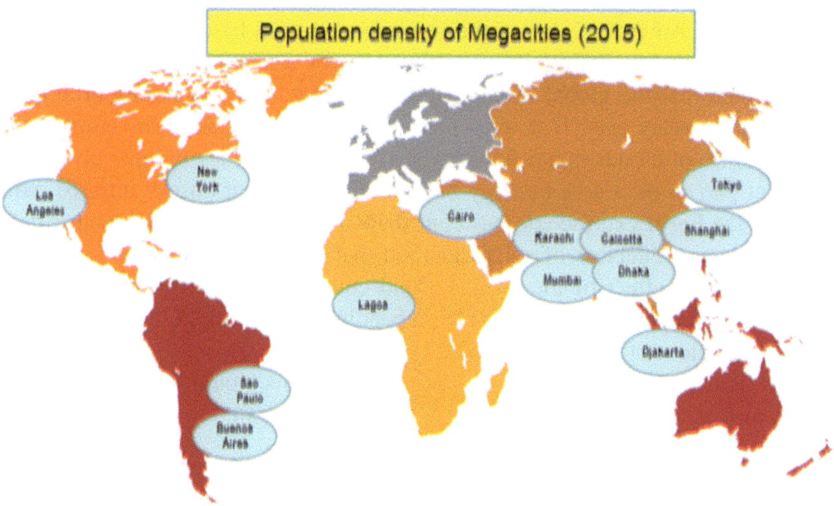

Fig. 2.1 World map of megacities "2015" (Own graph)

is caused by emissions of greenhouse gases from human activity mainly from combustion of carbon and by the clearing of natural vegetation. The Earth's system consists of five major interacting components: the atmosphere, the hydrosphere, the cryosphere, the land surface, and the biosphere. The climate is largely controlled by the flow of heat from the sun, 50 % in the form of the shortwave part of the electromagnetic spectrum and the other half by the near-infrared light spectrum. Radiation is about 30 % reflected by clouds, the atmosphere, or the surface, and 60 % of the radiation enters the Earth's system and is thereafter stored in oceans, land, the atmosphere, or the ice shield, warming the Earth's surface.

In order to balance the Earth's energy system it would be necessary for the amount of heat entering the system to be in equilibrium with the radiation. The atmosphere contains several trace gases including carbon dioxide (CO_2), methane (CH_4), and nitrous oxide (N_2O), together called "greenhouse gases" (GHG) or "Kyoto-gases," and also contains a considerable volume of water vapor that altogether can absorb a good portion of the infrared radiation. Thus the greenhouse gases hinder the heat leaving the Earth's atmosphere, resulting in the retention of more heat near the Earth's surface, a phenomenon that was already assumed from the famous "Keeling curve" that monitored the CO_2 concentration in the atmosphere at the Mauna Loa volcano in Hawaii. The curve could prove that the CO_2 concentration in the atmosphere increased from 310 ppm in year 1960 to more than 380 ppm in 2010. Schellnhuber et al. (2006) reported that the earliest record on CO_2 was from 1750 indicating a CO_2 level of 280 ppm and points out that today the CO_2 equivalent is already at 430 ppm. Similar increases were reported also from nitrous oxide and methane concentrations over the last 30 years (Stern 2007). The "greenhouse effect" led to a significant increase in surface and air temperature in recent decades. The increase resulted in a rise in the global sea level since 1970 to the present of about up to 3–6 cm, largely due to the loss of ice from Greenland and Antarctica. And if the rise is not stopped by the year 2100 the sea level will increase according to new estimations by a meter or more. It is moreover projected that the global average surface temperature will hardly drop in the first thousand years even if it were possible to cut greenhouse gas emissions to zero. The role of clouds is not fully understood although they definitively play an important part in the Earth's energy balance. Clouds either absorb infrared radiation from the Earth's surface or thus contribute to the warming. On the other hand, most clouds are reflectors of solar radiation that tend to cool the climate system. The net average effect of the Earth's cloud cover at present is a slight cooling. However, this effect is highly variable, depending on the height, type, and optical properties of clouds (IPCC 2001).

If the actual trend in temperature rise continues this would lead an increase in Earth's surface temperature of 2–4 °C. This would result in:

- A rise of the sea level of three to 4 m and will double the number of people (400 million) exposed to coastal flooding and near coastal salt water intrusion.

Thus small islands (e.g., Maldives or Tuvalu) are already about to settle in other countries handing over their national sovereignty into host countries' hands.

- An increase in the number of "extreme" events of disasters with more heat waves and less cold weather is expected.
- A change in water resource availability. Up to 30 % decrease in runoff in lower latitudes is expected, leading to more droughts and more flood events in high latitudes. This would result in an even more disproportional distribution of water resources on the Earth than today. More social and economic conflicts are assumed to develop when an additional one to four billion people suffer from water shortages.
- A deficiency of irrigation water in lower latitudes will lead to a dramatic decline in agricultural yield. Up to 3 million people, most of them in Africa, are assumed to be exposed more to malnutrition than today.
- Melting of inland glaciers will have an extremely severe impact on the Himalayas and the Andes. When the waters are not contained in snow and ice and therefore drained off even during rainy seasons this would lead to flooding and also to droughts in winter.
- An increase in vector-borne deceases (diarrhea, malaria, dengue, and others) and a shift of such epidemics towards higher latitudes. The slum quarters of developing countries especially will be extremely exposed to such diseases due to lack of sanitation and clean water access. A strong increase in heat wave fatalities is expected for the Indian Plain and Africa.
- An additional 40 % extinction of species will occur with a 2 °C increase. The drying up of the Amazon region will reduce the tropical rainforest there, and that again will have a strong influence on the world climate.

All the factors resulting from climate change will exaggerate the discrepancy already existing between the high- and low-income countries. In this regard it should be noted that the main emitters of CO_2 are the high-income countries, but the poorer nations will be "hit earliest and most severely" (Stern 2007, p. 99).

2.2.2 Benefit and Risk—A Cause–Effect Relationship

Modern risk management started in the 1980s when American sociologists began asking how risky life was in the context of an increasing use of nuclear energy for power generation in the United States. The Three Mile Island nuclear accident of 1979, where a core melting was narrowly avoided, triggered a huge debate in those days.

The discussions of how risky life is for all of us can be a summarized as (Kaplan and Garrick 1981): "[R]isks are ubiquitous and there is no life without a risk. The only choice we have is that we can choose between different kinds of risk. Risk is never zero; risk can (in the best case) only be small."

Risk can occur in every sphere of human life as business risk, social risk, political risk, safety risk, investment risk, military risk, and so on. Moreover they pointed out that "risk is (entirely) depending on the standpoint of the affected or an observer and is thus a 'subjective thing'" and pose the question: "Why [do] people expose themselves or are exposed in-voluntarily[sic] to a risk?"

The basics of risk are "that risk is has a component of uncertainty and of a kind of loss and damage that might be received." Societies normally do not enter into a risk where they do not see a benefit. Mountain climbers seek the recognition of the public or personal satisfaction by taking the risk to climb dangerous mountains. Tightrope walkers even earn their living with walking on a high wire. But also quite normal activities are in general based on a benefit orientation. A risk–benefit assessment carried out in the United States at the end of the 1990s on the acceptance of the public to site a hazardous waste disposal landfill site revealed that the public was more inclined to accept the risk when compensated (benefited) for the risk with free garbage collection for the community, although acceptance for the installation of a waste incinerator was higher when compensation for medical costs, financial reimbursements, and property value guarantees were offered by the implementing agency. In general risks are not taken without a rational justification. Generating nuclear energy might serve as an example given by Weinberg (1981). A nuclear power plant is erected because the energy consumer is asking for a cheap and constantly available energy supply. The energy company reacts to this by constructing nuclear power plants. The advantages and risks for the company and for the energy consumer are summarized in Fig. 2.2.

Although the benefit–risk relationship can be calculated by a mathematical algorithm, the result will not lead to general acceptance by society. A decision on

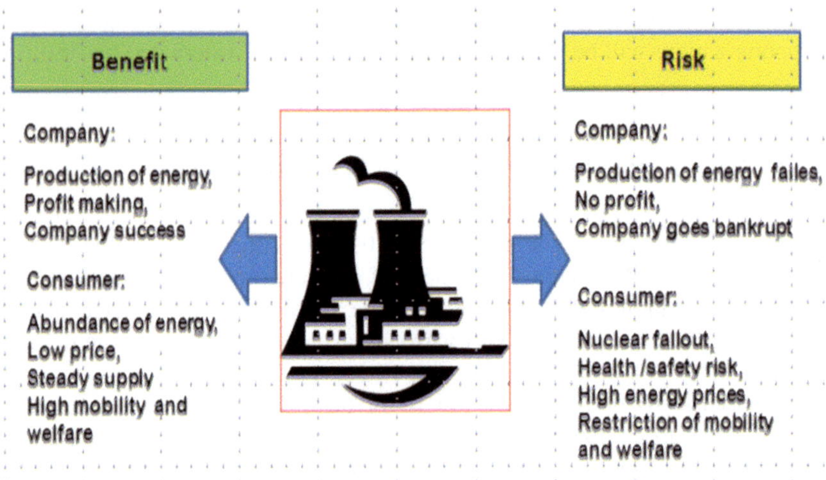

Fig. 2.2 Benefit—risk relationship of power generation from nuclear energy (Based on Weinberg 1981)

what risk we take for what benefit is basically a matter of a societal and common understanding, on how much risk a society is willing to accept in order to gain a profit, for example, of the supply of cheap energy by nuclear power plants. The decision is not a question of "right or wrong," but rather an outcome of all stakeholders involved in the decision-making process. The moment we deny the use of nuclear energy to generate power, we have to accept that the use of coal-fired power plants is for many countries unavoidable as long as the technology does not provide a cheap and sustainable way to store energy produced from renewable sources. Therefore the benefit–risk relationship is in practice not that simple to understand. There are risks with a quite easily identifiable cause–effect relationship. Natural disasters in general follow such simple linear relationships (see Sect. 5.3). But many risks have a very complex cause–effect relationship, originating in a multitude of potential causal agents resulting in multiple causes and effects that often have no or low identifiable interdependencies. The Fukushima earthquake cum tsunami nuclear accident caused Japanese car production to deteriorate and simultaneously led to high sales quantities for a German car manufacturer and an increase in the employment rate. Such complex relationships are called by the stochastic mathematics "black swan logic."

Shortly after the debate evolved in the United States a likewise discussion started on the same aspects in Western Europe as a reaction to the Chernobyl nuclear catastrophe when Europe, for the first time since the Second World War, was exposed to never-before experienced threat. In 1986 the German sociologist Beck (1986) published a book on how the individual and societies are nowadays under the paradigm of strong technically oriented economies exposed to risk and created the term "risk society." Beck pointed out that in our advancing modern economies societies steadily produce wealth and income but simultaneously that increase is systematically combined with an increase in risk exposure. He further pointed out that this increase results in an unequal distribution of risk among the different societal groups, leading to what he describes as a "paradigm of a risk society." From his point of view the term "risk" is mostly used as "a risk someone take[s] up voluntarily, demonstrating courage and automatically describing this person to be someone that is taking the challenge." The term is thus derived from the technological paradigm. In this regard it is a quite new phenomenon that man-made risks are different from natural ones in their outcome. Earthquakes can be identified more or less as occurring along fault zones. Volcanic lava, lahars, and ashfall can be located quite precisely, just the same as floods that will definitively occur in the river basins and as landslides that will occur at the foot of steep hills. In contrast, many of the man-made catastrophes can often not be seen, can't be smelled, or felt. Nuclear fallout will cover large parts of continents and chemical incidents may transport toxic substances into the oceans. Moreover the populations at risk will be totally differently exposed to such impacts, leading to a distinct disparity of risk in the society (Beck ibid), which means that in a society different social groups were "so to speak unavoidably assigned to civil risks." At the end of the twentieth century, when social ranking was much more dominated by social status, the saying, "Convictions are a result of the social status," described the situation quite well. In today's risk society however, Beck argues, "[T]he convictions (understanding) are defining actions." With this rationale, he pointed out that in the course of technological progress, modern societies arrived at a comprehensive understanding about the cause–effect relationship between the natural or man-made origin of a catastrophe and its social, economic, or ecological impacts. Only by understanding these dependencies will modern societies be able to define, work out, and implement effective countermeasures to increase their disaster resilience. Beck emphasized that "[K]nowledge, understanding and science is thus getting a political dimension that has to be developed further in natural and political sciences as well as in sociology."

This book wants to go even a step further by broadening the "risk context." In industrialized countries but especially in developing countries, disaster risk management strategies are often set up and brought into being without the participation of those who are affected by catastrophes. It has now become evident that without including the experience of the population at risk, no risk mitigation strategy will be effective. Participation can only be achieved by an early and comprehensive inclusion of the often decades-long experiences of the affected ones. But risk experience is not only the number one topic, but experiences worldwide have clearly shown that any risk mitigation strategy cannot be implemented and will not function against the will of the population. They are the ones who have to take up the recommendations. But often the measures turn out to be too technical and thus contradict their traditional beliefs. The root cause of this problem is that, according to law and administration procedures, those who are mandated with risk mitigation come from national authorities and from the scientific sector, whereas those who are affected have no functioning relationship with the institutions in charge. The main and indispensable task to bridge this social and technical gap is an open discussion forum such as a roundtable has to be established, giving the affected population a fully respected mandate in the decision-making process at the same eye-level sight. Once equal conditions are established it will be possible to reach a much higher state of resilience. Beck indicated that this will only be achieved when the society at risk succeeds in attributing the normally "un-political" natural disaster a "political" dimension (Beck 2011; Rayner 2006) Thus far all respective discussions and elaborations are still highly dominated by scientific and technical categories. This vision is also backed by Hans von Storch of the Max Planck Institute for Meteorology who emphasized (Spiegel Magazin 25/2013) that "[M]ore than generally perceived, natural sciences are a process highly affected by the actual socio-economic environment."

The upcoming awareness of social effects in the changing environment resulted in a new scientific branch: environmental sociology that deals with the "man–nature relationship" in the broadest sense (Diekmann and Preisendörfer 2001). But even in the early 1990s the integration of social elements of risk definition gained more and more acceptance. Up to that time, risk was mainly seen as a technical, scientific, and operational paradigm "to predict physical consequences" for the population of risk and its welfare systems "by extrapolating past experiences to the future," but the physical database for such extrapolations is permanently changing, very often making it impossible to "draw meaningful statistical inferences to predict future effects" (Zwick and Renn 2001). The way people perceive a risk is more a question of their anxiety, their cultural values and traditions, and their status in the societal hierarchy than their physical experience. Therefore only when the social agenda is "internalized" into the risk analysis and made an integral (if not central) part of the risk assessment (Luhmann 1990), will the exposed demand from the risk be covered comprehensively. The different way of perceiving a risk can go so far that a certain risk is seen by an individual or a societal group as absolutely unacceptable. The discussion on the use of nuclear energy in many countries is significant regarding this. The debate amplifies the risk perception that then is called "stigmatization" of certain risk and eclipses the (normally acceptable) impacts thus "determining the perceived seriousness of risk." And the "more stigma relevant elements a person links with a specific risk source, the more likely he will find this risk non-acceptable" (Renn 1989).

In the course of the climate change debate and the increasing world population a discussion arises as to how losses and victims will develop from natural disasters. It seems obvious that the population increase itself will not be the factor that matters, but the trend of poverty migration into the big megacities will be getting stronger. In the search for work and improved living conditions these migrants wear down the already difficult living conditions. Being the last in the chain in the search for living quarters they are forced to settle areas that are from their geological and geomorphological pattern not suitable for living, a behavior that just increases the hazard exposure. A similar outcome is envisaged from the changing climate. Holzer and Savage (2013) in a comprehensive study came to the conclusion on the future risk from earthquakes for people and their living environment that more people will die from earthquakes, even when the statistical occurrence of earthquakes remained more or less constant over the centuries. The study analyzed earthquakes with death tolls of more than 50,000 in the time span since 1500 AD. Comparing those estimates of world population history, they found that the number of catastrophic earthquakes has increased as the population has grown. After statistically correlating the number of catastrophic earthquakes in each century with world population, they predict that total deaths in the century to come could more than double to approximately 3.5 million people if world population grows to 10 billion by 2100 from 6 billion in 2000. The study underscores the need to build residential and commercial structures that will not collapse and kill people during earthquake shaking.

This example shows how much our daily life is governed by social and economic factors that are superimposed by the natural conditions to which we are exposed. Thus the general increase in disaster impacts that was experienced all over the world made clear that the risks are increasing and in future will be even higher. This finding alerted politicians and scientists and in May 1994 representatives of all nations assembled in Yokohama at the World Conference on Natural Disaster Reduction (UNIDNDR 1994) to adopt the "Yokohama Strategy for a Safer World." The strategy initiated Guidelines for Natural Disaster Prevention, Preparedness and Mitigation, accompanied by the Hyogo Plan of Action to be endorsed by the General Assembly of the United Nations as an internationally binding regulation. The Conference reacted to UN Resolution 44/236 to address the increasing casualties and damages from natural disasters on a global scale. Yokohama called for an integrated approach for disaster management in all its aspects and to initiate a process towards a global culture of prevention. The strategy aims to support efforts of national governments in the implementation of the program although acknowledging that each country bears the primary responsibility for protecting its own people, infrastructure, and other national assets from the impact of all kinds of natural disasters and moreover emphasized that each national government has the responsibility to enforce the law accordingly.

In the aftermath of the Yokohama Conference many states formulated national programs and action plans in order to ensure a higher level of resilience (UNISDR 2004). The great advantage of the conference was that it initiated the inclusion of all sectors of public life: the populations at risk, natural scientists, and engineers to benefit from their experience. The decision making was no longer seen as an exclusive task of the executive. Thus the discussion became broadly based on multiple stakeholders and was opened for implementing polycentric mitigation strategies. Germany already has reacted to the change by founding the German IDNDR-Committee that later developed into the German Committee for Disaster Prevention.

Since Yokohama natural scientists have also been called to take their part in the "formative actions" of the state, especially to initiate and organize a shift in paradigm from a "culture of risk" to a "culture of prevention." The inclusion of scientists and social groups in risk management led to an integration of all stakeholders involved in problem analysis, decision making, and implementation of mitigation efforts as well as in the final evaluation of achievements. By this a "formative state" in the best tradition of liberalism and democracy will increase its legitimation and strengthen public acceptance (WBGU 2011). Also in the sector of natural disaster management a state, when acting accordingly, can demonstrate that increasing disaster resilience is not a rationale to curtail individual freedoms or a call for abstaining from a self-nominated life, but is an opening chance for multistakeholder cooperation to increase the security of society in general. The best experience was made with such an inclusion approach when in Indonesia in 2006 the National Law on Disaster Management was enacted. Since then national disaster management has been nationwide and gives a robust mandate with well-defined responsibilities. With the newly formulated law it became possible to initiate a risk mitigation culture with a socially equal reorganization of decision-making mandates and implementing responsibilities, breaking the monopoly of science in the risk discussion.

2.2.3 *Population at Risk*

The daily impression of natural disasters makes us believe that disasters in general increase in number and severity and that nobody will be excluded from the impacts. The statistically proven reality on disaster from natural hazards according to information by Guha-Sapir et al. (2011) is that:

- The number of disasters is increasing.
- But the death toll from the disasters is decreasing.
- The economic impact is vastly different between low-, middle-, and high-income countries.

In order to assess how the populations in the different countries are really exposed to natural hazards, a closer look at the global risk distribution pattern is required. Only when the scope is widened to a larger than a local or regional perspective on the frequency and severity of natural disasters can the overall capacities of the people to withstand the risks be assessed and compared. Such a fact-based insight will improve the understanding of the general risk patterns and will lead to the identification of the root causes and their related threats. Based on fact-based findings the necessary technical, social, and financial mitigation measures can be identified and consequently implemented to reduce the risks. Such a generalized insight will moreover offer the chance to transfer knowledge and experience made in one part of the world to another, in order to increase local resilience there. Identifying and communicating global risk patterns proved to be the most appropriate approach to increase resilience, as the geotectonic setting, as well as the world weather situation are global phenomena both defining the risk patterns. In addition, changing climate conditions are posing a further global moment to risk distribution, the same as the exponential urbanization that results from a growing world population. For all areas at risk over the world a most realistic reliable and robust prediction on likely losses, magnitudes, and frequency of disaster events is indispensable. To be able to reduce disaster impacts efficiently, the linkages between the geotectonic, hydrometeorological, and climatic root causes and the social and economic development processes, such as urbanization and environmental change must be understood in addition to "invisible" risk factors such as gender bias, social inequity, sociopolitical conflict, and poor governance (UNISDR 2007). Although investigating the overall risk patterns enables us to understand the general distribution of risk, risk identification for specific social groups or certain regions requires a specific insight to local hazard conditions.

There are several international statistics on natural disaster occurrence, frequency, and severity assembled and regularly assessed mainly by the NATCAT Service, the download center for statistics on natural catastrophes of the Munich ReInsurance Company (Munich), the Sigma of Swiss ReInsurance Company (Zurich), by the UN organizations UNISDR, UNDP, and UNU-EHS, and also by the US Foreign Office of Disaster Assistance (US-OFDA) and many others especially the International Federation of the Red Cross. The most comprehensive database on natural and epidemic disasters has been collected by the Centre for Research on the Epidemiology of Disasters at the Catholic University of Louvain (CRED-EMDAT), Brussels, Belgium. The organization has been mandated by the United Nations as the central organization to collect and assess data on natural

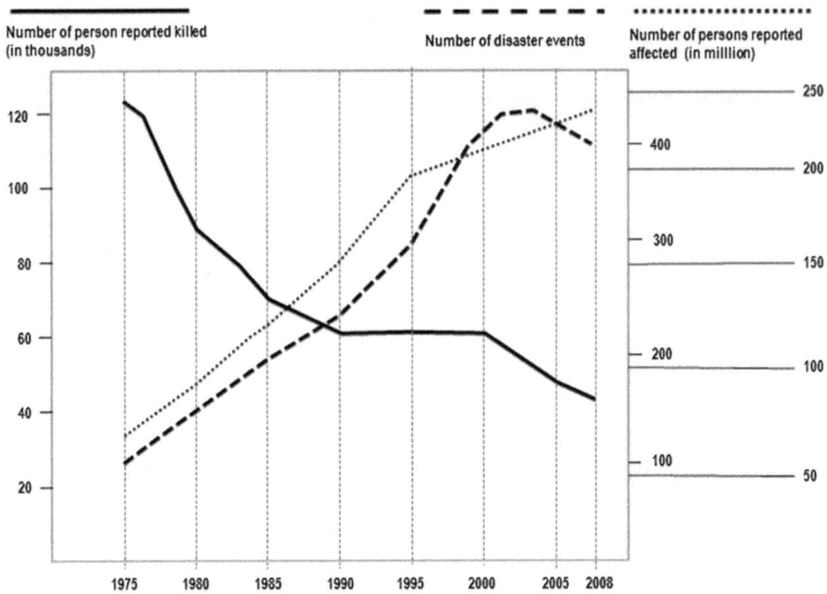

Fig. 2.3 Comparison of the number of disaster events, killed and exposed people (1975–2008) (*Courtesy* Guha-Sapir et al. 2011, 2013)

disasters. CRED is strongly supported by all disaster/emergency management organizations worldwide, especially by the USAID Office for Foreign Disaster Assistance. In total 8900 disaster events have been listed for the time span 1975 until 2008 in the EMDAT-Natural Disaster Database (CRED). The data collection could prove that there have been 600 disaster events yearly, that have killed in total 2.3 million people and have injured and made homeless more than a billion.

Today's ratio of the number of events to the people affected, respectively, killed by natural disasters has changed significantly in the last 35 years (Guha-Sapir et al. 2011, 2013). A clear trend in the ratio has been proven by the CRED-EMDAT database. For the analysis the number of disaster events was juxtaposed to number of persons killed and affected by disasters (Fig. 2.3). In 1975 about 100 events claimed 120,000 lives and affected 70 million people. Since then the number of yearly disaster events has risen to more than 600 disasters globally each year. But at the same time (only) 40,000 people lost their lives. Similar to the number of events the number of persons affected has also risen to more than 200 million. The graph clearly indicates that in the time under investigation although the number of events quadrupled, the number of people killed in disasters has been lowered to 30 %. When in 1975, 100 events claimed 120,000 lives, then—when the trend just would have been extrapolated—this would give a fourfold higher death toll. Instead the death toll in 2008 has been (only) 40,000 persons. This can be interpreted that by a preventive and effective disaster emergency management it is possible to drop the number of people killed

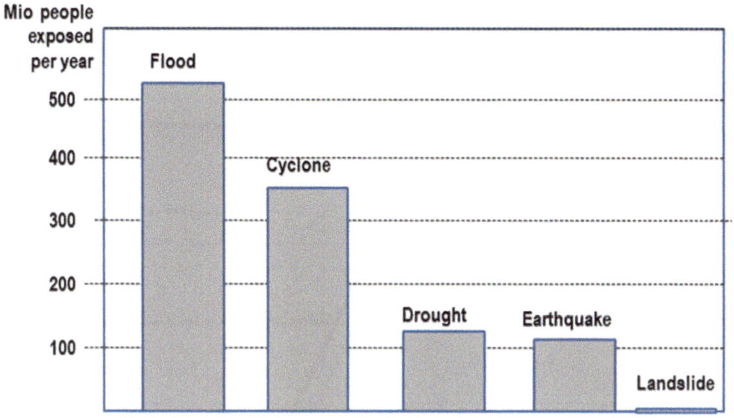

Fig. 2.4 People exposed per year to risk from different natural hazards (*Courtesy* Nadim et al. 2006)

(statistically) to 1/10. This graph reveals a much better insight to the "real" risk impact from natural disasters than when simply based on the number of people killed alone. This graph is furthermore a pledge that societal perception and from media coverage to political decision making on natural disaster severity should no longer solely be linked with the number of killed persons alone, but rather should take the number of those affected into consideration. The suffering of these societal groups has not been addressed properly in the past and the "fortunate" diminishing of the death toll ratio should not be taken as a justification to reduce the efforts in disaster mitigation.

An information CRED-Emdat stated that reliable information on disasters at a global level can (reproducibly) only be given for the times from 1985 onwards. They were able to prove that for the time before 1985 the data delivered to them were often very scarce, overrepresenting certain disaster hotspots, and were mostly lacking well-monitored timelines. When regarding only the time span from 1990 until today, the world disaster risk exposure gives a similar outcome to that previously stated. About 200 disaster events killed 60,000 people and exposed 250 million people. Compared with today's figures on the death toll, number of events, and exposed peoples, this would present the assumption that the impact of the today's disasters is (only) half of that in 1990. International statistics furthermore point out that most of the people worldwide are exposed to floods, followed by cyclones, droughts, and earthquakes and indicate that for everyone killed by a natural disaster about 3000 are exposed to hazards (Fig. 2.4).

Another striking feature from the CRED-EMDAT statistics highlights that the risk from natural disasters on a world scale is not at all distributed ubiquitously. Most disasters occurred in China with 35 events, followed by the United States of America with 26, and Indonesia and the Philippines with 20 each. Next come India (17), Afghanistan and Vietnam (14), Australia, Burundi, and Pakistan

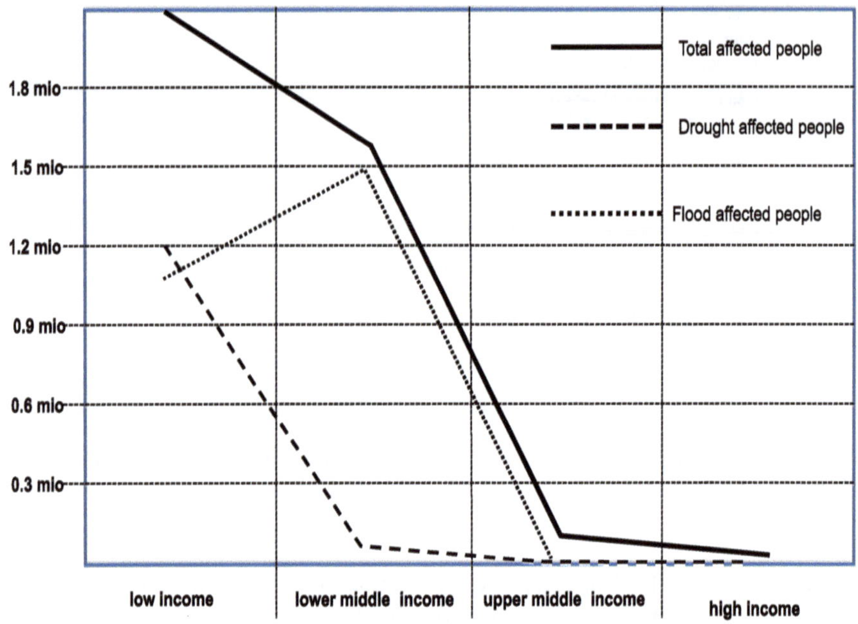

Fig. 2.5 World overview from 1975–2000 of number of people affected, categorized by income classes and disaster type (Compiled from Guha-Sapir et al. 2011, 2013)

(19 each), and Ethiopia, Germany, Mexico, and Romania (7); also Bangladesh, Canada, Japan, Kenya, and Malaysia share one group (6) and Papua New Guinea, Russia, and Somalia (5) another one. This compilation clearly shows that disasters occur in high-income countries at almost the same frequency as in the least-developed countries. But when the number of people affected by natural disasters is regarded for the time span 1975–2000, categorized according to income classes, it becomes obvious that low-income groups of societies worldwide are extremely overrepresented as can be seen from Fig. 2.5, which shows that about 2 million low-income people were hit compared to "only" 200,000 in high-income classes. Of the 13 biggest disasters since 1970, the low- to middle-income countries have claimed high death tolls, whereas disasters in high-income countries in general caused the highest economic losses (Fig. 2.6).

Figure 2.7 from Cred-EMDAT in which the economic damages are plotted according to the level of country income, further underline the above-given findings. The earthquake of Kobe, Japan in 1995 with an economic loss of about US$100 billion and Hurricane Katrina (United States) in 2005 with a loss of US$130 billion, have been the world's most costly disasters ever. Both disasters occurred in high-income countries, and middle-income countries face a much lower risk of economic damage. The maximum was reached in 1999 from the earthquakes in Turkey (Kocaeli) and Taiwan and floods in China. The risk of economic losses in low-income countries is comparably low with a ratio that

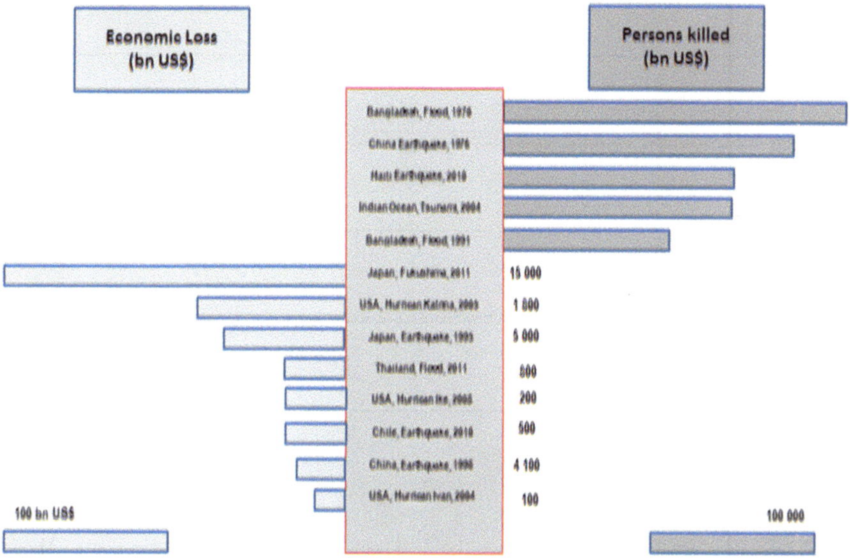

Fig. 2.6 Economic losses versus persons killed from natural disasters 1970–2010 (Compiled from UNISDR 2007, Guha-Sapir et al. 2011, 2013; Munich Re 2013)

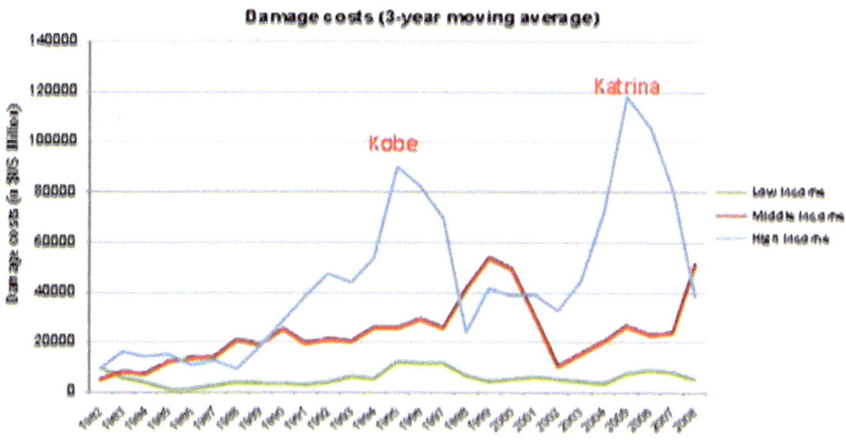

Fig. 2.7 Disaster losses according to country income levels (*Courtesy* Guha-Sapir et al. 2013)

has not changed in the last decades, a fact indicating that the amount of valuable assets accumulated is still very small. Nevertheless the biggest economic loss thus far ever was caused by the earthquake/tsunami/nuclear power plant accident of Fukushima (US$300 billion), but it is not possible to distinguish between the losses from the natural disaster and the losses caused by the power plant failure (Fig. 2.8).

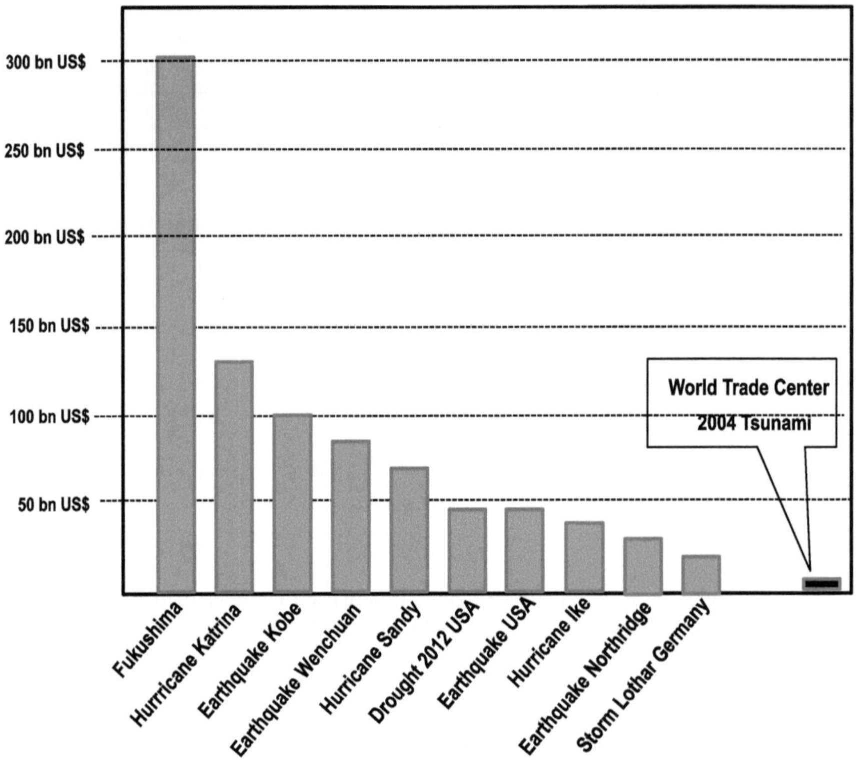

Fig. 2.8 The 10 biggest losses from natural disaster events compared to the losses from internationally best known disasters (tsunami, World Trade Center)

Regarding the economic losses from disasters since 1975, it seems that the losses have increased significantly from about 1990. There are at least two reasons for this impression: one is according to CRED information that the loss figures are "somehow" distorted owing to scarce and unproven data reported for the times before 1980, and second that because in the last 20 years, even in the least-developed countries, the accumulation of valuable assets (factories, office buildings, etc.) has increased significantly, a fact that is also indicated in the Munich Re World Map of Insured Losses (MunichRe 2013).

Moreover the impact of the different types of disaster is quite different (Fig. 2.9; redrawn from MunichRe 2012). Although geotectonic disasters make up more than 50 % of all fatalities, their share of loss is ("only") about 30 %, whereas weather-related events make up 70 % of all loss events; their fatality ratio is much lower (about 50 %). That means the lesser occurring geotectonic disasters claim comparably higher casualties whereas the many weather-related disasters claim significantly few lives. The ratio of economic losses due to these two disaster types shows another remarkable picture. The economic losses from weather-related disasters make up 70 % compared to geotectonic disasters (30 %),

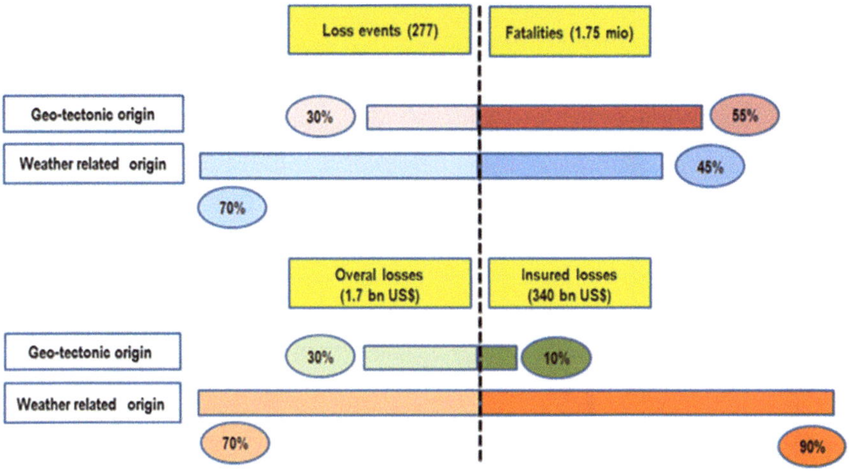

Fig. 2.9 Casualties versus economic losses of the 40 biggest disaster events since 1980 (Compiled from MunichRe 2012, 2013)

however, they are also responsible for more than 70 % of all economic losses (geotectonic 30 %). Regarding the insured losses, the picture is even more different. Although 90 % of all assets were insured against floods and wet mass movements, only 10 % of the assets were insured against geotectonic disasters. That means floods and wet mass movements occur mostly in regions where large volumes of economic assets have been accumulated, and geotectonic disasters occur mainly in regions with limited aggregated economic values, of which only some assets were insured.

This picture at a first impression contradicts the findings of Munich Re (2012, 2013) according to which 60 % of the casualties were victims of disasters of climate and weather origin, and 40 % of geotectonic. This is because there is no correlation between the number of events and the casualty ratio. Statistics revealed that annually there are about 50 geotectonic events (earthquakes, volcano eruptions, tsunami, mass movements, etc.) that claim 0.8 million people (40 %), and there are 300 hydrometeorological events that claim 1.2 million people (60 %). From the analysis it becomes clear that it is inadvisable to draw the simple assumption that the higher the frequency of a disaster type is, the higher will be the number of victims. If just a one-to one equation were rational, the amount of weather-derived disasters would claim about 5 million people.

The social dimension of natural disasters becomes even clearer when we compare the fatality ratio of developing countries with that of industrialized countries. According to data from SwissRe (2010) 95 % of the 1.8 million people killed by the 40 biggest natural disasters occurred in developing countries, whereas "only" 5 % in industrialized countries. The ratio of economic losses (here given as the amount of insured losses), however, show the opposite: of the more than US$300 billion losses, more than 90 % occur in industrialized countries, and

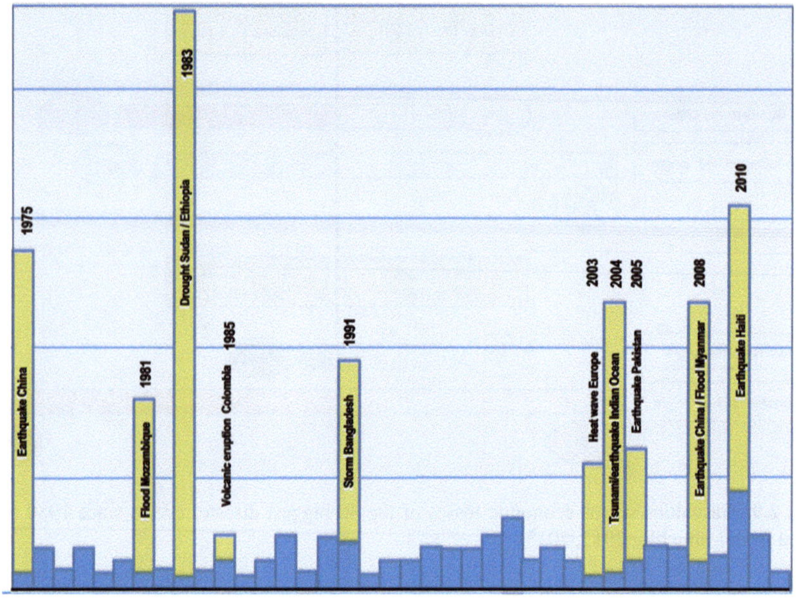

Fig. 2.10 World natural disaster events attributed to specific events, 1975–2000 (Compiled from Guha-Sapir et al. 2011, 2013)

"only" 5 % occurred in developing countries. When adapting the findings of SwissRe to the two biggest disasters of the last decade, Hurricane Katrina and the Indian Ocean tsunami, exactly the same relationship appears.

Another remarkable fact becomes obvious from Fig. 2.10. Most of the death toll in the time span from the years 1975 to 2000 derived from about 20 major disasters.

The figure clearly proves that most of the victims are attributed to the many severe droughts in the early 1980s that, for example, in the Sahel region killed about 400,000 just in the year 1983. In this context it should be noted that the statistical evidence on drought victims is not without ambiguity. Therefore the data should be treated with some caution. A similar death toll resulted from the large earthquake that hit China in 1975 (>200,000) or by the 2004 tsunami in the Indian Ocean that also killed more than 200,000 people in one event alone. In recent years one of the most disastrous events was the earthquake in Haiti that also claimed more than 200,000 lives. On all of these megadisasters as they were called by UNISDR (ibid; see also: Sect. 3.2.3) about 1.8 million people were killed. When taking out the megaevents from the statistics, a baseline of death toll risk of about 40,000 people per year seems to be the actual yearly fatal risk from natural disasters worldwide.

When a natural disaster strikes the most vulnerable groups are the poor, disabled, elderly, and the young. According to information from the World Health Organization (WHO) older adults are that fraction of a vulnerable society who are especially more likely to experience greater risks and adversity than others in any disaster. Most elderly (>65 years of age), children under 15, women, and the disabled experience the negative impacts of natural disasters the most, partly because of age-related disabilities but also because of social circumstances, such as social isolation. And WHO pointed out that worldwide the demographics are developing so that the population all over the world is aging. The projections suggest that there will be an almost threefold increase in the global population over 65 within the next half century (Tuohy 2011). But not only are the elderly disproportionally affected, the younger ones below 15 years of age are also. This group of society is mainly not able to get realistic insight to the problems and (especially in developing countries) moreover are often lacking technical and operational capacities to cope with accordingly. For instance, investigations on age and gender impact of Hurricane Katrina revealed the older adults were the fraction of the New Orleans population that faced disproportionately high adverse impacts compared to other population groups. The Indonesian tsunami of 2004 saw the highest death rates among the over-60 s and the deaths during the 2003 Paris heat wave killed more people over 70 years of age than any other group; and more than half of all casualties in the 1995 Kobe earthquake were older adults, with 90 % of deaths in this group. The same holds true for the mortality from the Tamil Nadu flood in India in 2006, where the under-10 years and the over-50 s had a five to ten times higher death toll ratio than the group between 10 and 30, or the occurrence of leptospirosis epidemics that increased 20 times in the days after the 2008 flood in Jakarta and that mostly affected the population directly dwelling near the rivers and canals in the city (Guha-Sapir et al. 2006). A disaster will amplify both personal and social challenges the older adults are facing already and as a result, older adults become more vulnerable to negative outcomes during disasters. Emergency preparedness planning must therefore take more care than before on the special age-related needs of older adults.

The above-given data on how different societies react to natural disasters clearly reveal that coping capacities differ highly: high-income countries have, due to their financial, technical, and managerial capabilities, better chances to withstand a catastrophe than many of the developing countries that are often lacking such ability. Their personal and social vulnerability hinders at all levels from national political decision making down to the individual to prepare, respond to, and recover from such events effectively.

Table 2.1 and Fig. 2.11 on the World Risk Index (WRI) compiled by the United Nations University, Bonn (UNU-EHS 2012), highlight that risk exposure is strongly dependent on the developing status of a country, In Fig. 2.11 the WRI has been correlated with the Human Development Index (HDI) to give statistical evidence of the correlation of social and economic living conditions and disaster exposure (UNDP 2013) for a selected group of countries. The boundaries of low to very high were set arbitrarily in order to make the indices comparable. The World Risk Index is based on indices reflecting exposure to natural hazards (earthquakes, floods, volcanic eruptions, etc.) as well as the vulnerability of a society by indicators describing the frequency of disaster occurrence and the deficiencies in coping with the impact and how a society has developed effective mitigation strategies.

Table 2.1 World Risk Index (WRI) of selected countries

Country	Abbreviation	WRI	HDI	Country	Abbreviation	WRI	HDI
Afghanistan	AFG	1	175	Congo	CNG	23	142
Ethiopia	ETH	2	173	Laos	LAO	24	139
Australia	AUS	3	2	Lesotho	LES	25	158
Bangladesh	BAN	4	146	Nepal	NEP	26	157
Benin	BEN	5	166	New Zealand	NWZ	27	6
Bolivia	BOL	6	108	Nicaragua	NIC	28	129
Brazil	BRA	7	85	Niger	NIG	29	187
Burkina Faso	BFA	8	183	Norway	NOR	30	1
Chile	CHI	9	40	Pakistan	PAK	31	147
China	CHN	10	101	Papua New Guinea	PNG	32	156
Germany	GER	11	5	Peru	PER	33	77
Dominican Republic	DOM	12	97	Philippines	PHI	34	114
Finland	FIN	13	21	Russia	RUS	35	55
Haiti	HAI	14	161	Samoa	SAM	36	99
India	IND	15	137	Switzerland	SWI	37	9
Indonesia	INO	16	121	Turkey	TUR	38	90
Iran	IRA	17	76	Hungary	HUN	39	37
Island	ISL	18	14	Vanuatu	VAN	40	124
Italy	ITA	19	25	Venezuela	VEN	41	71
Japan	JAP	20	10	United States	USA	42	3
Cameroon	CAM	21	150	United Kingdom	UK	43	27
Columbia	COL	22	91				

Courtesy UNU-EHS (2012)

Figure 2.11 shows that countries with a low risk exposure and a high HDI are almost all located in Western Europe, but even countries such as the United Kingdom (UK) and Germany both face regular floods. Although the flood events are often perceived by the affected population as "extreme events" (see Sect. 3.2.3) their death toll is normally very small and the impacts from the disasters do not greatly affect the national economies. On the opposite side of the graph countries such as Bangladesh, Afghanistan, Niger, and Haiti are located, all facing frequent and strong disasters and are low developed in status. Although having achieved partly significant improvements in their disaster resilience the large populations of these countries outnumber the achievements every time thus resulting in the very high risk figures. Nevertheless there are exemptions to this finding. One is Japan that on one hand has one of the highest HDI while its risk exposure due to its geotectonical exposure is also very high. And another is the Pacific island of Vanuatu that is according to UNU-EHS the most risk-prone country of the world. There a comparatively small number of earthquakes and cyclones affect a country that has due to its geographical situation almost no chance to develop effective

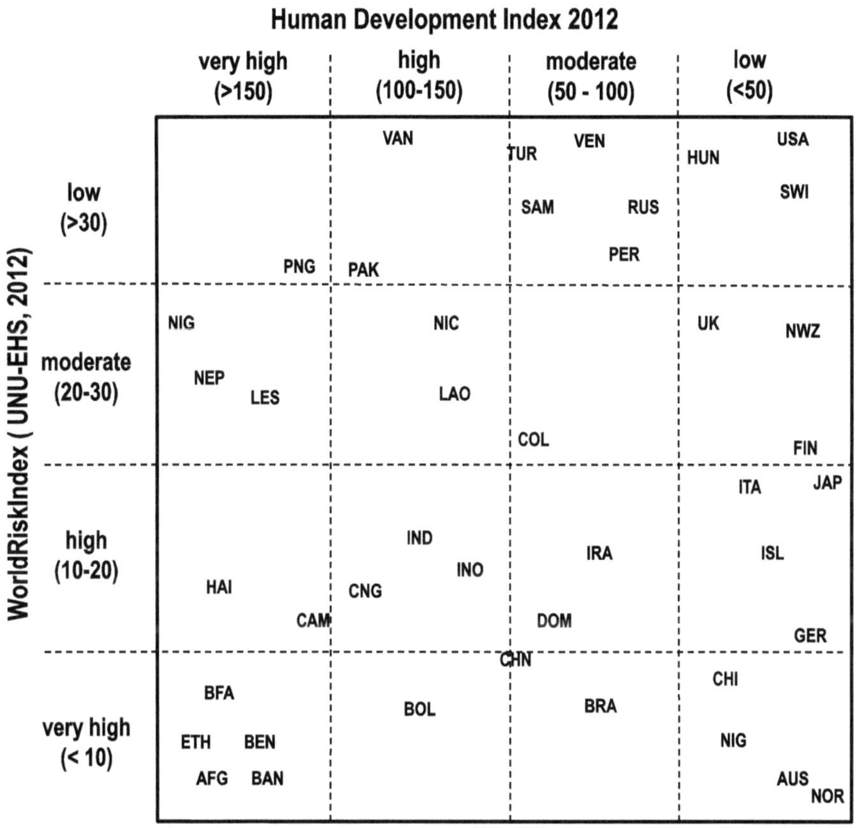

Fig. 2.11 Country risk exposure for the year 2013 (Based on UNU-EHS World Risk Index 2012 and Human Development Index, 2013)

countermeasures. This situation also holds true for many other small island states including Tonga, Kiribati, Fiji, and Maldives, among others. Summarizing the findings, it can be stated that there is a clear dependency between risk exposure and social and economic development.

Both indicators follow an opposite trend. This emphasizes the often-discussed finding that disaster risk management is more a matter of poverty alleviation than technically oriented emergency management. Such an analysis on reported death tolls and economic losses are of great importance to political decision making. In order to make societies more resilient in general against any kind of disaster, a political decision-making process has to answer the questions regarding the level of security that should be achieved and for what kind of hazard, social group, where, and to what extent it should be prepared. Or will society be safeguarded against each and every risk and at the highest level possible and is a society then willing to pay for such prevention.

2.2.4 Gender Relation to Natural Disasters

The tsunami of December 26th, 2004 killed about 230,000 people all around the Indian Ocean and claimed the lives of more than 170,000 alone in the Indonesian city of Banda Aceh at the northern tip of the island of Sumatra. However, it was not possible to count the death toll from the earthquake that triggered the tsunami. According to information of the Indonesian Ministry for Rehabilitation and Reconstruction (BRR 2007) six times more people were killed than injured in the province, a ratio that was, for instance, in Sri Lanka less than 1.5:1 and that dropped further towards the East African coast. And the tsunami killed more men than women in Indonesia. The ratio of killed men to women was between 1.2:1.0 in the entire Aceh province, whereas in the city of Meulaboh (West Aceh district) just opposite the earthquake epicenter the ratio was 2.1:1.0. Such a ratio is typical for tsunami and storm surges as men have a higher physical ability to use rescue opportunities, whereas on other hand the tsunami hit at eight o'clock on a Sunday morning when many men had already left their houses for market business. And the chance to survive the tsunami has been higher in the cities than in the rural areas. The death toll was also higher for children under 15 years and adults over 50, resulting in a death ratio of about double the amount of children and elderly than of adults.

This short description clearly shows the typical outcome of a natural disaster in developing countries. The victims are different according to their age, sex, and social status. It is the gender bias that creates the vulnerability. In general women are poorer than men, and disproportionately employed. And if employed they are mainly working in the informal sector, often unpaid or at least underpaid. Inherited laws and social patterns such as arranged marriages or the male- dominated banking system, superimpose women's dependence on fathers, husbands, and sons, thus limiting their access to resources and increasing their inability to change things (Anderson 1994). Moreover, health dangers as a result of multiple births also contribute to their low social status. Traditionally assigned responsibilities to home-based duties limit women's mobility and also hinder their chances for education and access to information as well as participation in political decision making. These factors push them deeper into the cycle of vulnerability. As women in developing countries work mainly at subsistence farming, the global shift to export-oriented agriculture undermines their economic base. This forces many of them to migrate into the big conurbations thus exposing them to rather unsafe living conditions on the fringes of the cities, moreover to urban environmental pollution but also to disasters such as flooding or landslides. As long as males dominate traditionally organized societies and as long as ideological constraints still prevail in many industrialized countries, women will still be more vulnerable to disasters. Moreover demographic trends put women increasingly at risk.

According to many studies and investigations on the social dimension of risk mainly by the United Nations (2000), IPCC (2001, 2007), UNIASC (2006), Birkmann (2006), and others, the World Bank Group and international donor agencies such as OXFAM (2000) have proved that gender vulnerability in general is a matter of poverty, or as stated before, "a lack of opportunities and capacities." Therefore the Millennium Development Goals (MDG3) underline the necessity to increase gender equality especially of the people at risk. Gender inequality is seen as an archetype that again produces further inequalities with negative consequences for women, their families, and their communities. MDG3 emphasizes that addressing gender disparities and empowering women is an important development objective. But the demand for gender equality does not necessarily mean equal outcomes for males and females. Gender inequality occurs significantly in three domains: the household, where it defines the distribution of household tasks, often limiting women's ability to work outside the home, as well as women's control over fertility decisions; in the market issue it reflects the unequal access to land, credit, and labor markets; and concerning society, it expresses restrictions on women's participation in civic and political life. Using the definition in the World Development Report, "Equity and Development" (World Bank 2006),

> [G]ender equality means equal access to the opportunities that allow people to pursue a life of their own choosing and to avoid extreme deprivations in outcomes that is, gender equality in rights, resources, and voice, as it appears that economic growth and social stability is positively correlated with gender equality of a society.

2.2.5 *Traumatization*

Natural disasters and other catastrophic events, such as traffic accidents, plane crashes, or a terrorist attack are extraordinarily stressful to the survivors. Although such traumatization occurs with many disaster events such kind of psychological impact is often not considered in emergency risk management practice. Stressful situations can harm a human population in a way in which the adverse psychological exposure exceeds the coping capacity of the affected population especially that of children, the disabled, or other socially deprived groups. Through the 2004 tsunami almost 10,000 children lost both parents according to information given by the National Indonesian Planning Commission (BAPPENAS 2005). Such disasters shatter one's sense of security, making one feel helpless and vulnerable in a dangerous emotional state and unable to rebuild a stable life. And such traumatization can last many years, if it can be cured at all. There is a clear difference between developing and industrialized countries in dealing with traumatization. In industrialized countries curing such impact is generally seen as the task of institutionalized medical services, whereas in traditional societies, for instance, Islamic societies, numerous kinds of social networks exist, helping the victim to a cure. There are also a number of programs that deal especially with orphans. Also in these countries the International Red Cross and the International Crescent Moon run specific programs that are oriented towards helping traumatized persons. Another problem of many societies in developing countries originates from the role of women in the society. According to tradition the men lead the household and represent the family. All legal contracts (house rent, plot documents, etc.) are signed by the males. In the case where a disaster has killed the husband, the surviving wife has few opportunities to claim her interest, as women still often do not hold their own passports or ID cards, depriving them from appealing a law case.

A lack of coping capacity for such incidents is especially symptomatic in poor countries and led to fact that more than 90 % of deaths due to natural disasters occur in such countries. The poor residents of New Orleans had to bear the heaviest loss of life, health, and property due to Hurricane Katrina. But such an event would have most likely caused a much higher death toll in a developing country. The "disparity in disaster outcomes between rich and poor can be understood as a function of both pre-event vulnerability and post-event response" (McCarroll et al. 2013). Socioeconomic factors such as individual technical and financial resources, the social and communal infrastructure, and overall political stability all affect the risk and consequences of natural disasters. Moreover, poverty is a well-known determinant of poor physical health, and the poor may therefore be more vulnerable to adverse physical health outcomes in the wake of a disaster. Malnourished, nonimmunized, and chronically ill persons are from experience less able to withstand the physical and emotional stress of a disaster. The impact of such disasters or traumatic events often goes far beyond physical damage. Injury is a leading cause of post-traumatic stress disorder (PTSD).

People react in different ways to disasters and traumatic events called PTSD. Most people who go through a trauma have some symptoms at the beginning, whereas others develop them over time. They may also come and go over many years. From medical experience it isn't clear why some people develop PTSD and others don't. Whether a victim develops PTSD depends mainly on how intense the trauma was, how long it lasted, and whether he or she received professional help and support after the event. The emotional distress in the aftermath of a traumatic event can result in a wide range of confusing and sometimes frightening emotions, with shock and disbelief in accepting the reality of what has happened and in fear that the same thing will happen again. Many people show symptoms of anxiety, that one might lose control and break down or helplessness on the unpredictable nature of a disaster. The symptoms usually start soon after the traumatic event, and can cause great distress; PSTD symptoms generally concern the emotional sphere, the cognitive situation of the patient, and his or her physical abilities. US-VA (2014) identified four major types of stress symptoms:

- Reliving the event (also called re-experiencing symptoms, or flashback). This becomes manifest in bad memories or nightmares that can come back at any time. Other examples are feeling on alert and on the lookout for danger, having trouble concentrating or sleeping, having a pounding heart, cold sweat, rapid breathing, or stomach tightening.
- Avoiding situations that remind one of the event. This becomes manifest in escaping stress-forming situations or crowds of people that can trigger memories and that can lead even to avoid talking or thinking about the event (e.g., avoiding driving after a car accident).

- Negative changes in beliefs and feelings. People may feel fear, sad, and depressive. They show grief and anger, and feel guilty that they were not able to prevent the situation. Moreover many are ashamed because they cannot control their feelings.
- Feeling keyed up (also called hyperarousal). This becomes manifest in always being on the alert and on the lookout for danger. Even harmless situations may arouse anger and irritation, like a sudden loud noise next door.

When mass casualties occur in a disaster, not only adults are affected. Children are disproportionately put at risk of being injured and traumatized. When children are exposed early to the death of parents, brothers, sisters, or close friends they discover that even parents and close relatives are susceptible to harm. The loss of important, care-giving relationships in their daily lives can generate long-standing traumatic experiences. Children suffer not only from the premature loss of a family member, but also from exposure to the cruel and violent nature of the death which may create feelings of on-going insecurity and exposure to danger or threat. Although children are generally exposed to the same spectrum of hazards as adults, they are still maturing physically, emotionally, cognitively, and socially. Thus, the "impact of perceived threat or physical harm must be put in relation to the child's developmental level and also within social context the child lives" (Shaw et al. 2007). For children, individual factors such as age, gender, race, educational level, medical and psychiatric history, and the child's level of functioning before and during the disaster are the main factors defining the trauma history. Family cohesiveness, the parent–child communication patterns, how the parents respond to the disaster impact, or post-disaster family functioning are powerful factors helping the child to rehabilitate. Moreover, some definable groups of children will require additional, specifically customized assistance for their protection and to facilitate their recovery from the event. Children with special needs include those who are developmentally disabled, children who are medically or psychiatrically ill, children living in poverty, foster care children, and children who have suffered from repetitive exposure to violence or maltreatment. After a disaster has occurred victims experience different kinds of stress reactions that may continue for a significant period of time, for instance, grieving and mourning. After all disasters, the experience of the loss of safety, security, and lack of predictability as to how life will go on, makes a sense of uncertainty become a part of life.

Observations specialists (FEMA 2013) who assist survivors in the aftermath of a disaster had successful and encouraging experiences with the following steps to reduce stress symptoms and to promote post-disaster readjustment and to rebuild emotional well-being and regain a sense of control following a disaster:

- Provide a "safe haven" that gives shelter, food and water, sanitation, allows privacy, and open opportunities to mourn the losses and adjust to the adverse situation.
- Immediately establish direct personal and family contacts to regain a sense of hope, purpose, and self-esteem.
- Establish a self-help group of victims under the guidance of medical assistance in order to talk about the experiences ("tell the story") and to share grief with others.

- Identify key resources such as national or international organizations for debris management, health services, shelter, and basic emergency assistance.
- Identify local cultural or community supports to help maintain or re-establish normal activities such as attending religious services.
- Understand the root causes and consequences of disaster occurrences.
- Change social and health behaviors to enhance ability to cope with excessive stress.
- Establish daily living routines.

2.2.6 Social Connotation of Disaster Impact

There is no better indicator for the social connotation of natural disasters than the fact that 90 % of all death casualties occur in developing countries, and 90 % of all economic losses (most of them insured) occur in industrialized countries. Even more it is anticipated that 90 % of all rescue operations are carried out by the affected people themselves. Figure 2.12 delineates this distinct difference between income and poverty, where 95 % of the death toll of the 40 biggest natural disasters found 95 % of the deaths in developing countries, and 90 % of the economic losses occurred in industrialized countries, 75 % in the United States of America alone. This is a ratio that can be transferred directly to the situation of the catastrophes of Hurricane Katrina and the 2004 tsunami in Indonesia, where the tsunami claimed 90 % of the victims, and the hurricane was responsible for 90 % of the economic losses.

The international disaster statistics mainly from Guha-Sapir et al. (2011, 2013) and UNISDR (2007) confirm this significant difference between developing

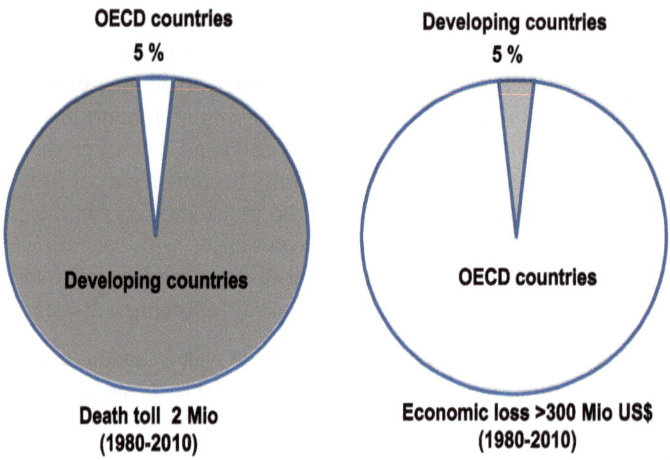

Fig. 2.12 Death toll and economic losses distinguished between industrialized and developing countries (Compiled from Guha-Sapir et al. 2011; MunichRe 2012, 2013; UNU-EHS 2012; UNISDR 2009)

countries and industrialized countries. Although the numbers of casualties are still regarded as unacceptably high, great strides were made in the last decades on the "survival ratio" from natural disasters in developing countries. However, the differences are still dramatic. To give an example: in 1980 there was a series of drought events striking the Sahel Zone, claiming a death toll of more than 400,000. Although the exact death toll figure is not confirmed, casualties were by far the largest in history. Furthermore it should be noted that 1980 was not the only year when Northern Africa was that badly hit by a drought. But the disaster initiated a multitude of national and international help and assistance initiatives. In the aftermath considerable achievements have been made by many developing countries and so-called "threshold countries" in order to safeguard their populations from disasters. These initiatives are highly subsidized by international donor agencies, resulting in a drop of the death toll figure especially from drought disasters from about 30,000 per year to less than 5,000 today. But the achieved reduction should not camouflage two other distinctive aspects in disaster exposure. The death toll itself fortunately dropped considerably, however, the number of the people exposed to a drought hazard has more than doubled, the same as the values of assets prone to damage. This is mostly attributed to the fact that high birth rates and poverty have driven migration into the large conurbations bringing more people to the brink of disasters, thus undermining many of the mitigation achievements.

2.3 Risk to Economy

2.3.1 Eyjafjallajökull, Iceland

In April 2010 the Eyjafjallajökull, one of Iceland‘s mountain glacier volcanoes erupted explosively. Although the impact from the eruption was quite a local phenomenon and did not have a serious impact on Iceland itself (about US$3 million in damage), the economic impact on international and European air traffic was enormous. The volcano erupted twice after more than 100 years of rest before the April 2010 eruption. But this time the eruption was 10–20 times more powerful and caused the cancellation of thousands of flights across Europe and to Iceland. At the time of the eruption the prevailing winds transported the ash clouds first over the North Sea towards England and then turned east to Central Russia and days later shifted south until it reached the Alps. Immediately upon the eruption and as a consequence of the large amount of ashes that were ejected into the atmosphere the International Civil Aviation Organization (ICAO) and the European Air Control Agency ordered a complete stop of all flights over northern and central European airspace for more than five days. Sixty percent of the daily flight connections were cancelled and hundreds of thousands of passengers were forced to stay on the ground. Even the German Chancellor Angela Merkel on her way from the United States had to land in Italy and was forced to take a car back to Germany. The Eyjafjallajökull eruption caused the biggest international air traffic disruption since the World Trade Center attack in 2001. The event affected 10 million passengers and claimed economic losses of about US$2.0 billion according to information from the European Commission. The airline companies complained that the disruption order was overexaggerated as the order was based on recommendations of the Volcanic Ash Advisory Center (see Annex B) but not on aircraft producers given a threshold value of ash concentrations. But the European Commission reiterated putting the safety of passengers first and insisted on the flight moratorium.

2.3.2 International Impact of Local Events (Fukushima Nuclear Power Plant Failure)

The devastating magnitude 9.1 earthquake of March 11th, 2011 along the northeastern Japanese coast was not anticipated to be of such great magnitude. From seismic records, seismologists (Geller 2011) were of the opinion that such a strong earthquake could not occur on this subduction zone. Earthquakes with a magnitude of 8 were expected and accordingly planned for, either for the nuclear power plant of Fukushima at Sandai–Daiichi or for the tsunami protection facilities along the east coast of Japan. The giant magnitude 9 earthquake, which released 30 times more energy than an 8 magnitude earthquake overtopped the 10-m seawalls, causing enormous damage to the coastline and destroying the four power plants of Fukushima–Daiichi. Thus Fukushima is an example of a technically intrinsic and well carried out natural disaster assessment that was toppled by reality. The reason that the seismologists formerly, "did not see the possibility of an earthquake of such a magnitude, was that the historic record on earthquakes along plates boundaries was very scarce," as pointed by Stein and Okal (2011). Instead the record fostered the opinion that "[E]arthquakes with a magnitude of nine and greater will only occur where the lithosphere is younger than 80 million years old and that is moving with a speed of faster than 50 mm per year." This assumption made intuitive sense, as it seemed understandable that both, "[T]he young age of the plate and its high speed favor strong mechanical coupling at the interface between the two plates." At the interface "The strong coupling was therefore assumed, to give rise to larger earthquakes when the interface eventually slipped in a great thrust fault earthquake." Furthermore it was anticipated by Stein and Okal (ibid) that the "rupture-process is performed in segments" as could be demonstrated for the 2004 tsunami (over a length of 1100 km) and thus "the more segments are generated the more energy is released."

The Fukushima–Daiichi nuclear accident was a major catastrophe that had a serious impact on the Japanese and world economy. It turned out that the earthquake itself and the damage from the tsunami could to be rated "quite" low, although with undoubtedly serious impact on the people living in the area and on the national economy. In Japan private houses are insured by a national insurance pool that covers most of the losses against such kinds of disasters. The losses due to evacuation, resettling in other regions, and medical costs are also borne by the government. No private insurance is liable for events like this. If only these two disaster aspects are considered, than Fukushima can be classified a "medium class" catastrophe, which would have resulted only in a small impact on the world economy. It is a fact that highly industrialized nations such as Japan or the United States generally quickly recover from such disasters. Moreover very often the money invested to recover from a catastrophe leads to modernizing the social and economic infrastructure at a higher reliance level. In industrialized countries the losses from disasters normally lower the gross domestic product (GDP) by only about 1–2 %, allowing for a recovery within a year or two (Hurricane Katrina, New Orleans), whereas such disasters in developing countries can have impacts on the GDP of more than 15 %, according to information given by the World Bank in 2004. The Fukushima catastrophe, however, had a great impact on international economies as well as on global ecology. The release of much radioactive contaminated cooling water will result in an increased radionuclide exposition of the offshore regions. Moreover the failure of the power plant led the Japanese government to shut down all nuclear power plants temporarily, resulting in power supply restrictions for the private and industrial sectors. This again resulted in a dramatic drop in industrial productivity especially of the world's leading car manufacturers. For the first time in decades Japanese carmakers suffered high losses, while on the other hand, the car manufacturers in Europe and America gained much profit.

Natural disasters such as the Eyjafjallajökull eruption not only strike people at the location of the disaster, but can also severely affect the living conditions of people far away. In this example the volcanic eruption affected the international air traffic sector and hindered many people from running their businesses or to connecting with others. The eruption thus has an impact on conditions essential for private as well as public sector life. Although the private sector (houses, household, family organization) is a matter of personal disposal, there are quite a number of technical, administrational, and managerial, physical, or virtual systems that are indispensable to provide essential services to maintain the functioning of a society, called critical infrastructure. Critical infrastructure refers to technical assets as well as to organizational systems that can be especially at risk from natural hazards, the consequences of climate change, or nowadays from terrorism that are essential to sustain societal functioning during a catastrophic emergency.

Critical infrastructure disruptions thus can have direct impacts on social welfare and business. Whereas in many societies, critical infrastructure comprises all kinds of technical and social assets and their operational setups that can be at risk, Norway distinguishes particularly between the challenges to enterprises that are responsible for critical infrastructure and critical societal functions. They define "critical infrastructure" as power generation and supply, electronic and satellite-based communication, water supply and sewage, and the road/rail/air and waterway traffic system; and critical societal functions are the banking and finance sector, food supply, health services, social services and social security system, law enforcement including the police and military services, as well as emergency and rescue services and crisis management (NOU 2006). The critical infrastructure is diverse and complex. It includes distribution networks, highly varying organizational structures and operating models, and interdependent functions and systems in both the physical space as well as in the recently increasing cyberspace. It comprises governance constructs that involve authorities, responsibilities, and regulations from the local up to the national and international levels. Critical infrastructure can be at risk from various natural, man-made, and technological hazards that can result in human casualties, property destruction, adverse economic stability, and public health and safety, and that can consequently damage public morale and confidence in the national problem-solving capability. The risks are heightened by the complex system of interdependencies, which can produce cascading effects far beyond the initially affected sector and physical location of the incident. Securing critical infrastructure-related functioning is a national task whereas the specific mandates, roles, and responsibilities at the national and the local levels and among the public and private owners and operators must be clarified. In Europe the national governments are responsible for the development of a situational awareness and mitigation capability during incidents, whereas in the United States the Secretary of Homeland Security provides strategic guidance assigned in the USDHS (2002).

2.3.3 The Great Flood of 1993 (United States)

Every year the United States sees an extraordinary impact from natural disasters and atypical weather situations. The economic losses from these events have been considerable. In only half a century (from 1989 to the mid-1990s) insurance companies have paid out more than US$45 billion in damage claims stemming from blizzards, hurricanes, earthquakes, tornadoes, floods, droughts, mudslides, wildfires, and other calamities. Altogether, these disasters have affected the economy deeply in terms of property damage, lost wages, utility disruptions, industrial and agricultural production failure, in addition to claiming hundreds of lives. The effect on the economy varies considerably. Some natural disasters, such as tornadoes, hurricanes, and earthquakes are more or less short-term events, lasting several or a few hours, but causing substantial destruction in a concentrated area, whereas others, such as droughts or floods, tend to be of a longer duration, spreading their damaging effects over a relatively larger expanse for days or weeks. Any type of disaster, however, can leave an economic imprint that may persist for years. A major flood has the capacity to affect numerous sectors of the economy from agriculture to manufacturing to transportation. In addition to the obvious damage to public and private structures, other damages are not so obvious, for instance, a reduced fertility of farmland, weakened structural foundations of buildings, or waterlogged roads. There are other factors, such as transportation delays and adversely affected crop and livestock markets.

The damage from the "Great Flood of 1993" in the United States, which primarily hit the states along the upper and middle Mississippi River basin, were so widespread that for more than 500 counties in nine states, including the entire state of Iowa, a "state of emergency" was declared. In the St. Louis area, the 1993 flood topped the previous record flood of 1973. The flood was in those days ranked one of the costliest natural disasters of all time, just behind Hurricane Andrew in 1992. The overall costs were estimated to be up to US$20 billion, with a large percentage of uninsured losses (Kliesen 1994). According to the Insurance Information Institute, insured nonagricultural losses were about US$800 million, and insured crop losses were put at US$250 million. Although the flood affected several important sectors of the economy, the disruptions to transportation were the greatest, especially on railroad connections in the Midwest. Numerous disruptions forced many railroads to lay emergency tracks to reach a sustained delivery of the production, especially of the car manufacturers upstream. The Association of American Railroads (AAR) at that time calculated direct losses of US$130 million primarily on physical destruction of rail lines, bridges, and signalling equipment, and another US$50 million as indirect losses from rerouting of trains. The AAR believes that other indirect losses, for example, from business interruptions and lost revenue could reach another US$100 million. As the Upper Mississippi River is an important transportation lifeline, moving a significant percentage of the nation's grain, coal, chemicals, fertilizers, and other goods, the Maritime Administration estimates that indirect flood losses totaled nearly US$280 million. Agriculture also incurred significant losses, with US$530 million in disaster assistance disbursed to nearly 150,000 farmers and another US$500 million in crop insurance. Of this nearly US$1 billion disbursement, 50 % was received by the farmers in Iowa and Minnesota alone. In total the US federal government spent over US$2.5 billion, a financial injection that was intended to support the economic recovery of the region.

As with flood impact everywhere, the largest effects from the great flood were on physical damage, production, employment, wages, and the capital stock at the local or regional level. The flood, moreover, resulted in multifold impacts across the country especially as it came in addition to the big Northridge earthquake and the winter storms in the South, Midwest, and East. Altogether, these events in 1993 affected about one half of the entire US population, disrupted construction in the housing industry, and caused significant reductions in the output of automotive, steel, and appliances, yet on the other hand the adverse weather conditions boosted output of nation's coal mines (Kliesen ibid). But economically the overall effect of these temporary disruptions did not really put the American economy under serious pressure, an assessment that also was anticipated for the US east coast that suffered from a series of blizzards and storms.

At the beginning of 1993, most economists were expecting the US economy to grow at about 3 %. But the first quarter 1993 GDP was only at 0.8 %. This significant drop was attributed by many economists to the adverse weather conditions. But when the second-quarter real GDP growth rate was also below expectations this made it apparent that the first quarter's weakness was not entirely weather-related. As the economic effects of a disaster on a national economy are often superimposed by other than natural factors, calculating the impact often may result in misleading pictures of the economy's overall performance. The many experiences in the disaster–economy relationship tend to assume that a disaster has often a less serious rather than a challenging impact on the economy than the overall national or international economic situation itself. This is due mainly to the fact that the disaster impact influences a multitude of economic sectors that are highly intertwined in innumerable and unseen ways, making a calculation of the real economic effect of a natural disaster a difficult task (Kliesen ibid). In a paper on economic effects from natural disasters Chang (1984) confirms a finding of Dacy and Kunreuther (1969) that "Although a society as a whole suffers from a net economic loss, the recovery efforts in a disaster area may be more than sufficient to replace old roads, bridges and other community assets. If so, disaster areas may be said to benefit from disasters even if the benefit, if any, is a transfer benefit from other areas."

References

Anderson,M. (1994) : Understanding the disaster-development continuum. - Focus on Gender, Vol. 2, Part 1, Oxfam Publications, Oxford.

BAPPENAS (2005): Indonesia: Preliminary damage and loss assessment - The December 2004 Natural Disaster. – Technical Report prepared for BAPPENAS and the International Donor Community, The World Bank, Jakarta.

Beck, U. (1986): Risikogesellschaft - Auf dem Weg in eine andere Moderne. – Suhrkamp, Berlin.

Beck, S. (2011): Zwischen Entpolitisierung von Politik und Politisierung von Wissenschaft: die wissenschaftliche Stellvertreterdebatte um Klimapolitik; in: Schüttemeyer, S.S., (Hrsg.): Politik im Klimawandel: keine Macht für gerechte Lösungen? Nomos, Baden-Baden, p. 239 – 258.

Birkmann, J.(ed) (2006): Measuring vulnerability to natural hazards – Towards disaster resilient societies. United Nations University Press, Tokyo.

BRR (2007): Tsunami recovery indicators package (TRIP) for Aceh and Nias.-National Indonesian Agency for Rehabilitation and Reconstruction of Aceh and Nias, Government of Indonesia.

Chang. S. (1984): Do Disaster Areas Benefit from Disasters? - Growth and Change, Vol, 15, Issue 4, Wiley Online Library (online: DOI: 10.1111/j.1468-2257.1984.tb00748.x).

Dacy, D.C, & Kunreuther, H. (1969): The Economics of Natural Disasters: Implications for Federal Policy. - The Free Press, New York NY.

Diekmann, A. &Preisendörfer, P. (2001): Umweltsoziologie - Eine Einführung. – Rowohlt, Reinbeck.

FAO (2013): Resilient Livelihoods – Disaster Risk Reduction for Food and Nutrition Security Framework Program, Food and Agriculture Organization of the United Nations, Committee on Agriculture, Rome (online: www.fao.org/publications).

FEMA (2013): Coping with disasters. - Federal Emergency Management Agency (FEMA), Disaster Survivor Assistance, Washington DC (online: www.fema.gov/coping-disaster).

Geller, R. J. (2011): Shake-up Time for Japanese Seismology. – Nature, Vol. 472, p. 407-409, Macmillan Publishers Limited.(online: www.nature.com/nature/journal/v472/n7343).

Guha-Sapir D., Parry L., Degomme O., Joshi P.C., Saulina Arnold J.P. (2006): Risks factors for Mortality and injury : Post-tsunami epidemiological findings from Tamil Nadu, CRED Working Paper. - The OFDA/CRED International Disaster Database, Brussels.

Guha-Sapir. D., Vos, F., Below, R. with Ponserre, S. (2011): Annual Disaster Statistical Review 2011: The Numbers and Trends.- The OFDA/CRED International Disaster Database, Brussels.
Guha-Sapir, D., Below, R. & Hoyois, Ph. (2013): EM-DAT: International Disaster Database (2013). - The OFDA/CRED International Disaster Database, Brussels.
Haub, C., & Yanagishita, M. (2011): World population data sheet, Population Reference Bureau, Washington DC (www.prb.org/2011world-population-data-sheet).
Holzer, T. & Savage, J. (2013): Global earthquake fatalities and population.- Earthquake Spectra, Vol. 29, Issue S1, Oakland CA.
Horwell, C.J. & Baxter, P.J. (2006): The respiratory health hazards from of volcanic ashes: a review for volcanic risk mitigations.- Bulletin Volcanology, Vol. 69, p.1-24, Springer-online.
IPCC (2001): Third Assessment Report - Climate Change 2001 - The Scientific Basis, Section 2, - Special Report of the Intergovernmental Panel on Climate Change, Cambridge University Press, Cambridge MD.
IPCC (2007): Fourth Assessment Report - Climate Change 2007 - The Physical Science Basis, Contribution of Working Group I. - Special Report of the Intergovernmental Panel on Climate Change Cambridge University Press, Cambridge MD.
Jonkmann, S.N., Maaskant, B., Boyd, E. & Levitan, M.L. (2009): Loss of life caused by the flooding of New Orleans after Hurrican Katrina: Analysis of the relationship between flood characteristics and mortality. – Risk Analysis, Vol. 9, No. 5, Wiley Online Library.
Kaplan, S. & Garrick, B.J. (1981): Risk Analysis, Vol. 1, No. 1, Wiley Online Library.
Kliesen, K.L. (1994): The Economics of Natural Disasters. - A Quarterly Review of Business and Economic Conditions, The Regional Economist, The Federal Reserve Bank of St. Louis, MS.
Luhmann, N. (1990): Technology, Environment and Social risk – A Systems Perspective. - Industrial Crisis Quarterly, Vol. 4, p.223-231, Sage Publisher, New York NY.
Macfarlane, A.(2012): Fukushima lessons: The disconnect between geology and nuclear engineering.- Elements, International Magazine of Mineralogy, Geochemistry, and Petrology, Vol. 8, No. 3, p.165 (online: www.elementsmagazine.org).
McCarroll, J. E., Vineburgh, N.T. & Ursano, R.J. (2013): Disaster, disease and distress – Resources to Promote Psychological Health and Resilience in Military and Civilian Communities. - Center for the Study of Traumatic Stress, Department of Psychiatry, University of the Health Sciences, Bethesda, USA (www.CSTonline.org).
MunichRe (2012): Topics Geo online – 2012. - Munich Reinsurance Company, Munich.
MunichRe (2013): Topics Geo online - 2013. - Munich Reinsurance Company, Munich.
Nadim, F., Kjekstad, O, Peduzzi,P., Herold,C. & Jaedicke, C. (2006): Global landslide and avalanches hotspots. - Landslides, Vol. 3, No 2, p. 159-173, Springer Link.
NOU (2006): Protection of critical infrastructures and critical societal functions in Norway. - Report NOU 2006:6 submitted to the Ministry of Justice and The Police by the government appointed Commission for the Protection of Critical Infrastructure on 5th April 2006, Oslo.
OXFAM (2000): The Oxfam Poverty Report. – Oxfam, London.
Pirsching, M. (2006): Good Bye New Orleans - Der Hurrikan Katrina und die amerikanische Gesellschaft.- Leykam, Graz.
Rayner (2006): What drives environmental policy ? - Global Environment Change, Vol. 16, Elsevier (online: www.sciencedirect.com).
Renn, O. (1989): Risikowahrnehmung und Risikobewertung in der Gesellschaft – in: Hosemann, G. (ed): Risiko in der Industriegesellschaft, Analyse, Vorsorge und Akzeptanz. - Erlanger Forschungen, Vol. 19, p. 176-192, Erlangen.
Schellnhuber, H.J., Cramer, W., Nakicenovic,N., Wigley, T. & Yohe., G. (eds) (2006) *Avoiding Dangerous Climate Change.* - Cambridge University Press, Cambridge MD, ISBN 9780521864718.
Shaw, J.A., Zelde Espinel, & Shultz, J.M (2007): Children: Stress, Trauma and Disasters. - Center for Disaster & Extreme Event Preparedness, Department of Epidemiology & Public Health Clinical Research Building (DEEP), Disaster Life Support Publishing, Tampa, Florida (online: www.umdeepcenter.org).

Soderblom, J. (2008): Climate change: is it the greatest security threat of the 21st century?. - *Security Solutions, No. 52, March/April, Canberra.*

Stein, S. & Okal, E. (2011): The size of the 2011 Tohoku earthquake need not have been a surprise. – EOS, Transactions American Geophysical Union, Vol. 92, No. 27, Wiley Online Library.

Stern, N. (2007): The economy of climate change - The Stern Review.- Cambridge University Press, Cambridge MD.

Swiss Re (2010): Natur und Man-made Katastrophen 2010. - Sigma, Nr. 1/2011, Swiss Reinsurance Company, Zurich.

Tuohy, R. (2011): Exploring older adults' personal and social vulnerability in a disaster. – International Journal of Emergency Management, Vol.8, p.60-73, Geneva.

USDHS (2002): Homeland Security Act of 2002. The Department of Homeland Security, (online: http:// www.whitehouse.gov/deptofhomeland/sect1.html).

UN (2000): Millennium Development Goal. – United Nations General Assembly, United Nations Millennium Declaration, New York NY.

UNDP (2013): Human Development Report 2013 - The Rise of the South: Human Progress in a diverse World. - United Nations Development Programme (UNDP), United Nations, Geneva.

UNFCCC (2009): Climate Change- Global risks, challenges & decisions. United Nations Framework Convention on Climate Change (UNFCCC), Synthesis Report Copenhagen 2009, 10-12 March, 2nd edition, University of Copenhagen (online: www.climatecongress.ku.dk).

UNIASC (2006): Women, Girls, Boys and Men: Different Needs – Equal Opportunities. – The UN-IASC Gender Handbook. - Inter-Agency Standing Committee Inter-Agency Standing Committee (IASC) United Nations General Assembly Resolution 46/182 (online: www.humanitarianinfo.org/iasc/gender).

UNIDNDR (1994): Yokohama Strategy and Plan of Action for a Safer World - Guidelines for Natural Disaster Prevention, Preparedness and Mitigation.- World Conference on Natural Disaster Reduction, 23-27 May, Yokohama, International Decade of Natural Disaster Reduction (IDNDR), United Nations, Geneva.

UNISDR (2004): Living with Risk - A Global Review of Disaster Reduction Initiatives". – United Nation International Strategy of Disaster Reduction (UNISDR), Geneva.

UNISDR (2007): Global risk reduction: 2007 Global Review. - United Nations International Strategy for Disaster Risk Reduction (ISDR), ISDR/GP/2007/3, United Nations, Geneva.

UNISDR (2009): Global assessment report on disaster risk reduction – Risk and poverty in a changing climate.- United Nations International Strategy for Disaster Risk Reduction (ISDR), United Nations, Geneva.

UNU-EHS (2012): WeltRisikoBericht 2012. – United Nations University (UNU-EHS).

USVA (2014): Effects of Disasters: Risk and Resilience Factors. – United States Veteran Affairs (USVA), Washington DC (online: www.va.gov).

WBGU (2007): Welt im Wandel: Sicherheitsrisiko Klimawandel.- Wissenschaftlicher Beirat der Bundesregierung Globale Umweltveränderungen, Hauptgutachten 2007, Springer Heidelberg.

WBGU (2011): Welt im Wandel, Gesellschaftsvertrag für eine grosse Transformation. -Wissenschaftlicher Beirat der Bundesregierung Globale Umweltveränderungen, Hauptgutachten 2011, Springer Heidelberg.

Weinberg, A., (1981): Reflection on risk assessment. - Risk Analysis, Vol. 1, No 1, Wiley Online Library.

Winchester, S. (2003): Krakatao – The day the World exploded, August 27, 1883. – Harper & Collins Publishers.

World Bank (2006): Global Facility for Disaster Reduction and Recovery – Reducing Vulnerability to Natural Hazards. – The World Bank, Washington DC (online: www.gfdrr.org).

WRI (2007): Annual report 20076-2007. - World Resources Institute, Washington, DC.

Zwick, M.M. & Renn, O. (2001): Perception of risk – Findings in of the Baden-Württemberg Risk Survey 2001.- Joint Research Report, Center of Technology Assessment, University of Stuttgart.

Chapter 3
Natural Disasters: Definitions and Classification

3.1 Natural Disaster Classification: General Aspects

Thinking of disasters and catastrophes that are threatening our daily life, automatically impressions of volcanic eruptions, earthquake-destroyed houses, tsunami devastating coastal villages, or pictures of starving children in tent camps of the Sahel Zone come to mind. But the different manifestations of disasters are different in their origin and differ enormously in their impacts. Moreover it has been proved that certain disasters result in specific impacts. Although the term natural disaster expressing "processes made by nature" (with no interference of human beings) is quite in use, the term nevertheless only describes the outcome of the natural processes, and does not include the origin/trigger level of the process. Thus natural disasters are by definition the outcome of the process, and the triggering elements of disasters are called "natural hazards."

Therefore in order to better assess the cause–effect relationship, first of all a systematic classification of natural hazards is needed. Second, as all natural hazards pose a threat to human beings, natural disasters actually only occur when the potential threats interfere with human life. At third there is quite a large group of hazards existing that comprise all hazards derived from technical and human activity. And moreover we have to acknowledge that all three categories of hazards can interact with each other resulting in disasters the origin of which sometimes can hardly be identified (traced back) anymore.

3.2 Natural Hazards Originating from the Solid Earth

Regarding natural hazards there are quite a lot of different classification schemes in use worldwide that all have their advantages. As long as there is no one system agreed upon worldwide, this book wants therefore to emphasize that a classification scheme should be used that is simple enough to be understood everywhere

© Springer International Publishing Switzerland 2016

U. Ranke, *Natural Disaster Risk Management*,
DOI 10.1007/978-3-319-20675-2_3

and that is transparent enough to open ways to harmonize with other systems. Generally natural hazards are divided in these main categories:

1. *Geotectonic hazards*: Natural processes that have their origin in Earth's crust and mantle resulting in convectional movements that cause lithospheric plates to be permanently in motion; this motion lets mountain ranges build up, oceanic plates subduct under continental plates, or oceanic ridges develop. These movements are the triggering elements for earthquakes, volcanic eruptions, mass movements, or the uplifting or subsidence of land.
2. *Hydrometeorological hazards*: Natural processes that have their origin in the Earth's atmosphere. They are responsible for climate variations that create flash floods, droughts, storms, and/or extreme weather.

In order not to come up with just another system, this book wants to follow the classification scheme that has been introduced by CRED and Munich Re (Fig. 3.1) and that has proved its comprehensive applicability in many cases.

3.2.1 Geotectonic Hazards

3.2.1.1 The Earth Structure

The Earth is composed of three main layers that exhibit quite different chemical and physical composition (Fig. 3.2).

The outermost layer is called the Earth's crust and is comparatively thin, with thickness ranging from 5 to more than 70 km. This layer can be described as the outer shell of the Earth and is chemically composed of an upper layer made up mainly of silicate and aluminum (SIAL) and a lower layer mainly made up of silicate and magnesium (SIMA). The crust and the upper part of the lower lying earth mantle form the hard and rigid outer layer of the Earth called the lithosphere. The lithosphere is underlain by the asthenosphere constituting the weaker, hotter, and deeper part of the upper mantle. Two types of lithosphere are distinguished:

- *Oceanic lithosphere*, which is associated with the oceanic crust and which exists under the ocean basins (density of about 2.9 g/cbcm)
- *Continental lithosphere*, which is associated with the continental crust (density of about 2.7 g/cbcm)

This division should not be confused with the chemical subdivision of these same layers comprising both the asthenosphere and the mantle portion of the lithosphere and the crust. A piece of mantle may be part of the lithosphere or the asthenosphere at different times depending on its temperature and pressure. The asthenosphere is the ductile part of the Earth just below the lithosphere, including the upper mantle. The asthenosphere is about 180 km thick.

The oceanic crust forms the ocean basins. This crust type is rich in silica, iron, and magnesium. According to this chemical composition most of the volcanic

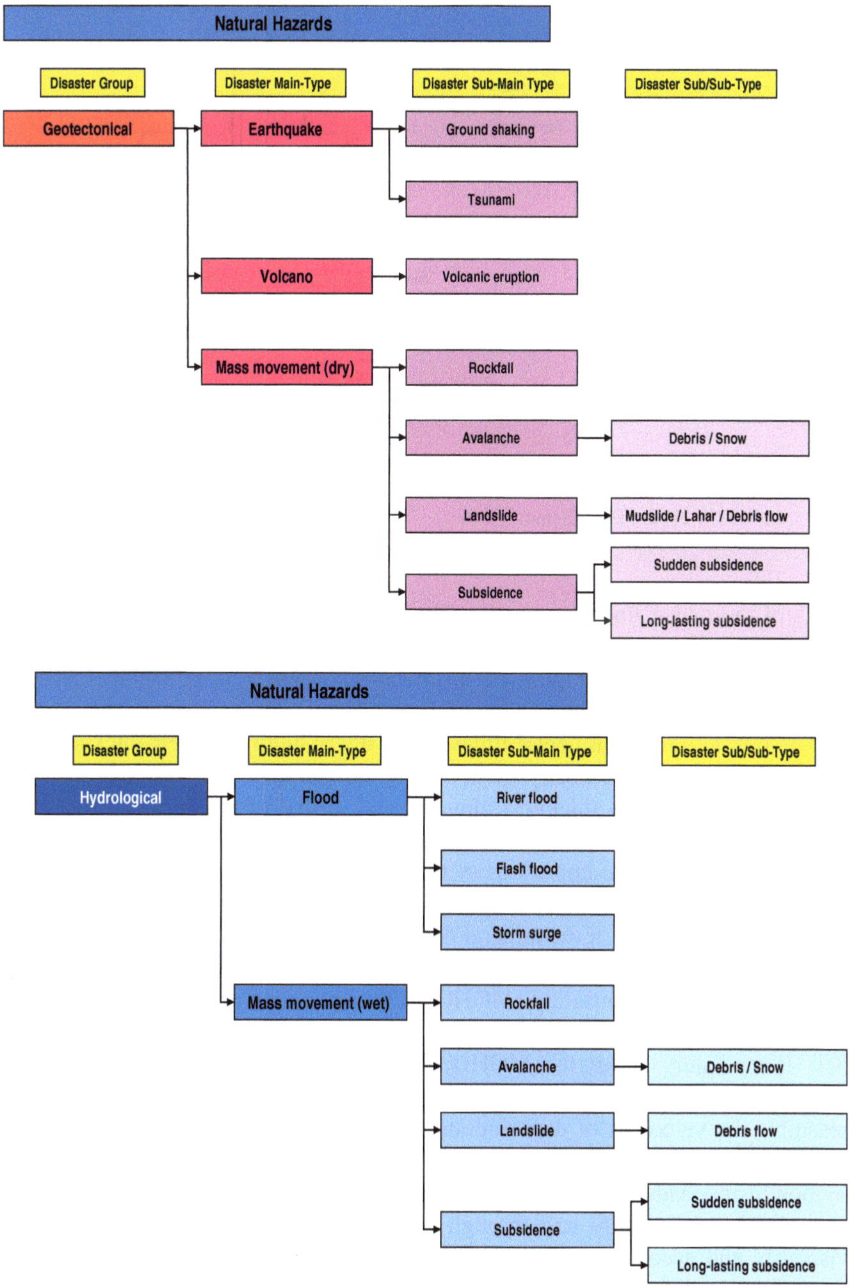

Fig. 3.1 Disaster type classification proposed by CRED and Munich Re

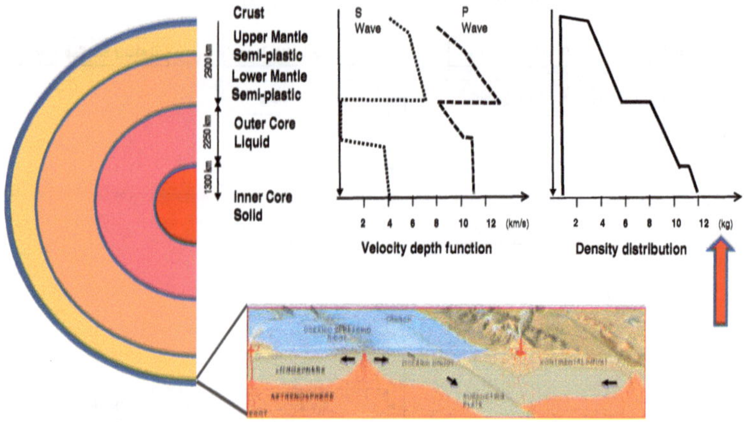

Fig. 3.2 Structure of the Earth and respective seismic velocity and density distribution (Compiled from USGS 2008a; Berckhemer 1990)

rocks of the ocean floor are basalts. The oceanic crust is much thinner than the continental crust that is made up of igneous granitic, sedimentary, and metamorphic rocks. About 40 % of the Earth's surface is covered by continental crust and it makes up about 70 % of the volume of the earth crust.

The oceanic and continental lithospheres differ highly in their thickness. The oceanic lithosphere is typically about 50–140 km thick, but only directly under the mid-ocean ridges is its thickness almost equal to that of the oceanic crust there, whereas the continental lithosphere has a range in thickness from about 40 km to more than 250 km. The upper part of the continental lithosphere is defined as the continental crust, typically from 30 to 50 km thick. The mantle part of the lithosphere consists largely of the mineral peridotite.

The boundary between the crust and the lower lying upper mantle is called the Mohorovicic discontinuity (MOHO), although it lies mainly within the lithosphere. The discontinuity is characterized by a sudden increase in seismic velocities. Immediately above the MOHO, the velocities of primary seismic waves (P-waves) are equal to those of basalt (6.7–7.2 km/s), whereas below it they resemble the velocities of the minerals peridotite and dunite (7.6–8.6 km/s). The boundary between the lithosphere and the underlying asthenosphere is defined by how it responds to external stress. Whereas the lithosphere remains rigid and only deforms elastically and through brittle failure, the asthenosphere deforms viscously and accommodates strain through plastic deformation. The base of the lithosphere is defined to be the temperature regime of about 1000 °C where the mineral olivine begins to deform viscously. When the stress exceeds the elasticity modulus of the lithosphere it breaks, forming the tectonic plates.

As the upper part of the Earth mantle, the asthenosphere is about 500 km wide (the lower boundary of the mantle is defined at depths of 2900 km) and exhibits a semi-plastic state that permits the above-lying lithosphere to "float" on the mantle material. As the lithosphere has a highly variable composition and thickness, the floating of the lithospheric plates is controlled by "isostasy", a physical phenomenon that can be compared to buoyancy. The lithospheric plates are thus floating on the asthenosphere like wooden logs in water. The thicker the continental crust rises, the deeper the roots sink into the mantle. The roots of ocean basins, however, are low lying because the oceanic crust is thin and dense. Under the Himalayas, for instance, the crust sinks to about 70 km whereas under the oceanic plates depths from 8 to 10 km occur.

The Earth core lies at depths beyond 2900 km and represents about 30 % of the planet's mass. The core again is divided into two layers. The "outer core" is about 2100 km thick. This layer is characterized by rocks that are mostly made up of iron and nickel (NIFE). The outer core has the highest specific gravity of the Earth material and is responsible for the Earth magnetisms. The "upper core" has a molten state due to excessive heat that is thought to originate from radioactive decay producing temperatures that lie between 2200 and 2750 °C. In contrast, the "inner core" (about 1400 km thick) shows a solid sphere due to the combined factors of high pressures and temperatures that could partly range between 4300 and 7200 °C. The density of the inner core is estimated between 12.8 and 13.1 tons/m^3. The pressure in the inner core of the Earth is between 3.3 and 3.6 million atmospheres.

Plate Tectonics

The similarity of both the coastlines of West Africa and South America was for a long time seen as just a curiosity of nature. It was in 1929 when Wegener (1929) a Germany geographer, made these striking shapes the basis for a new idea on how the continents may have been formed. He named his theory "continental drift". Up to that time most geoscientists were convinced the Earth is contracting due to cooling over the billions of years of its existence. Wegener's theory was mostly rejected because he could not explain the driving mechanisms of the continental drifting and what happen if drifted continents collided. Fifty years later the analytical possibilities had much advanced and opened the chance for a revolutionary change in the vision of Earth's development. When in 1965 Bullard et al. (1965) presented their fit of the African and South American coastlines—this time based on the geophysical evidence that the continental shelves are part of the continents—they could show that along the contour line at about 1000 m water depth, the two continents fit together almost perfectly. It was in the 1960s when the first leading papers were published by Hess, Wilson, Vine and Matthews, and many others, indicating that Earth's internal forces are the driving mechanisms for the plate movements. In particular the following scientific developments in the 1960s strengthened the formulation of the plate tectonics theory and seafloor-spreading hypothesis:

- The detection of mid-ocean ridges and the young age of the ocean floors
- The evidence of repeated reversals of the Earth's magnetic field
- The documentation that the world's earthquake and volcanic activity is concentrated along oceanic trenches and submarine mountain ranges

As about two thirds of the Earth's surface lies beneath the oceans, the knowledge that developed significantly after World War II of oceans helped significantly to understand the process that moves the Earth's plates as originally stated by Alfred Wegener 50 years earlier. With echo-sounding systems measuring the sea bottom morphology and using magnetometers recognizing odd magnetic variations across the ocean floor, the scientific base for modern plate tectonic theory has been laid. The understanding of the magnetic patterns ("normal" and "reversed") of ocean floor basalts reflect the reversing of the Earth's magnetic field with time that later became known as magnetic striping (Fig. 3.3). The striping pattern that was identified as running exactly parallel to the (discovered at the same time) mid-ocean ridges especially led to the assumption that the ridges mark a structurally weak zone in the oceanic crust where new magma erupts to the surface and creates a new oceanic crust, a process that was called "sea floor spreading." The magnetic reversals in the oceanic rock can only be dated back about 180 million years (Early Jurassic) and indicated that the older crustal parts had already been subducted. This evidence was a further indicator that enabled answering the question of why the sediments of the oceanic crust are older when departing from the mid-ocean ridges.

Plate tectonics is basically a kinematic phenomenon making the plates move with respect to one another. Today the driving force behind tectonic plate motion is assumed to be generated by large-scale heat convection currents in the upper mantle that dissipate heat through a process called "mantle convection." When mantle material close to the radioactive core is heated up it becomes less dense than the comparatively cooler upper mantle rocks. The "warmer" rocks rise while the "cooler" rocks sink, creating steady vertical convection cells within the mantle. These convection cells are assumed to be the driving force for mantle material. Although the movement is just only a few centimeters a year it provides a powerful source of energy that makes the plates move with a velocity ranging from 10 to 40 mm/year along the mid-Atlantic ridge up to about 160 mm/year at the Nazca plate.

Although this theory has gained wide acceptance, there is still a debate as to how mantle convection can directly and indirectly be related to plate motion as

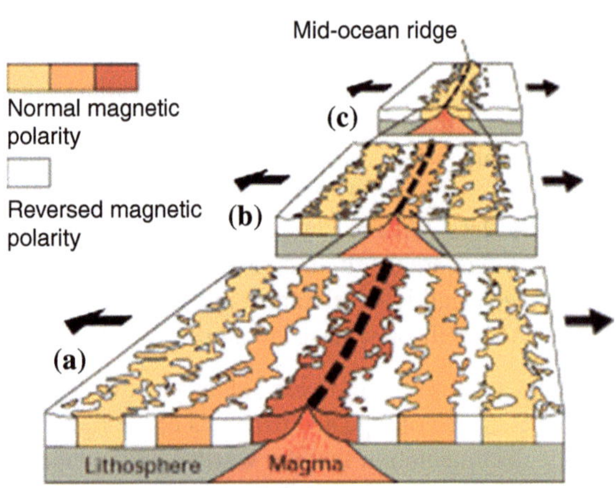

Fig. 3.3 Magnetic striping pattern of the oceanic crust as a result of the reversing of the Earth's magnetic field (*Courtesy* USGS 2008a)

even modern techniques (seismic tomography) still were not able to discover the predicted large-scale convection cells. Somehow, the energy must be transferred to the lithosphere for the plates to move. Another source of energy for moving the plates is driving forces related to gravity, although such forces are seen as secondary phenomena within the framework of mantle convection. At the mid-ocean ridges the uprising magma forms higher elevations from the hot mantle material along the spreading ridges. As in the course of plate movement, this material is gradually cooling and thickening with age. The cooling of the oceanic lithosphere makes it increasingly denser than the hot mantle material. Thus the lithosphere gradually subsides into the mantle to compensate the greater load. The result is a slight lateral incline with increased distance from the ridge axis, a driving mechanism that is often referred to as "ridge push," although it is a gravitational sliding rather than a push movement. A very significant driving force occurs when the oceanic plate converges with a continental plate and the plate on its way down into the mantle pulls the plate by its density, a force that is called "slab pull" and is widely thought to constitute a great force acting on the plates.

Plate Boundaries

As the lithospheric plates of the Earth are either moving apart (diverging) or coming together (converging) they permanently change the Earth's surface. Although all the plates appear to be moving at different relative speeds and independently of each other, a worldwide puzzle of plates results in a multitude of highly interconnected and interrelated plates (Fig. 3.4). No single plate moves without affecting others and the activity of one can influence another thousands of kilometers away. For example, as the Atlantic Ocean grows wider with the spreading of the African plate away from the South American plate, the Pacific sea floor is being consumed in deep subduction trenches over 10,000 km away (USGS, ibid).

There are three primary types of tectonic plate boundaries as shown in Figs. 3.5, 3.6, 3.7 and 3.8 that all have been provided by the USGS (ibid).

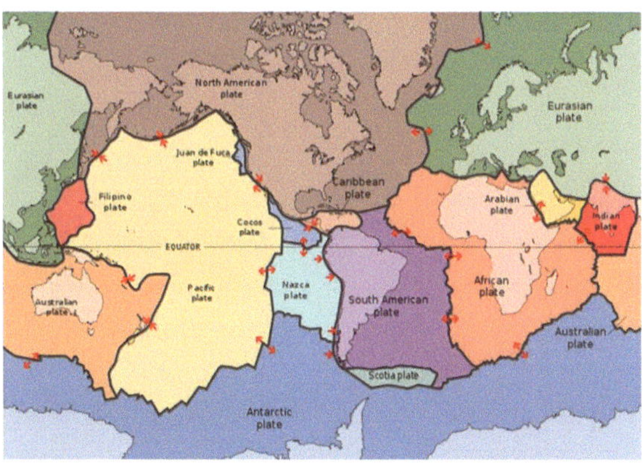

Fig. 3.4 Pattern of global lithosphere plates (*Courtesy* USGS, ibid)

Fig. 3.5 Divergent plate margin of Iceland, exhibiting that Europe and America are drifting apart (*Courtesy* USGS, ibid)

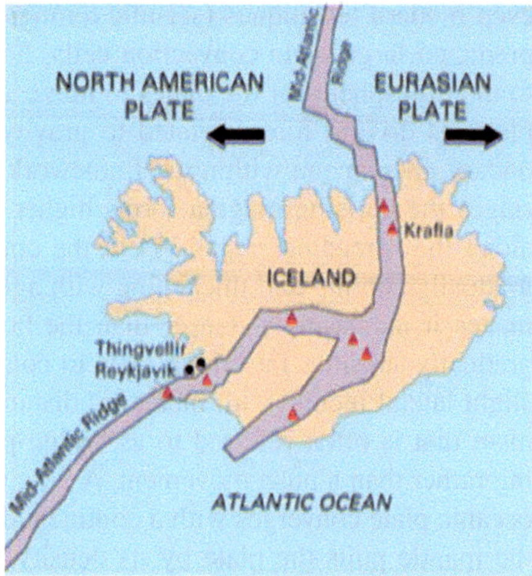

Fig. 3.6 Oceanic–continental plate convergence (*Courtesy* USGS, ibid)

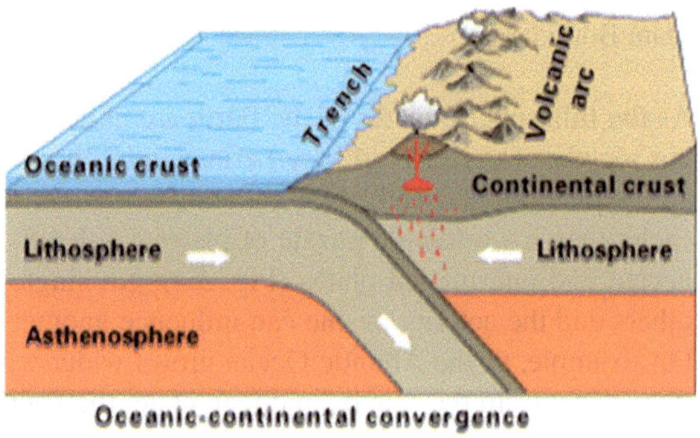

Fig. 3.7 Oceanic–oceanic plate convergence (*Courtesy* USGS, ibid)

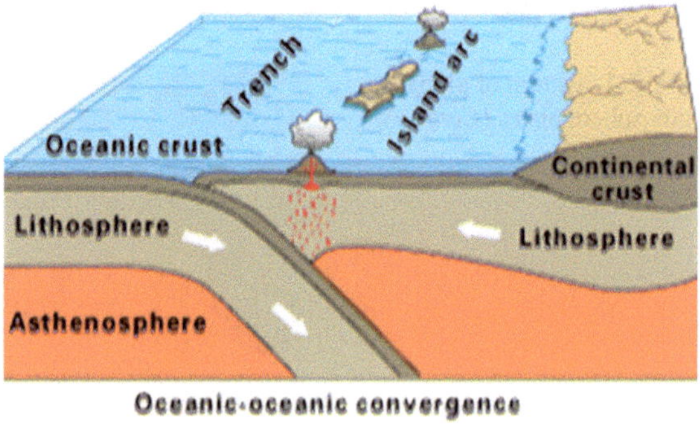

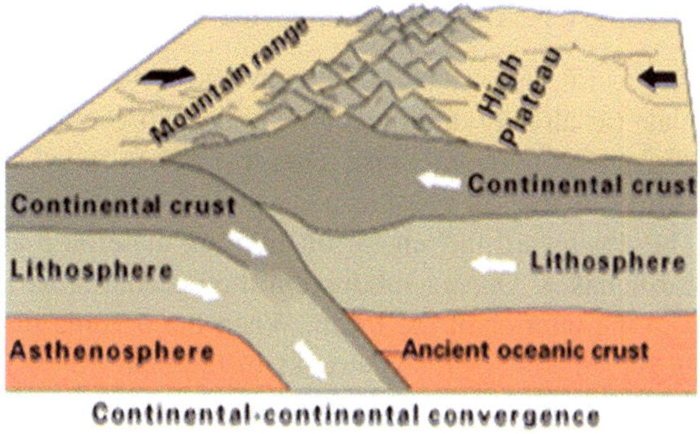

Fig. 3.8 Continental plate–continental plate collision (*Courtesy* USGS, ibid)

Divergent Boundaries

Places where plates are moving apart are called divergent boundaries. When the Earth's brittle lithosphere is pulled apart, it typically breaks along parallel faults (Fig. 3.5). While continuing to separate along the boundary, the material between the two sides of the fault is plunging down into the soft plastic interior (asthenosphere). The sinking of the block thus forms a central valley called a rift. Magma seeps upward to fill the fractures and allows new crust to be formed. Earthquakes occur along the faults, and volcanoes form where the magma reaches the surface. Oceans like the Atlantic are born this way and grow wider when the plates pull apart. Where a divergent boundary occurs at the ocean floor a rift valley is formed, generally about several kilometers wide as can be seen along the mid-Atlantic ridge. They normally rise a kilometer above the ocean floor and form a global network of rifts that can be traced all over the globe. The plates' separation is quite slow. For example, the speed of divergence along the mid-Atlantic ridge is only about 2 cm/year. When a diverging boundary occurs on land a rift, or separation, will arise and over time that mass of land will break apart into distinct land masses and the surrounding water will fill the space between them. This can be seen in the East African Afar triangle, where the African and the Arabian plates have diverged since 2010 at a local speed of up to a meter per year. Eastern Africa will inevitably break apart into two separate land masses in the future, and Indian Ocean waters will eventually rush into fill the widening and deepening space between. On land divergent boundary rift valleys are formed that are typically 30–50 km wide. Examples include the East Africa rift from Ethiopia down to Mozambique or the Rio Grande rift system in New Mexico.

Convergent Boundaries

The locations where plates collide or "crash" together are called convergent boundaries. Three types of plate convergence occur.

Oceanic Crust: Continental Crust Convergence

In general the oceanic plate converges under the continental plate as the oceanic plate (as already described above) has a higher density and thus plunges under the "lighter" continental crust, a process that is called subduction (Fig. 3.6). At the sea bottom where an oceanic plate moves downward, deep sea trenches are formed. The downward movement of the rock material makes the plate break up causing earthquakes. During subduction the oceanic crust is destroyed and recycled back into the interior of the Earth. When the oceanic plate continues to slip down into the Earth's interior, some rocks of the plates melt. The subducting plate is heated up at a depth of about 150 km and at temperatures beyond 1000 °C. Magma chambers are produced as a result of this melting, and as the magma is lower in density than the surrounding rock material it begins ascending by melting and fracturing its way through the overlying rock material. Magma chambers that reach the surface break through to form a volcanic eruption. Moreover, the water content in the oceanic crust is released due to temperature and pressure increase. The vapor reduces the solidus temperature, for instance, of the basalt. On its way up through the continental plate the melted rock causes volcanic-related earthquake tremors, and finally forming volcanic eruptions where it reaches the surface. As the average travel speed of the plates differs from 2 to 17 cm/year, the collisions generally last millions of years, a speed that is from a geological history viewpoint rather fast. The majority of global mountain chains and volcanoes are found where plates converge; the best example for a chain of volcanoes is found around the Pacific Ocean, called "the Ring of Fire." Furthermore with the bending down of the oceanic plate the overriding continental plate is lifted up and a mountain range is created like the Andes Mountains or the Indonesian Archipelago.

Oceanic Plate: Oceanic Plate Convergence

When two oceanic plates converge the plates—similar to the oceanic–continental plate situation—subduct under the other (Fig. 3.7). Normally the older plate will subduct because of its higher density. This type of plate convergence forms undersea volcanoes. And when over millions of years the erupted lava and volcanic debris pile up on the ocean floor, a submarine volcano rises above sea level to form an island volcano. Such volcanoes are typically strung out in chains called island arcs. In the course of the subduction the well-known deep oceanic trenches are formed, such as the Marianas Trench as a result of the Philippine plate subducting under the Pacific plate. With continued development the islands can grow larger, and merge to real landmasses like Japan, the Aleutian Islands, or the Eastern Caribbean islands.

Continental: Continental Convergence

When two continents meet head on they do not subduct because both continental plates have a density that is much lower than the mantle, which prevents them from subduction; although there may be a small amount of subduction when the heavier lithosphere below the continental crust might break free from the crust and subduct. Fragments of crust or continental margin sediments might be caught in the collision zone between the continents forming a highly deformed melange of rock. Instead of subducting the plates tend to buckle and to be pushed upward or sideways (Fig. 3.8). The Himalayan mountain range is the best active example of this type of plate boundary, where the collision of the India continental plate 3.50 million years ago caused the Eurasian continental plate to crumple up and override the Indian plate. After the collision, the slow continuous convergence of the two plates over millions of years pushed up the Himalayas and the Tibetan plateau to their present heights. Most of this growth occurred during the past 10 million years. The Appalachian mountain range is an ancient example of this collision type. The continental–continental plate collision process is still poorly understood when compared to the other types of plate boundaries. Nevertheless the huge global mountain chains prove that by this type of convergence a powerful collision can occur. The intense compression can also cause extensive folding and faulting of rocks within the two colliding plates. This deformation can extend hundreds of kilometers. Moreover the effects from collision include shallow earthquake activity, and shortening and thickening of the plates.

Transform Boundaries

Transform boundaries are characterized by two plates sliding horizontally past one another with sideways and not vertical displacement (Fig. 3.9). These boundaries are also known as transform fault boundaries or faults. Transform faults differ from normal strike-slip faults because the sense of movement is in the opposite direction. A strike-slip fault is a simple offset, however, a transform fault is formed between two different continental plates, each moving away from the spreading center. Most transform faults are found on the ocean floor. They are best known from the active spreading ridges—like the mid-Atlantic ridge—producing "zigzag plate margins" that are also known as shallow earthquake locations. A few of such continental–continental plate margins are also known to occur on land and are in general marked by linear valleys along the boundary where rock has been ground up by the sliding. The best examples are the Alps, the North Anatolian fracture zone in Turkey or the famous San Andreas Fault zone in California. The San Andreas is one of the best monitored transform faults on Earth. It is about 1300 km long and in places tens of kilometers wide. Along it, the Pacific plate has been moving past the North American plate for 10 million years, at an average rate of about 5 cm/year. This would lead to the fact that in about 10 million years' time both cities San Francisco and Los Angeles will be located side by side. Along the San Andreas fault the land on the Pacific side of the fault zone is moving in a northwesterly direction relative to the land on the east side of the fault zone on the North American plate. Transform faults are locations of recurring earthquake activity and faulting. The earthquakes are usually shallow because they occur within and between plates that are not involved in subduction. Volcanic activity is normally not present because the typical magma chamber convection or a melting subducting plate is not present.

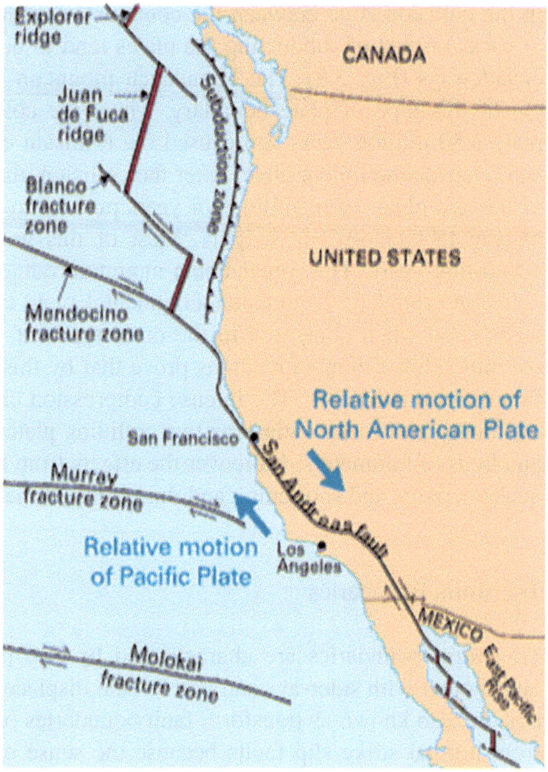

Fig. 3.9 Transform fault (*Courtesy* USGS, ibid)

Earthquakes

About three billion people making up almost 50 % of the world population are supposed to live in areas that are today classified as earthquake prone. And since 1900 it is supposed that more than 1.2 million people were killed in earthquake disasters; most of the victims were claimed in China (500,000), Japan (200,000), and Italy 100,000. Nevertheless the strength of earthquakes differs very much from region to region. Thus in California, the region with the highest earthquake risk of the United States, there have been in the last 100 years along the San Andreas fault the same amount of victims claimed as in the Romanian capital, Bucharest (1500), on only one occasion in 1977. Four catastrophic earthquakes have already struck since the beginning of the twenty-first century, including the 2004 Sumatra–Andaman earthquake and tsunami and the 2010 Haiti earthquake that each may have killed over 200,000 people. Guha-Sapir et al. (2011) listed that on average since 1990 every year 27,000 people are killed by earthquakes worldwide. Worldwide about several millions of earthquakes occur every year. More than 90 % of them are unidentifiable without technical devices. Only some thousand are strong enough to be recognized and able to create minor damage. It is supposed that fewer than 1000 earthquakes occur every year that we perceive as tremors, of which only a couple really create devastating damage. Surface manifestations of earthquakes are the most impressive signs people perceive and that makes them fear. By far the biggest impression comes from the ground motions that the people feel directly or from the damage of the buildings and infrastructure installations and rescue operations where people try desperately to dig out victims using only their hands. But also distorted rows of trees or fences are often recognized as well as a shift along fault zones, best seen in aerial photos or satellite imagery (Fig. 3.10).

With the continuous moving of the lithospheric plates by pulling apart, converging, or steadily transforming the boundaries a tremendous amount of energy is unleashed that often results in earthquakes, tremors, and volcanism. In a very general sense, earthquakes describe any seismic event, whether natural or caused by human activity, that generates seismic waves. Earthquakes are caused mostly by rupture along geological faults or fractures, but can also be generated by other events such as volcanic activity, landslides, mine blasts, or nuclear explosions. An earthquake's point of initial rupture is called its "focus" or "hypocenter", and the point at ground level directly above the hypocenter is called the "epicenter".

The energy is released in the form of a pressure front propagating from the location where the plates collide or drift away from each other. The energy is released in the form of elastic seismic waves, as the Earth's material behaves in general elastically. The degree of elasticity determines how well the waves are transmitted through the Earth's interior. The released energy brings the material either by compression, tension, or shearing under external strain that means it

Fig. 3.10 San Andreas Fault zone, California and its impact on the Earth surface (*Source* Wikipedia, file: kluft-photo-Carrizo-Plain-Nov-2007-Img 0327.jpg; accessed: 24 May 2014)

changes its volume and/or its shape. The material reacts to this strain either elastically, meaning after the external pressure is over the material returns to its former volume and shape, or it reacts inelastically (ductile, plastic) meaning the external deformation remains. When the compression, tension, or shearing (stress) and the resulting strain overstep the limits of the material elasticity (elastic moduli), the material begins either to crack and crush (brittle) or to react (ductile).

For long time the "elastic rebound theory" was seen to best explain earthquake generation. This theory describes earthquakes as a result of the elastic rebound of previously stored elastic strain in the rocks. For a couple of years now there have been indications that this theory alone will not answer all the questions concerning earthquake generation (Chui 2009). If elasticity is the main trigger this would imply that earthquakes might be a regular phenomenon as the strain that has been built up must be released either in the form of earthquakes or through a slow quiet alternative called aseismic slip. Another view was that faults generally exhibit rupturing in large characteristic earthquakes of about the same magnitude again and again. Another model asserts that big quakes are most likely to strike in seismic gaps that haven't suffered major jolts in a long time. But recent research reveals that earthquakes don't occur on simple fault structures, rather on fault systems that are very often involved in complex interactions among faults. In some cases, quakes come in clusters such as the 9.1 magnitude earthquake that hit Sumatra in 2004 and that led to a series of aftershocks (even until 2014). In other instances, the earthquakes are rupturing large sections of faults that had not been known to quakes before. Furthermore there are indications that earthquakes can trigger tremors, geyser eruptions, and other seismic activity thousands of kilometers away.

The seismic waves that travel the Earth are characterized by the way they move the rock particles: either they move it in the direction of the wave propagation or perpendicular to it.

Four different seismic wave types are distinguished:

- Compressional waves have a longitudinal polarization that means the displacement of the rock is back and forth in the direction of the wave propagation. These waves have a high travel speed of 5.5–11.5 km/s. Accordingly they arrive first in the seismogram and are therefore called primary waves (P-waves). P-waves are able to propagate in solid, gaseous, and liquid media and thus are able to travel through the liquid core of the Earth. The P-wave exhibits a change in volume and shape of the rock material.
- The other underground wave types are shear waves (S-waves). This wave type propagates horizontally or vertically perpendicular to the wave propagation direction and only shear the rock material with no change in its volume. Due to this different mode of wave propagation S-waves travel much more slowly (3.5–6.0 km/s) and thus they "arrive" in the seismograms after the P-waves; therefore they are also called "secondary waves." An S-wave can only propagate through solid materials.
- Love waves result from the interference of many horizontal shear waves. They travel with a slower velocity than P- or S-waves, but faster than Rayleigh waves.

The particle motion of a Love wave is transverse and forms a horizontal line perpendicular to the direction of propagation. Because Love waves travel on the Earth's surface, the strength of the waves decreases exponentially with the depth of an earthquake. Large earthquakes may generate Love waves that travel around the Earth several times before dissipating and their amplitude decays comparatively slowly. Love waves are the most destructive forces outside the immediate area of the earthquake's focus (epicenter).

- Rayleigh waves are a type of wave that travels near the surface of solids. Rayleigh waves include both longitudinal and transverse motions that decrease exponentially in amplitude as distance from the surface increases. Rayleigh waves are generated by the interaction of P- and S-waves at the surface of the Earth, and travel with a velocity of about 3 km/s that is lower than the P-, S-, and Love-wave velocities.

The seismic waves propagating from an earthquake spread out in all directions through the Earth's interior. The waves can either travel through the Earth's interior (body waves) or exclusively along the Earth's surface (surface waves). They are measured by seismometers that take advantage of a pendulum transferring seismic waves into a horizontal or vertical movement. The wave propagation from the hypocenter can be described as a mechanical wave or vibration that leads to a compression and extension of the rock particles. In this regard a mechanical wave is comparable to a sound wave. In general earthquake waves can either run through the Earth's interior or along its surface. Seismic waves propagate faster in hard, solid, and uniformly made rock sequences, whereas in soft sediments or heavily fractured rock sequences they travel much more slowly as they lose their energy. Such a loss in energy is called "wave attenuation."

Seismic stations that are located all over the world record the waves according to the time difference the waves need to travel from the epicenter. The travel velocity depends on material properties such as composition, density, mineral phase, temperature, and so on of the rock material through which seismic waves pass. Seismic waves travel more quickly through denser materials and therefore generally arrive earlier with respect to the travelled distance. Moreover, seismic waves move more slowly through a liquid than a solid. Therefore molten areas within the Earth slow down P-waves and stop S-waves because their shearing motion cannot be transmitted through a liquid. When seismic waves pass between geologic layers with different seismic travel velocities the waves are reflected, bend, or can even produce new wave phases.

From the seismograms clear indications can be drawn of where the earthquake has happened and what travel paths the seismic rays have taken (Fig. 3.11). When the incoming wave signal is identified as originating at a distance of less than 30° from the epicenter it has mainly travelled through the upper mantle. Between 30 and 100° the P- and S-waves have travelled through the lower mantle and beyond 100° only P-waves can be recorded, as they have travelled through the outer core. From the time differences of the P- and S-waves recorded at a seismic station the internal structure of the Earth has been identified. The outer core especially is due to its liquid phase which is very indicative as it does not allow S-waves to pass. Seismic ray paths show that at distances between 100° and 180° a zone occurs on the surface where no S-waves can be recorded, the so-called S-wave shadow. In contrast, the reflection of P-waves at the mantle–outer core boundary reveals a P-wave shadow between 103 and 143°.

A theoretical impression of how seismic waves propagate through the Earth's interior is given in Fig. 3.11. The waves travel either along the surface or through the body of the Earth (body waves). When the waves encounter a lithospheric discontinuity a part of the waves is reflected or refracted and another part deviates like light when passing from one medium to another. Thus the seismic waves are able to cross the entire Earth and can be recorded even on the other side of the globe.

According to information given by the USGS (2014) there are about 5 million earthquakes per year of which:

50,000	have a magnitude between 3–3.9
6000	have a magnitude between 4–4.9
800	have a magnitude between 5–5.9
120	have a magnitude between 6–6.9
18	have a magnitude between 7–7.9
1	has a magnitude of higher than 8

The largest earthquakes in historic times have been of magnitude slightly over 9 (Valdivia 9.3, Chile 1960), although there is no limit to the possible magnitude. The most recent large earthquakes of magnitude 9.0 or larger were the 2004 Sumatra quake and the 9.0 earthquake in Japan in March 2011.

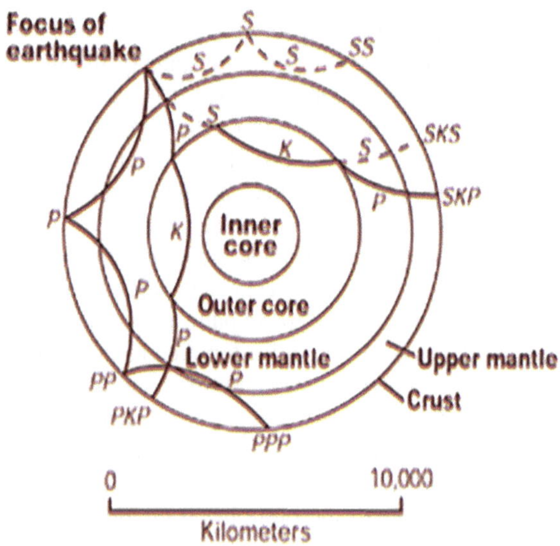

Fig. 3.11 Seismic ray paths through Earth's interior (*Courtesy* USGS, ibid)

3.2.1.2 Measuring Earthquakes

People who once or often were affected by earthquakes have worried about why such a tragedy happened and whether it might happen again. This information demand can be taken as the origin of modern seismology. From the very beginning people were interested in having an instrument at hand to provide an objective assessment of the earthquake's strength. Therefore in historical documents a multitude of information on natural disaster events such as flood, volcanic eruption, droughts, and also earthquakes can be found. The problem in reconstructing natural disaster events from historical records is that the information given is generally very inhomogeneous, too scarce, and very subjective. What was needed was a measure that is objective, reproducible, and could be used all over the world under the same conditions. With the seismograph developed by Wiechert and followers the technical basis for such independent measures was laid. From the records gathered by these instruments in the course of time two different assessment methodologies to define the strength of an earthquake were developed: the "magnitude" (Richter magnitude scale) which measures the amount of energy released (ES) by the event and the "intensity" (Mercalli intensity scale).

Information on the first earthquake in man's history was given by the Chinese about 1100 BC (Fig. 3.12), and it was mentioned in an earthquake catalogue that listed several dozen large earthquakes during the next few thousand years. In Europe the first descriptive records on earthquakes date back to the mid-sixteenth century. The earliest known earthquake reports in the Americas came from Mexico where events in the late fourteenth century and in Peru in 1471 were recorded

Fig. 3.12 Model of the first earthquake identification instrument (Han Dynasty), China, about 132 AD (*Source* file: EastHanSeismograph.JPG, Wikipedia; "Zhang Heng"; access: 20 May 2014)

although not very well documented. By the seventeenth century, descriptions of the effects of earthquakes were being published around the world, with still an often highly questionable degree of reliability.

The first seismograph of modern times that was able to measure ground motions reproducibly was built by Emil Wiechert at the University of Göttingen in 1900. He invented a recording system based on the concept of a reversed and air-damped pendulum. A solid mass hanging independently over the moving Earth takes up the amplitude of the ground motion with some delay. The normal amplitude of a free hanging pendulum is according to the volume of the Earth one second and therefore pendulums like this are called "second pendulums." As ground motion not only has horizontal amplitude but also a vertical component, Wiechert also invented a pendulum that is able the measure the vertical ground motion component.

Thus developing a reliable and homogeneous dataset, for instance, of earthquake distribution and damage patterns requires the use of as many different data sources as possible and makes cross-checking of the results indispensable. Nevertheless it has to be acknowledged that the older an event dates back the less reliable is the information on patterns of earthquake location, strength, and destruction pattern. Furthermore it has to be acknowledged that before 1950 there were no instruments available that guaranteed an accurate and reproducible record of an event. Meanwhile the technology of seismographs has been much developed and today even the smallest earthquake anywhere on the globe can be identified in all three directions (X/Y/Z). The most important and reliable measure to identify how strong an earthquake was, is the energy released by seismographs/accelerographs expressed by its peak ground acceleration (PGA).

Richter Magnitude Scale

From the seismograms not only the epicenter distance and the travel path can be derived but also the strength of an earthquake as the amplitude of the seismogram is clearly related to the quantity of energy released.

The best-known measure to rate the strength or total energy of earthquakes is the famous Richter magnitude scale, invented by Charles Richter and Beno Gutenberg (Richter 1935). Gutenberg proposed to use the maximum amplitude of a wave group to be taken as an indicator of the total energy. The scale those days opened the possibility for a worldwide comparison of earthquake events. It has no upper limit but usually ranges from 1 to 9. The scale uses a logarithmic scale (base 10-log) which defines the magnitude as the logarithm of the ratio of the amplitude of the seismic wave. Because it is logarithmic scale, an earthquake rated as 5 is ten times as powerful as one rated as 4. Physically this value corresponds to a 31.6 times greater release in energy than a magnitude 4 earthquake. Richter and Gutenberg developed the method by putting the seismic travel time and amplitude of a seismic registration into an empirical relation, based on the evidence that the farther the distance of an earthquake the longer is its travel time and the higher the amplitude in the seismogram, the stronger are the damages.

The Richter magnitude scale was later abolished as the instruments once used by Richter and Gutenberg were replaced by worldwide networks of seismic stations. Today the seismic moment of an event is calculated that is proportional to the physical size of the event as it is derived from the area of the rupture times the average slip that took place in the earthquake. The millions of seismic data from all over the world collected every day enable the seismologists to quite precisely calculate the magnitude even in the range higher than magnitude 9. Moreover, as seismographs are recording the ground motions as a function of time, a comparison of different signals for known seismic monitoring stations enables the calculation of the distance between the seismic source and instrument.

Macroseismic

Even in ancient times, the first impressions the people had of earthquakes were the destructive impact on their houses, the number of persons killed, and the amount of people made homeless. For more than 1000 years the effects of past earthquakes were therefore written down in historical records mainly describing the death toll and the date of the event. According to the technical capability of those times the people mostly reported in short notes, often scarce and imprecise in the description of the damage pattern and often unclear concerning the date. Therefore since the Middle Ages all over Europe—the first records can even be traced back until 1000 AD—natural disasters events (earthquakes, volcanic eruptions, and floods) were recorded.

As the severity of destruction was the only outcome people could easily recognize, this fact became the basis of a classification system. Geophysically the impact on the Earth's surface is related to the strength of the shaking that again is related to the energy released. This part of seismology is called "macroseismology" and has developed into a useful and well-accepted part of seismology, although macroseismology is more a classification of observed effects in a limited area, rather than a measure of the strength of the shaking of an earthquake. The classification provides an "idealized" description of the effects generally called "macroseismic intensity." Such an assessment nevertheless is exclusively based on a subjective impression made by an unequipped observer rather than on a physical parameter. Macroseismic is the most important cornerstone to formulate a seismic risk reduction strategy. The basic concept is to provide information on the earthquake severity from such observation of the damage at the local and regional levels. By transferring such information into a vulnerability model of an area under investigation it is possible to define the probable damage distribution of future earthquakes.

The first steps towards an earthquake damage assessment method were based on visual impressions stated in ancient times. But it took until the 1880s when the Italian geophysicist Guiseppe Mercalli introduced his famous "Mercalli scale" which soon became the internationally agreed-upon classification system. Since then the scale underwent a couple of modifications. It originated on the times widely used 10-degree Rossi–Forel Scale. In 1902 the 10-degree Mercalli scale was expanded to 12 degrees by Italian physicist Adolfo Cancani and later redefined by the German geophysicist August Heinrich Sieberg. From that time on the scale became known as the Mercalli–Cancani–Sieberg (MCS) scale. In 1956 the scale was completely overhauled by Charles Richter (MM56). Today the Mercalli scale is known as the modified Mercalli scale (MM) or modified Mercalli intensity scale (MMI). In order to make the scale more practically adoptable quantitative aspects of the damage were introduced by Medvedev, Sponheuer, and Karnik in 1964 (MSK) and used in Europe for almost half a century. Today the European Macroseismic Scale (EMS-98) has been agreed upon to set the basis for evaluation of seismic intensities in European countries (Grünthal 1998). The scale is divided into 12 sections and has meanwhile proved to function as an encouragement to the interdisciplinary cooperation between engineers and seismologists. Since then the scale has undergone several modifications and in 1998 the EMS-98 was declared the sole measure to compare seismic impacts at least in the European region.

By the EMS-98 scale the effects of an earthquake on the Earth's surface, humans, objects of nature, and man-made structures can be quantified and thus compared on a scale from I (not felt) to XII (total destruction). The lower degrees of the Mercalli scale generally deal with the manner in which the earthquake is felt by people. The higher numbers of the scale are based on observed structural damage. The effect depends upon the distance to the earthquake, with the highest intensities being around the epicentral area. Data gathered from people who have experienced the quake are used to determine an intensity value for their location. The advantage of the method is that it relates directly to the natural phenomenon, with the damage to buildings, people, and their living environment. The disadvantage is that the Mercalli scale is quite openly formulated, leaving much space for interpretation. Moreover, the assessment is in general carried out by those affected and even when carried out by experts is seldom exact enough to be used for nationwide or international comparison.

The essential feature that distinguishes the MMI from the EMS98 is according to IASPEI (2002, Chap. 12) "that:

- MMI attempted to distinguish between the effects of earthquake shaking on buildings of different construction types, using type as an analog of strength,
- EMS 98 employs a series of six vulnerability classes which represent strength directly and involve construction type, but also other factors such as workmanship and condition."

Moreover the EMS 98 distinguishes between structural and nonstructural damage and the different forms of damage defining five classes of destruction: slight, moderate, heavy, very heavy, and complete destruction.

In Table 3.1 the European magnitude scale (EMS-98) is given with its respective vulnerability classification (Description) and in Fig. 3.13 the vulnerability classes according to building/construction type are also based on EMS-98.

Table 3.1 European macroseismic scale (EMS-98)

EMS	Definition	Description
1	Not felt	Not felt, even under the moat favourable circumstances
2	Scarcely felt	Vibration is felt only by individual people at rest in houses, especially on upper floor of buildings
3	Weak	The vibration is weak and is felt indoors by a few people. People at rest feel swaying or light trembling
4	Largely observed	The earthquake is felt indoors by many people, outdoors by very few. A few people are awakened. The level of vibration is not frightening. Windows, doors and dishes rattle. Hanging object swing
5	Strong	The earthquake is felt indoors by most, outdoors by few. Many sleeping people awake. A few run outdoors. Building tremble throughout. Hanging objects swing considerably. China and glasses clatter together. The vibration is strong. Top heavy object topple over. Doors and windows swing open or shut
6	Slighty damaging	Felt by most indoors and by many outdoors. Many people in buildings are frightened and run outdoors. Small objects fall. Slight damage to many ordinary buildings e.g., fine cracks in plaster and small pieces of plaster fall
7	Damaging	Most people are frightened and run outdoors. Furniture is shifted and objects fall from shelves in large numbers. Many ordinary buildings suffer moderate damage: small cracks in walls; partial collapse of chimneys
3	Heavily damaging	Furniture may be overturned. Many ordinary buildings suffer damage: chimneys fall; large cracks appear in walls and a few buildings may partially collapse
9	Destructive	Monuments and columns fall or are twisted. Many ordinary buildings partially collapse and a few collapse completely
10	Very destructive	Many ordinary buildings collapse
11	Devastating	Most ordinary buildings collapse
12	Completely devastating	Practically all structures above and below ground are heavily damaged or destroyed

Courtesy IASPEI (2002)

Other than earthquake magnitude that is recorded from instruments, the intensity is assessed exclusively by visual inspection. In general, data collection on seismic intensities is based on first-hand surveys of seismologists, second from interrogation among the people that were exposed to the quake, or third by using historical data often found in old church documents. The main obstacle to overcome by interviewing is that memory fades very quickly and that often the same event is highly differently perceived by different people. Questionnaires should therefore be easily understandable, simple to answer, but as precise as possible regarding the factual base of the event. The interviewee should either be chosen randomly or unselected. Another fraction of interviewees are the mandated officials and experts. The questionnaire must also be designed for that group of people. There is a variety of questionnaires in use all having in common that they assess the earthquake shaking, its sound, the effects on the people and animals, the

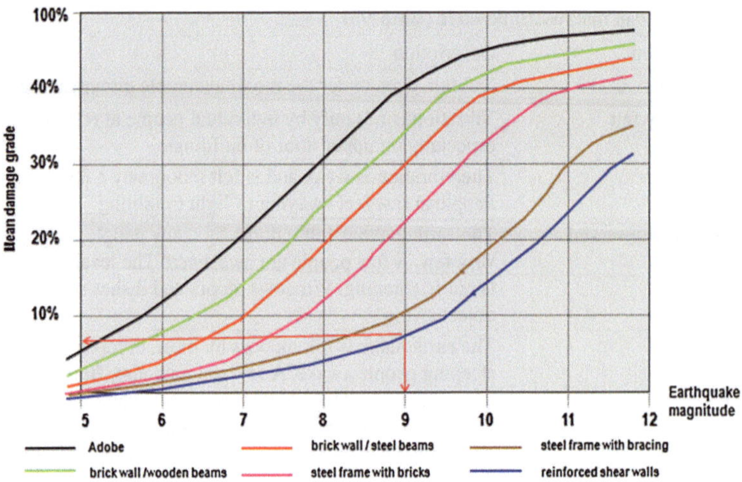

Fig. 3.13 Vulnerability classes according to building/construction type (Based on EMS-98)

Table 3.2 Macroseismic intensity assessment questionnaire

Personal data	Effects on person	Effects on objects	Effects on animals	Effects on building
Name Address Date/Time Location	Position during the earthquake If inside, which floor Activity before the earthquake began How strong the earthquake was felt What type of motion by the earthquake was felt Did other people feel the earthquake also Immediate personal reaction • No reaction • Surprise • Awakening • Fear • Problems maintaining balance • Running outside • Panic	Hanqing objects Glass ware, China Windows, doors Easily movable objects Small stable objects Light furniture Large objects overturned Liquids (jar, containers) Tombstones (moved) Wave in the ground Monuments, columns	Animals indoor Animals outdoor	Type of building • No idea • No coment • Stone masonry • Brick masonry concrete • Steel • Wood Use of building • No idea • No comment • Private • Public Degree of damage • No damage observed • Hairline cracks • Cracks in walls • Falling plaster • Chimney collapsed • Individual walls and columns collapsed • Total collapse

Courtesy Lang (2002)

effects of the household, and the building substance. An example recommended for Switzerland is given in Table 3.2.

Today such records are used to trace back the disaster history of a region as demonstrated for all of Switzerland or by Barbano et al. (2001) for the earthquake events of the town of Nicolosi, Sicily, or for an entire country such as Italy by the "Catalog of Historical Earthquakes in Italy" (CPTI; Camassi 2004). Once all the information is transferred into numbers, they can be plotted on a map. The resulting intensity map gives a comprehensive picture of the pattern of effects of the earthquake in contours that are equal to the intensity values that have been observed. The contours are called isoseismals and will be highest near the epicenter and gradually decrease with distance. The decay of intensity with distance is called attenuation. Mostly the contours have an elliptical shape with the long axis directed according to the orientation of the fault that produced the earthquake.

Transforming locally observed damage into a regional earthquake intensity distribution has been carried out for many earthquake-prone regions of the world, for example, on the 2009 L'Aquila (Italy) earthquake damage using a Web-based survey. With the help of an online questionnaire 65,000 people reported on their individual impressions. The huge amount of answers allowed a broad range of statistical assessments that revealed good agreement with the magnitude recording of the area. A similar assessment has been carried out by the USGS on an earthquake along the San Andreas fault system. Macroseismic assessment from all over the world proved that it is generally possible to derive information from earthquake magnitudes from intensities. This is especially useful as it helps to extend earthquake catalogues back into historical times. Nevertheless it has to be acknowledged that local geology strongly influences the peak ground velocity, or that strong but short-time ground motion has lower damaging potential than accelerations that are less in strength but last for a longer period.

Volcano Seismology

Not only plate movements but also volcanic activity can cause seismic signals. Moreover, it was recognized that nearly all volcanic eruptions have been accompanied by seismic signals such as the Mt. Pinatubo eruption in 1991 (Newhall and Punongbayan 1996). Since its first detection in the early 1900s, extensive seismic networks were deployed, today collecting a variety of information at a great number of volcanoes worldwide (Scarpa and Tilling 1996).

Nevertheless a huge number of volcanoes are still unmonitored to the extent necessary (USGS 2005a). Almost all volcanoes showed increased levels of earthquake shaking before an eruption took place. It is assumed that rising of the magma in the vent and the degassing of the magma are the origin for the low-frequency shaking as the material must find its way through the volcanic structure either by using existing fractures and cracks or by breaking up new pathways. The shaking can be monitored as high-frequency earthquake signals indicative of rock breaking caused by the fracturing of brittle rock as magma forces its way upward. These short-period earthquakes signify the growth of a magma body near the surface, and low-frequency waves occur when existing smaller cracks are used. Often the entire volcanic system is under constant shaking called "volcanic tremor". In most of the volcanoes worldwide volcanic earthquake shaking is identified at depths below 10 km and released earthquakes of magnitude 2–3. Furthermore volcanic seismic signals often do not occur as single events but rather as swarms made up of a multitude of seismic events over a longer time period.

The installed seismic arrays at many volcanoes today revealed that identification of source mechanisms of magma uplift can be an effective tool for a volcano early warning. Although the elastic wave propagation is generally the same as that of earthquakes, the volcanic seismic signals differ somehow and therefore also the measuring devices and the seismic station's geometry can be different. The main problem in volcano seismology is that it is still unclear how and to what extent the magma movements, fluid flows, degassing of the magma, or cracking in the volcano's structure can be identified from the seismic signals. Furthermore the appearance of continuous seismic signals (volcanic tremor) is seen as the result of rockfalls, lahars, landslides, pyroclastic flows, and minor volcanic explosions.

Measuring seismic volcanic activity today represents one of the major steps of an effective early warning system and is often the only tool to forecast volcanic eruptions and to monitor the eruption process (see Sect. 3.2.1.7). A higher rate of volcanic tremors is seen as a signal of increased volcanic activity that manifests long before the actual eruption takes place. Together with the other signs of enhanced volcanic unrest (fumaroles, ground deformation from doming-up of the structure, thermometry, emanation of radon and other gases, hot water occurrences) seismic activity is one of the major components of a volcano prediction. On December 18th, 2001, seismologists and volcanologists successfully predicted an eruption of Mt. Popocatepetl (Mexico City) hours before the volcano underwent its most violent explosion in centuries. The early warning gave Mexican officials enough time to evacuate some 40,000 people living in the shadow of the volcano.

Reservoir-Induced Seismicity

Since 1940 when the first extensive study of the correlation between increased earthquake shaking and the volume of impoundment in a reservoir was made for Hoover Dam (United States), it became obvious that large dams can trigger earthquakes. Today evidence of a causal relationship between earth tremors and artificial water reservoirs was found for more than 100 dams worldwide. The most powerful reservoir-induced seismicity (RIS) with a magnitude 6.3 destroyed the Indian village of Koynanagar on December 11th, 1967, killing 180 people, injuring 1500, and rendering thousands homeless. The dam was seriously damaged by the quake that was felt up to 230 km from its epicenter. During the 1990s, two seismic events greater than M5 and a series of smaller earthquakes occurred in the vicinity of Koyna following the impoundment of the Warna Reservoir. These new quakes gave rise to a unique experiment that drilled 21 boreholes to monitor the water levels continuously. The measurements revealed coseismic changes of several centimeters occurring in the wells associated with seismic shaking (Gupta 2002). RIS is also suspected to have contributed to one of the world's most deadly dam disasters, that of Diga del Vajont in the Italian Alps in 1963 killing 2600 people. The 261-m dam—at the time the world's fourth highest dam—was completed in 1960 in a limestone gorge at the base of Mount Toc. Tremors began as soon as the filling of the reservoir started. Later the reservoir was partially drained, and the seismic activity and slope movement almost stopped. When the reservoir was then filled again an increase in tremors was recorded. In 1963 heavy summer rains swelled the reservoir and in September, 60 shocks were registered and the movement on Mount Toc started to accelerate. On the night of October 9th, 350 million cubic meters of rock broke off Mount Toc and plunged into the reservoir.

The most widely accepted explanation of how dams cause RIS is thought to occur in two ways (Jauhari 1999):

- By adding millions of tons of stored water that increases the original stress to faults, causing them to rupture
- By water seeping into the rock and changing the fluid pressure in microcracks and fissures underground, thus "lubricating" the faults that are already under tectonic strain, but were prevented from slipping by the friction of the rock surfaces.

Although the load effect of the first case is immediate, the pore pressure effect is delayed because it requires the flow of the water through rock. This delay can cause some reservoirs to begin triggering earthquakes years after the first impounding.

Most of the strongest cases of RIS have been observed for dams over 100-m depth. Nevertheless dams with just half the filling height are also believed to be prone to quakes. Filling reservoirs can both increase the frequency of earthquakes in areas of already high seismic activity but can also trigger earthquakes in areas previously seismically inactive. Seismologists point to the fact that just the latter effect is the most dangerous as structures in areas thought to be quiescent are generally not designed to withstand even minor earthquakes (Gupta, ibid). He therefore recommended generally to consider RIS for reservoirs with filling depth heights of more than 100 m as the depth of the reservoir is the most important factor, but the volume of water also plays a significant role. RIS can be immediately noticed during filling periods of reservoirs or after a certain time lag. The RIS cases investigated revealed that the seismic patterns are unique for every reservoir. Moreover, the intensity of seismic activity is generally increased within around 25 km of the reservoir as it is filled. The strongest shocks normally occurred within days, but sometimes within several years. After the filling of the reservoir has reached its maximum, RIS events normally continued but usually with less frequency and strength than before. Thus far the actual mechanisms of RIS are not well understood therefore seismologists will not be able to predict accurately which dams will induce earthquakes or how strong the tremors are likely to be.

Lately the 7.9-magnitude Wenchuan earthquake in May 2008, which killed an estimated 80,000 people has been linked to the construction of the Zipingpu Artificial Reservoir. An intensive debate began in the public on the cause–effect relationship generated due to the fact that the newly impounded Zipingpu reservoir was located just about 20 km east of the earthquake epicenter. But intense seismological investigations on pore pressure diffusion in relation to the variation of water level in the reservoir revealed that the reservoir operation did not cause an increase in local stress at the earthquake hypocenter and the surrounding region, leading to the assumption that the reservoir probably did not play a role in generating these disastrous earthquakes (Galahaut and Galahaut 2010).

Geothermal Power Plant at Landau, Rhine-Pallatine (Germany)

The location of the city of Landau in close vicinity of the Upper Rhine Graben Structure that exhibits a significantly increased geothermal heat flow, made the idea of using low enthalpy geothermal power for heating purposes technically and economically feasible. The power plant started operation in 2007 to produce 3 MW of energy from 160 °C hot water from Triassic limestones at 3000 m depth. The total investment was about €20 million mostly for the injection and the production wells. Two years after starting operation, a number of small earthquakes were registered with a magnitude of M2.4–2.7. The earthquakes were located just below the city of Landau about 2 km distant from the wells and at a depth of 2.8 km. Some streets were deforming, some houses showed cracks in the walls, and arsenic was detected in shallow ground water wells. The State Pallatine commissioned the German Geological Survey (BGR 2010) to analyze the situation. The experts came to the conclusion that there is a causal relationship between the seismicity and energy production. Although small damage to nonstructural parts of exposed buildings may occur, damage to the supporting structure can be ruled out. The main reason for the quakes was seen in the reinjection of the thermal waters that led to an increase in pore pressure and resulted in a change of the local tectonic stress regime. The experts proposed to introduce a high-sensitive seismic grid around the plant to enable a real-time identification of an increase in seismicity one or two days before the stress releases. Based on such information the reinjection could be reduced accordingly. Moreover the state government ordered the plant operator to increase the risk insurance coverage. A mediated participation was established where all those affected were asked to present their specific concerns. The increased costs from reduced energy production and for the monitoring of the seismicity and the coverage of the damages led the operating consortium in 2013 to halt the operation.

Earthquakes Generated from Producing Gas Field (Groningen, Netherlands)

In August 2012, an earthquake with a magnitude from 1.8 to 3.2 on the Richter scale occurred in the village of Huizinge in the province of Groningen in the Netherlands (Kraaijpoel et al. 2011). The tremors were recorded in the vicinity of the Groningen gas field, the tenth largest gas field in the world, from which NAM, a consortium of Royal Dutch Shell and ExxonMobil, has produced natural gas since 1963. The field has about 300 wells across 20 production sites. Forty percent of the shares of the operation are held by the Dutch government. Seismological evidence revealed the tremors to be triggered from the change in reservoir pressure through production. Gas extraction is known to lead to tectonic subsidence along the naturally occurring faults and that the energy released results in local tremors. In the case of the Groningen field, the hypocenter is relatively close to the surface at about three kilometers underground, where the gas is situated. It should be acknowledged that NAM is producing the gas from conventional production wells not by fracking. The geotectonic situation of the northeastern Netherlands on the surface exhibits a flat landscape yet the subsurface terrain is highly complex. Near Groningen the stratification is distorted severely mainly due to salt tectonics. The reservoirs are cut into compartments by vertical fault systems. The earthquakes are associated with differential compaction due to gas extraction and reactivation of the existing faults. The seismicity pattern shows some lineation, especially for the larger earthquakes. The catalogue of earthquakes induced by gas production in the north of the Netherlands contains 688 events to date.

During the initial decades of production the surveying organization KNMI (Royal Netherlands Meteorological Institute) did not detect any earth tremors from the field. This changed in the mid-1980s when about 10–20 tremors were recognized per year. In 2003 there was a significant increase in tremors monitored and KNMI found evidence that the tremors could directly be linked with the gas production.

After the earthquake of August 2012 the Dutch Government started an initiative to revise the existing gas production plan. Subsequently NAM submitted a revised production plan on which the Dutch government will make a decision in January 2014. The plan comprises numerous actions to be taken to increase local resilience including:

- Enhancing knowledge of the regional seismic fault patterns.
- Assessing of the likelihood of the occurrence of higher intensity earthquakes.
- Identifying measures suitable to prevent structural damage to buildings.
- Strengthening communication between NAM and the population at risk.
- Introducing a "Groningen gas website" and other public information campaigns.
- Implementing a program to provide information on structural reinforcement for vulnerable houses.

Moreover the compensation allowance for damage has been substantially increased and broadened by NAM in order to compensate the victims in a timely manner for the 2500 damages claimed since August 2012.

Induced Earthquakes from Waste Water Reinjection

Induced seismicity is earthquake shaking that results from human activity that is beyond the normal regional level of historical seismic activity and that can lead to damage to the surrounding communities. In addition, if the seismic activity returns to background activity after the human activity stops, that would be another sign that the seismic activity was induced. One of the causes of induced seismicity can be fluid injection. Fluids in the pores and fractures of the rocks (pore pressure) play a major role in controlling the pressure in a reservoir. If pore pressure is increased, earthquake activity can be accelerated along existing fractures (shear failure). Injecting fluids into the underground generally results in an increase of reservoir pore pressure and that can cause faults and fractures to "fail" more easily. However, seismicity can also be induced by extracting fluids. In order to improve understanding of the triggering mechanisms of wastewater reinjection, knowledge of the local earthquake distribution pattern, the reservoir characteristics, and the time and amount of waters injected should be analyzed. The epicenters of earthquakes induced by fluid-injection activities are not always located close to the point of injection. It is assumed that the injected fluids can migrate for larger horizontal and vertical distances from the injection location. Therefore induced earthquakes commonly occur several kilometers below the injection point. In some cases, induced earthquakes could be located as far as 10 km from the injection well.

One of the many applications for wastewater reinjection is carried out by the oil industry. All over the world oilfield wastewater, after removal of the oil and the solids, is injected into a deep well for permanent storage underground. In the United States oilfield wastewater reinjection has been carried out for decades and it was reported that within the central and eastern United States, the number of earthquakes has increased dramatically over the past few years in areas that are under oil exploitation license. More than 300 earthquakes above a magnitude 3.0 occurred in the years from 2010–2012, compared with an average rate of 21 events per year observed from 1967–2000 (EERI 2014). These earthquakes are fairly small, although large enough to have been felt by many people, yet small enough rarely to have caused damage.

USGS scientists analyzed changes in the rate of earthquake occurrence recorded since 1970 in these areas (USGS 2013). The increase in seismicity has been found to coincide with the injection of wastewater in deep disposal wells in Midwest locations from Texas to Ohio. Hydraulic fracturing, commonly known as "fracking," does not appear to be linked to the increased rate of magnitude 3 and larger earthquakes. Although wastewater injection has not yet been linked to earthquakes larger than M6 scientists cannot eliminate its probability as it appears that wastewater disposal was the cause of the M5 earthquakes at Raton Bazon (Colorado) and Prague (Oklahoma), that both have led to a few injuries and damage to more than a dozen homes. Evidence from some case histories suggests that the magnitude of the largest earthquake tends to increase as the total volume of injected wastewater increases. Injection pressure and rate of injection may also be factors, although more research is needed to determine answers to these important questions. Nevertheless, according to EERI (ibid), the overall risk of induced seismicity from wastewater disposal can be rated to be—although not zero—either minimal or able to be handled in a cost-effective manner.

3.2.1.3 Liquefaction

The damaging impact from earthquakes is different from location to location and is mainly the result of two different phenomena. One is, how strong a building is to withstand seismic motion. The other is how the geology (rocks and sediments) reacts to the seismic energy released. It turned out from the very beginning of seismology that the geology surrounding the epicenter has a significant influence on the earthquake impact. Thus it was noticed that unconsolidated sediments tend to react much differently when exposed to seismic energy. Hard crystalline rocks stay stable, but soft sediments tend to react like a liquid. Seismograms reveal that in soft unconsolidated sediments the amplitudes can reach three times the height of hard rocks. Furthermore the time spans the acceleration holds on are considerably longer. The upper 30 m of building grounds especially often consist of unconsolidated soils, loose sands of river terraces, or gently inclined slopes, and therefore react significantly to ground motions. This phenomenon is called liquefaction. Physically liquefaction can be explained as soft sediments with low or no cohesion being exposed to dynamic acceleration; the energy introduced into the system increases the pore pressure. And when the pore water cannot be drained off, the soils react like a liquid. The energy input into the system is then amplified by the soft medium as it has a different elasticity modulus that leads to a long period of shaking of the ground, thus the intensity of a magnitude 5 earthquake in liquefaction-prone sediments can be much higher than if it strikes hard ground. Furthermore the elasticity modulus of soft sediments is much different from that of steel-reinforced masonry buildings, with the result that both materials work against each other and massive large-scale and widely distributed damage occurs. Many regions on Earth are exposed to this type of natural hazard, especially former dried-up lake sediments that have been settled extensively afterwards. The Kathmandu basins in Nepal with its 700,000 inhabitants today is one of the famous examples of this. Or in Turkey when the building itself remains structurally intact but the building ground reacted like a "liquid" and made the entire building sink into the ground (Fig. 3.14). In contrast, the island of Manhattan (New York) is built of granitic ground making the construction of high skyscrapers possible at all.

3.2.1.4 Earthquake Prediction

Although numerous efforts have been undertaken in the last decades to improve earthquake prediction, a reliable prediction tool still does not exist (Geller et al.

Fig. 3.14 Liquefaction of the subsurface strata from the earthquake event in 1999 made this masonry structure collapse (Adapazari; Turkey)

1997; Kagan 1997). And this unpredictability makes earthquakes one of the most lethal natural disasters. Even in the areas best known for earthquakes in the California San Andreas Fault system or the Adana Fracture Zone in Northern Turkey, no evidence was ever found that would enable geophysicists to predict the exact date for an earthquake to occur. For the San Andreas Fault the United States Geological Survey recently released information that they expect a major earthquake in the area of 6.7 Richter scale to occur until the year 2038 with a 99.7 % probability and an earthquake of magnitude 7.5 with a probability of 46 % for the same time span. But they also declared that all monitoring instruments and interpretation methods are still not reliable enough to predict the onset at a sufficient level of reliability, although the area is plugged with thousands of seismic monitoring devices.

USGS statistics on earthquake occurrences (USGS 2014) regularly report on earthquake events from all over the globe. More than 10,000 events occur yearly with a magnitude of M4.0 and higher, although it is thus far not possible to predict a future event with a reliable level of certainty. There are geophysicists who are convinced that a robust forecast will also not be possible in the near future (Geller et al., ibid), although another school of thought believes that prediction is "inherently possible" (Bakun and Lindh 1985).

The aim of earthquake prediction is to warn the people of a potentially damaging earthquake early enough and specific concerning the time, location, and magnitude to allow appropriate response to the expected disaster. There have been a number of predictions of earthquake events in the last decades, but none of them succeeded in identifying the proper parameters for the expected events.

In many cases such predictions did not specify all the required parameters, but often left out one or more (Bakun and Lindh, ibid). Such an approach is mostly called trivial as it normally only describes the earthquake parameters in such a generalized and unspecified manner that the message holds many possible events true. His experience with earthquake prediction led Charles F. Richter in 1977 on the occasion of the presentation of the honor, "Medal of the Seismological Society of America," on earthquake prediction to state: "Since my first attachment to seismology, I have had a horror of predictions and of predictors. Prediction provides a happy hunting ground for amateurs, cranks, and outright publicity-seeking fakers."

The most reliable way to assess potential future earthquakes is still by studying the history of large earthquakes in a specific area and the rate at which strain accumulates (USGS 1995a; Kagan, ibid). The assessments are based on millions of events, listed in international statistics covering the last centuries, as well as from seismological research in the field but also in laboratories and on theoretical investigations. For example, the USGS has been monitoring the strain accumulation for years along the fault segments of the San Andreas Fault zone, and measures the time that has passed since the last earthquake and calculates the strain that was released during the last earthquake. This information is then used to calculate the time required for the accumulating strain to build to the level that may result in an earthquake event. This simple model is complicated by the fact that such detailed information about faults is rare. In the United States, only the San Andreas Fault system has adequate records for using this prediction method (Segall et al. 2007). The information collected thus far allows the assessment of most earthquake mechanisms and to describe the major fault zones. On studying the frequency of large earthquakes in the past it is possible to determine at least the future likelihood of similar large shocks. For example, if a region has experienced four magnitude 7 or larger earthquakes during 200 years of recorded history, and if these shocks occurred randomly in time, then scientists would assign a 50 % probability to the occurrence of another magnitude 7 or larger quake in the region during the next 50 years (200:4 = 50).

To achieve meaningful predictions the assessment should define the range of possibilities concerning the time the event will set in, the proper location, and the expected magnitude. For a long time, earthquake prediction was mainly carried out in a deterministic way, based on various kinds of more or less clearly identifiable precursors. Meanwhile earthquake prediction is aiming to quantify the uncertainty of an earthquake event to occur. To achieve a short-time prediction a reliable forecast requires a statistical approach to simulate the probability of an earthquake risk. The many forecasting models at hand and the tremendous capacity of computing today allow us to apply a broad range of parameters and then to check the results against real observations. Kagan and Jackson (1991) have applied the likelihood method for many earthquake occurrence studies. Their stochastic modeling describes "seismicity as a random process, for which a continuous space-time density distribution of the earthquake occurrence can be defined." The use of stochastic models is achieving increasing success in answering the question of what the chance is that another large earthquake will occur in the near future.

A different approach is used when forecasting the next earthquake in the long term. This approach assumes that on the same seismogenic source, earthquakes have similar rupture areas, similar mechanisms, and similar magnitudes, and are defined by a remarkable regularity. They are often assumed to have similar hypocenters, similar displacement distributions within the rupture area, similar source time functions (leading to similar seismograms), and quasi-periodic recurrence. This probabilistic approach can be applied when assuming that the elastic strain energy accumulates over a long period of time after the occurrence of one earthquake before the fault is prepared to release in the next earthquake.

Bilham et al. (1989) proposed a method to predict potential future earthquakes by space geodesy measuring the speed at the plate boundaries. Fundamental to the assumption that geodetic studies can be helpful to forecast, is a model in which the strain that is released by an earthquake is equal to the strain developed in the following interseismic period. They point to the fact that plate motions constitute a 1–10 cm/year displacement input signal to the earthquake process plate movement and that most of the earthquakes occur within plate boundaries that are typically less than a few hundred kilometers wide. But as the internal strain rates of plates are normally less than 1 microstrain per year and the strain at failure is typically between 10 and 100 microstrain, the measurements can only be successful when a narrow spaced measuring array is established with spacing less than several tens of kilometers to understand the mechanisms of plate boundary deformation and rupture. They furthermore point to the fact that the earthquake rupture zone (the region where the strain is released) is proportional to the magnitude of the earthquake. Thus if one knows the regional extent of a potential future earthquake its magnitude can be estimated. Space geodesy has three significant advantages over terrestrial methods: three-dimensional relative point positions are obtained from a single observing pair, site intervisibility is not needed, and baselines of any length can be included in a network. These advantages permit great flexibility in field operations Nevertheless they stated that at present it does not appear promising for space geodesy to provide predictions of precursory deformation that may be manifest in the days and weeks preceding earthquakes.

3.2.1.5 Examples of Earthquake Occurrences

The First "Well-Known" Earthquake (Lisbon Earthquake, 1755)

One of the biggest earthquakes in Europe and the one that led to the birth of modern seismology was the earthquake of Lisbon in 1755. In combination with subsequent fires and a tsunami, the earthquake almost destroyed the entire city of Lisbon and the adjoining areas. Lisbon was in those days one of the richest and wealthiest cities of the world. As the historic record is not very reliable it is anticipated that most probably 30–40,000 people lost their lives in the event (Pereira 2009). Today the Lisbon earthquake is estimated to have had a magnitude in the range of 8.7–9.0 on the Richter scale with its epicenter 300 km west-southwest of Lisbon in the Atlantic Ocean.

The earthquake was triggered along the Azores–Gibraltar transform fault that marks the boundary between the African plate and Europe and that runs just south of Portugal and Spain east–west into the Mediterranean Sea. The fault system is seen responsible also for many earlier earthquakes, such as the ones that occurred in 1724 and 1750. Historical record proves that the earthquake lasted about 9 min (one of the longest according to earthquake records) and caused a 5-m wide fissure that crossed the entire center of the city. Forty minutes after the earthquake a huge tsunami occurred that rushed into the harbor and destroyed those parts that thus far had not been affected by the earthquake. Afterwards a fire broke out that destroyed all the houses that were neither affected by the earthquake nor by the tsunami and that resulted in about 50 % of the houses being burned down. But the tsunami did not only affect the city of Lisbon but almost the entire Algarve coast and also heavily struck the islands of Madeira and the Azores. Even the Caribbean islands of Martinique and Barbados were hit by a 10-m wave. Shocks of the earthquake were recorded all over Europe.

The earthquake not only destroyed 85 % of Lisbon's buildings but had enormous economic and social impact on the entire kingdom of Portugal. A large amount of assets was accumulated in the city, for example, from a high influx of gold from Brazil and the African colonies. Pereira estimated the economic losses due to the earthquake of about 50 % of the gross domestic product, a figure that is enormous when seen in respect to the overwhelming richness of the Portuguese empire in those days. In the aftermath wheat and barley prices rose by more than 80 % (Pereira, ibid) for many years to come. In addition the prices for wood and other construction materials increased considerably, as did labor costs. This was also due to the effect that many craftsmen refused to work in Lisbon as they feared to be hit by another earthquake. Nevertheless 10,000 huts were reconstructed in the following months. Historically earthquakes and their following cascade of disasters were seen as a distortion of national sovereignty. It also became the trigger moment for a radical change in the general policy of the former Portuguese Kingdom away from close trade relationships with England to a stronger orientation on colonialism.

Transform Fault Earthquake (Loma Prieta, California, 1989)

The Loma Prieta earthquake was one of the major earthquakes of the world. It struck the San Francisco Bay Area (California), one of the most densely populated regions of the United States, although luckily it did not strike the San Francisco metropolitan area itself. The earthquake occurred on October 17th, 1989, at 5 p.m. and was the first earthquake that was broadcast live on television as it happened during a basketball sports game (Bakun and Prescott 1989).

The earthquake was generated by a strike-slip motion along the San Andreas Fault. The quake lasted 10–15 s and measured 6.9 on the Richter scale. The quake killed 63 people throughout Northern California, injured 3757 and left up to 10,000 people homeless. A total economic loss of US$6 billion was claimed.

The slip occurred over 35 km of fault at depths ranging from 7 to 20 km with a maximum offset of 2.3 m. Although the earthquakes along the San Andreas Fault system are in general strike-slip movements, the Loma Prieta earthquake had a significant uplift component along its southwest dipping fault plane. The jolted crustal segment mostly coincided with a fault segment that had already ruptured in 1988. Although the San Andreas Fault occurs as a through-going fault in the epicentral region, the Loma Prieta rupture surface forms a separate fault strand. "Seismological evidence revealed that the earthquake may not have released all of the strain stored in rocks and therefore may have a potential for another damaging earthquake in the Santa Cruz Mountains in the future" (Bakun and Prescott, ibid).

Holzer (1998) describes in detail the main effects caused by the Loma Prieta earthquake: strong ground shaking and liquefaction of both floodplain deposits in the Monterey Bay region and on the sandy artificial fills along the margins of San Francisco Bay and by landslides in the epicentral region. The strong ground shaking was amplified by a factor of about two by soft soils (liquefaction) and caused damage even 100 km away from the epicenter. Liquefaction alone is seen responsible for about US$100 million in economic loss. Landslides caused US$30 million in earthquake losses, damaging at least 200 residences. Many landslides showed that they had already moved in previous earthquakes. Post-earthquake studies provided one of the most comprehensive case histories of earthquake effects ever made in the United States. A comparison of the liquefaction and landslide impacts from 1906 with those of 1989 lay the base to work out new methodologies to map liquefaction and landslide hazards. Seismographs installed in 28 buildings recorded the respective building responses to the earthquake shaking and provided physical evidence to understand how different building structures interact with their foundations when shaken and how liquefaction can amplify ground motion. It furthermore prompted the California legislature in 1990 to pass the Seismic Hazards Mapping Act that required the California Geological Survey to delineate areas potentially susceptible to these hazards and communities to regulate development in these zones (Holzer, ibid).

Approximately 16,000 housing units were no longer habitable after the earthquake; 13,000 alone in the San Francisco Bay region, and another 30,000 units

Fig. 3.15 Collapse of the MacArthur Maze (Cypress Viaduct) in Oakland from the 1989 Loma Prieta earthquake

were even moderately damaged. It turned out that rented houses and low-income residents were particularly hard hit. Structural failure of the many highway systems was the single largest cause of loss of life during the earthquake and led to a death toll of 42 of the 63 earthquake fatalities; all of them died when the Cypress Viaduct in Oakland collapsed (Fig. 3.15). The cost to repair and replace highways damaged by the earthquake was US$2 billion, about half of which was to replace the Cypress Viaduct. Major bridge failures were the result of antiquated designs and inadequate anticipation of seismic loading. The gas distribution lines and water pipelines showed a multitude of leaks and breaks all over the region.

In 1989 Mileti (1989) summarized how people and the economy reacted to the event. Most people responded calmly and without panic to the earthquake and acted to get themselves to a safe location. Economically the earthquake resulted only in a minimal disruption to the regional economy that resulted in maximum losses to the gross regional product of US$3 billion in the months after; but the loss was 80 % recovered during the first six months of 1990. Approximately 7000 workers were laid off for a certain time period. For the local and national emergency managers, the Loma Prieta earthquake provided the "first test of the newly established post-earthquake review process that places red, yellow, or green placards on shaken buildings. Its successful application has led to widespread use in other disasters including the September 11, 2001, New York City World Trade Center attack" (Mileti, ibid).

Transform Fault Earthquake (Haiti, 2010)

On Tuesday, January 12th, 2010 an earthquake with magnitude 7.0 (Richter scale) struck the entire region of the capital, Port-au-Prince, Haiti. The epicenter was located near the town of Léogâne approximately 25 km west of the capital. Within the next two weeks almost 60 aftershocks were recorded with magnitudes of 4.5 or greater. The two largest aftershocks were of magnitude 6.0 and 5.9.

An estimated three million people were affected by the quake. According to official estimates (UNEG 2010) the death toll was estimated to be at least 220,000, another 300,000 people were injured, 1.3 million people were made homeless, and almost 100,000 houses were completely destroyed. The earthquake caused major building damage in Port-au-Prince and other settlements around the capital. Many notable landmark and infrastructure buildings were significantly damaged or totally destroyed, including the Presidential Palace, the National Assembly Building, and the Port-au-Prince Cathedral. Nearly the entire communication system; air, land, and sea transport facilities; hospitals; and electrical networks had been damaged or destroyed totally by the earthquake. The lack of a functioning infrastructure complicated almost all rescue and relief efforts, resulting in heavy confusion over who was in charge in air traffic and land transport congestion. A major problem was the long unsolvable problems on prioritization of relief flights. Port-au-Prince's authorities were overwhelmed with tens of thousands of bodies. These had to be buried in mass graves. Delays in food and water supplies, medical care and sanitation, and an unprofessional aid distribution management led to angry protests by the survivors and locally to looting and sporadic violence (IFRC 2011a, b).

The earthquake occurred along the plate boundary that separates the Caribbean plate and the North America plate. This plate boundary is dominated by strike-slip motion and compressional tectonics that moves the Caribbean plate eastward with respect to the North America plate with a speed of about 20 mm/year. The earthquake did not, although consistent with former seismic events, result in a significant surface displacement on the morphologically well-expressed main-strand of the Garden Fault system, but instead appears to have caused a considerable uplift of the Léogâne delta. The Garden Fault zone is supposed to have triggered the historic earthquakes in 1860, 1770, and 1751. Interferometry analysis of the crustal structure (Calais et al. 2010) revealed that the quake resulted in a vertical uplift of more than 2 m of the northern block of the fault zone and that it was laterally displaced by 50 cm in the northwest direction. The southern block instead moved 50 cm in the northeast direction. The plate boundary along Hispaniola Island (Haiti and Santo Domingo) is partitioned into two major east–west trending, strike-slip fault systems: one that occurs in northern Haiti and the other where the earthquake happened runs along the Garden Fault system in southern Haiti. The steep inclination of the deep earthquakes (Wadati–Benioff zone) indicates the subduction of the oceanic lithosphere of the Caribbean plate along the Central American and Atlantic Ocean margins, a subduction that is accompanied with deep ocean trenches and volcanic arcs volcanoes. On the other hand shallow seismicity and focal mechanisms of major shocks in Guatemala, northern Venezuela, and the Cayman Ridge and Cayman Trench indicate transform fault and pull-apart basin tectonics at the backside.

The earthquake hit a society that has been living ever since in an extremely poor social and economic situation. Even in times of no catastrophe, Haiti is one of the poorest nations on Earth with an almost nonfunctioning public sector. Even the minimum requirements for a sustainable living were missing. The high birth rate, poor labor opportunities, and many people with no income, a poorly functioning medical system, and a highly rotten technical and communication infrastructure together with a weak political system lay the basis for the total negligence of any disaster preparedness. The quake thus hit a society that was only oriented to make their daily living. Even during the years before the earthquake, the country was exposed to several natural disasters (hurricanes, floods, landslides, etc.) but did not pave the way for any emergency management.

Immediately upon the disaster the United Nations increased its presence in the country and supported the recovery, reconstruction, and stability efforts to restore a secure and stable environment, to promote the political process, to strengthen Haiti's government institutions and rule-of-law structures, as well as to promote and to protect human rights. More than three years after the devastating earthquake tens of thousands of families were still living in shelters. The majority of them are women and children. Those who been made homeless have to struggle to make a living for themselves and their families with little access to safe drinking water, sanitation, health care, schools, or other essential services. The number of internally displaced people and the number of makeshift camps has been decreasing since July 2010, from a peak of some 1.5 million people living in 1500 camps to 320,000 people living in 385 camps as of the end of March 2013, according to the International Organization for Migration (IOM). Thousands of families have left the camps for other accommodation provided through different projects and programs. However, forced evictions appear to have become an important factor leading to the reduction in camp numbers.

Even despite massive support from international donors, Haiti remains one of the poorest nations in the world, with significant needs in all basic services. Over half of its population lives on less than US$1/day. And Haiti is a country with a most unequal distribution of income (Gini-Coefficient 0.59). Nevertheless the Haitian economy has been recovering slightly since the earthquake. After a decline in the GDP of 5.4 % in fiscal year 2009–2010, it grew by 5.6 % in FY 2010–2011. Although small, the macroeconomic situation today is relatively stable with domestic revenue rising slowly but steadily. However, the momentum of economic recovery slowed in 2012, due to drought-induced declines in agricultural output and higher food prices, as well as due to damage from Hurricanes Isaac and Sandy.

Earthquake Hit a Megacity (Kobe, Japan 1995)

On January 17th, 2005 an earthquake hit the center of the megacity of Kobe, Japan. It was the first time in human history that a megacity was struck directly. Even the big earthquakes along the San Andreas Fault all spared cities like San Francisco, Oakland, or Los Angeles.

The earthquake lasted for only 20 s but was the worst in Japan since the Great Tokyo Earthquake in 1923 (magnitude 7.9 on the Richter scale) that killed 140,000 people and the greatest disaster in Japan since World War II. The reason why the disaster so heavily influenced the Japanese soul was:

- It wasn't anticipated by the public that the Kobe area, not located on one of the central faults, would be affected by a major earthquake and therefore only minor prevention measures were undertaken.
- Many Japanese were of the opinion that their country was one of the technically most advanced, and would be able to master any disaster whatever.
- Kobe was considered one of the nicest cities in Japan and where (ironically) many people moved to escape earthquakes elsewhere.

Therefore the earthquake of January 17th was more than a natural disaster; it almost twisted the national sobriety and it happened ironically in the Hyogo Prefecture.

The earthquake was generated by a strike-slip lateral movement of the rock masses along the Nojima Fault, which thus far was not considered a dangerous fault, shifting the two sides of the fault 20–30 m in opposite directions. The epicenter was estimated to be 60 km away from Kobe City between Awaji Island and Honshu. The surface along the fault moved five feet in one place. This movement could be seen in a rice field on Awaji Island. One of the major reasons for the extensive damage was that the quake was generated at very shallow depth. The earthquake claimed 6400 dead, injured 25,000, displaced 300,000 people, damaged or destroyed 100,000 buildings, and caused at least US$132 billion of economic loss, constituting about 2.5 % of Japan's GDP. Only US$3 billion was covered by insurance. More than 35,000 people were pulled from collapsed buildings by neighbors or rescue workers.

As the earthquake took place early in the morning most of the people were caught asleep. Therefore many bodies were found at or near their homes, although this later led to the identification of most of the victims in only 10 days. In addition the damage pattern was very spotty, leaving some areas almost untouched whereas others were destroyed completely. Even some of the new houses were badly damaged and some old ones were undisturbed. Nearly 80 % of the victims died from being crushed by the collapsing buildings. Many people were even killed when the heavy typhoon-resistant tile roofs collapsed on top of them. Sixty percent of the victims were older than 60 years and had lived in traditional wood frame structures that were built shortly after World War II. Many of these houses caught fire from toppled stoves and kerosene cookers as the earthquake happened in the very cold Japanese winter season. New, even tall, buildings that were constructed to be seismic resistant according to the 1981 building codes all remained standing. Electricity and water supplies were badly damaged over large areas. This meant no power for heating, lights, cooking, and so on. Moreover a clean, fresh water supply was short until April 1995.

Heavy damages were identified along all the major freeways, mostly as they were not designed to withstand such a strong earthquake. Therefore many of them collapsed, streets were uplifted, and road tracks buckled and twisted. It took much time to clear the streets from the rubble. But the immediate rescue and relief operations just cleared traffic routes. This became the biggest complaint of survivors to the authorities, that they failed to manage this problem first. Emergency experts later pointed out that the management of this disaster symbolizes a weak point in Japanese society after World War II. Public administration even on the local level turned out be completely inflexible (an attitude that was also recognized in how TEPCO handled the Fukushima nuclear catastrophe). Any political development decision in Japan is usually made by consensus and strongly follows the hierarchy principle. This made the authorities in Kobe reluctant to make decisions on their own. For instance, the broken telephone lines between the offices and ministries were not repaired immediately, as such a decision was not in the responsibility of the local administration. As a result the prime minister's office received the information on the disaster next morning on television. An order to send military forces for rescue was not issued until nine hours after the quake.

The cost for rebuilding the social and technical infrastructure was estimated at about US$130 billion of which the government allocated only 50 %, including supplementary budgets to carry out "reconstruction" work such as debris removal and for a period of 10 years.

About 70,000 people were still living in provisional shelters two months after the earthquake, some of them even until January 2000. It was furthermore learned that providing new housing did not solve all the problems. Many of the victims who lost their homes became depressed after being disconnected from their traditional social networks. Many of homeowners who lost their homes were not insured and were later reimbursed according to a classification of their damaged houses as "partially damaged" or "totally damaged." But this classification was carried out rather restrictedly and left many homeowners to the rigidity of the authorities. Nevertheless by January 1999, about 150,000 housing units had been constructed, new laws were passed to make buildings and transport structures even more earthquake proof, and a significant amount of money was invested to install more instruments in the region to monitor earthquake movements.

Earthquake in Istanbul (1999)

The northern part of Turkey as well its eastern flank are characterized by two major fault zones (North Anatolian Fault zone and the East Anatolian Fault zone) that both experience earthquakes. Along the North Anatolian Fault zone plate movements of up to 25 mm/year of right lateral motion, running parallel to the Marmara Sea were measured (Ilkesik 2002), making northern Turkey highly vulnerable to natural disasters, particularly earthquakes. Historic records reveal a statistical recurrence of one destructive earthquake hitting Istanbul each century.

In 1999 the Marmara/Kocaeli earthquakes of magnitude 7.4 claimed a death toll of over 17,000 people and caused direct economic losses estimated at about US$5 billion, or around 2.5 % of GDP along the earthquake occurrence. Even in Istanbul approximately 1000 people were killed and the damage to buildings was rather serious, although the epicenter was more than 110 km away.

The city of Istanbul and its surroundings have in Turkey the highest earthquake risk context. Istanbul is most vulnerable because of its location in the western continuation of the seismic fault zone and due to its population of roughly 15 million people and its commercial industrial densities. In 2002 JICA (Ilkesik, ibid) carried out a risk assessment for Istanbul that revealed the probability of a major earthquake affecting the city in the next 30 years was more than 60 %, and the likelihood of such a devastating event within the next decade is calculated to be more than 30 % (Erdik and Durukal 2008). Compared to the cities of Los Angeles or San Francisco, both experiencing a comparably high risk, the damage potential to Istanbul is significantly higher because of its greater structural vulnerabilities. If a seismic event of the same magnitude as that in 1999 occurred near Istanbul, the human suffering as well as the social, economic, and environmental impacts would be dramatically higher than in the Marmara region. JICA estimated that an event similar to the Marmara earthquake could result in up to 90,000 fatalities, 135,000 injuries, and heavy damage to 350,000 public and private buildings resulting in an economic impact of more than US$50 billion. The number of injured and affected people was estimated to be around 150,000, but 30 % of the hospitals are located in the potentially risk-prone areas south of the city.

The government of Turkey clearly sees that an interruption of Istanbul's social, economic, and financial life would heavily affect the national economy and the social sector for many years to come. Managing natural disasters therefore requires a multirisk approach and the government fosters the development of a comprehensive hazard risk-management strategy, not only for Istanbul but for the entire country. It has identified a serious need for improved knowledge, methods, and integrated framework for the assessment of hazards, vulnerability, and risks. Furthermore in order to manage the potential earthquake disaster in Istanbul, it is necessary to prepare disaster prevention and mitigation plans, emergency rescue plans, and a restoration plan for the earthquake stricken. Therefore detailed geological, geotechnical, and geophysical studies of the surface strata down to 250 m have to be carried out in order to forecast likely earthquake motions realistically. A three-dimensional model should be established to explain the earthquake trigger elements, the kind of probable tectonic processes, and the geologically originated amplification features. Recently the Istanbul municipality has run a microzoning project at the southwestern part of the city to detail information of local ground conditions. This will later be used to establish the appropriate design parameters for a city building code, which should be adopted for the more than 1.3 million buildings in Istanbul (Erdik and Durukal, ibid).

Earthquakes in Western Europe

On April 13th, 1992 the strongest earthquake since 1756 hit the German–Dutch boundary region, a region that is well known as seismically active. The quake lasted for 15 s, had a magnitude of 5.9 Richter scale, had its epicenter 4 km south of the Dutch city of Roermond, and the hypocenter was identified at 18 km depth. According to the information of the Geological Survey of North Rhine-Westphalia (GSNRW 1992) the quake was felt all over Western Europe and caused damages of €80 million. About 30 people were injured, most from falling tiles and stones; 150 houses were damaged in the nearby city of Heinsberg and even the Cologne Dom was partially damaged. The earthquake had no precursors but caused a series of about 150 aftershocks, the strongest with a magnitude of 3.6. The entire area west of the river Rhine along the Niederrheinische Bucht has long been known as seismically active. The area is part of the north–south oriented fault zone that crosses Western Europe from the Rhone River delta up to the Oslo Fjord. Along this fault zone Europe is pulled apart as a result of the ongoing pressure from the African plate hitting the European plate. The pulling apart has also formed the Rhine Graben structure and is responsible for a multitude of normal faults and deep-reaching thrusts west of Cologne.

But the Niederrheinische Bucht (shown in Fig. 3.16) is not the only region in Germany that is seismically active. The Earthquake Hazard Map of Germany

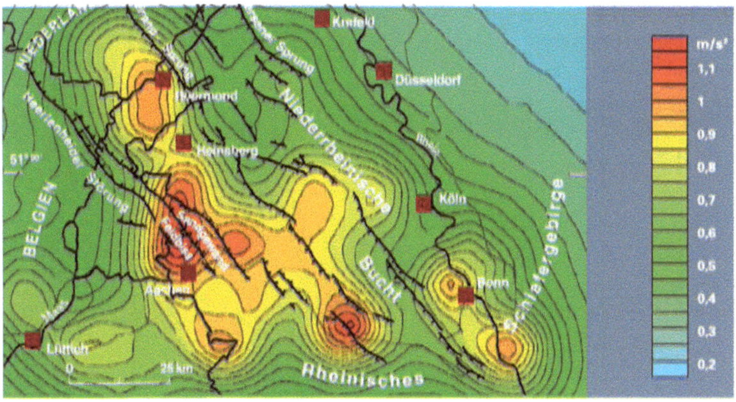

Fig. 3.16 Earthquake hazard of the Niederrheinische Bucht based on maximal horizontal acceleration exceeding one in 500 years (*Courtesy* Geological Survey of North Rhine-Westphalia; Germany)

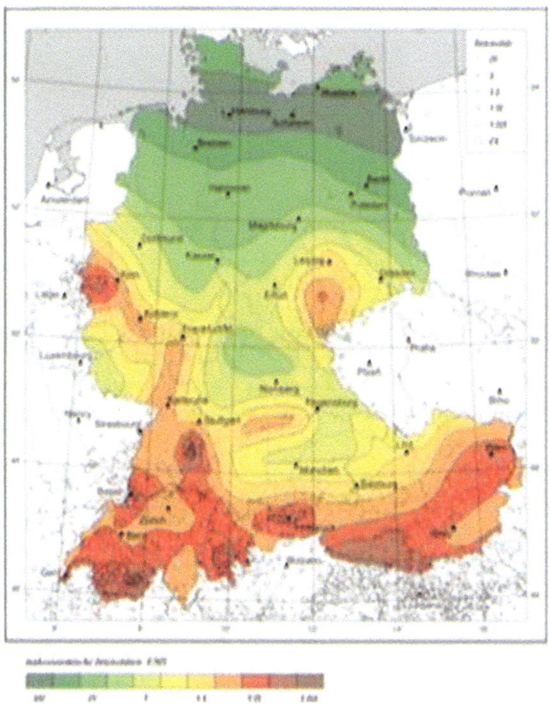

Fig. 3.17 Earthquake Map of Germany, Austria, and Switzerland (D-A-CH), based on earthquake intensities that will not exceed 90 % probability in the next 50 years (*Courtesy* Helmholtz Centre Potsdam—GFZ German Research Centre for Geosciences, Annual Report 2011)

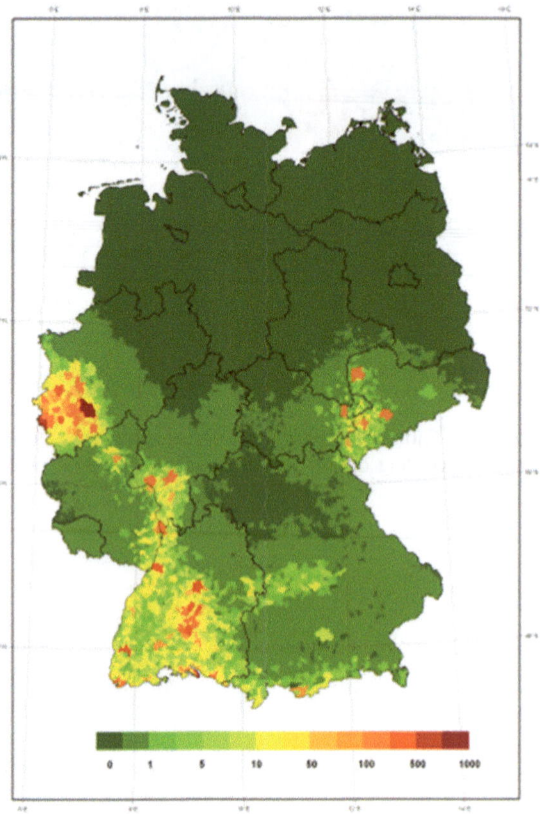

Fig. 3.18 Seismic risk zones of Germany (*Courtesy* Tyagunov et al. 2006)

shown in Fig. 3.17 identified at least four regions that experience a higher seismicity:

- Upper Rhine Valley/Swabian Alb
- The Alps
- Thüringer Wald
- Lower Rhine Valley

The classified seismicity zones from III (low risk) to VIII (high risk) give a measure of the probability for seismic activities at a certain location, a defined time span, and for a defined seismic intensity or ground acceleration. The map supplies the basis for introducing seismic building codes in Germany to reduce infrastructure damage.

Based on the earthquake hazard map, the distribution of seismic risk was assessed by CEDIM and laid down in a special map (see Fig. 3.18) expressing about a 90 % probability for an earthquake event in Germany in the next 50 years.

3.2.1.6 Tsunami

Tsunamis are water waves that occur due to mechanical disturbances in water bodies such as oceans, ocean bays, or lakes but also in artificial reservoirs (dams). The triggering mechanism for the disturbance of the water can be of seismic origin, landslides, rockfalls, and volcanic eruptions or may even be of an extraterrestrial source, such as meteorites.

Tsunamis are found to be one of the most threatening and life-demanding natural disasters of the world. Most tsunamis are definitively generated by earthquakes. But it should be borne in mind that not every earthquake generates a tsunami (Bryant 2001). Sedimentological indicators of geological coastal strata revealed that tsunami events occurred all over the world and cover the entire geological history. Such records could range from sudden onsets of coarse-grained sand layers in normally fine-grained coastal sediments to the occurrence of several meters large limestone blocks, as found in southern Italy and Greece (Soloviev et al. 2000). In the Mediterranean region alone about 300 tsunamis in the last 4000 years could be distinguished by their sedimentary record.

The term "tsunami" became known to the world with the tsunami that hit the Indian Ocean on December 26th, 2004. Up to that day the term "tsunami" was known to just a handful of specialists and disaster managers, but from that day, the term "tsunami" and its related term "early warning system" entered many spheres of social and economic politics. The 2004 tsunami triggered unprecedented help and relief assistance from all over the world. It provided US$8 billion in aid money, a sum that was never raised before and never after. See Fig. 3.19.

The term "tsunami" derives from the Japanese words *tsu* which means harbor and *nami* that means great wave (*tsunami* = great harbor wave), as the phenomenon was first observed by Japanese fishermen when coming home from the sea and found their harbor and houses destroyed by sea waves, although not having realized any significant rise in the sea level far offshore.

Fig. 3.19 The 2004 tsunami moved a ship onto a house in Banda Aceh. Today the ship serves as a tourist spot (Own photo)

Most tsunamis are generated by:

- Tectonic acceleration of the sea floor as a result of an earthquake
- Submarine volcano eruptions
- Landslides or rockfalls that occur either at the flanks of submarine volcanoes or of rockfalls and landslides from near shore cliffs
- Impact of large extraterrestrial meteorites

The above-mentioned effects cause a disturbance of the entire water body from the sea floor up to sea level. Thus tsunamis are different from normal, wind- and tide-induced sea waves that are defined as orbital waves and occur mainly in the upper 30 m of the water column. In a tsunami the entire water column is agitated; that means in the open ocean a water column of 4–5 km can be in motion.

From the source of generation the tsunami propagates in all directions. In the open ocean the travel speed may reach 800–900 km/h with an only marginal rise in the sea level but wave amplitude of more than 200 km. While approaching the shelf, wave amplitude and travel speed diminish strongly to about 60–80 km/h and even less on the shelf (about 10–30 km/h). This reduction in speed causes the wave heights to increase dramatically, so that they may reach 30 m or more at

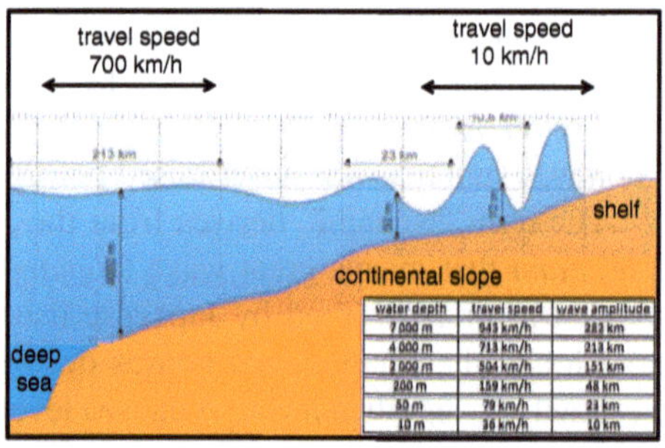

Fig. 3.20 Tsunami travel speed, wave height, and amplitude (*Courtesy* Franzius Institute, University of Hannover)

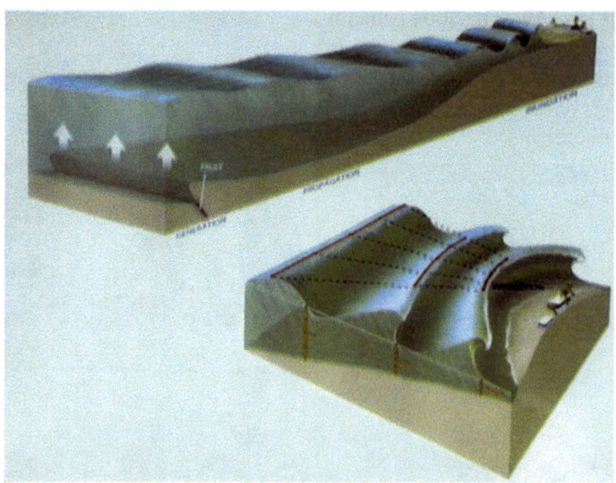

Fig. 3.21 Model of a tsunami wave train

landfall (Fig. 3.20). Travel speed, wave amplitude, and wave heights can be calculated according to the mathematical equation (the travel speed of a tsunami is the square root of the earth gravity times the water height (Mader 1974).

Like an acoustic signal, a tsunami is made up of a series of wave troughs and wave crests forming what is called a wave train (Fig. 3.21). Thus a tsunami is not made up of one single wave but comprises a series of waves. The tsunami of 1960 on the island of Hawaii that devastated the city of Hilo was made up of a total of 11 waves, most of which had wave heights between 4–8 m above sea level, and the fourth wave exceeded 14 m, the one that destroyed large parts of the city and harbor (USGS 2009a).

The agitation of the water column floor can either result in a wave crest or a wave trough, automatically defining the way the tsunami reaches land. Wave troughs lead to a significant retreat, withdrawing the water body, and can reach many kilometers, as recognized in the Indian Ocean tsunami, December 2004, at Khao Lak, Thailand; or in a rapid and strong increase in the water level (wave crest). Both indicators are the first and most reliable signs that a tsunami is approaching, still leaving 5–10 min for an immediate evacuation of the area.

Because in a tsunami the entire water column is in motion, the path a tsunami takes in the ocean is next to the concentric propagation from its epicenter, strongly controlled by the sea floor morphology. Furthermore the tsunami reacts more strongly on the sea floor morphology the more it enters shallower waters. Thus bottom morphology and coastal configuration highly influence where and in which direction the tsunami makes landfall. The coastal morphology also defines the impact. Steep cliffs lead to a "piling up" of the water masses, whereas gently inclined coastlines give ample space for a wide inland intrusion. Thus the impact of a tsunami is described by the water height (run-up height) and the distance the water penetrated the land (inundation distance; Fig. 3.22). How strongly morphology can influence tsunami impact was seen with the tsunami of 1992 in the city of

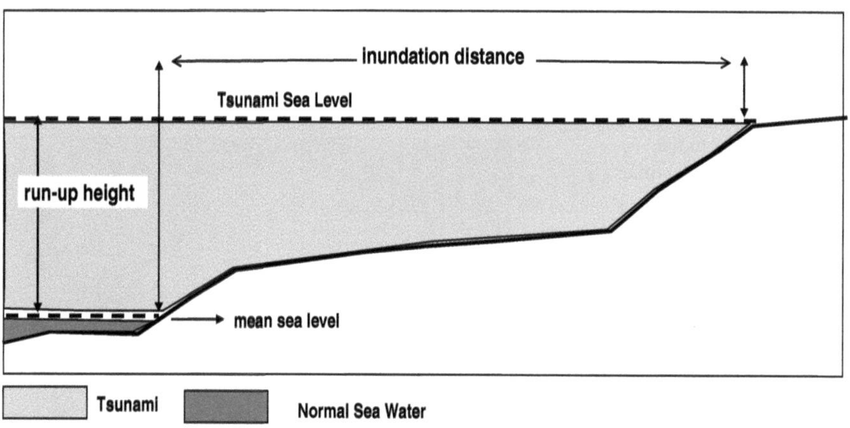

Fig. 3.22 Relation between run-up height and inundation distance (Own graph)

Maumere (Flores Island, Indonesia) where the tsunami flooded a gently seaward inclined limestone more than 12 m high, where in the city itself, the water run-up height was only 3–5 m. According to worldwide evidence the run-up height is normally twice the amount of the fault slip amount. The 2004 tsunami fault slip was about 10 m, resulting in a run-up height of more than 20 m. Another rule of thumb is given by Bryant (ibid), who stated that the run-up height in general equals 10 times the open ocean wave height (a 0.5 m wave height in the ocean would result in a run-up height of 5 m). But nevertheless it should be kept in mind that local coastal morphology plays a significant role in defining how far inland the wave may travel. The inland penetration of a tsunami can be roughly estimated as follows. On a very smooth, flat-lying terrain with low topographic roughness, inland penetration can be up to several km (e.g., at a run-up height of 10 m). In contrast, a steep cliff exposed to a tsunami will only be inundated by several tens of meters and the run-up height may reach more than 20 m. A coastal area that is densely populated with buildings close to the shoreline or coastal rim covered with trees and bushes both hamper the flood from penetrating far inland. The more densely a tsunami-exposed terrain is covered by buildings and/or trees, the less the waves are allowed to penetrate inland.

A further indicator for assessing a tsunami impact is that the wave crest/propagation is constantly orienting itself perpendicular to the sea bottom morphology. Thus tsunamis are subject to sea bottom refraction and bending around higher sea floor topographies. There is an impressive example of this. In 1996 a tsunami hit the island of Biak (Irian Jaya, East Indonesia), an almost perfectly round shaped island. The tsunami approached the island from the east with wave heights of 1–2 m, it traveled on both sides around the tiny island and the waves added themselves on the leeward side to heights of 4–5 m, flooding two small villages and killing 107 people (Matsutomi et al. 2001). The diffractions of the 2004 tsunami making it travel around the southern tip of Sri Lanka or enter into the Strait of Malacca are also impressive signs for this phenomenon. Following this, there is also the possibility that the tsunami, instead of being deviated from a critical area, may also be focused on a critical region.

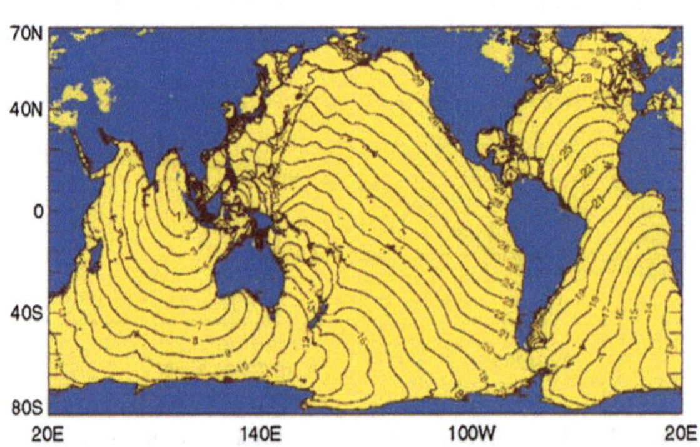

Fig. 3.23 2004 Indian Ocean tsunami travel times (*Courtesy* Kowalik et al. 2005)

The high speed of wave propagation of a tsunami in the open ocean results in travel times of less than a day to pass the most vulnerable tsunami area of the world: the Pacific Ocean. According to the mathematical equation $V = \sqrt{(g \cdot h)}$ a tsunami that was generated in southern Chile takes about 20–24 h to reach the Japanese Islands. The same holds true for a tsunami generated along the Eurasian/Pacific plate boundary to reach Chile (Fig. 3.23). Also a tsunami generated offshore Alaska or along the Aleutian Trench will cross the ocean in almost a day to reach Australia and New Zealand. Regardless of which tsunami generation location is taken, each will pass the islands of Hawaii on their way across the Pacific Ocean. Hawaii is reached from the east, west, or north within just 12 h and is thus the most tsunami-vulnerable spot in the Pacific Ocean. Following the 1932 tsunami that hit the Hawaii capital, Hilo, the world's first Tsunami Warning Center (Pacific Tsunami Warning Center, PTWC) was established there, followed by the Alaska Tsunami Warning Center (WCATWC), located in Anchorage, that monitors the northern Pacific region.

Less frequent than earthquake triggered, but nevertheless important for the generation of many tsunamis, are submarine or near coastal induced mass movements. The biggest tsunami ever recorded was that of Storrega (Norway; identified in seismic lines) where a submarine slump 12,000 years BC released a large amount of sediment from the shelf that today covers large parts of the Norwegian shelf between Bergen and Tromsö. The impacts of the tsunami could be traced all along the northwest English coast and was even identified along the North American coast. Another well-known record of a tsunami generated by a mass movement is that of Lituya Bay, Alaska (Fritz et al. 2009). There in 1958 in a small estuarine, north of Anchorage, a landslide was triggered (most probably by an earthquake) that released a huge mass of sediment. The slump formed a 30 m high tsunami that overran the 1.3 km wide estuary and climbed up the opposite coast more than 200 m and devastated the entire coastal rim. Still today the scar of the tsunami, its pathway, and the eroded coasts can be seen in satellite imagery.

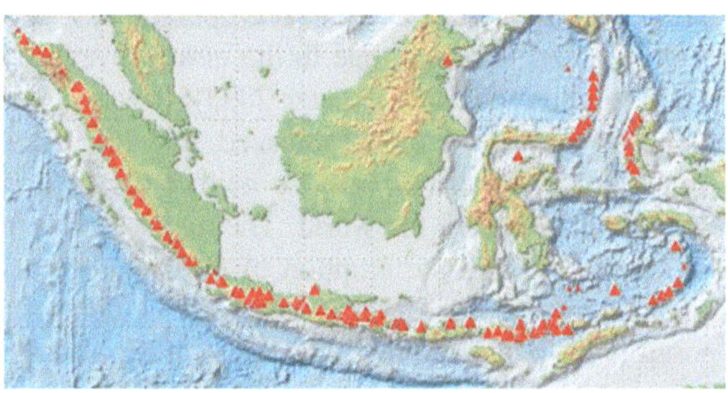

Fig. 3.24 Active volcanoes of the Indonesian archipelago are lined up like the pearls of a necklace (*Courtesy* BG/Georisk 2005)

Fig. 3.25 Basaltic lava streams from Mauna Loa, Hawaii, reaching the Pacific Ocean

3.2.1.7 Volcano

Sharing the same plate tectonic origin as earthquakes, the volcanoes of the world are also a result of the above-described plate motion. More than 90 % of the worldwide active and inactive volcanoes are situated along these boundaries and are thus called plate boundary volcanoes. In contrast, volcanoes that occur within the plates are called intraplate volcanoes (Pichler 2006). The Indonesian archipelago impressively demonstrates how volcanos mirror the boundary of the Indian and Eurasian Plates (Fig. 3.24). Roughly 200 active volcanoes are lined up like pearls on a necklace from the westernmost tip of Sumatra (Aceh Province) to the east (Iryan Jaya) and even continue eastward into the Philippines and the Japanese Islands.

Volcanoes are the most impressive manifestations of the Earth and are thus known to everybody. There passes almost no day without impressive pictures disseminated in the news on an erupting volcano, on high lava fountains putting the landscape in glowing light, on clouds of ashes that rise up kilometers in the atmosphere, or lava streams that pour into the ocean forming distinct lava forms (Fig. 3.25). The geology of volcanoes is one of the oldest scientific subjects in geoscience. Hundreds of volcanologists meanwhile have studied the nature of volcanoes, among them famous scientists including Robert Tilling, Tom Simkin, Haroun Tazieff, Chris Newhall, and the German volcanologist Ulrich Schmincke; all of them contributed significantly to the state of knowledge.

Nevertheless in addition to photos on erupting volcanoes, are volcano disasters that are remembered most by the people. The big eruption of the Krakatoa volcano in 1863 in Indonesia was the first eruption that was witnessed (Verbeek 1885) in last centuries, and that was (due to the newly invented telegraph) reported all over the world within a day (Winchester 2003). It was also the Krakatoa eruption that triggered the first government-financed scientific survey worldwide (Royal Society, London). But Krakatoa was the neither the biggest nor the most severe volcano disaster that affected the world. The biggest volcano eruption is, according to geological record, still the eruption of the Toba volcano about 75,000 years BC. The eruption formed the famous Toba Lake caldera of 100 km by 30 km and is supposed to have had an energy release that equaled 40 million H-bombs (Info: online www.ajb-hennings.de, 2008). The biggest volcano eruption of the last 20,000 years was the Tambora volcano eruption in 1815 which claimed a total of 90,000 lives. This eruption produced a crater of 7 km in diameter with an estimated volume of at least 30 km^2 (some estimates even say it might be 100 km^2) of ashes and bombs. The ash clouds and the huge amount of sulphur emitted traveled the stratosphere several times around the Earth and resulted in a significant change in world climate the year after (Stommel and Stommel 1983). Historic records reveal that in Switzerland there was a snow cover of 20 cm even during the summertime, resulting in a drastic loss of crops and a large famine was thus called, "the year without a summer."

Although daily present and partially severe in impact, volcanic eruptions are less lethal than many other natural disasters (flood, earthquakes, landslides). This is based on the fact that a volcano can be localized directly and seen physically. The uppermost flanks are normally sparsely settled and the people around have years of long experience in coping with the volcano. Furthermore the volcano most often announces an eruption days or months earlier. Therefore the death toll, as well as the economic losses due to eruptions is comparably small. It is estimated that about nine percent of the world's population lives in a radius of about 100 km around a historically active volcano (Small and Naumann 2001). Five hundred million people are exposed to volcano hazards and 10 million people live within a distance of up to 30 km of an active volcano. Volcano eruptions claimed the lives of 30,000 people, affected about 6 million, and caused damage of US$2.5 billion in 230 events in the time span of 1900–2013 (Guha-Sapir et al. 2013).

There are records of more than 1500 volcanoes that have erupted since Holocene times according to the *Catalogue of the Active Volcanoes of the World* published by the International Association of Volcanology, using this definition, by which there are more than 500 active volcanoes. About half of them are classified as being active and have erupted in recorded times. On average 50 volcanoes erupt each year, a frequency that was found by Simkin et al. (1981) to not have changed since historical times. Inactive volcanoes even when found at repose status for many centuries may erupt one day. Volcanologists distinguish between active volcanoes and inactive volcanoes. Those that erupt regularly are called active, and inactive volcanboes are defined as either dormant, meaning they are in an inactive status but can still erupt, or as extinct, meaning that they are expected never to erupt again. Those that have erupted in historical times but are now quiet are called dormant or inactive and those that have not erupted in historical times are called extinct. However, there is no consensus among volcanologists on how to define an "active" volcano as the lifespan of a volcano can vary from months to several million years. Simkin et al. (ibid) stated that in general, "The longer the period of interruption reposes is, the more energetic the next eruption" can be (see Sect. 3.2.1).

Generally three types of volcanoes are distinguished:

- Shield volcanoes (Plateau volcano)
- Strato volcanoes
- Rift volcanoes

The best examples of basaltic shield volcanoes are the Kilauea Iki and Mauna Loa (Hawaii) or the Icelandic volcanoes. Extinct shield volcanoes are the Southwest African volcano formation of Etendaka that was formerly connected with the Brazilian Parana formation and the Indian Deccan Trap of the massive volcanic formation that formed Siberia.

Strato volcanoes are in general connected with converging plates. About 1000 volcanoes are known to be active along such margins, but in general occur more than 150 km in the direction of the subducting plate and have an angle of 30–60°. This situation was exactly that experienced at the Indian Ocean 2004 earthquake cum tsunami in Indonesia. The best examples of strato volcanoes are the Nevado del Ruiz (Columbia), the Merapi (Indonesia), and the Fujiyama (Japan).

About 250 active rift volcanoes are counted worldwide, most of which are located along the East African Rift system. But by far most of the rift volcanoes are definitively generated in geological history along the submarine rift systems. The best examples of rift volcanoes are the Nyiragongo (DR Congo) or the Erte Ale (Ethiopia), which both allow insight into the lava/magma chamber.

Volcano eruptions are generated as the magma in the Earth's crust has a lower density than the surrounding rock material. This leads to an uprising momentum that seeks structurally predetermined faults or material weaknesses in the Earth's crust to make its way upward. The uprising magma is called lava. The higher the lava rises, the more the solved gases are set free, finally resulting in a massive expansion of the gas volume that then drives the molten lava out of the vent.

The generation sequence becomes quite different when meteoric water penetrates into the upper layers of the Earth. An overlying rock column of even several hundred meters increases the pore pressure drastically so that the boiling point of the water is raised to about 200 °C. When such a water–lava mixture then rises upward, the pressure is released suddenly and the overheated water changes in the vapor, leading to a hydrothermal explosion.

The molten uprising magma produces lava, ashes, rock fragments (bombs, lapilli), or gases. What kind of material the volcano is producing depends on the composition of the magma. In general two different types of magma/lava are distinguished:

- Basaltic magma
- Dacitic magma

Basaltic magma has a rather low SiO_2 content (46–52 %), low gas content, and is in general up to 1200 °C hot. Such magma forms effusive lava of a low viscosity: it flows easily and quite fast. There is a distinct relationship between SiO_2 content and the effusion capacity. The lower the SiO_2 content in the magma, the higher is the flowing capacity of the lava. Basaltic magma does not produce ashes nor does it emit large amounts of noxious gases. These facts make such volcanoes the ones with a low potential hazard. When the low content of solved gases is boiled out of the lava before it reaches the surface, the lava shows a red stream of molten rock material (Hawaiian volcano type).

Andesitic or dacitic magma is characterized by SiO_2 content of 52 % up to 68 % and a temperature range from below 1100 °C. Andesitic lava has SiO_2 contents of 52–58 %, and dacitic lava is characterized by SiO_2 content of 58–62 %. The lower temperatures and the higher SiO_2 content give these lava types a higher viscosity and make them less liquid. Therefore these lavas tend to accumulate in the volcano vent and when the internal pressure exceeds the mechanical stability of the surrounding rock material it comes to an eruption. Ashes, rock fragments, and/ or molten lava are than explosively produced. In the case where meteoritic water comes in contact with the lava, a phreatomagmatic explosion may be generated. The andesitic or dacitic lava often contains many gases (CO_2, NHN, H_2O and others at a minor quantity). The severity of such an explosion is indicated by the Volcano Explosivity Index (VEI) established by Newhall and Self (1982; see Sect. 5.2).

Gaseous emissions are the most powerful and dangerous threat from volcanic eruptions. Although lava emissions, bombs, lapilli, and ashes can be localized quite precisely and the area they damage is comparatively small, easily identifiable, and often the eruptions are a slow process, the gaseous emissions can spread over square kilometers rushing downhill with a speed of up to several hundred kilometers per hour. Moreover, such gases can reach temperatures of up to 400 °C and are often highly toxic. In nongaseous volcanic eruptions the vapor–gas mixture emissions can rise up into the sky for several kilometers. On their way up, the ashes fall off according to particle size. The gases, however, reach the stratosphere and are transported by the wind. So disaster managers while assessing the vulnerability of a certain volcano need to consider the main wind direction. The gases when pushed up into the sky are, when they reach the stratosphere, traveling a couple of times around the globe as many examples from Tambora volcano (Indonesia), El Chichon (Mexico), or Krakatoa (Indonesia) have revealed. At Mt. St. Helens the ash cloud traveled at a speed of 100 km/h, drawn eastward by the prevailing winds into the midwestern United States.

The biggest threat from such gaseous vapor emissions are generated when, as at Mt. St. Helens or Mt. Pelée (Martinique) a flank of the volcano collapses and the uprising oriented thermodynamics of the ash cloud collapses, reverses, and rushes downhill. The famous disaster at Mt. Vesuvius (Italy) has given at the city of Pompeii an impressive example of what happens when a city is covered by a gas cloud within a couple of minutes. At Pompeii the ashes preserved bodies of citizens who are concentrated in the uppermost of the eight ash layers, a clear indicator that only the last ash eruption was accompanied by high gas content. A high risk for the population exists when large-size conurbations such as Mexico City are located in the vicinity of active volcanoes that are made up of SiO_2-rich magma that are highly prone to paroxysmal eruptions.

In general volcanic emissions are not composed of only one volcanic material alone: they normally comprise a mixture of ash, vapor, and gas and even larger rock fragments. When in the course of the eruption rock fragments break off the dome flank, the gases dissolved in the lava are explosively set free and fragment the lava into small pieces. Such an ejection type is called a pyroclastic surge as it is characterized by a suspension of super-heated gas, ash, and rock fragments.

Once released the pyroclastic surge increases its volume while rushing downhill. The experience revealed that the steeper the volcanic flanks are, the higher is the probability that a pyroclastic surge (also called *nuée ardente*) is generated. A pyroclastic surge can be generated when the thermodynamics of the uprising ash–gas–vapor cloud is interrupted, often due to a collapse of the volcano flanks. This eruption type is the most deadly of all volcano eruption types. When the gas column collapses the ejected particles fall down and rush downhill at speeds up to 400 km/h. At the base of the ash cloud, temperatures of up to 800 °C can be reached, making the ashes practically glow. The volume of the gas increases when the pressure is released by 1 atmosphere for every 4 m and can thus reach a 10 times larger volume than originally in the magma.

When volcano emissions enter the atmosphere the aerosols emitted are anticipated by the public to add strongly to the greenhouse gas content. However, it was calculated that all world active volcanoes generate about 250 tons of SO_2 annually, whereas automotive and industrial activities are responsible for a volume of 25 billion tons of SO_2 per year. Thus it becomes obvious that the volcanos are "only" contributing to just 1 % of the world's SO_2 concentration; the SO_2 emission of Mt. St. Helens was about 1 million tons of SO_2 released within a couple of weeks. Another indicator that volcanic aerosols have only a negligible impact on the world's atmosphere is the fact that CO_2 measurements that are continuously recorded by the Scripps Institution of Oceanography since 1960 at Mauna Loa (Keeling Curve: 1960 = 310 ppm CO_2; 2010 = 390 ppm CO_2) did not reveal a significant increase after major volcanic eruptions. Instead the total CO_2 concentration has increased smoothly and steadily with no erratic increase or decrease (USGS 1995b). If the increased CO_2 concentrations were of volcanic origin, then CO_2 concentrations should rise following volcanic eruptions; the opposite is true: CO_2 concentrations decline for decades continuously, smoothly, and steadily. Assuming that all 200 most active volcanoes of the world emit a similar amount of aerosols to that Mt. Pinatubo once generated, the yearly total emission will be "only" less than 20 million of CO_2. Another figure clarifies how small the input from volcanoes is on the Earth's atmosphere: the emission of the Mauna Loa volcano of Hawaii of about 2 × 15 kg CO_2 equals an emission of CO_2 of 200 MW coal-fired power stations, and contributes a volume of just 1 % of the entire anthropogenic contribution. Nevertheless even when the annual contribution of aerosols of about 370 ppm is comparatively small and thus does "not directly impact the greenhouse effect, over the lifetime of the Earth, these gases have been the main source of the planet's atmosphere and ocean" (Robock 2000, p. 193). Therefore even small volcanic eruptions emit gases and these emissions mainly occur along the plate boundaries. Of the many different gases emitted, H_2O in the form of water vapor, CO, SO_2, CH_4, and N_2 are the major contributors. Although not the biggest constituent, SO_2 is seen to be the biggest contributor to the greenhouse gas concentration. The SO_2 reacts with other gases to H_2SO_4 (Newhall and Self 1982), forming aerosols that have a strong radiative capacity, capable of withholding the sun's radiation from penetrating the atmosphere and thus cooling the Earth's surface. In contrast to that, CO_2 has the effect of increasing global

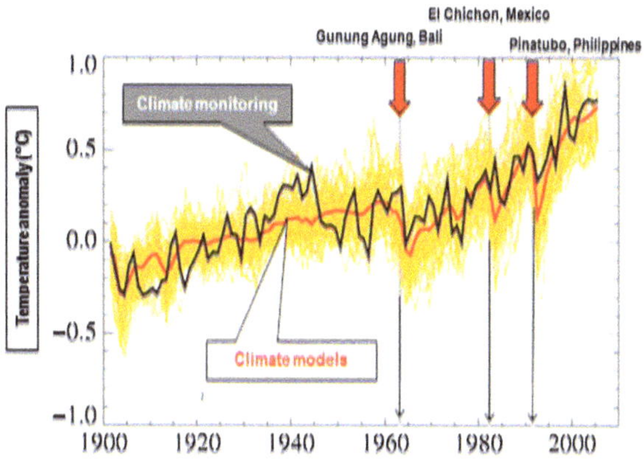

Fig. 3.26 Drop in atmospheric temperature due to volcanic ash emissions from volcanoes Gunung Agung (Bal), El Chichon (Mexico), and Pinatubo (Philippines) (Based on: IPCC 2007)

warming (USGS 1995b) as it destroys the ozone layer. Many of the volcano events of the last decades, most prominently the eruptions of Mt. St. Helens, Gunung Agung, Pinatubo, and El Chichon proved that their gas emissions had a significant cooling effect (Fig. 3.26). From measurements of the gas concentrations in the atmosphere it was calculated that the El Chichon eruption emitted about 20 million tons of aerosols (Bluth et al. 1992) that traveled the globe in less than three weeks, similar to the gases of the Pinatubo eruption. The emissions from Krakatoa were reportedly visible for months all over the globe and were documented in many art drawings and photographs.

The ashes and gases emitted not only have an impact on global cooling or warming, but also can have a considerable impact on public health. In fact not only human health but international air traffic can also be highly at risk from volcanic ash. It was on June 1982 when a British Airways flight from Kuala Lumpur to Perth (Australia) flew into a cloud of volcanic ash at 12,000 m. The ashes were emitted from the Indonesian volcano Galunggung (Western Java, Indonesia) that erupted a couple of days earlier. All of a sudden the windows were hit by particles that sent them into a sparkling light (like that emitted by a welding machine). Simultaneously the wings were set with bright glowing light and the cockpit filled with sulfur-smelling smoke. Finally all four engines failed. A similar experience occurred on a KLM flight bound for Anchorage, Alaska. That time the aircraft entered an ash cloud of the Alaskan Redoubt volcano, 200 km away. The route across Alaska is one of world's busiest jet airline routes over the North Pacific. As the ash particles are of volcanic origin, they have a solidus temperature of about 1000–1100 °C. In contrast to that, aircraft turbines run at a temperature of about 1400 °C. These higher temperatures make the glass particles melt and cover the turbine blades with a thin coating of silica, making it impossible for them to work. The cool volcanic particles outside the turbine are tiny, jagged, very hard particles of sand and silt size (>2 mm in diameter), splintery with sharp edges that have an enormous abrasive potential to diminish visibility.

Fig. 3.27 Autosiphon degassing carbon dioxide from Lake Nyos, Cameroon

Another hazard accompanied with volcanism is carbon dioxide and methane accumulations in water bodies, such as in the African Lake Nyos (Cameroon) and Lake Kivu (Rwanda). In both lakes gases (CO_2/CH_4) have accumulated in the water column and were released suddenly to the surface. At Lake Nyos on August 21st, 1986 a great amount of CO_2 was released from the water body that is supposed to be generated from underwater volcanoes at a depth of several hundred meters. At about 200–300 m water depth the CO_2 is in a liquid state due to the overlying load of the water column. About two million tons of CO_2 were supposed to have accumulated over a time period of 300 years. All of a sudden the gas was released, and bubbled up like the gas in a bottle of mineral water when opened. The trigger is supposed to be either an earthquake or a seismic tremor generated from uprising magma in the volcano vent. As CO_2 is under an atmospheric condition heavier than air, the CO_2 accumulated at the surface and was flowing downwards following the morphology. It flooded the next city 27 km away and killed about 1700 people within a couple of minutes. In order to prevent such CO_2 from bubbling again an auto siphon system was installed

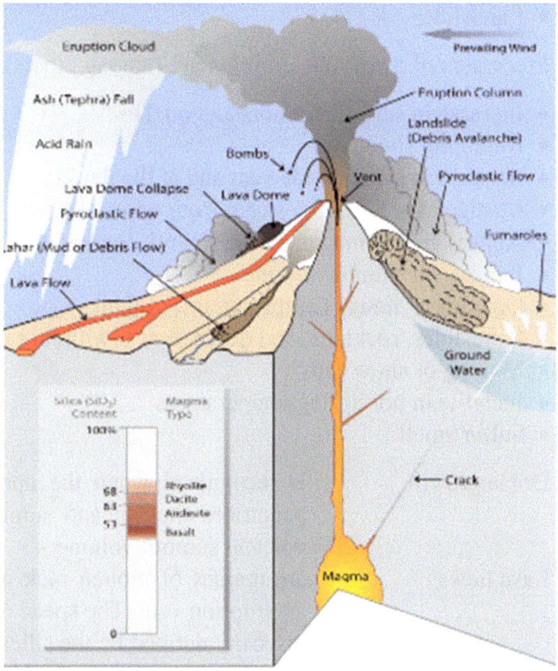

Fig. 3.28 Simplified sketch of typical hazards from strato volcanoes (*Courtesy* USGS 2008b)

to de-gas the deeper water layers constantly. In Fig. 3.27 the water fountain is displayed, the result of the gas driving the water up to 50 m in the air. A similar auto siphon system was installed at Lake Kivu where it now steadily releases the methane accumulation and thus protects the people from being at risk from a methane outburst.

The multifold hazards that can derive from volcanic eruptions are impressively summarized by Fig. 3.28 from the USGS.

Directly associated are:

- Fumaroles, hot water pools, solfatares, geysers
- Dome growth, microseismic tremors
- Lava flow
- Ash fall, tephra fall, acid rain
- Volcanic bombs, lapilli
- Lateral blasts
- Pyroclastic flows, gas surges
- Dome collapses
- Acid rain

Indirectly associated are:

- Mud flow, debris flows, lahars
- Hydrothermal explosions

- Landslides

Precursors of a volcanic eruption are (among others):

- Increase in seismic tremors and quakes
- Rumbling sounds
- Increase in steam at the vent and at the flanks
- Change in color of the steam from white to gray
- Appearance of magma at the summit (glowing magma)
- Increase in summit volume
- Opening of fissures at the summit of flanks
- Landslides, rockfalls, and debris flows (without rain)
- Melting of snow caps
- Increase in hot Spring temperatures
- Sulfur smell

Dome growth	is recognized when the uprising lava (magma chamber) penetrates the volcano summit leading to an increase in volcano summit volume.
Lava flows	are masses of molten rock material cascading down from an eruption vent. The speed of the lava flow depends on the viscosity defined by the silica and water content. Lava high in silica and water is highly viscous and flows at 3 km/days, whereas lava with a low viscosity can flow up to 50 km/h. Speed and geometry of the lava flow depend on the local topography: steep slopes channel low viscous lava to form elongated lava streams.
Pyroclastic surges	are a turbulent density current made up of a mixture of hot and often toxic gases and fragmented volcanic materials, that follows the slope topography with speeds of up to several hundred km per hour and with temperatures that can reach 800 °C and thus can cover large areas within a couple of minutes.
Volcanic gas	volcanic gas is a basic component of magma and lava. It can contain water, SO_2, SO_4, CO, CO_2, HCL, and also HF, all components that might react toxically when released in large quantities.
Hot and lateral blasts	occur when hot gases cannot make their way directly to the atmosphere due to the impermeable material around the volcano summit. The pressured system is often released by a rapid and explosive escape. Such blasts are among the most dangerous volcanic eruptions. They can blast off the entire summit or can explode obliquely to the vent as lateral blasts.
Mud flow/lahars	are a mixture of fragmented volcanic debris and water that rushes down the flanks during a volcanic eruption following the morphology. Lahars can also be generated long

Table 3.3 Volcanic materials that may be emitted during an eruption (Own compilation)

Lava	Lahar (mud flow)	Ash	Gas
Ejected magma Liquid Temperature up to >1100 °C Lava travel speed 2–3 km/h Liquid follows morphology Viscous Temperature of 700–800 °C Slowly moving downhill	Mixture of volcanic debris and water (rain, condensed vapor, ice) Highly turbulent stream (similar to liquid concrete) Travel speed up to 100 km/h Travel path follows morphology Highly destructive	Solid material of different size: Bombs (up to 7 cm in diameter), Lapilli (<2 cm) Ash (<2 mm) Pyroclastic surge up to 800 °C Up to 400 km/h fast Covers a large area within minutes Ash particles sharp-edged (<2 μm), Threat to respiratory organs Abrasive to jet turbines	CO_2 SO_2 SO_4 HF HCL H_2S

after an eruption when the unconsolidated ashes and debris accumulated at the flanks are wet by heavy rainfall.

Table 3.3 gives a general overview of the kind of volcanic materials that are emitted during an eruption and what threat can be derived from this.

3.2.1.8 Examples of Volcanic Eruptions

Paricutin, Mexico (Birth of a Volcano)

The birth of a volcano was for the first time in history witnessed in Mexico when in February 1943 a peasant observed that the Earth opened on his crop field and smoke rose up from a hole that was normally used to dispose of debris or other material. The hole was there for many years and had astonishingly never been filled up. That time the hole rapidly developed into a crack a couple of meters long. Several hours later the crack produced ashes and began to pile up. The ash production increased and also rock fragments started to get unearthed. At the end of the day, the ash mount had risen to about 2 m. The very next day the area began to tremble and glowing ashes and lava occurred, piling up to 50 m. A day later the first seismic tremors set and became more and more intensive, reaching a magnitude of 4.5 on the Richter scale. Andesitic lava was produced having a temperature of more than 1000 °C. Two months later the newly developed volcano had reached a height of 300 m. By June 1943 the village near the volcano had to be evacuated completely. Due to continuously ongoing lava production the volcano easily reached a height of 450 m within the next months and covered an area of 25 km^2. About 10 years later the activities ceased and the clock tower of the former village was left as the sole visible remnant.

The increased density of the worldwide volcano and seismic monitoring networks today makes it possible to identify and follow the birth of new volcanoes on our planet earlier, more directly, and much more precisely than ever before. Therefore in 1963 in Iceland the birth of a new volcano (Surtsey) was recorded (Jakobsson 2007). The volcano rose within its first day of appearance to a height of 10 m above sea level and to a length of 500 m. Half a year later the new volcano was 175 m high (in total 300 m above the sea floor) and about 2 km long. From satellite imagery it could be seen that in the vicinity of the volcano Erte Ale in Ethiopia, the Earth was opening for decades at a speed of about a couple of millimeters per year, and from January 2010 on, rifting speed was recorded to have accelerated to several meters per year; indicating the birth of a new ocean. Further evidence on volcanism and rifting is reported from Central Africa, where in Cameroon a 1500 km long belt of intraplate volcanoes occurs, also showing here that volcanism is a direct indicator of plate movements (Fitton 1980). In 2012 at the southern tip of the Spanish island of Hierro (Canary Islands) the birth of a new volcano was announced by a bubbling sea surface and volcanic gases. The news spread all over the world within hours and made everybody a witness to this event. In the next days the volcano produced much hot water, gas, and vapor. On underwater photos and on seismic records the size and shape of the volcano crater could clearly be identified.

Explosive and Lahars (Mt. Pinatubo, Luzon, Philippines)

The eruption of Mt. Pinatubo was one of the infrequent events in modern volcanology (in ancient history it was Plinius the Younger who witnessed the eruption of Vesuvius in Italy) that opened the chance to monitor a volcanic eruption long before it began until the last mitigation operations. And Mt. Pinatubo is the only example where it was possible by detailed geoscientific analysis to predict a volcanic eruption successfully and to prevent larger casualties. The investigations were comprehensively documented in a voluminous report (Newhall and Punongbayan 1996). Mount Pinatubo is a strato volcano that is located on the island of Luzon, quite near the Philippine capital, Manila. At its western flank the American air base (Clark Air Base) and at its southern rim the American Naval Base for East Asia (Subic Bay) were also located. The volcano had in those days a height of 1750 m; 30,000 people were living around the volcano and about a million around its foot. At the top of Mt. Pinatubo there was a small ethnic minority, Aeta, living who still had animistic beliefs. Until 1991 the volcano was assumed to be extinct with no signs of volcanic activity for over 500 years. It was in April 1991 when the monitoring volcanologists registered the first signs of activity, small seismic tremors accompanied by the first powerful steam explosions, producing molten rock and lava. Many thousand tons of noxious sulfur dioxide gases were also emitted. Ten weeks later the volcano erupted in one of the biggest volcanic events in modern history. Immediately upon the first eruptions being registered, a team of geologists and volcanologists of the Philippines Institute of Volcanology (PHIVOLCS) and the Geological Survey of America (USGS) installed additional monitoring equipment around the volcano that was able to identify during its first day about 200 smaller seismic tremors. The center of Mt.

Pinatubo monitoring was established at Clark Air Base. The scientists made a volcanic hazard map of the area and identified the extent and severity of former volcanic eruptions. The gas emissions, the shape of the volcano, and its temperature were measured directly at the volcano as well as from satellite images and aerial photos. The scientists came to the conclusion that an eruption of Mt. Pinatubo lay ahead.

On June 12th PHIVOLCS ordered the evacuation of all people in a radius of 10 km around the volcano, a radius that was soon extended to 30 km. About 60,000 people were evacuated within two days. On the 15th of June Mt. Pinatubo erupted cataclysmically. The entire summit exploded reducing the height by 150 m, forming instead a caldera of 2.5 km in diameter. The volcano ejected 5 km^3 of material, making Mt. Pinatubo the second largest volcano eruption in the last century, VEI 6. The ash cloud rose 34 km into the air and an incredible amount of gas and ashes were released. Giant mud flows and pyroclastic surges were formed, running over 50 km downhill at high speed. These surges filled former deep valleys up to the top and fresh volcanic deposits accumulated locally up to 200 m. The pyroclastic surge at one place was five years later easily identified as its deposits were still approximately 500 °C hot. The ashes covered an area of about 4000 km^2 with a blanket up to 1 m high. Many houses in the area (including many of the airbase barracks) collapsed under the load of newly fallen ash, especially as during the time of the eruption a tropical cyclone hit the area. The water soaked the dry ashes and thus doubled the weight. The (typical of the region) quite flat metal roofs were not able to withstand the increased load and broke, causing the biggest fraction of the 700 deaths. The living environment of many people was completely destroyed and after the eruption no vegetation was left and the predominant color of the region was gray. Quite a number of aircraft were traveling the area on the day of the eruption and 16 of them came in contact with the ashes; with two of them one of the engines failed.

USGS and PHIVOLCS have estimated the costs and benefits for forecast and evacuation of the people living around the volcano. There were 5000 people registered to be evacuated but it was estimated that in total up to 40,000 people could be brought into security, among them 20,000 of the indigenous Aeta people who settled directly at the top of the volcano. Moreover, 20,000 servicemen of the US airbase were also brought to safety. Altogether property losses of conservatively estimated US$250 million with an upper range of U$500 million had been saved by the eruption forecast. And USGS gave an estimate of the monetary value of the lives saved of the 5000 refugees registered refugees: it was "estimated" to be a value between US$0.1 and US$1 million per life (see Sect. 7.2). The total costs for safeguarding lives and property have been estimated at about US$50 million, including US$1.5 million for the volcano monitoring and eruption forecast, resulting in a cost–benefit ratio of about 1:10. Altogether the Mt. Pinatubo eruption is a story of a successful volcano disaster management. The severity as well as the area of impact was precisely predicted by the scientists. "Only" 700 people lost their lives, most of them members of the Aeta minority who for a long time denied the evacuation order. What were the factors that made the disaster and emergency management so

Fig. 3.29 Mount St. Helens, Washington State

successful at Mt. Pinatubo, where it failed at so many other volcanoes? The reasons are:

- The Philippines have long-lasting in-depth experiences with volcano eruptions.
- They very early established a well-functioning and skillful disaster management.
- They were working in close cooperation with international well-experienced volcanological institutes along with the USGS.
- The population at risk was early and comprehensively involved in risk prevention.

Lateral Blasts and Lahars (Mt. St. Helens, United States)

The eruption of Mt. St. Helens, in Washington State in 1980, is today the best recorded and analyzed volcano eruption of the world (USGS 2005b). It is also a synonym for a lateral blast: an eruption that runs obliquely to the central vent and explodes at the flank (Fig. 3.29). Mount St. Helens is still today one of the most active volcanos in the Cascade Range located in the northwestern part of the Rocky Mountains near the Canadian border. The volcano has proved to have a great potential to erupt explosively. Activity of Mt. St. Helens can be traced back over more than 250,000 years. In this time the volcano has produced both violent eruptions of ash and debris as well as quiet outpourings of lava.

Volcanic monitoring proved that months before the eruption, magma was intruding into the volcano leading to an increase in dome volume. The higher level of activity of Mt. St. Helens dated back to early 1980 when a series of small earthquakes occurred, followed by hundreds of smaller quakes. A strong steam explosion on March 17th, 1980 opened a crater through the volcano's ice cap that reached a diameter of about 300 m within several days. Two giant crack systems crossed the entire summit area. More than 10,000 earthquakes have been recorded and the northern flank of the volcano had grown outward by at least 150 m to form a significant bump. Such dramatic deformation of the volcano was interpreted by the volcanologists from USGS as strong evidence that molten rock had risen high into the volcano.

At the time of eruption, the volcano dome had reached its point of instability (USGS, ibid). On May 18th, an earthquake of magnitude 5.1 occurred that lasted only about 20 s and that made the entire volcano's bulge and summit slide away in a huge landslide. Film recordings reveal that the landslide removed the volcano's entire bulge and summit within a couple of minutes, the largest landslide ever recorded in Earth's history. This reduction in material released the pressure from the magma chamber of the volcano abruptly and allowed the hot water in the system to flash to steam and to expand explosively. The decreasing pressure even went through the volcanic conduit down to the magma chamber. The lava began to rise, formed bubbles, and then degassed explosively. For nine hours a hydrothermal lateral blast produced rock fragments, ashes, and gas water vapor, an eruption that is called cataclastic.

The eruption produced in the first 15 min an ash and gas column that reached 25 km high up in the sky. An hour later a second eruption took place, this time from the newly formed crater that was followed by the first smaller avalanches of hot ash, pumice, and gas. The eruptions blasted off the entire summit of the volcano lowering it by 400 m from its previous 2950 m. Then the first real pyroclastic surge occurred running at 120 km/h more than 10 km down the northern flank. USGS estimated the speed of the ash column to be about 400 km/h that then slowed down considerably and led to the fall of the huge amount of ash, debris, and rock fragments. Over the day the volcano released more than 500 million tons of ash north and eastward across Washington State; and even caused complete darkness in the city of Spokane (Washington State) more than 400 km away from Mt. St. Helens. Major ashfalls were also registered to the north as far as central Montana and covered large parts of the Great Plains to the east. Within the next three days the ash clouds spread across the entire United States and circled the Earth in 15 days.

The hot rocks and gas made large parts of the snow and ice cap of the volcano melt, creating surges of water that mixed with loose rock debris to form massive mudflows (lahars). The largest and most destructive lahar was formed by water seeping over days from inside the huge landslide deposit. This sustained flow of water eroded material from the landslide deposit. While flowing down the lahars increased in size and as they traveled downstream, flooded the two arms of the Toutle River, destroying everything (houses, bridges, trees, etc.). Impressive pictures show massive trees ripped off by the eruption all lying aligned in the direction away from the volcano. The region still today is only covered by small bushes and loose vegetation. Since 1982 the area has been under a natural conservation program and was given the status of National Monument.

In the years since the eruptions of 1980 the volcano showed a multitude of different volcanic activities. Its most recent series of eruptions began in 1986 with several periods of increased seismicity. Between 1989 and 1991 there were about 30 events of short but intense seismic activity lasting minutes to hours, accompanied by small explosions from the dome. The explosions formed a new vent on the north side of the dome and produced small eruption columns that rose several kilometers above the volcano. A few explosions also hurled hot rocks a meter in diameter from the dome, generated small pyroclastic flows in the crater, and formed small lahars. During 1995 and 1998 seismicity increased for several months, but there were no accompanying explosions. The first signs of recovery of fauna and flora were seen about six years later (Dale et al. 2005) when a small rat was seen that survived the catastrophe in hollows under the volcanic debris and spread lupine seeds on which it lived. Successively other plants and animals invaded the area. Today the former devastated and grey ash-covered area is covered by bushes, single trees, and widespread grass, flowers, and colors. For a couple of years now Mt. St. Helens is again showing higher signs of seismic and volcanic unrest. In 2004 a new lava dome inside the already newly emerged dome continued to grow. Thermal imaging confirmed that lava extrusion increased noticeably. In addition, the areas of uplift and those with intense deformation continued to move southward, nearing the crater wall. The new lava extrusion, which occupies the western part of the uplift, had a volume of almost 1.5 million cubic meters. And temperatures of more than 600 °C were measured. The many different types of indications of unrest clearly document that Mt. St. Helens will also be an active and highly hazardous volcano in the Cascade Range in the future.

Mount Pelée (Pyroclastic Surge = Nuée Ardente)

The paroxysmal eruption of the Mount Pelée on the island of Martinique in the year 1902 laid the base for an up to that day unknown type of volcano eruption (Pichler 2006, pp. 125–127). Since that time such an eruption type has been called a pyroclastic surge or in French *nuée ardente*. Mount Pelée was for a long time a dormant resting volcano for the 26,000 inhabitants of island's capital, St. Pierre, situated 7 km distant at the foot of the volcano, less a risk than seen as a leisure destiny. Although pyroclastic surges were recorded in the decades before the great eruption, no one really felt at risk. But in April 1902 the volcano started to produce ashes, gas, and white smoke. Sometimes eruption clouds were seen that rose a couple of hundred meters into the sky. And ash fell on the city. At the end of April the first seismic tremors were registered and the ashfalls became more intense. It was May 6th when according to historic evidence lava penetrated into the crater and produced a molten lava dome there. Simultaneously the heat generated the first lahars from the molten icecap that rushed downhill and killed 25 people in an adjacent valley. The people became frightened and sought refuge in the city of St. Pierre. The governor in his intention not to disturb the upcoming election appeased the people with the words that no one was at risk. But Mt. Pelée activities increased and on the top burning volcanic flames were seen. On May 8th the volcano erupted and a dark cloud of ashes ran downhill.

Later, evidence revealed that the cloud had reached a speed of up to 500 km/h and was up to 800 °C hot. The cloud was accompanied by a sound shock wave that traveled at a speed of 450 m/s and that covered an area of about 30 km^2

from the top of the volcano 1300 m down to the seaside. The pyroclastic surge reached the capital, St. Pierre, within a couple of minutes and killed all but two of the inhabitants. Nearly all the houses were completely destroyed. The thus far unprecedented witness of this paroxysmal eruption was given by sailors who just at the moment the volcano was erupting were about to enter St. Pierre harbor. The French Academy of Science immediately sent a team of volcanologists to Martinique to investigate the disaster. They found that the magma of Mt. Pelée was very rich in SiO_2 (dacitic magma) and that the tectonically highly fractured volcano allowed much phreatic water to penetrate into the volcano vents, leading to a highly explosive situation. These explosions destroyed the crater flanks and the normally thermodynamically controlled uprising ash cloud collapsed and the pyroclastic surge then rushed downhill (Lacroix 1904, 1908 cited in Pichler 2006, p. 127).

Such types of volcanoes spreading hot and toxic gases over large parts of the volcano flanks are known from many volcanoes of the world. The best examples are the city of Pompeii (Italy) whose entire populace was killed within a couple of minutes by a pyroclastic surge of the Vesuvius eruption in 79 AD. Or from Mt. Merapi (Indonesia) whose many eruptions often produced highly dangerous pyroclastic surges, most recently in June 2006. As this volcano type was intensively investigated during the eruption of the Surtsey volcano (Iceland) this type is also called Surtseyian.

Nevada del Ruiz (Mudflow, Lahar)

The volcano eruption of the Nevada del Ruiz in Columbia in 1985 is a world-famous example of a volcano-triggered mudflow (lahar). The term "lahar" comes from the Indonesian language and describes a mixture of water with sandy material, gravel, and larger boulders that suspended in a fluid–liquid mixture has a physical density like floating concrete.

The Nevada del Ruiz volcano is about 5400 m high and is one of the volcanoes of many covered with an icecap. The glacier at the top of the volcano those days spread over an area of about 23 km^2 over the almost flat-lying summit with ice of up to 200 m thickness. Furthermore under the icecap there was a large crater lake (Arenas Crater). For more than 150 years the volcano was in repose status. But on November 13th, 1985 the volcano suddenly produced ashes and increased fumarole activity with vapor that rose up to 2 km high (Pichler 2006, pp. 116–118). Some minor but significant seismic tremors had been recognized for weeks indicating uprising magma in the volcano vent. Consequently a group of Columbians from the national Committee of Volcano Monitoring investigated the volcano, but due to limited and rather outdated equipment the investigations were not found serious enough to convince the national and local authorities about the potential risk. It was nevertheless understood that in case of an eruption this would definitively melt the ice cap and would generate a massive lahar. In 1595 and 1845 similar lahars had already been recorded.

The initial blast occurred on November 13th, 1985 producing a large amount of ash and pumice that showered down on the area and that even reached the city of Armero about 70 km away from the volcano. Several hours later molten rocks and lava erupted from the summit for the first time and also produced a local pyroclastic surge. Together with the eruption a heavy storm with torrential rain set in forcing the people to stay in houses for the next hours. The heavy downpour led the electricity system in the region to fail and there was no radio communication in operation (see Sect. 5.3). The uprising magma led to a melting of 10 % of the ice-cap and filled up the Arenas Crater lake with melt water. Once the surrounding ice barrier was melted a highly destructive mudflow (lahar) made up of water, ashes, rock fragments, gravel, and sand was generated that rushed downhill at a speed of up to 50 km/h and that reached Armero two hours later. The city of Armero was built on top of a former mudflow that occurred in 1845.

There is no clear evidence that the inhabitants of Armero ever received a general evacuation order (Tilling 1989). Even attempts of the local Red Cross (Voight 1986) to evacuate the people failed, as most of the citizens did not have confidence in the technicians' statements and instead preferred to follow the advice of the local priests and city mayor, both disseminating their messages about three hours before by radio and loudspeakers that "the city is not at risk." An earlier attempt to consult USGS volcanologists was stopped by the US government as they saw a certain risk for their people by terrorist attacks. A total of 23,000 inhabitants of Armero died that night when the lahar devastated the entire city with a 3-m high flood causing an economic loss of about US$1 billion (in US$ 1985 currency). Summarizing the tragedy of Nevada del Ruiz it was "purely and simply a human error caused by misjudgment and bureaucratic shortsightedness" (Voight, ibid).

Island (Rift Volcano)

A special form of plate boundary volcanoes are the so-called rift volcanoes that occur along the divergent plate boundaries. The best-known examples of that volcano type are Iceland or the volcanoes in the East-African rift valley. Nevertheless by far the most rift volcanoes occur under the sea; it is estimated that there are more than several thousand, although only 250 of them are active and identified. In Iceland the divergent plates formed in historic times (since 1783), including the world-famous 27 km long Laki rift valley where the Eurasian and North American plates diverge with a speed of 1–2 cm/year. Thus one half of Iceland is traveling with the American plate westward while the other goes east. Next to the Laki rift there is a series of different almost parallel running rifts that have separated Iceland for more than 20 million years (Hjartarson et al. 2009). At the time of the initial rifting of the Laki rift a 25-km wide lava mass was formed covering an area of 560 km^2 with a volume of almost 15 km^3, the biggest lava production during one eruption phase ever recorded (Pichler 2006, p. 64). Similarly, about 100 years later the eruption emitted a high amount of sulphur aerosols into the atmosphere that dropped the average temperature in the northern hemisphere in the year after

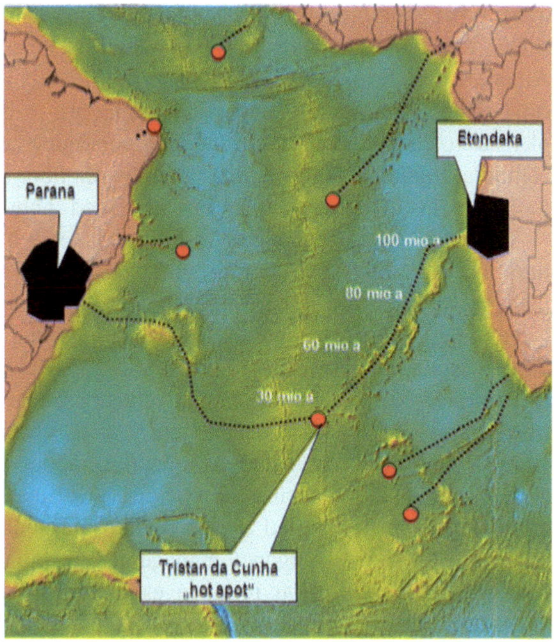

Fig. 3.30 Hotspots in the South Atlantic and their travel paths (Graph based: Duncan and Richards 1991)

by about 1.3 °C and made the summer very cold, resulting in a massive crop failure. In those days according to historic records Iceland's population diminished by about 25 % and the livestock losses amounted to 70 %.

Hawaii (Hotspot)

Hotspots are accumulations of magma in the Earth's crust that are not directly attached to plate boundaries. Like magma of other chamber types the lava continuously protrudes through the plates thus forming a volcano. The characteristic feature of a hotspot is that the overlying plate is steadily moving while the magma chamber is stable in its position. Thus in geological times a chain of volcanoes is built. Hotspots are identified all over the world. Well-known examples in Europe are the hotspots located along a line from the French Cevennes in the west and the German Eifel volcanoes towards the east (Schmincke 2000), all of them generated by the northward movement of the African continental plates. The Hawaiian Island chain is acknowledged to be the most famous example of a hotspot. The westward movement of the Pacific plate results in the oldest Hawaiian volcano lying in the west having been generated about 30 million years ago and the youngest (2 million years old) located in the east (Lockwood and Hazlett 2010). The Hawaiian volcano ridge is more than 75 million years old and has produced more than 200 volcanoes. The ridge itself can be traced for more than 5000 km to the north, where it found its outermost extension in the Emperor Seamount chain. There many other examples of hotspots in the Pacific, for instance, the Marquesas Island ridge, the Salomon Islands, or the Marshall Islands. A famous hotspot in the South Atlantic is the Island of Tristan da Cunha which is located just at the mid-Atlantic ridge. The precursors of Tristan da Cunha are to be found in Brazil, the Parana plateau volcanoes, and on the African continent the basaltic layers of Etendaka, Namibia. Figure 3.30 shows that from the sea floor morphology the way the plates took over this hotspot can clearly be identified (Duncan and Richards 1991).

The Yellowstone volcano is also a result of a hotspot. The magma protruded into the mountain long before, forming the biggest caldera on Earth. Since then the volcano did not give significant signs of an ongoing effusion process. Nevertheless there were numerous indications that revealed that the Yellowstone hotspot is not extinct. Since 1950 the area has lifted up less than a meter, as could be identified by radar interferometry imagery. Since 2004 the Yellowstone volcano rose at a record speed of up to 7 cm/year as investigations of the University of Utah (United States) revealed. At a depth of about 10 km the geologists identified a huge accumulation of molten magma (www.unews.utah). This uplift frightens many people in the United States, therefore a geologist in 2009 made an urgent request to the local emergency management to immediately evacuate Yellowstone National Park as according to his investigations a major eruption is about to occur very soon. Since than the volcano did not show any significant unrest. The Utah geologists stated that uplift must not automatically be followed by an eruption. Uplift can rather be followed by a descent due to cooling of the magma that rose up in the vent.

Lake Nyos (CO_2 Accumulated in Volcanic Crater Lakes)

Another phenomenon attached to volcano eruptions is submarine gas emissions, either offshore or on land in crater lakes. Lake Nyos in Cameroon is a prominent example for a submarine gas emission, and caused a major tragedy in 1986 (Walker et al. 1992). The lake is located in the Oku Volcano region within the Mbere-Graben structure and is a former explosion crater that for centuries has been filled with water. As the ongoing volcanism continuously produces CO_2 the gas accumulates in the deeper parts of the water in liquid form. According to the thermodynamic equation a water column of 200 m can take up 10 times more CO_2 than water at the surface. Thus it was assumed that at a depth of more than 200 m a considerable amount of CO_2 was accumulated. In August 1986 that CO_2 suddenly was released from the lake and the CO_2 having a lower density than water was jetted to the surface and resulted in a change in the water–CO_2 pressure equilibrium and resulted in a degassing of more and more CO_2. As at the surface CO_2 on the other hand is heavier than air, the CO_2 accumulated in the adjacent river system and morphological depressions where many people were living. It was calculated that more than 1.6 million tons of CO_2 had been released that time, an amount of gas that took about 300 years to accumulate. In that night 1700 people lost their lives and thousands of cattle and other livestock were killed. Although

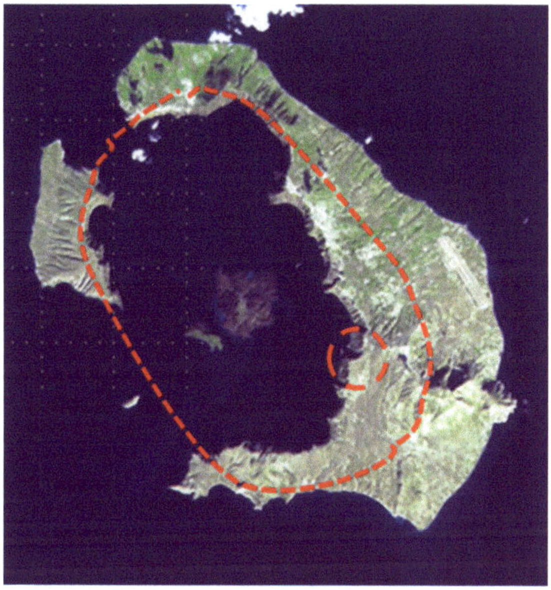

Fig. 3.31 The Island of Santorini and the position of the former volcano

the actual trigger for this tragedy is not clear up to now, it is assumed that it was either an earthquake or submarine slum that turned the normal water layering upside down. A similar tragedy was reported from Lake Kivu in Ruanda (Walker et al., ibid), one of the largest lakes in eastern Africa that is 2000 times the size of Lake Nyos. There it is assumed that about 250 cbm of CO_2 and 55 billion cbm of CH_4 have accumulated.

Caldera (Santorini Volcano)

The volcanic island of Santorini (Greece) today encompasses four islands named Thera, Therasia, Aspronisi, and the central Kameni Islands. This group of islands makes up what is called a caldera (Fig. 3.31). The Santorini volcano is part of the Cyclades Island Chain in the southern Aegean Sea, located halfway between Greece and Turkey. The Cyclades Island Chain itself is generated by the northward subduction of the African plate along the Hellenic trench system under the Eurasian plate. Its outstanding appearance, its perfect circular shape, and its mere size, make the Santorini volcano the type locality of a caldera. Other well-known calderas are the large caldera of the Krakatoa volcano in the Sunda Strait (Indonesia) or the beautifully shaped caldera of Crater Lake in the Rocky Mountains, although this one is much smaller in size. The vision that the eruption of the Santorini caldera was responsible for the decline of the Minoan culture is often cited in archeological literature but could not be verified by geological evidence (Pichler, ibid, p. 40).

Calderas are former strato volcanoes whose central parts, after having emptied the shallow lying magma chamber through a ring of volcano vents, are collapsed into the depleted reservoir, thus forming a bowl-shaped depression. In this depression water ingresses to fill up the former volcano summit (Gudmudsson 2008). This caldera type is normally generated after the main phase of ash-producing eruptions.

The island of Santorini wasn't the result of a single event, but rather of several eruptions. The geologic record reveals a long history of eruptions, all consistent with its subduction-zone setting. At least 12 eruptive phases have occurred over the last one million years. One of the ash clouds is taken to have reached an altitude of more than 30 km (Pichler, ibid, p. 42) and Santorini aerosols could be traced up to Greenland. The latest Bronze Age event was a Plinian eruption with an estimated "Volcanic Explosivity Index" (VEI) of 6.9; an explosivity that was only surpassed by seven other terrestrial eruptions in the past four millennia. Before this Bronze Age catastrophe, Santorini was a large circular island with a water-filled embayment whose central highland collapsed again to generate the modern caldera. From archeological records it can be followed that Santorini has been inhabited by numerous civilizations going back to the thirteenth century BC, contemporaneous with the most recent eruptive events. Historic eruptions may well have provided dramatic events for these various civilizations. Archeological excavations indicate that the island of Thera was colonized by the Minoans, a Bronze Age civilization named after the legendary King Minos of Crete, and that appears to have had a thriving economy based on an intensive trade throughout the eastern Mediterranean. The exact date of the last eruption is still controversial. Most radiometric studies show that it might date from 1615–1645 BC, consistent with a pronounced acid-ice layer from the Greenland cores, dated at 1636 BC. Since the late Bronze Age eruption, two new islands have formed in the center of the caldera by numerous eruptive events over the past 2000 years. Santorini thus appears to be particularly active compared to its geologic past. There have been several eruptions in the nineteenth and twentieth centuries, with the most recent occurring in 1950. Still today the island group exhibits ongoing seismic activity, and both fumaroles and hydrothermal springs are common features on the islands.

3.2.2 Hydrological Hazards

This section describes disaster events that were caused by deviations in the normal water cycle and/or overflow of bodies of water caused by wind setup:

- Flood
- Mass movement (wet)
- Land subsidence
- Avalanche

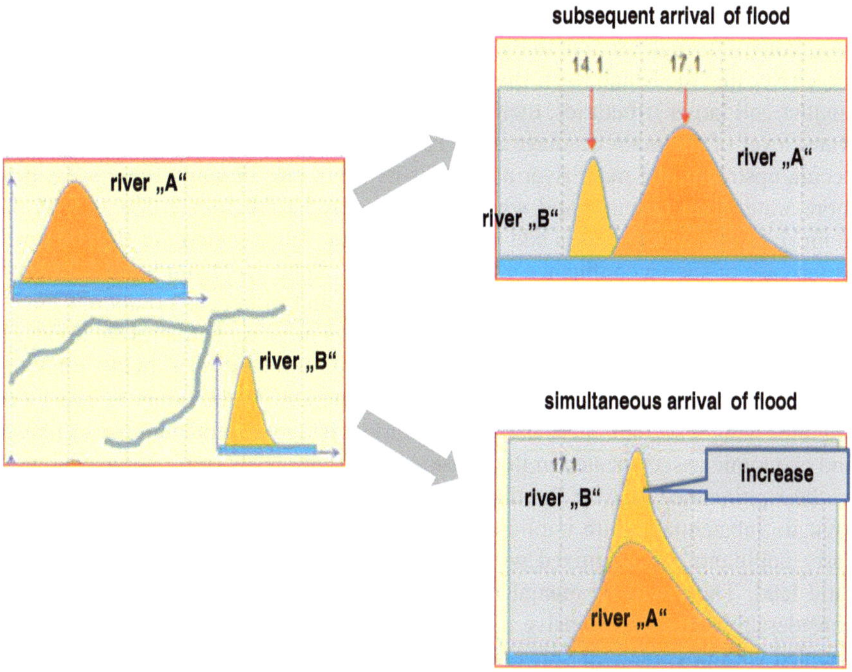

Fig. 3.32 Flood level increase due to subsequent or simultaneous flood arrival (Own graph)

3.2.2.1 Flood

Different types of floods occur, such as river floods and floods from the sea in coastal areas. Flash floods and urban floods have been described in Sect. 3.2.3, however, this section is concerned with the flooding of riverbeds and lowlands.

Flooding occurs when the normal water level in the rivers is increased, in general due to excessive rainfall, but also when in times of snow melting the water level is increased or due to failure of technical infrastructure. A higher than normal rainfall can occur either as a result of precipitation over a longer time period or as an abrupt and short-term heavy downpour from thunderstorms. In general about 30 % of precipitation is drained at the surface, 30 % enters the top soil, and the rest is taken out of the system by evapotranspiration. The moment the soil pore space is saturated with water the additional water masses can only drain off, thus increasing the normal percentage of drained water and it comes to a flood. The responsibility to define what is a "flood" lies in general with the government administrations that in Western Europe, according to centuries of hydrological data, in general define the highest water level of the last 100 years (HQ100) as the threshold value. Nevertheless higher and lower water levels can be defined for areas with lower risk or areas where, for example, industrial assets or historical buildings have to be protected.

The water of a river normally flows downstream according to the river valley morphology from the source to the river mouth, a journey that can take weeks with large river like the Rhine river. But as rivers are normally composed of a system of smaller and larger tributaries, their water masses can have a great influence on the water level of the main drainage area. In the case of a regional rainfall event that occurs upstream, the main river as well as its tributaries draining these areas bear more water than normal. And where one or several tributaries conjoins with the main river, the water masses add up to a level that can overstep the HQ100 level. In the case of such a confluence a flood situation can be avoided. Figure 3.32 gives an impression of how water masses can add up and moreover points out how a flood management comprising the entire catchment area can help prevent floods.

There are two trends that point to an increase of flood risk worldwide. First, the magnitude and frequency of floods are likely to increase in the future as a result of climate change; second, there has been a marked increase in the number of people and economic assets located in flood risk zones. There is also a growing awareness of the significance of river flooding on human health, both physical and psychological. Substantial health implications can occur, for example, when floodwaters carry pollutants, or are mixed with contaminated water from drains and agricultural land. There will be mental health consequences as well: in addition to the considerable stress of extensive damage, the threat of repeated floods, sometimes coupled with possible withdrawal of insurance cover can make properties impossible to sell. All experts in flood management see clear indications that the risk from floods will increase considerably during the coming decades and this holds true for developing as well as industrialized countries. The challenge for flood managers as well as for regional planners but also for the populations at flood risk is to anticipate these changes and to work out strategies to protect society and the environment from the negative effects of floods.

Swiss Re (2010) listed 675 floods all the over the world in the time span 1998 to 2010 that claimed 130,000 deaths and resulted in US$30 billion in insured losses. The country in Western Europe most affected by floods is the United Kingdom, where in 2007 the biggest flood ever in the last 200 years of history occurred. The flood devastated large parts of middle England when 360 mm rain fell in the months from May to July 2007, more than double the amount of normal rainfall, and covered the country with water for weeks causing damages of up to US$8 billion, of which about US$6 billion was insured (MunichRe 2012). The country is regularly struck by floods that have caused damage of another US$2.5 billion in the time from 1994 to 2000.

More than 200 flood events occur worldwide yearly in the last 30 years of which about 10 are classified "great and disastrous" (MunichRe 2013). From 1975 to 2001 the number of flood-related disasters increased from 20 (1975) to more than 150 in 2001, and the number of people killed by floods in general was quite stable between 5000 and 10,000 per year. CRED-EMDAT listed the 10 most severe floods of the last 20 years (Guha-Sapir et al. 2013) and it came out that although China did not occur among the 10 most deadly events, it made up all 10 most severe in regard to the people made homeless. Of the 10 most severe floods

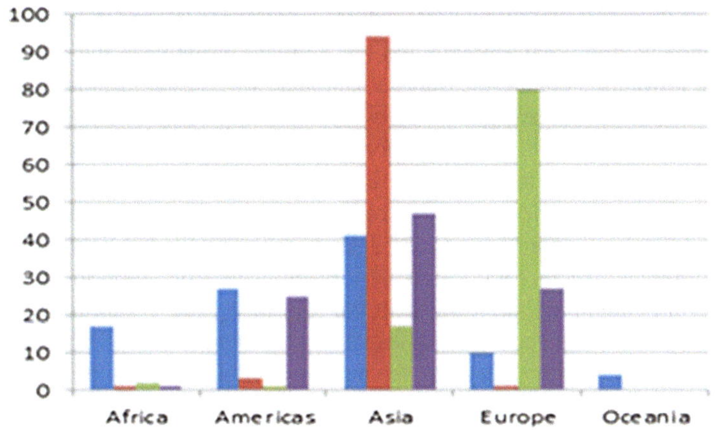

Flood	1900-2012 events (%)	CRED-Emdat killed (%)	affected	damages
Africa	17	1	2	1
Americas	27	3	1	25
Asia	41	94	17	47
Europe	10	1	80	27
Oceania	4	0	0	0
in total	1280	2.5 Mio	5.1 billion	89.3 billion

Fig. 3.33 Statistical evidence on worldwide flood risk (Compiled from MunichRe 2012; Guha-Sapir et al. 2013; UNISDR 2011)

in regard to economic losses, 80 % occurred in Asia with 20 % in the United States and Western Europe.

In the year 2012 floods claimed 1600 lives worldwide, made more than 50,000 homeless, and caused an economic loss of more than US$20 million. Similar to other natural disasters, floods have also primarily affected the low- and middle-income classes. It must nevertheless be pointed out that although the number of people exposed to flood hazards has risen significantly from about 100 million in 1980–1990 to more than 150 million by the end of 1990, the overall death toll from floods is decreasing, a trend that reflects the general death toll from natural disasters worldwide. By far most of these victims live in Asia, where 90 % of all the people that were ever affected by natural disasters live. Between 1998 and 2002 Europe suffered over 100 major damaging floods, including the catastrophic floods along the Danube and Elbe rivers in 2002. Since 1998, floods have caused some 700 fatalities, and were responsible for the displacement of about half a million people and at least €25 billion in insured economic losses (EEA 2003). Floods in developing countries often have wiped out the investments made in infrastructure of the previous 50 years. Especially in Asian developing countries where poverty and social indifference are widespread and basic needs are often not secured is there a great need to build societal resilience to floods. Such a higher level of resilience will only be achieved when society in general acknowledges the importance of flood management and the political levels pave the administrative ways for prevention countermeasures at national as well as at local levels.

Statistics on flood damages given by MunichRe (2012), CRED-Emdat (Guha-Sapir et al. 2013), and the United Nations (UNISDR 2011) reveal that floods are in general responsible for more than 30 % of the total economic losses and for more than two thirds of the people who have been in total affected by natural disasters worldwide in last century. Figure 3.33 gives more detailed insights to the flood risks worldwide. About 1300 events have taken place from 1900 to 2012; in total 2.5 million people have been killed in the said time span, 5 million people have been affected, and the economic losses total up to more than US$3 billion. But the geographical distribution of events, the number of claimed lives, the affected ones, and damages vary significantly. Although more than 80 % of the events occurred in Africa, Asia, and the Americas, 90 % of the persons killed were in Asia alone and 80 % of the affected ones live in Europe. Another remarkable fact is that in America only 1 % of the worldwide flood affected people's lives, but the damages (mostly derived from the Mississippi/Missouri flood plain) sum up to about 25 % of the worldwide flood damages.

The EU-EWFD (2000) stresses the need for an integrated flood management evaluating flood probabilities and flood consequences in a risk-based assessment. Methods for the analysis of flood risk should include the following steps: (a) determination of the probability of flooding, (b) simulation of flood characteristics, and (c) an assessment of the consequences from flooding (Apel et al. 2006).

In order to reduce the flood risk either of the individual as well as of society there are three factors that define the main mitigation strategy:

- Reducing flood probability by giving the rivers more space to take up the rainwater by rebuilding the often aligned river courses, by widening the riverbeds, by using natural morphological depressions as areas to be flooded (retention areas, flood polders), and by increasing the dam heights
- Reducing the consequences of flooding by making households at risk flood resistant, by increasing the flood resilience of critical infrastructure (hospitals, water and power supply installations, bridges, etc.)
- Establishing organizational measurers such as faster and comprehensive flood risk probability assessment, a dense and rapid flood level monitoring, reliable flood simulation programs that cover all types of flood scenarios, an improved early warning system, an early involvement of the individuals at risk and a strengthening of the individual coping capacities as well as an early and comprehensive involvement of those societal groups that are highly at risk in the official flood management

Risks from flooding predominantly occur along rivers and the coastal regions. River flooding is mainly defined by the water retention capability of the river catchment area as a function of its size and morphology. In this area, when exposed to rainfall, the amount and intensity of rain define how strongly the river is able to drain off the water and prevent adjacent areas from flooding. In Europe the economic values that are exposed to flood risk are estimated to be more than €150 billion and along the river Rhine alone, and along its German part more than 2 million people are supposed to be at risk (ICPR 2010, p. 8). By human intervention in the natural environment the original capacity of the catchment areas to withhold floods in many regions of the world is already highly affected: flood plains were sealed by settlements, river courses straightened, and deforestations led to higher runoff.

Flooding in coastal areas, however, mostly depends on the morphology of the coastal regions that are exposed to storm-induced sea-level increase. The total value of economic assets located within a 500-m perimeter of the European coastline is estimated to reach €1000 billion (EU-EWFD ibid). As with river flood plains the coastal regions are home to more than 50 % of the world's population as these regions are highly settled and used agriculturally and industrially. Thus flooding leads to high economic and social damages, but also has severe environmental consequences, for example, when coastal aquifers become saline or polluted waters enter into the sea, with negative impacts on the wetland areas and the biodiversity.

(a) When during a heavy rain a river extends its bed and covers the adjacent plains this is already a flood.
(b) Mass movement (wet)/mass movement (dry).

3.2.2.2 Mass Movements (Landslides, Debris Flows, Avalanches)

Landslides contribute to major disasters every year on a global scale, and the frequency of occurrence is on an upward trend. The increasing number of landslide disasters can be attributed in large part to the new reality of changing climate resulting in more extreme weather conditions combined with overexploitation of natural resources and deforestation, increased urbanization, and uncontrolled use of land (Nadim et al. 2006a). Recent examples are the mudflows of December 1999 in Venezuela, involving over 20,000 deaths, and the El Salvador earthquakes of 2001, which caused 600 deaths in just one landslide. In total more than 600 landslide events have taken place since 1900 according to the CRED-EMDAT database, claimed about 70,000 lives, affected more than 10 million people, and caused economic damage of more than US$10 billion.

The landslide hotspots in the world are located in general along the mountainous fold belts in the west along the Rocky Mountains and the Andean mountain range and along the European–Asian mountain belt that is formed by the Alps, the Caucasus, the Zagros Chain, and the Himalayas all along through Indonesia down to New Zealand and northwards along the Philippines, and Japan up to Kamchatka. In general the landslide risk of these mountain regions has been identified by Nadim et al. (2006b) to be moderate to medium, although—as landslide and mass movements are preferably local phenomena—can be locally of very high risk.

The term “landslide” describes a wide variety of processes of slope-forming materials including rock, soil, artificial fill, or a combination of it, that result in downslope movement of soil, rock, and organic materials under the effects of gravity. Based on the trendsetting classifications made by Varnes (1978), Cruden and Varnes (1996), and Hutchinson (1988), the term “landslide” meanwhile has become widely accepted especially by engineering geologists and civil engineers. The term “landslide” describes all types of gravitational slope failures: rotational and translational slides, slow moving earth flows, and fast-moving debris flows composed of mud, gravel (up to boulder-sized material), and organic debris that often mobilize from slides (Pierson et al. 1996). In response to periods of intense rainfall landslides often turn into debris flows. They initiate as rotational or translational slides that turn into muddy slurries, or from concentrated erosion of surface material by runoff. As they travel downhill slopes and channels, the slurries can substantially increase in volume by incorporating additional material. Addition of sufficient volumes of water relative to the sediment content can also result in dilution of the debris flow to the consistency of normal mudflow. The term also describes the landform that results from such movement (USGS 2008c). It comprises many different ways the materials can move (see Fig. 3.34): either by falling, toppling, sliding, spreading, or flowing.

Nevertheless there are quite a number of other terms in use that are interchangeable with the term landslide (mass movement, slope failure, flow-like mass movements, debris flow, debris avalanche and mudslide, rockfalls, soil creeping, and many other variations). A definition that can provide a more precise description of gravity-induced mass movements by emphasizing the formation process has been proposed by Hungr et al. (2001). They presented a “new division of landslide materials based on genetic and morphological aspects rather than arbitrary grain-size limits.” The proposed definition distinguishes between:

- Slow, nonliquefied sand or gravel flows
- Extremely rapid sand, silt, or debris flow slides accompanied by liquefaction
- Clay flow slides involving extrasensitive clays
- Peat flows
- Slow to rapid earth flows in nonsensitive plastic clays
- Debris flows that occur in steep established channels or gullies
- Mud flows considered as cohesive debris flows
- Debris floods involving massive sediment transport at limited discharges
- Debris avalanches that occur on open hill slopes and rock avalanches formed by large-scale failures of bedrock

Regardless of the definition actually in use, the definition of the different type of landslides (dry and wet mass movements) as given above outlines the basic understanding of landslides and is summarized in the eight pictures in Fig. 3.34.

Landslides can occur virtually anywhere in the world, a finding that contradicts the traditional viewpoint that landslides are restricted to extremely steep slopes. And there is almost no country on Earth that is not exposed to this kind of natural hazard. Although the primary driving force for a landslide is gravity, there are

Fig. 3.34 Generalized overview of landslide types (*Courtesy* USGS 2008c)

other factors that influence and affect the original stability of a slope, for instance, the specific subsurface conditions that make an area or a slope prone to failure, whereas a landslide often requires a trigger before being released. Among the different triggering effects the major causes for landslide are:

- Precipitation
- Earthquakes
- Volcanoes
- Permafrost
- Forest fires
- Erosion
- Flooding
- Human activities

Extremely dry areas as well as very humid areas are highly prone to slope failure. Moreover not only steep slopes are a necessary prerequisite for landslides to occur but also gentle slopes with an inclination of only 1–2° have been observed to fail. Concerning the impact of human activities it turned out that many of them superimpose the natural preconditions and lead to more severe triggering situations. But landslides can not only occur in bedrock or on soils, cultivated land, barren slopes, or on natural forests; landslides can also occur under water and are also recorded from extraterrestrial planets such as the Ophir Chasma landslide on the Moon that experienced a total height of about 5000 m. New studies revealed that also thermal expansion for daily temperature fluctuation contributes to slope movement.

Although landslides have been recorded from everywhere around the world, three major triggering mechanisms can be distinguished that can occur either singly or in combination. The impact of all of these root causes can vary widely and depend on geomorphological factors such as steepness of slope, shape of terrain, geological factors such as soil type and underlying geology, and on the human factor including agricultural activity, settlements, or technical infrastructure. Landslides typically occur when rainfall infiltrates a relatively competent mass of soil making the soil become gradually saturated. This leads to an increase in the pore-water pressure while simultaneously decreasing the shear strengths. The more water infiltrates, the more the initial landslide changes into muddy slurry transforming the landslides gradually into a debris flow. Such a phenomenon is preferably observed on hill slopes steeper than 15° (Iverson et al. 1997). Landslide-generated debris flows can move rapidly downslope and frequently incorporate significant volumes of sediment along their way down, thus increasing in volume. Landscapes disturbed by wildfire, foresting, construction of roads or dams, or volcanic eruptions reduce transpiration rates as a result of the loss of vegetation and to root decay associated with decreases in soil cohesion, that can result in a higher landslide hazard potential as the increase in rainfall triggers the soil moisture content (Schmidt et al. 2001). In areas burned by wildfire, for instance, it was found that debris flows caused by landslide could occur during the first rainy season immediately after the fire and that hazard potential can last about 10 years after the fire, especially when such areas are exposed to prolonged, but infrequent rainfall events often in combination with rapid snowmelt.

Young mountain ranges that are generally subject to a comparably higher level in earthquake activity consequently increase the likelihood in vulnerability to landslides. Earthquakes in such areas experience a comparably significant higher amount of landslide events due to ground shaking, liquefaction, or just that the ground motion allows the infiltration of large amounts of water into the subsurface. Furthermore rockfalls and rock toppling can also be generated by loosening the rocky formations. There is also a great danger of landslides forming dams of debris in streams and rivers. These landslide dams often can block the water completely from flowing, causing water to hold up behind the dam. While the water level is increasing these dams often erode and can fail completely, releasing large amounts of water in a flash flood.

Landslides due to volcanic activity represent some of the most devastating types of failures. Volcanic lava may melt snow rapidly, which can form a deluge of rock, soil, ash, and water that accelerates rapidly on the steep slopes of volcanoes, devastating anything in its path. These volcanic debris flows (also known as lahars) can reach great distances after they leave the flanks of the volcano and can damage structures in flat areas surrounding the volcanoes.

Volcanic edifices are young, unconsolidated, and geologically weak structures that in many cases can collapse and cause rockslides, landslides, and debris avalanches. Many islands of volcanic origin experience periodic failure of their perimeter areas (due to the weak volcanic surface deposits), and masses of soil and rock slide into the ocean or other water bodies, such as inlets. Such collapses may create massive submarine landslides that may also rapidly displace water, subsequently creating deadly tsunamis that can travel and do damage at great distances, as well as locally.

Landslides also can cause tsunamis and seiches, can overtop dam reservoirs, and/or reduce the capacity of reservoirs to store water. Steep wildfire-burned slopes often are landslide-prone due to a combination of the burning and resultant denudation of vegetation on slopes, a change in soil chemistry due to burning, and a subsequent saturation of slopes by water from various sources, such as rainfall. Debris flows are the most common type of landslide on burned slopes.

Additionally such landslides—which usually occur in small, steep stream channels—can have the same impact as a flash flood whereas they also can cause flooding by itself when the bulk debris rock material blocks a stream channel thus holding back large volumes of water behind such a "dam".

3.2.2.3 Landslide Dams

Rivers dammed by natural processes mainly from landslides, glacial ice, and volcanic debris present a great threat to people and property in mountainous areas. The most common initiation mechanisms for dam-forming landslides are excessive rainfall and snowmelt or earthquakes that form dams from rockfall, debris avalanches, debris flows, mud slides, and the like. The mass movements block the flow of a river, causing a lake to form behind the blockage. The backwater will pile up to certain level and when the "dam" fails will subsequently flow downstream. Moreover, the solid debris material increases the density of the slurry that can easily reach a density as high as concrete slurry. Most of these dams are short-lived as the water will eventually erode the dam. If not destroyed by natural erosional processes or modified by human action, this blockage creates a new lake. Such lakes can last for a long time, although they can suddenly be released and cause massive flooding downstream, but they may also cause upstream flooding as the lake rises. Although data are few, there are clear indications that flooding from a glacier lake dam failure is generally smaller than those from landslide, moraines, or volcanic debris. Moraine dam failures appear to produce some of the largest downstream flood peaks due to their normally large extensions whereas in contrast, dam failures triggered in more or less steep valleys are more "local" events.

Costa and Schuster (1988) classified landslide dams worldwide based on their areal distribution in relation to the valley floor. A great number of dams are just covering a part or span the entire valley floor. Most dams worldwide, however, fill up the valley over a considerable distance both upstream and downstream from the landslide failure. Seldom are dams created when a single landslide sends a tongue of debris into a valley forming a dam. There are some examples of massive rock failure that extends under the stream or valley and emerges on the opposite valley side. Most landslide-triggered dams fail within a short time after formation. The investigations of Costa and Schuster (ibid) revealed that a third of the landslide dams failed less than 1 day after formation, and about 50 % failed within 10 days. Overtopping of the water by increasing the water level in the lake is by far the most common cause of dam failure. The timing of failure and the magnitude of the resulting floods are controlled by dam size, its geometry, and the material characteristics of the blockage.

In addition to the direct risk from landslides and debris flows, the deposition of volcanic material in valleys after a volcanic eruption often forms unstable natural dams that cause blockage of the former drainage system. Such landslide types vary in size from small mass movements of loose debris on the surface of a volcano to massive failures of the entire summit or flanks of a volcano. But volcanic landslides are not always associated with eruptions. Heavy rainfall or a large regional earthquake can also trigger a landslide on steep slopes. Volcanic material is highly susceptible to landslides because it is composed of layers of loose, fragmented, volcanic material that is piled up on top of the generally steeply inclined topography. Furthermore, some of these rocks have been altered to soft, slippery, clay minerals.

Volcanic debris-generated dams have formed all over the world. One of the best-known examples is that of Mt. St. Helens, where the eruption of 1980 produced the largest debris avalanche on Earth in recorded history. The material was spread over large parts on the north of the volcano covering the area with several meters of ashes and debris along the South and the North Fork Toutle River. The debris blocked the water level of Spirit Lake and raised the water level by 20 m. Spirit Lake has been repeatedly dammed by volcanic material, than filled to up to the spill point with water, and at least partially drained due to dam failures causing several major floods and lahars down the Toutle River. It is believed that pyroclastic flows around 3350 BC first dammed the river to form the lake.

Engineering operations to release water for the lakes comprise digging artificial spillways, water diversions, and tunnels, but can also be achieved by blasting the rock material or by conventional excavation. An impressive example for engineering dam rehabilitation is that of the Xinjiang earthquake of 2008. An earthquake with magnitude of M7.2 formed a series of 35 lakes, among them one that was only 3 km away from the next provincial capital, the area worst hit in the devastating quake. The lake held about 130 million cubic meters of water and was inaccessible by road. The impounded water mass was at risk of breaching its banks. An emergency plan was disseminated among the 1.3 million inhabitants in 169 communities of the entire region to make them aware that if the dam broke they had to be evacuated within four hours. Meanwhile, hundreds of workers with 40 heavy-duty bulldozers and excavators and other earth-moving equipment worked non-stop on top of the barrier to construct a diversion channel for the water. At least 50,000 m^3 of debris had been removed to build the diversion channel. Finally the workers succeeded in releasing the water without any further damage to the people and their livelihoods.

3.2.2.4 Glacial Lake Outburst Flood (GLOF)

Glaciers are nature's most effective renewable storehouse of fresh water. However, in the course of climate change, which can most impressively be observed just as the world's glaciers retreat, the natural fresh water storage capability of the glaciers is changing. Already today their storage capacity is reducing steadily and will in future give way to a higher melting rate. Much of the melt waters are cached in the high mountainous valleys behind very young and mostly unstable terminal moraines. The more water is accumulated the higher is the risk that the walls breach. From many alpine mountainous regions of the world, such as the Columbian Andes but especially from Nepal and Bhutan (Himalayas) such breaches are known. The breaching results in a sudden dramatic discharge of huge amounts of water and debris, the so-called "glacial lake outburst flood" (GLOF), often with catastrophic effects downstream (Mool et al. 2001). In Nepal more than 3000 glaciers were identified and 2300 glacial lakes, of which 20 were considered potentially dangerous. In Bhutan, 600 glaciers were identified and 2600 glacial lakes, of which 24 were considered potentially dangerous.

There are several methods available for mitigating the impact of glacial lake outburst flood surges. The most important mitigation measure is to reduce the volume of water in the lake. But also downstream in the GLOF-prone area, prevention measures can be taken to protect infrastructure against the destructive forces of the surge. Careful evaluation of the lake, the glaciers, the damming materials, and the surrounding geological, geomorphological, and social living conditions are essential in choosing the most appropriate mitigation measures. Monitoring the dam stability, lake volume, and geometry as well as the glacier retreat are indispensable tasks during and after the mitigation process.

The prevention of a peak surge discharge of the glacier lakes can technically be achieved by:

- *Controlled breaching*: A controlled outflow over the terminal moraine can be achieved either by blasting or the excavation of a drainage pass. Several successful attempts have been reported from Lake Bogatyr, Kazakhastan (Mool et al., ibid), where an outflow channel was excavated using explosives and 7 million cubic meters of water were successfully released in a period of two days, although such a fast lowering of the lake level can lead to a strong and uncontrollable regressive erosion of the moraine wall. Also by opening cuts in the moraine dam during the dry season when a lake's water level is lower, a reduction of the water level can be achieved. However, such a method is risky as any displacement of material from the dam may give way to surges from snow and ice avalanches and can result in a complete breach of the moraine.
- *Construction of an outlet control structure*: A solid structure made of stone, concrete, or steel can be used to install a permanent dam through which the water can be controllably released. However, such constructions need comprehensive maintenance works at high elevations, in difficult terrain conditions, and under extreme logistic situations. Therefore, preference should be given to constructing the dams with locally available materials such as boulders and stones that can be held in place by gabions or appropriate anchors.
- *Pumping or siphoning out the water from the lake*: Today turbines, propelled by the water force at the outside of the moraine dam, are seen as the most effective means to release water from critical glacier lakes by steadily pumping off a defined amount of water. Such siphons that can be adjusted to the specific situation of a particular lake are cost-effective as they are easy to transport and to install. Nevertheless, such siphons also need steady maintenance at high altitudes, especially as the debris load of the water may erode the turbine blades.
- *Tunneling through the moraine barrier*: Digging a tunnel through moraines or debris barriers is another measure to lower the water in glacier lakes, although the method is highly dependent on the geological and geomorphological situation especially of the type of material blocking the lake. Tunneling is best applied through competent rock formations beneath or beside a moraine dam. The cost of such a method is very high. Also problems have been reported from a tunnel through a moraine dam that had been severely affected by an earthquake in Peru. A successful tunnel was installed at Spirit Lake after the Mt. St. Helens volcanic eruption (Sager and Chambers 1986). In the Himalayan region tunneling will most probably not be efficient as there is no secure electric power supply and transporting fuel for the diesel engines is too costly.

But not only remedies such as the outflow methods described before are seen as options to reduce the risk from GLOFs. Prevention measures by geological mapping of the potential sources of snow and ice avalanches, landslides, or rockfalls around the lake area that may have a direct impact on the lake and dam were carried out at many at-risk GLOF areas. Such studies were helpful in defining preventive measures against slope instabilities by, for instance, removing masses of loose rock. Additionally bridges were built at levels higher than the expected GLOF surge levels and gabions were installed to protect the base of the river embankments. Settlements have already been moved to higher places in general to the upper river terraces in order to increase people's resilience.

Factors Influencing the Onset of Landslides

Slope saturation by water is a primary cause of landslides and is closely related to precipitation, runoff, and the saturation of the ground. Water saturation of the soil can occur from intense rainfall, snowmelt, changes in groundwater levels, and surface-water level changes along coastlines, lakes, reservoirs, or rivers. In the last decade many comprehensive investigations have been carried out to define the onset value for rainfall triggering landsides: the so-called "global rainfall intensity–duration threshold" (Hong et al. 2006) which revealed that rainfall of more than 20–200 mm rain/day is assumed to trigger a landslide. This threshold value can be overstepped when, for example, it rains about 6–8 mm/h for one day, or 4–5 mm/h for 2–3 days but also when it rains only about 1 mm/h for more than 7 days. Therefore not only a heavy downpour (the rainfall during the Wenchuan earthquake 2011 revealed 350 mm in 24 h) but also longer-lasting "mild" precipitation of many days can lead to landslides.

NOAA/USGS investigations came to a similar result on the onset of debris flows in the San Francisco Bay (USGS 2005c). There an experimental debris-flow prediction and warning system was operated from 1986 to 1995. The model relied on rainfall forecasts and measurements of precipitation linked to empirical precipitation thresholds to predict the onset of rainfall-triggered debris flows. The investigations revealed different rainfall intensity–duration thresholds that are indicative of onset values for landslides (Fig. 3.35) and that can be used as a debris-flow warning system. The model identifies a safe area at rainfall levels below 20 mm in 3 h and

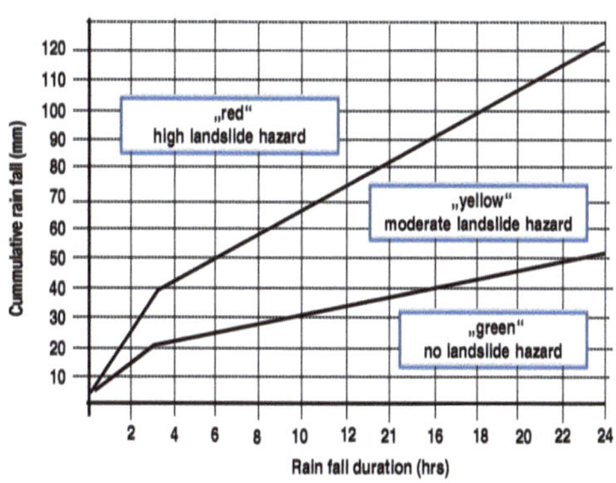

Fig. 3.35 NOAA/USGS empirical "rainfall intensity–duration threshold" for the San Francisco Bay area (Redrawn from USGS 2005c)

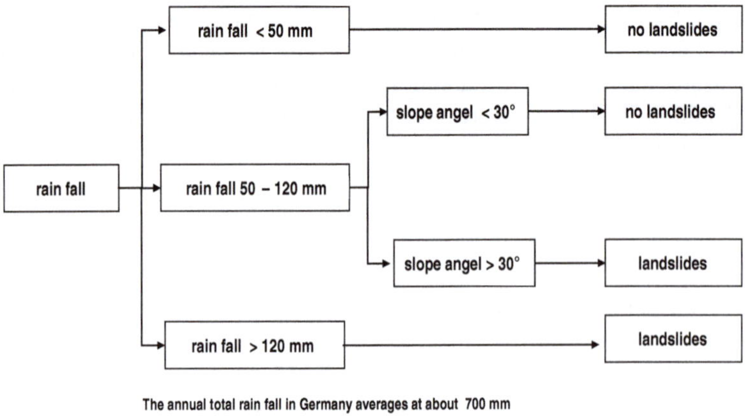

Fig. 3.36 Decision tree for the occurrence of landslides (Generalized after Hamberger 2007)

60 mm in 24 h (green), a sector with moderate hazard to debris flows as between 50 mm in 3 h and 120 mm/in 24 h (yellow), and in the red area where rainfall volumes exceed 50 mm in 3 h and 24 mm in 24 h, debris flows are likely to occur.

Next to the water factor the factor of slope inclination is also of significant importance for defining the hazard potential for a landslide. Following a heavy rainfall (150 mm in 2 h) in the area of Sachseln (Switzerland) Hamberger (2007) was able to define that most of the 700 shallow landslides that occurred in the area were triggered at slope angles of between 20° and 40°. The investigations further revealed that a combination of several parameters mostly morphology, geology, soil property, the general hydrological regime, as well as vegetation all define the

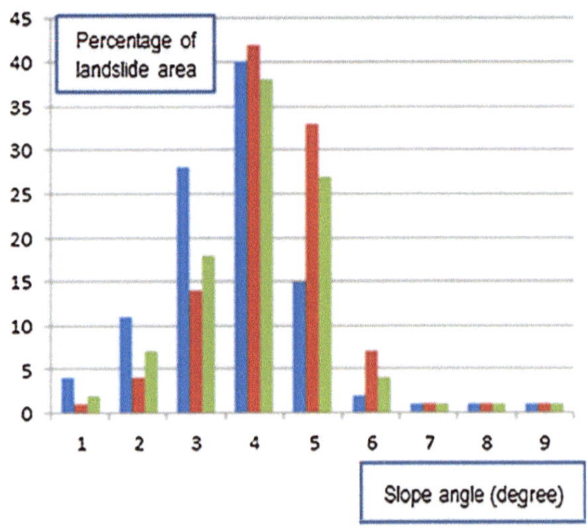

Fig. 3.37 Percentage of landslide areas following the 2008 Wenchuan earthquake (Redrawn from Tang et al. 2011)

major causes of the landslides there. Following his investigations Hamberger gave a "decision tree" that combines slope angle and precipitation that can be used to assess the landslide hazard potential at least as a very general approach (Fig. 3.36).

Other investigations on rainfall volume and critical slope angle reveal even more differentiated relationships. According to investigations by Tang et al. (2011) carried out on landslides that occurred before, during, and after the Wenchuan earthquake (China) in May 2008, the critical slope angles were different for the different types of landslides (Fig. 3.37). Slope angles between 20° and 40° were identified for most of the landslides that occurred before the earthquake stroke, whereas coseismic landslides were mainly triggered at slope angles of 30°–50°. Based on aerial photographs and remote sensing imagery 40 pre-earthquake landslides and 2200 coseismic landslides were identified. As the area was subject to a strong rainfall even four months after the quake about 1000 new landslides were triggered as a result of the earthquake tremors and the subsequent rainstorm that severely weathered the topsoil strata. The earthquake triggered at first massive landslides and that subsequent strong rainfall prompted the development of new landslides as well as reactivated pre-existing slides. An almost identical frequency distribution on slope angles has been published by Ruff (2005, p. 80) who investigated landslide susceptibility in the Austrian Alps. The landslide frequency distribution showed its maximum (40 %) between 20° and 40°. But about 10 % of landslides also occurred in young valley fills along the river Lech at angles between 10° and 20°. The landslides identified by Tang and coauthors and also by Ruff, occurred in geologically young mountain ranges, although the situation in China is much more critical as the region is subject to many earthquakes.

Next to water and slope angle the rock material is also a denominating factor that defines the onset of landslide events. Investigations carried out by Carrara et al. (1977) in the Calabrian mountains of southern Italy revealed that different lithologies also have an influence on critical landslide-generating slope angles. Although with sand, clay, and marls the critical slope angle lies between 10° and 20°, hard rock (gneisses, etc.) sees 25° to 40° as critical. All lithologies together in the area under investigation provided the maximum critical slope angle between 10° and 20°.

All the given examples show that the corridor of critical slope angles in general ranges between 20° and 40°. But it should be taken into consideration that water saturation, vegetation, lithology, or external factors such as an input of energy into the system such as from an earthquake, may lead to a more different trigger mechanism. As described above, gentle slopes (1°–2°) can also be subject to landslides, a phenomenon that is best described as soil creeping.

The best prevention measures against erosion, wet mass movements, or debris flows in hilly regions are terraces. Especially in Asia terraced hill slopes are normally not subject to serious erosion, as long as they are well maintained. Moreover from volcanic trass regions it is known that slope angles of more than 80° are also not prone to landslides. It will most probably not be possible to give the "one" critical slope angle applicable for all landslide-prone areas worldwide. Thus it is recommended to analyze the region under investigation carefully to find out what variety of slope angles occur, what substrate and vegetation are dominant, what is the exposure to rain and storm, and how far the distance is to the next drainage path.

3.2.2.5 Green Lake Landslide (New Zealand)

The Green Lake landslide in New Zealand is a very large ancient rockslide that is considered to be the largest documented landslide of its type on earth (although larger submarine slides are known). The landslide was generated from gneisses and granodiorites located in the deeply glaciated Hunter Mountains (Hancox and Perrin 1994). Geology and geomorphic evidence suggest the slide occurred just after the end of the last glaciation about 13,000 years ago. The landslide has an estimated volume of about 27 km^3 with a surface area of 45 km^2. The main features of the Green Lake landslide include a large area of hummocky, bush-covered slide debris up to 1000 m thick. Within the debris a number of large, semi-intact blocks up to 2.5 km long occur as well as a prominent v-shaped head scarp that extends for about 14 km, and four large pull-apart basins. Three of these basins contain large landslide ponds, the largest being Green Lake and Island Lake.

The landslide area is actually the in-filled part of the former Lake Monowai, which was cut in half when the landslide occurred and gradually filled with glacial sediments and swamp deposits. Geomorphologic evidence indicates that the landslide probably was triggered by a rapid rock failure, possibly occurring in two phases. First a 1500 m high mountain ridge collapsed on the east side of the former Monowai valley resulting in the destruction of a 9 km long section of the southern Hunter Mountains. Then the slide debris was transported up to 2.5 km laterally, and fell about 700 m vertically into the deeply glaciated former Monowai valley, which at the time of the landslide was probably filled with a glacial lake. The enormous debris volume formed a landslide dam about 800 m high in the valley, which cut the original Lake Monowai in two, impounding a lake of 11 km length, which was then gradually filled with glacial sediments, and later with peat and swamp deposits. Radiocarbon (Delta 14C) dating of lake sediments indicated that the final infill occurred about 11,000 to 11,500 years ago. Dating of peat deposits moreover revealed that the lake was later drained about 9000 years ago after recession of the glacier. The geomorphological investigations showed that flooding of the landslide failure surface although having reduced the slope stability, would not be sufficient to cause such a large mountain mass to collapse. An earthquake simulation revealed that ground motion of an earthquake magnitude of 8–10 most probably must have triggered the landslide. This earthquake may also have triggered some of the other old large landslides also identified in the region. Today, the landslide dam remains essentially intact, and apart from local failures around the steep head scarps surrounding the landslide, there is little potential for reactivation of the Green Lake landslide. However, when tectonic movements might occur again in the future, there are many other slopes in the area that might cause very large catastrophic landslides. Although experience from many glaciated mountain ranges revealed that glaciated mountain slopes are most vulnerable to collapse just after ice withdrawal, it is unlikely that these will be on the same scale as the Green Lake landslide.

New Zealand geologists see the Green Lake landslide as a good example of a hazard type of landslide generated by deglaciation in a mountain area of high seismicity. Such landslides can move an enormous volume of debris with extensive catastrophic effects resulting from a collapse of high mountain ranges.

3.2.2.6 Mt. Rainer Landslide (Lahar)

Water-saturated landslides originating from volcanic eruptions (lahars) pose a significant hazard to downstream environments. The flows are characterized by their long travel distances and their high speed, thus reaching areas that are normally far away from the debris flow source in a couple of hours. Such lahars have caused more than 20,000 deaths on one occasion at the Nevada del Ruiz volcano (see Sect. 3.2.1.7) and pose a risk to many other volcano locations on Earth (Witham 2005).

Mt. Rainier located in Washington State (United States) at the northern part of the Rocky Mountain chain is the highest of the active volcanoes that make up the Cascade Range and erupted the last time in 1895. The volcano carries a voluminous ice cap and at its western foothills are located many smaller towns and also the megacity of Seattle at Puget Sound. Although most of the Mt. Rainier volcanic materials of lava, ash, and bombs are known to concentrate despite the great topographic relief but due to their high viscosity in a radius of a few kilometers around the volcano, there are quite a number of lahars in the sedimentary record that have been deposited along the valleys that drain the volcano. The steady increasing urbanization of the lowlands downstream of Mt. Rainier makes the area one of the most risky in the United States from a lahar flow (USGS 2009b).

In order to assess the risk from debris flows and lahars to the people in the area, USGS geologists investigated the volcano and its potential socioeconomic impact. The geologists were able to identify a series of large, generally clay-rich debris flows that originated as landslides, to occur on average every 500 to 1000 years during the last 6000 years at Mt. Rainier. The Osceola debris flow about 5000 years ago was found to be the biggest: about 3 km^3 of material were removed by a huge landslide from the summit of Mt. Rainier by that time. Although large blocks of the landslide have formed numerous sand and clay mounds along the White River valley before spreading the flow over a wide area of the Puget Sound Lowland, the town of Orting is actually directly situated on this debris flow. Based on the lahar sequences identified, Driedger and Scott (2008) calculated that there is roughly a 1-in-10 chance of a lahar reaching the cities around Puget Sound in the next 50 years, an assumption that holds true especially for the Puyallup and the adjacent Carbon River that both conflate at the small town of Orting (USGS 2000). The travel times for lahars from Mt. Rainier to reach Orting are assumed to be about 40 min after a lahar is detected.

In the 2008 the local government together with the United States Geological Survey presented the "Mt. Rainer Volcanic Hazard Plan." The plan identified the areas that are prone to lahars and classified the lahars into three categories:

- Debris flows that are relatively small in size and of minor destruction potential
- Lahars that develop from debris flows and that give ample time to warn the people
- Lahars that occur without prior warning, that travel at high speed, and that have a large destruction potential

The plan was based on the assumption that almost 80,000 people are exposed to risks from the dangerous lahars, 5000 of them in the city of Orting. As in many other communities in the valley Orting also has a relatively small amount of developed land in the lahar hazard zone, which likely represents single-family housing and associated buildings such as garages and sheds. Next to the residents of the area, Mt. Rainier and its national park host more than two million visitors every year, making the volcano a significant source of income and labor in the region. In order to increase the population's resilience to volcanic threats, the local government is operating a lahar-warning system along the Puyallup and Carbon River. Ground motion is recorded constantly and the data are telemetered to the local emergency management center. The early warning system is complete with regular monitoring of the geochemistry of the springs around Mt. Rainier, thermal monitoring, and a visual inspection of its volume. Furthermore the local government enacted the "Orting Hazard Response Plan" that defined evacuation as the major instrument in order to safeguard the people. But the hazard plan also indicated that there are only three smaller bridges to the south, west, and north available to leave the area at risk.

This scenario was taken up by a master's thesis at the Geological Institute of the University of Goettingen (Germany) that proposed another way to reduce the risk from flooding (Friess 2010). The paper proposed that instead of making evacuation the first priority of emergency management, it should be considered whether the lahar could rather be hindered from entering the town of Orting. The paper proposes to install a series of Sabo dams placed at critical positions several kilometers upstream along Puyallup and Carbon River in order to diminish the lahar's suspension load by removing the bulk rock material from the water. Such a measure has proven to be able to reduce the destruction potential of a lahar considerably at many risk areas worldwide and it can help safeguard the infrastructure and the buildings.

3.2.2.7 Landslides Triggered by Tropical Storm

Torrential rains that accompanied Hurricane Mitch in October and November of 1998 triggered thousands of landslides in the moderate to steep terrain in eastern Guatemala. For five days the hurricane swept over Central America with windspeeds of up to 300 km/h, generating torrential rains that flooded large parts of the areas with mud and debris. The rainfall was exceptional because it was geographically widespread in the central and eastern regions, lasted over a week, was moderate to heavy in intensity, and occurred at the end of the rainy season, when the ground already had high moisture content.

Guatemalan Hurricane Mitch triggered more than 10,000 landslides in an area of more than 10,000 km^2; on average, this is about one landslide/sq km, a ratio that could range up to as many as 120 landslides/km^2. The main concentrations of landslides were found on moderate-to-steep hill slopes underlain by diverse geologic units. The landslides were of two general types:

- Relatively small, translational, and rotational landslides that mobilized into debris flows and covered less than several hectares in area
- Large, commonly translational, landslides that sometimes generated debris flows and covered between 15 and 25 ha

Between 10,000 to 15,000 people were assumed to be killed by Hurricane Mitch in Central America, 1.5 million affected, and the total economic losses were estimated to be about US$6 billion (USGS 2001). Like all the countries in the region the small country of Guatemala was also seriously affected, although the death toll was much less: 268 dead and 100,000 affected of whom at the end of the year many thousands still had to live in shelters. Damages were estimated by the government of Guatemala and the United Nations of US$550 million of which the agricultural sector claimed the greatest damages (US$350 million), and the damage to highways, infrastructure, and private buildings was about US$150 million. Damage to infrastructure included 100 damaged or destroyed bridges, 100 road sections, and 2000 houses completely destroyed and 20,000 damaged. There was extensive damage to productive agricultural areas and farm-to-market access roads. The agricultural sector was affected most with 90,000 ha of losses in basic grains, coffee, vegetables, and bananas. Facilities for small production coffee processing were also seriously affected.

In order to assess the geological dimension of Hurricane Mitch's impact, the USGS (ibid) together with the Guatemala Institute of Seismology, Volcanology, Meteorology and Hydrology (INSIVUMEH) undertook an assessment of the regional landslide susceptibility. Based on information on the frequency and severity probability distribution of historic and Mitch-triggered landslides, it was possible to identify the critical landslide-prone areas and to discover the basic onset mechanisms responsible for the gravity movements. To create a landslide susceptibility map of the region under investigation, a susceptibility threshold equation based on elevation and slope gradient was developed. For this the ratio of the elevations of each landslide taken at its point of initiation and the elevation of the particular grid cell of the topographic map was calculated. About 96 % of the landslides were initiated from elevations between 500 and 2500 m. At the next step, landslide frequency was tabulated for each 100 m interval in elevation and revealed the maximum of the landslides (12 %) occurred between 2000 and 2100 m and were initiated from slope gradients between 15° and 45°. When tabulated for every 5° slope interval, the highest percentage of landslides (27 %) occurred between 25° and 30°. The ratio between the slope angles where landslides were initiated to the entire population of slope gradients indicated that in the areas between 1200 and 2800 m, the ratio is greater than 1, indicative of a high susceptibility to landslides. The analysis furthermore indicated that in certain areas no landslides occurred at all, whereas in the surrounding region many of them were generated. These areas experienced slope angles below 9°, a slope angle found obviously not susceptible to landslide formation during the hurricane.

All the landslide-prone areas received between 200 and 600 mm of rain over the period from October 25th until November 6th, a precipitation equivalent to the amount the region normally receives during one year. The highest rainfall amounts (400–600 mm) occurred in the Upper Polochic valley and in the central Sierra de las Minas. Lower rainfall amounts (200–400 mm) occurred in the hills surrounding Sierra de las Minas and along the border region with Honduras. One rain gauge located near the La Lima landslide recorded a precipitation of 275 mm over the six days of the hurricane before the landslide occurred. As Mitch occurred at the end of the rainy season, USGS and INSIVUMEH (USGS, ibid) assumed that the rainy season already had saturated the soils and that the 275 mm additional precipitation in six days (an average of 46 mm/day) was enough to overstep the threshold and trigger the La Lima landslide. And as documented in the report, this type of rainfall (hurricane), on already saturated or nearly saturated ground has the capability to trigger both shallow as well as deep-seated landslides over a large area. The study furthermore points out that areas susceptible to rainfall-triggered landslides are not necessarily the same as those susceptible to earthquake-triggered landslides as USGS (1981) reported on landslides triggered by the 1976 M 7.6 earthquake in Guatemala.

The characteristics of rainfall-triggered landslides found with Hurricane Mitch in Guatemala can serve as a practical guideline to assess future landslides triggered by rainstorms. The data revealed that landslide susceptibility is highest on moderate to steep hill slopes. But also less steep areas directly below channels draining the hill slopes and on alluvial fans at the mouths of drainage fronts can be highly susceptible to landslides. The investigations further emphasized that records of historic landslides can be the best indicators for future landslide activity and provide very useful information to determine the level of future hazard. The study proved that landslide inventory maps showing historic and modern landslides are of critical importance.

3.2.2.8 Submarine Landslides (Grand Banks, Canada; Storegga, Norway; Lithuya Bay, Alaska)

Landslides do not only occur on solid earth but are also well known in offshore regions. The so-called "submarine landslide" describes the downslope mass movement of geologic materials from shallower to deeper regions of the ocean. Such events can transport a high amount of sediment from the outer shelf down to the foot of the continental slopes, but not only there. Submarine landslides are also known to occur in rivers and lakes. The biggest submarine landslides ever recorded were either triggered by an earthquake or by erosion, but all of them have caused deadly tsunamis (see Sect. 3.2.1), such as in 1929 at the Grand Banks (Newfoundland, Canada), the Storegga landslide offshore Norway in 7950 BP, or the 1958 Lithuya Bay rockfall in Alaska (United States).

On November 18th, 1929 an earthquake of magnitude 7.2 occurred at the Grand Banks with its epicenter about 400 km at the outer shelf edge south of Newfoundland. The earthquake was felt as far away as New York and Montreal and was generated along two fault zones 250 km south of the Burin Peninsula. The quake triggered a large submarine landslide of a volume of more than 200 km^3 that led to the generation of a large tsunami. The tsunami arrived at the Newfoundland coast in three waves, each 3–4 m high, about three hours after the earthquake occurred. The waves traveled at speeds up to 130 km/h and were recorded as far away as South Carolina and Portugal. The tsunami destroyed many south coastal communities, causing US$400,000 in economic losses, left 10,000 homeless, and killed 30 people, the highest death toll attributed to an earthquake in Canada ever since. The tsunami destroyed the entire communication infrastructure and moreover the relief efforts were hampered by a blizzard that struck the day after.

The submarine slide snapped 12 submarine transatlantic telegraph cables connecting America and Europe. From the time sequence of the cutting off of the cables, the submarine travel speed of the slump was calculated. It was the first time in history that such a phenomenon was ever recorded. The knowledge gained by this event laid the basis for a marine science study on submarine landslides, turbidity currents, and tsunamis by scientists from Columbia University. Since then geologists have been looking at layers of sand for indicators on tsunamis that can be originated from earthquakes. Most of the economic loss of the Grand Banks mudslide was the cost to repair the damaged transatlantic cables.

From evidence in submarine seismic profiles of the oil industry all over the Norwegian shelf and from geological evidence along the North Atlantic coast from Norway to Greenland, a large submarine landslide has been assumed to have originated at the Norwegian coast off the city of Alesund, the so-called Storegga landslide. The landslide is supposed to have occurred about 7300 years BP (Bondevik et al. 2005) and to have generated exceptionally large waves that inundated most coastlines around the North Atlantic. Even today the submarine morphology offshore Alesund gives a clear indication on the location and the dimension of the slides as it can be seen by Goggle Earth "Offshore Norway." As the trigger mechanism for the Storegga slide a strong earthquake in the North Atlantic was assumed but also methane hydrate emissions can be seen that might have destabilized the shelf edge.

In total five landslides are supposed to be broken off the Norwegian coast, of which the first was the largest in volume (about 3000 km^3) covering an area of around almost 100,000 km^2. The area coverage was about the size of Scotland and the mobilized sedimentary volume was large enough to generate a large tsunami. The materials were transported north and westwards over thousands of kilometers across the North Atlantic to be deposited at the coasts of Scotland, Iceland,

and Greenland and also along the coast of Norway. In fjords in Shetland and the Faeroe Islands deposits show that the waves reached elevations at least 20 m above the contemporary sea level, although at the Norwegian coast the tsunami waves reached heights of more than 10 m. Numerical simulations on the Storegga slide revealed that it was about 400 m thick in the upper part of the slope and its wave-front crossed the North Atlantic within 3 h, with maximal sea- level elevation on the open ocean of 3 m (Bondevik et al., ibid). The simulation further revealed that along the Norwegian coast the arrival of the first wave might have been associated with a major water withdrawal, dropping the sea level by 20 m.

The tsunamogenic sediments comprise fine- to coarse-grained sand layers containing fish bones, mollusk fragments, and even eroded diatoms that are found intercalated into shallow marine, tidal flat, and swampy sediments. Even at higher topographical levels along the Norwegian coast tsunami-inundated fresh-water bodies were observed, again leaving behind characteristic sand layers. These deposits contain redeposited lake mud, rip-up clasts, and marine fossils. The sand layers furthermore show a distinct erosional base although the top is transitional, properties characteristic of modern tsunami deposits.

On the night of July 9th, 1958 an earthquake of magnitude 8.3 (later revised to 7.7) and with an intensity XI (MMI) along the Fairweather Fault zone in Alaska triggered a rockfall at Lituya Bay that generated the highest tsunami wave run-up in recorded history (USGS 1960, 1993). Lituya Bay is located in the Glacier Bay National Park at the southern end of Alaska, directly on the Pacific coast. The bay was formed by one of the glaciers of Mt. Crillon that shaped the Gilbert Inlet, a valley following the north–south trending highly seismic fault zone that parallels the coast. The bay is about 10 km long and 2 km wide and opens westwards towards the Pacific, thus forming the landlocked nature of the bay. At the mouth of the bay a sill of about 200 m width was formed from glacial debris and moraines. The submarine contours show a pronounced U-shaped morphology with steep walls and a broad flat floor sloping gently downward from bay's end to the mouth: the depth reaches a maximum of about 250 m in the center of Lituya Bay, then rises again towards the entrance to a minimum of 10 m.

From geological and geophysical investigations there is good evidence that the Fairweather Fault is of lateral and oblique habit. The magnitude was determined to be M7.7 according to the report of the USGS (1960). The southwest side of the fault moved northwestward for at least 6 m and went up for 1 m. Landslides and other evidence of strong motion observed in the area indicated a total shift along the fault probably for up to 60 km. The rocks in the area are largely of diorite and slightly metamorphosed volcanic rocks, of slate and graywacke that are exposed on the southwest shore of Gilbert Inlet and the adjoining north shore of the bay. Bedded sedimentary and volcanic rocks from the Tertiary Age are exposed on Cenotaph Island, however, in the model of Lituya Bay most of the outer part of the bay boulder is still exposed at the surface or lies under a thin soil.

The rockfall occurred on steep cliffs above the northeast shore of Gilbert Inlet and loosened about 30 million cubic meters of rock at the eastern end of Lituya Bay (USGS, ibid). According to eyewitnesses the earthquake's shaking lasted for

3–4 min and 2 min later the rock mass plunged into the water. This mass of rock fell from an altitude of approximately 900 m down into the bay and generated a giant but very local tsunami. The tsunami crashed against the opposite lying shore with such power that it swept completely over the spur of land as high as 524 m above sea level that separates Gilbert Inlet from the main body of Lituya Bay. This is the highest tsunami wave that has ever been known, although the actual run-up height owed more to the inclination of 30–40° of the crystalline rock formation, making the climb by the tsunami wave rather easy, but not a result of a 524 m high tsunami wave. The wave then continued to flow down the entire length of Lituya Bay into the Gulf of Alaska. The force of the wave removed all trees and vegetation from the entire northern and southern shores of the bay and left completely barren rocks. Still today the new cover of vegetation (trees and bushes) can be seen on satellite imagery indicating the outer rim of the inundation level.

Moreover the United States Geological Survey was able to identify evidence for the occurrence of large waves in Lituya Bay prior to that of 1958. At least four previous large waves could be distinguished with estimated dates of 1936, 1899, 1874, and 1853/1854. All of these waves are supposed to be significant in size although the shoreline evidence for all of them was removed by the 1958 tsunami.

3.2.2.9 Snow Avalanche

An avalanche occurred on February 23rd, 1999 in the Alpine village of Galtür, Austria. Within a minute a powder avalanche, 50 m high and with a speed of estimated 300 km/h, hit the tourist center just at the winter recreation peak season (BMLFUW 2012). The avalanche buried 57 people, ruined many buildings, and overturned many cars. By the time rescue crews managed to arrive, 31 people—locals and tourists—died. At the time of the avalanche about 800 locals and about 5000 tourists were present in Galtür. The avalanche was considered the worst Alpine avalanche in 40 years, although before the accident the town was supposed to be safe. Three major low-pressure weather systems the days before accounted for large snowfalls totaling around 4 m in the area and led the snow to freeze and thaw. The weather conditions coupled with high windspeeds caused roughly 170,000 tons of snow to be deposited. In the decades before the accident, the Galtür area was subject to many severe snow avalanches with a considerable number of fatalities. Because of the snow masses in this part of Austria, the entire region was given the highest snow avalanche warning, blocking many traffic connections. The snow masses also buried the only road connection to Galtür, so that the victims could not be reached by road. The rescue was organized by military helicopters that flew out most of the tourists when weather conditions allowed flying. Since then Galtür has built up a series of countermeasures to increase local resilience. The central part is a 345 m long retention wall that is in part 20 m high. Behind the wall a museum and recreation center are incorporated that today give the town a unique ensemble of modern architecture and snow avalanche prevention.

Snow avalanches are a special sector of landslide hazards (mass movement, gravity movements) that occur exclusively in alpine regions, regions that are more or less permanently covered with snow and ice. Like landslides, snow avalanches also pose a significant threat to humans and infrastructure in mountainous regions. The worldwide annual average of snow avalanche fatality is estimated at about 250 fatalities, of which in the Swiss Alps alone snow avalanches cause an average of 26 fatalities per year (Tschirky et al. 2000). Of the Swiss fatalities, 90 % can be attributed to avalanches triggered by tourists (skiers, snowboarders, climbers) and again 90 % of which are triggered by the victims themselves. And moreover 90 % of all fatal avalanche accidents occur in uncontrolled avalanche risk terrain. In the United States on average 17 were killed annually by snow avalanches. Most of them were snowmobilers, whereas in the European Alps no victims occur from snowmobiling as such a recreation activity is not allowed there. Statistical evidence from fatal accidents revealed that most fatalities occur in areas that were rated of low to moderate danger levels. And most of the victims were involved in recreational activities. At higher danger levels, in addition to recreational activity, people were killed while driving, walking on paths, or residing in buildings. Fatalities in the high avalanche danger areas derive from the nonrecreational sector. When the extreme category is excluded, the normalized distribution of fatal incidents from international statistics is quite similar worldwide.

According to experience from many snow avalanche risk areas, avalanche victims do not have a chance to survive if buried for more than 45 min because of hypothermia. Medical data suggests a core body temperature cooling rate of more than 3 °C/h between burial and arrival at hospital will be fatal. And experience revealed that the cooling rate increases even immediately after extraction of the victim.

In general snow avalanches are divided into two distinct groups (McClung and Schaerer 1993, p. 61):

- Loose snow avalanches start from a point and move down the slope as a snow mass without internal cohesion of the snow particles spreading out on their way down to a triangular shape. They normally move only the upper part of the snow cover. As their material is not densely packed the hazard from this type is rather low.
- Slab avalanches start as a cohesive block of snow triggered by cracks in the upper onset line that propagates through the entire snow cover. Slab avalanches normally activate a large proportion of the snow cover. Slab avalanches can further be divided into dry and wet snow avalanches. Dry slab avalanches can be naturally triggered, for example, by new snowfall, or by artificial triggers such as tourists. Because of their high density, their regional extension, and their travel speed, they are normally highly dangerous to people and property. Dry snow slab avalanches are responsible for more fatalities and damage to property than wet snow avalanches.

Ninety percent of all avalanches occur on moderate slopes with an angle of 30°–45° indicating that snow does not tend to accumulate on steeper slopes.

Avalanches occur when the gravity of the collected snow mass at the top of the slope is greater than the internal cohesion forces of the snow cover itself. The critical onset of a slab avalanche is mainly defined by the existence of a weak layer within the snow cover (McClung and Schaerer, ibid) that moreover can also change horizontally. Deficit zones are areas on a slope where the snow slab is no longer supported at the base by the weak layer. Pre-existing cracks in the snow cover are seen as initiators for fracture propagation in the weak layer. The layers expand as the crack propagates outwards from the existing deficit zone through the weak layer (Kronholm 2004). A change in precipitation, temperature, or wind in combination with the local slope angle but also loud noise or vibrations can all trigger an avalanche at the "starting zone" at the top of a slope. The avalanche continues to move downslope steadily increasing in speed and in general following the slope morphology and ultimately fans out and settles in the "run-up zone." Internationally, the Alpine countries of France, Austria, Switzerland, and Italy experience the greatest number of avalanches and loss of life annually. The United States ranks fifth worldwide in avalanche danger and there the states of Colorado, Alaska, and Utah are the most deadly.

The spatial variability of the snow cover has long been subject to intense investigations. The Swiss Federal Institute for Snow and Avalanche Research (Davos), undoubtedly the leading research institute on snow avalanches, was able to identify distinct differences in the vertical layering and in the horizontal variation within individual layers at a regional scale making snow cover stability spatially highly variable (Schweizer and Jamieson 2003). The investigation further revealed that even at the slope scale, the density of the snow cover is not always constant and homogeneous (Kronholm, ibid).

As a great variety of factors determine snow quality, avalanche prevention and mitigation are very complex. Avalanche forecast centers in the United States, Canada, and Europe daily assess the danger from avalanche according to a five-level scale (see Sect. 5.2). They submit daily avalanche forecasts by print media, television, and the Internet. Snow avalanche bulletins are also published daily by all the winter tourist centers and give detailed warnings on which areas should not be entered. These bulletins typically describe important snowpack features as well as the current weather situation. In order to cover the entire country the Swiss Alps are divided into about 100 forecast areas. Nevertheless it has to be stated that much of the hazard assessment is still based on visual observation inasmuch as an automatic and quantitative verification of avalanche forecast is difficult to make and quite costly (Schweizer and Jamieson, ibid) and that more than 10 % of the avalanche fatalities occurred on days with no forecast.

3.2.3 Natural Disasters Versus "Extreme Events"

In the summer of 2003 a heat wave hit western and central Europe, an event that was seen in large parts of the affected population and the national administrations as an extreme event. The countries that were affected most by this disaster were France, Spain, Portugal, Germany, Italy, Austria, Switzerland, and Hungary (WHO 2008). As a result of a so-called "Omega" weather situation, the 2003 summer was the hottest ever recorded in Europe for at least 500 years, with a maximum temperature of 47.3 °C in the Portuguese Alentejo region. The temperatures for large parts of Western and Central Europe reached 3–5 °C above the long-term average. The heat was superimposed by dry weather conditions that in many parts of Europe already prevailed in the months before. Also during the night the temperatures were much higher than normal, hampering the normal cooling down. The heat wave led to health crises in several countries and combined with a long-standing drought created a crop shortfall in parts of southern Europe. Statistical evidence proved that the European death toll reached about 70,000 fatalities and caused economic damage of an estimated €10 billion. In all the countries, older people were more strongly affected than other social groups. But not all the European countries were hit ubiquitously.

France was hit exceptionally hard and experienced seven consecutive days with maximum temperatures above 35 °C and night temperatures above 2 °C. Here about 15,000 heat-related deaths were claimed, a death toll ratio 50 % higher than normal, whereas in other western European countries the excess mortality was about 10–20 %. Previously unexperienced high death toll figures were recorded from large capitals including Paris (>130 %) according to information given by the World Meteorological Organization (WMO 2010). Most of the victims were over 65 years of age and many of them died from dehydration, hypothermia, or cardiovascular system failure. The risk was overlain by the fact that the month of August is the holiday peak season in France, a time when most French pass their vacations at the seaside. Accordingly public life runs at a much lower intensity at that time. The same holds true for the medical system and public health assistance. The hospitals were thus understaffed and as such an event was never experienced before, the medical system especially in Northern France was not prepared to withstand such a disaster. In the cities in the south, for example, Marseille, the death toll—although higher than normal—increased by 20 %, an indication that the people there are more adjusted to higher temperatures. The analysis of the 2003 heat wave highlighted that the disaster resulted from the intricate association of natural and social factors: unusually high temperatures, as well as socioeconomic vulnerability, along with the social attenuation of hazards. In addition to age and gender, combinatorial factors included pre-existing disease, medication, urban residence, isolation, poverty, and, probably, air pollution.

The following discussion gives some insights to the state of knowledge of extreme events. People exposed to such events often have the impression of being completely helpless. Although most such events are seen to occur by chance, when examined in more detail, specific root causes can be identified that trigger such events. In the actual discussion the term "extreme" event is mostly used in the context of climate-related disasters such as the many flood events that occurred in Germany in the last 10 years or the 2003 European heat wave, but also earthquakes, volcano eruptions, mass movements, floods, or other natural phenomena that are often seen as "extreme" events. But such events vary considerably in their frequency and their regional distribution. Thus when does a disaster or a series of disasters occurring at a specific region fulfill the requirements of a normal event or when does it fulfill the criterion "extreme". Numerous climate models, for instance, of the ENSEMBLES research project of the European Union (Vander Linden and Mitchel 2009) and others, clearly show that events such as the 2003 heat wave most probably will occur in future more often as a result of the changing climate, with probably even higher temperatures. Meanwhile many Western European big cities have already started to adjust and improve their emergency management systems, thus such extreme events will in future without doubt be better managed than before. Thus automatically the question arises whether such a disaster then should be called an "extreme event" any more, rather than a strong but normal event.

The question arises as to what a "normal" disaster event is and what an "extreme" event is and consequently according to what definition can both event types be distinguished. As the definition is in practical terms quite unclear, there are only limited robust data available and "Scientists do not really understand what causes extreme events, how they develop and when and where they occur" (Jentsch et al. 2006). When "extreme" events are characterized as "normal" events of natural, social, or financial origin that take place very rarely, such disasters might be more appropriately named "rare events."

Metrics to quantify extreme impacts may include, among others:

- Human casualties and injuries
- Numbers of permanently or temporarily displaced people
- Impacts on property, measured in terms of numbers of buildings damaged or destroyed
- Impacts on infrastructure and lifelines
- Financial or economic loss
- Duration of the above impacts

In order to structure and to standardize the ongoing discussion on the definition of the term "extreme event," IPCC introduced the following definition, although knowing that such a generalized definition will not cover the full range of this hazard type. Extreme events are characterized as (IPCC 2007):

- To be rare to very rare events (low probability)
- Are of local extension (exceptional)
- Have a distinct social and economic impact (catastrophic)
- Have a high relevance to the society (socioeconomic consequences)
- Are very visible (traumatic)

The definition of extreme events of the United States National Science Foundation (Steward and Bostrom 2002) is backed by Sornette (2002), a specialist on the prediction of critical (natural/financial) events, who reported on the occasion of an "Extreme Event Congress" in Hannover (Volkswagen Foundation, 14 Feb 2013) that for him also an event such as the French Revolution in 1789 must be seen as an extreme event. Triggered by a large drought that occurred in 1788–1789 that took the crowd to the streets, it toppled the feudal system that (over time) opened the way to a democratic policy system that spread all over the world. This assumption followed an argument given by Grove (1998) who stated that the 1788–1789 drought in France was most probably an El Niño effect. A more recent example of what an extreme event might be, are the decisions taken by some small island states such as Tuvalu, Samoa, the Maldives, and others to buy land in Australia, India, Sri Lanka, and New Zealand in order to settle there after being forced to leave their countries by the rising sea level.

The problem of understanding the term "extreme event" may derive from the fact that "extreme" either is used to describe the frequency (how often a disaster occurs), or to describe the degree of severity (how relevant the event is for the socioeconomic situation), or to describe that this event got high media coverage (social visibility). The questions arise whether extreme events are defined by just one of the factors or do they require a certain (to be defined) relationship to each other. For example, people living for decades on one of the German North Sea islands while awaiting a winter storm will rate this hazard as normal, although the impact might become of an extreme severity. But when such people are on winter holiday in the Alps and are witness to a snow avalanche, they might get alarmed although the avalanche is of a comparably normal severity. Or do local administrations of a region that just experienced a fatal rockfall (as on the German island of Rügen where one child was killed and the event was covered for a long time in the media) rate this event extreme as it might have serious consequences for local tourism. Do insurance companies rate a hailstorm that produced hail of 8 cm diameter, an event that occurred statistically once in 20 years an extreme event just from the ball size or from the losses to be covered, especially when the losses from the hailstorm were much less, as a very local event, compared to a thunderstorm of a lower severity but with an extremely higher damage ratio.

In the climate change and adaptation discussion extreme events are generally considered on physical evidence, such as the increase in sea-level height, windspeed, or rainfall amount per hour. Such a definition will not fully cover the impact of natural disasters on the population at risk. Therefore in the disaster risk context, the term "extreme" should take the level of severity of a natural disaster event, based on its social connotation, into consideration—fatality, technical damage, and economic loss—rather than defining this event type on its natural phenomena. The following examples make this relationship more transparent. The extreme events, for example, the hot summer in Europe in 2003 with temperatures exceeding 40 °C for many days, or the heavy rainfall in Mumbai, India in 2005 that experienced 950 mm in one day (double the amount of the maximum rainfall thus far) were nevertheless perceived by different societies totally differently. The European heat wave was without doubt something never experienced before and was therefore named "extreme." Whether the heavy rainfall in Mumbai would also be seen by the people there as an extreme event might be questionable. Scheffran, on the occasion of the above-cited "VW-Foundation Meeting Hannover" (2013), gave an impressive example of this: "[A] month of daily temperatures corresponding to the daily maximum in Chennai, India, would be termed a heat wave in France; a snow storm expected every year in New York, USA, might initiate a disaster when it occurs in southern China" (see also: IPCC 2012, p. 53).

When extreme events are defined mainly by the classical means of type of disaster, its severity, frequency, economic losses, and fatalities, it will not arrive at a common understanding and a general acceptance of the term. When extreme events are targeted rather on the social impact a societal-oriented discussion is required that brings together natural scientists, sociologists, mathematicians, and political decision makers including the different populations at risk. An extreme event that is believed to change the entire global climate system is the Asian monsoon. A change in the monsoon rainy season pattern will lead to a melting of the ice caps in the Himalayas with the result that less water will arrive at the Indian subcontinent and can have even more of an impact on North Atlantic thermal circulation, the Greenland ice shield, or Amazon tropical rain forest, but should not be held responsible for the floods of the rivers Elbe or the Danube.

When the overall significance of extreme events is its rareness and not its extremeness then we have to define what we understand under the definition of "rare." The probability distribution function on frequency (variance) and severity (tolerance) of disasters (natural, social, financial, etc.) is in general well approximated by log-normal Gaussian distribution, as can be seen in Fig. 3.38. The blue line gives, for example, the temperature distribution for a certain region for a one-year period. Most of the temperatures experienced fall into the sector "normal" thus representing the temperature regime that has existed for the most of the year. The bell-shaped curve shows a distinct lower frequency for temperatures on both sides of the median characterizing slightly lower or higher temperatures. At the end of the curve the so-called "extreme" events are defined as the tail ends of the log-normal risk distributions, in this case describing "very cold" or "very hot" temperatures. Extreme events are the infrequent events at the high and low end of the range of values of a particular probability distribution. Therefore to understand

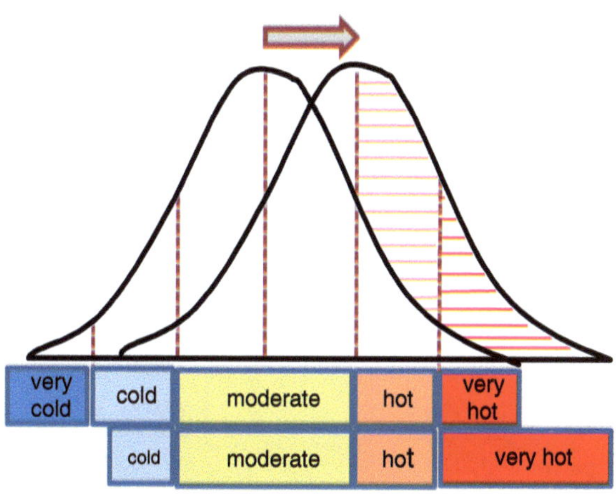

Fig. 3.38 Probability distribution function on frequency and severity of natural disasters in general is well approximated by log-normal Gaussian distribution (Redrawn from: IPCC 2007)

what is "rare," the IPCC definition needs as a further step the definition of the onset value of such an event. In order to have an internationally agreed boundary line that would be verifiable all over the world, the IPCC (2007) proposed to take the 10th and 90th percentiles of the observed probability density to serve as the threshold values.

The figure furthermore explains another phenomenon we have experienced for the last decade. In the course of the changing climate, the Earth's temperature regime is getting steadily warmer, with higher temperatures in summer and less cool in winter times. In the log-normal probability temperature curve this phenomenon is expressed in a shift of the bell-shaped curve towards the extreme "hot" tail end (red line). It "illustrates the effect a small shift (corresponding to a small change in the average or center of the distribution) can have on the frequency of extremes at either end of the distribution. And an increase in the frequency of one extreme (e.g., the number of hot days) will often be accompanied by a decline in the opposite extreme (in this case the number of cold days such as frosts)" (IPCC 2001).The figure shows quite impressively that the tail end of "extreme hot"

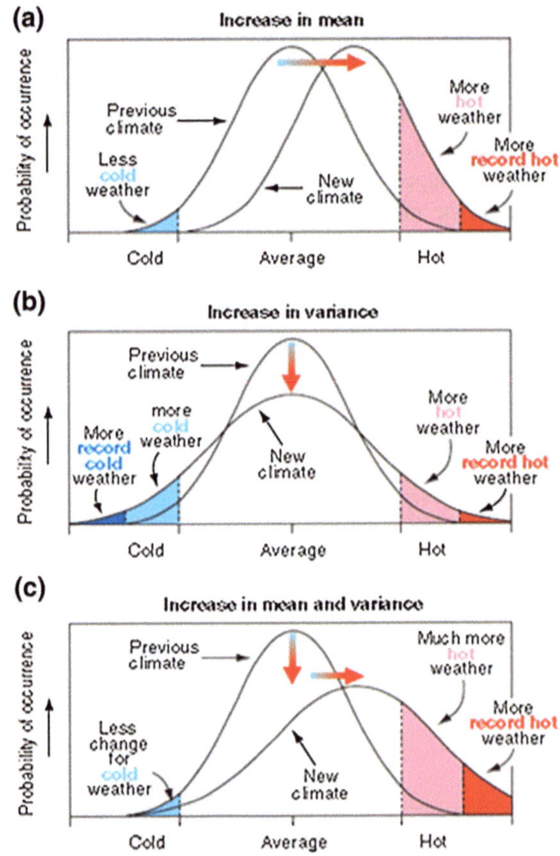

Fig. 3.39 Change in probability distribution function due to shift in the shape of the log-normal distribution: **a** shifting the mean, **b** broadening probability, **c** shift in skewness (*Courtesy* IPCC 2001)

increases significantly whereas on the other side the "cold" tail end is diminishing in its area. That the assumption of shifting from a colder to warmer climate reflects the changing climate conditions could be proved by an analysis of Jonas et al. (2005) in which they assessed the average temperatures in Germany from 1760 until today. In the time span 1760–1880 the average temperatures had their means at about 0 to −0.5 °C. Since then the average temperatures have risen so that today the mean temperature lies at about 1 °C and the "hot" tail end increased to a temperature of 3.4 °C. This simple statistical reasoning indicates that "[S]ubstantial changes in the frequency of extreme events can result from a relatively small shift of the distribution of a weather or climate variable" (IPCC 2001).

The third IPCC report (IPCC ibid) furthermore gave indications on how a small shift in the probability distribution function can lead to a comparably strong change in the climate distribution (Fig. 3.39). As described above, a shifting of the mean of the bell-shaped distribution towards higher temperatures increases the "hot to very hot" tail end considerably and the cold tail end is reduced (part "a"). But not only a shift but also a drop in the mean temperature, leading to a broadening of the variance, will result in an increase in the cold as well as the warm tail ends (part "b"). A change in the temperature regime can moreover result

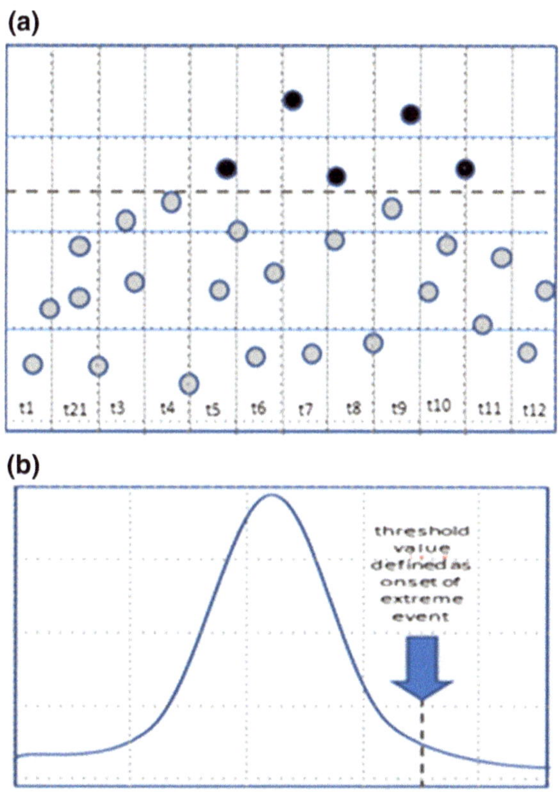

Fig. 3.40 Definition of extreme event threshold value based on **a** random sampling, or **b** from normal distribution function (Own graph)

in a change of median and variance that results in an increase of one of the tail ends significantly (part "c"). In this context it has to be considered that up to now for most of the climate change indicators it is not possible to define whether the changes result from a changed mean, variance, or both.

Typical indicators for a heat wave are related to the number of days above a pre-defined temperature value or on a given precipitation threshold (Fig. 3.39). The big advantage of using such pre-defined indices is that such a method allows an easy interpretation of the data record and can serve as a good comparison with other regions. Another definition is based on the threshold of exceedance, where the number of events, the percentage and fraction of days (i.e., with maximum or minimum temperature), the amount of days with temperature below the 1st, 5th, or 10th, or above the 90th, 95th, or 99th percentile for a given timeframe (days, month, season, annual) are given in a Gaussian normal distribution (Fig. 3.40). In meteorology it is common practice that temperatures for the 10th and 90th percentiles of Tmax/Tmin are referred to as "cold/warm days/nights" (IPCC-SREX 2011, p. 116). A large amount of the available scientific literature on climate extremes is based on the use of so-called extreme indices, which can either be based on the probability of occurrence of given quantities or on threshold exceedances.

To reach an internationally accepted definition one has to take into consideration that different social and economic systems will define "extreme" differently. The people in Bangladesh will arrive at another definition than those in New Zealand or in the Republic of Congo. Therefore the definition of extreme events is not an exclusive task of natural scientists but will not be successful without a comprehensive incorporation of the people who are exposed to the threat (see Sect. 8.6).

According to the definition, extreme events have to be defined from the impact they trigger rather than from the "extremeness" of the type of occurrence. Following this definition that extreme events lead to a fundamental change in the paradigm of social and economic life, the drying up of Lake Aral or the fire clearance of the Indonesian tropical forests with its regional impact on the climate are extreme events. Thus far extreme events have in general been attributed to climate and meteorological phenomena, but in risk science other types of extreme events also exist: the best known is the terrorist attacks on the World Trade Center that led to a complete change in the security architecture of most countries of the world improving the international antiterror networks.

The IPCC-SREX Report (see also IPCC 2012) warned that just by setting up probability distribution functions based on frequency and severity of disasters to assess the situation of future extreme events can be misleading, as it is recognized that climate-related events are mostly characterized by "non-stationary situations." Therefore "past experiences may turn out not to be a reliable predictor of the characteristic and frequency of future events as nature is more complex to be described by just these two variables." It has to be acknowledged that the information base on climate indicators is still very limited and thus does not yet allow us to draw generalized scenarios from it. Consequently the main interest of emergency managers, researchers, and engineers is get a better understanding of the physical processes that lead to extreme natural events by focusing on the cause–effect relationship of an extreme physical event and its impact. Extreme impacts are seen to depend strongly on the social and economic context, reflecting both the degree of vulnerability and susceptibility to which populations, the economy, but also the ecosystem and other elements at risk are located in the exposure path of the extreme event. Instrumental records of variability typically extend only over about 150 years, although since the year 2000 the worldwide monitoring networks on extreme events has developed substantially, so that on daily temperature and rainfall extremes there is now a worldwide comprehensive and verifiable database available for that time span.

3.3 Natural Disaster Distribution

3.3.1 Type of Disaster

Of the many types of disaster the world is exposed to every day, droughts proved to be the deadliest of all. But, as it not may be believed, drought-prone areas are not restricted to the well-known drought-stricken areas of the sub-Saharan regions but also comprise regions in the central United States, Brazil, China, India, and Australia. Moreover regions in the European Mediterranean are also regularly subject to drought events. According to Dilley et al. (2005) more than 1 billion people, that is, about 20 % of the world's population, living in about 10 % of the world's land areas, are regularly exposed to droughts. Other sources see this figure to be almost double (Misereor 2010). UNISDR (2010) citing data of the CRED-EMDAT database pointed out that the annual average death toll for the decade 2000 was almost 80,000, a number that was considerably higher than the previous decade's.

Droughts are not restricted to easily defined regions and do not occur at regular return periods like floods or hurricanes. "Droughts emerge slowly and quietly and lacks [sic] highly visible and structural impacts" (Below et al. 2007). But droughts are predictable as they do not occur overnight. Droughts affect societies more powerfully than many other natural disasters when the event is coupled with lack of financial means, with emergency management failure, and a lack of administrative power to enforce existing laws (UNFCCC 2012).

More than 90 % of the death toll from the 40 biggest natural disasters (1970–2008) occurred in countries that are in a developing state, including China and India. According to CRED-EMDAT (Guha-Sapir et al. 2011, 2013) the total number of people killed by natural disasters exceeded US$ 2.3 billion in the time span 1975 to 2005. As has been already given in Fig. (2.10), 80 % of this death toll is concentrated in only 20 major disaster events. Of these the 1983 drought in Ethiopia and Sudan claimed 450,000 lives, the earthquake of 1976 in China 240,000, the 1991 cyclone in Bangladesh 140,000, and the 2004 Indian Ocean tsunami 225,000.

Similar to the figures on risk mortality and on the economic losses, CRED data moreover reveal a distinct regional distribution. Of the 20 most costly disasters with damages exceeding US$10 billion from 1975 and 2006, 90 % were concentrated in the industrialized countries. Also the economic loss, although slightly less prominent than the death toll figures, was 40 % on 20 disaster events. The most costly natural disaster ever recorded with estimated economic damages of US$210 billion was the Tōhoku earthquake and tsunami in Japan, causing the failure of the nuclear power plant of Fukushima-Daiichi, followed by Hurricane Katrina in New Orleans causing damages of US$125 billion, and the Kobe earthquake in 1995 with a loss of more than US$100 billion. Of the 20 most costly disasters 14 were of hydrometeorological and climate origin, and 6 were caused by earthquakes (Guha-Sapir et al. 2011).

Over the last decade, China, the United States, the Philippines, India, and Indonesia constituted together the top five countries that are most frequently hit by natural disasters. Of the almost 300 disaster events analyzed by MunichRe (2005, 2012, 2013) only about 30 % were of geological–tectonic origin, responsible for about 60 % of the death toll, but only 30 % of the economic damages.

The above-given summary on disaster occurrences and impacts of the last 30 years shows a general trend in regional distribution, number of casualties, and economic losses that is just mirrored by the disasters of the year 2011, making this year somehow representative of the overall disaster distributions the world is facing today. According to CRED-EMDAT on the disasters of year 2011 there were:

- 332 natural disasters
  Asia was the continent most often hit by natural disasters (44.0 %), followed by the Americas (28.0 %), Africa (19.3 %), Europe (5.4 %), and Oceania (3.3 %).
- 30,773 people killed. 79.2 % of global reported disaster mortalities occurred in seven countries, classified as high-income or upper-middle income economies, a figure that is quite unusual due to the impact of the Tōhoku earthquake and tsunami (Fukushima-Daiichi) that alone caused nearly 19,000 deaths, representing 64.5 % of worldwide disaster mortality. The tropical cyclone "Washi" alone caused 1439 deaths, making it the most lethal storm worldwide in 2011.
- 244.7 million victims (injured/homeless). Asia accounted for 86.3 % of global disaster victims, followed by Africa (9.2 %). A total of 65.1 % (159.3 million) of the victims were in China stemming from two floods causing 87.9 million, a drought affecting 35.0 million, and storms with 22.0 million victims. In Ethiopia, Kenya, and Somalia droughts/famines caused 13.2 million victims. Natural disasters claimed 42.9 % of Somalia's population as victims in 2011, making the country the strongest hit in the world. The largest impact came from hydrological disasters that caused 57.1 % (139.8 million) of the victims.
- US$366.1 billion economic damages. With 75.4 % of the total disaster damages Asia suffered the most, followed by the Americas (18.4 %) and Oceania (5.6 %).The Tōhoku earthquake/tsunami (Fukushima-Daiichi) was the most

expensive natural disaster ever recorded (US$210 billion). The flooding in Thailand caused damages of US$40.0 billion, the earthquake in New Zealand US$15.0 billion, and a series of storms in the United States total US$25.0 billion. The economic losses increased by 235 % compared to the annual average damages of US$109.3 billion from 2001 to 2010. Among them, the damages from geophysical disasters (mostly earthquakes) increased the most (US$230.3 billion against US$24.1 billion (average 2001–2010). Geophysical disasters thus represented a share of 62.9 % of total damages caused by natural disasters in 2011.

- The Philippines experienced 33 natural disasters, the highest number ever registered in its history. The country was affected by 18 floods and landslides, 12 storms, two volcanic eruptions, and one earthquake.

3.3.2 Regional Distribution (Hotspots)

Section 2.2 focused on the victims of natural disasters and was thus mainly oriented on the socioeconomic aspects of natural disasters, however, this section focuses on the locations, respectively, regions, where such disasters occur. As stated earlier, natural disasters are very unevenly distributed over the world and have highly different generating modes. Therefore certain regions of the world

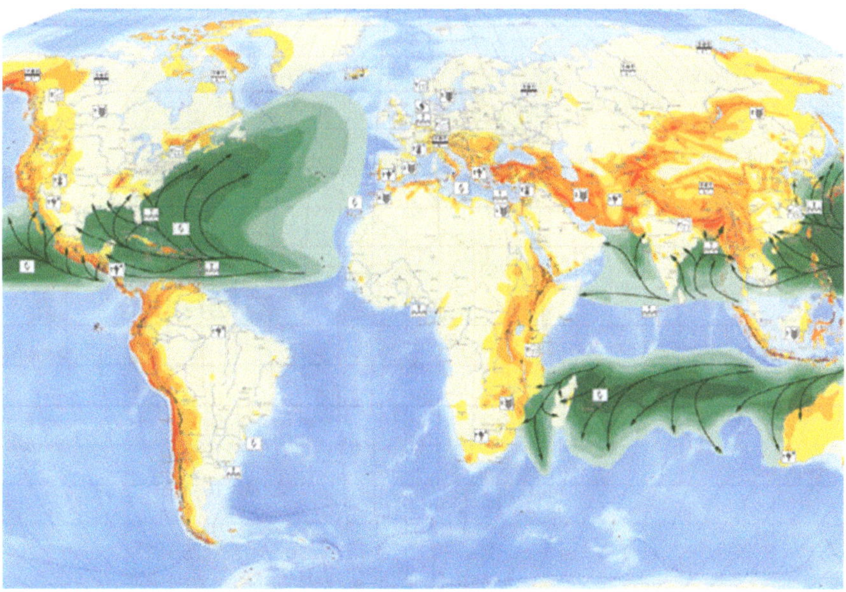

Fig. 3.41 World Map of Natural Hazards; excerpt (*Courtesy* Munich Re, NATHAN, 2011)

are exposed to a higher level of threats from disasters whereas others face only smaller risks. Such areas are called natural disaster hotspots and describe areas where large-scale disasters regularly claim a significant death toll and/or cause heavy economic losses.

In order to assess the regional aspects of disaster distribution there are a number of analytical instruments at hand. The causal relationship of hazard, vulnerability, and risk (see Chaps. 6 and 7) makes an assessment of the worldwide hazard distribution a substantial tool for such an assessment. Furthermore the regional distributions of vulnerability and risk, among others, also exist, all serving the same purpose. In the following some global perspectives of so-called disaster hotspots are presented. It should, however, be noted that there are more of these kinds of assessments worldwide in use.

The famous World Map of Natural Hazards of the Munich Re Insurance Company (2011) impressively displays where on Earth what kind of hazard is to be expected (Fig. 3.41). The author gratefully appreciates the permission to make use of the many data available with the Munich Re Insurance Company NATCAT Service (MunichRe, ibid). The map just intends to give a general understanding of the worldwide hazard distribution. Due to scale it is not possible to display each hazard in its local distribution; therefore the map presents only those hazards that are to occur with a probability of more than 50 %.

Another way to present an overview about the hazards exposure of a country or region is presented in Fig. 3.42. The different hazard types are given by using self-explanatory pictograms/icons symbolizing different hazard types. Such a presentation is best suited to give a generalized overview, simply to inform a broad public

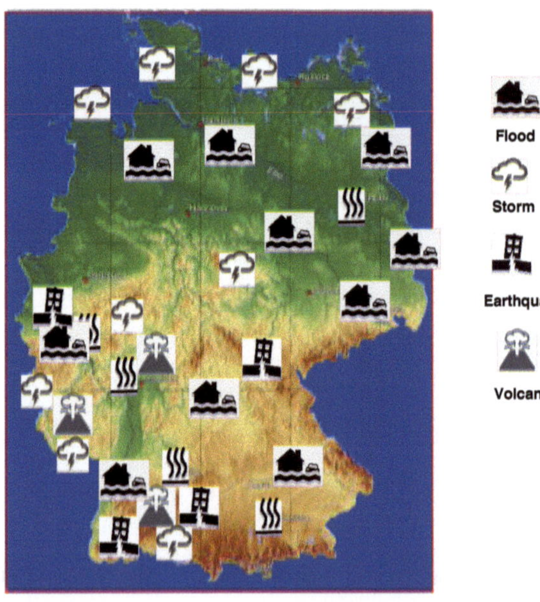

Fig. 3.42 Synoptic Hazard Map of Germany; example of a synoptic hazard map using symbols for indicating type and regional distribution of natural hazards (Own graph)

rapidly and in an easily understandable manner. Such synoptic maps can provide valuable information to the people at risk, to laymen, or other people who are not familiar with natural hazard/disaster/risk assessment. The icon distribution, moreover, delineates a potential agglomerated exposition of hazard types of a specific region.

In the following, a number of tools to define disaster hotspots are presented. First are those that assess the regional risk level by a country-to-country comparison based on a standardized and harmonized risk-defining algorithm, the so-called "risk index." Second are those that identify disaster hotspots based on disaster events (worst-case scenarios). The first tools are restricted in their significance due to the fact that the data availability is very different in number and quality, making a country-by-country assessment a real challenge, whereas the other type of hotspot assessment tool, based on real events, lacks the possibility to generalize the findings. This holds true especially for such types of disasters that are to occur frequently and severely, but have only a small regional extension, whereas other events are to occur quite frequently and cover large areas but only affect a small population. More on the methodology to assess disasters and risks at world and local scales is given in Chaps. 6 and 7.

One approach to give a perspective on disaster occurrence and impact on a world scale is presented by the Disaster Risk Index (DRI) of the United Nations based on the "Global Risk and Vulnerability Index Trend per Year Programme" (GRAVITY; Peduzzi et al. 2002, 2005, 2009). The disaster risk assessment presented here (Fig. 3.43) covers the risk of mortality exclusively. The assessment clearly revealed that the Asian and Eastern African countries are especially at the highest risk of mortality from natural disasters worldwide.

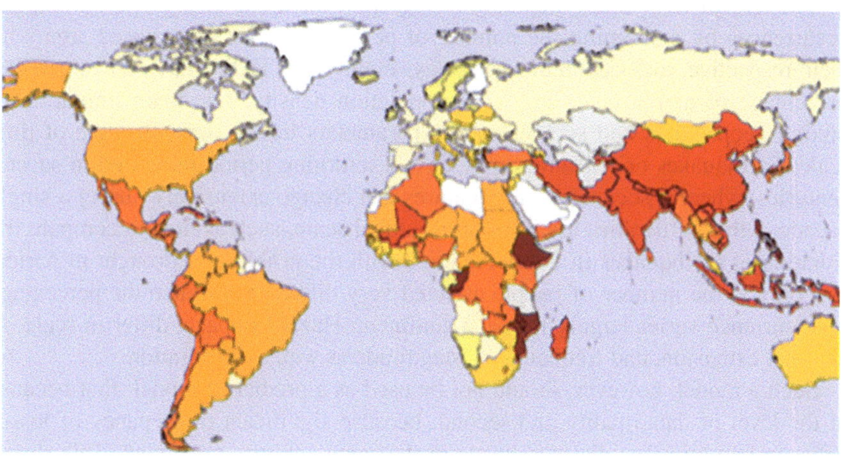

Fig. 3.43 Spatial distribution of risk mortality classes assessed by the Disaster Risk Index (DRI) of the UNDP-GRAVITY Programme (*Courtesy* Peduzzi et al. 2005)

UNDP has initiated the GRAVITY-Programme to assess worldwide vulnerability as a compulsory step to identify the countries' different risk exposure levels. The purposes of the GRAVITY research were to identify whether global datasets could be used for identifying populations living in risk-exposed areas. The program moreover was targeted to identify the links between socioeconomic parameters and vulnerability. With the GRAVITY-Programme UNDP was able to highlight the root causes leading to human vulnerability and provided substantial information identifying the populations at risk. The research was focused on the four natural hazards: earthquakes, volcanoes, cyclones, and floods based on data provided by the CRED-EMDAT database.

The four maps on vulnerability/exposure and on risk clearly indicate where on the globe the people are exposed to a higher risk. But maps at global scale like the DRI and its accompanying statistical findings should not be used as risk predictors. Local disaster risk reduction should always be based on detailed local assessments. By using GIS for spatial analysis a significant relationship between the number of casualties, physical exposure, and socioeconomic parameters was found. Now confirmed by statistical evidence it was possible to show the role of the development in the resilience capacity, a relationship that thus far was more intuitively understood. The analysis revealed that there is a clear relation, that a low development may lead to high casualties, while a high hazard exposure may also result in a low economic development. The statistical analysis demonstrated that physical exposure constitutes the major factor leading to casualties, but other socioeconomic parameters are also substantial variables that lead to high human vulnerability. The level of correlation achieved delineates that both physical exposure and socioeconomic variables are of significant importance and can be easily adopted from international statistics. All in all, the method used in this statistical analysis proved to be appropriate and allows the identification of the parameters leading to a higher risk and vulnerability.

The Global Risk Index was able to highlight the areas of high natural hazard occurrences by combining the number of people living in an exposed area with their respective socioeconomic variables, mainly the HDI, GDP, urban growth, percentage of arable land, and local population density. The main limitation of mixing geophysical and socioeconomic parameters lies in the difference of time scale. Earthquakes or volcanoes may have a returning period measured in several centuries, whereas socioeconomic features can change extensively during a single decade. Other difficulties are inherent to global scale, such as how to compare the situation of earthquakes in South America with the problem of drought in Africa. Not only is the number of people affected very different, but also the percentage of occurrence varies largely for each continent. Hazard impacts differ in scale, in regional extension, and frequency or magnitude as well as in duration.

Such a model, however, should not be used as a predictive model: first because of the level of data quality and second, because significant discrepancy of losses between two (similar) disaster events in the same country was found. This shows the high variability is often due to a temporal context. For earthquakes the number of those killed is highly dependent on where and at what time the disaster happened (during the night or during the day); it moreover depends on the type of habitat, type of soils, direction of fault lines, depth of epicenter, and so on. To bring such variables into a worldwide context is hardly possible.

Another approach for a generalized world disaster risk distribution assessment, called the World Risk Index 2012, was given by the United Nations University, Bonn (UNU-EHS 2012). The assessment was also mainly based on the CRED-EMDAT database. But other than the assessments of UNDP and World Bank (see further below), the UNU-World Risk Index (WRI) is not only restricted to risk exposure as indicated by the "frequency" of disaster occurrence and social vulnerability, but furthermore included the factor of "coping capacity." In this regard UNU-EHS further distinguishes between the coping capacity, defining the capability of a society to cope with adverse effects from natural disasters, and the adaptation capacity. Adaptation capacity in this sense sets in when "a society has already changed structurally before a disaster strikes in a sense that this makes much mitigation no longer necessary" (UNU-EHS, ibid, p. 17).

From the many World Risk Index Maps published by UNU-EHS, the one on "Coping Capacity" is presented here (Fig. 3.44). Although the map might at a first glance not be informative regarding the disaster cum risk distribution of the world, the factor "coping capacity" describes a substantial input societies may be provided in order to reduce disaster impact and is thus contributing to a better understanding of risk exposure of the world.

Like the other world maps on disaster distribution, the distribution of deficits in the coping capacity also revealed that the Asian countries are at high exposure to risk from disasters. Nevertheless the coping capacity shows certain differentiations: the highest deficits are identified in Central Africa but also for parts of Central America. When the assessment of World "Coping Capacity Deficit" is combined with "Hazard Distribution" and the DRI-Index maps (here on "Risk of

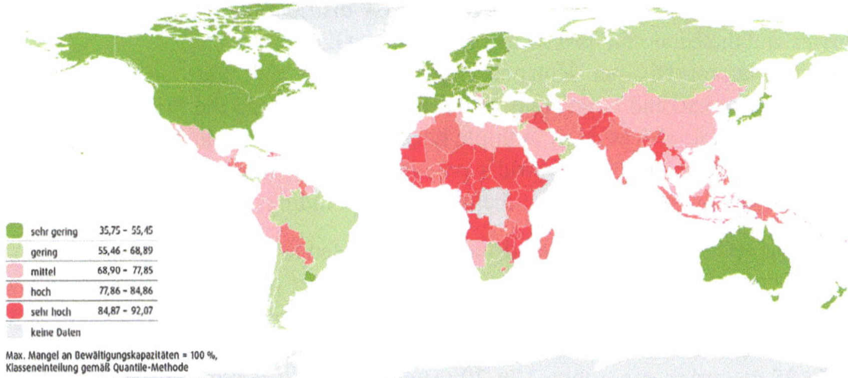

Fig. 3.44 World Risk Index Map showing the "Deficit in Risk Coping Capacity" (*Courtesy* UNU-EHS, Bonn)

Mortality") a realistic impression of the world risk from disasters can be derived. For example, Australia and Chile both are in the same high-risk exposure class, although their technical standards to cope with a disaster differ a great deal. The opposite holds true for Mongolia. There, the overall risk exposure is low, but the country has a very high deficit in disaster structural and socioeconomic capability, especially against risk from climate change, a situation Mongolia shares with Bolivia and Paraguay. Africa (with the exception of South Africa) is the region of the world that is at the highest risk in all categories, the same as Afghanistan and Pakistan, whereas the other Asian countries down to Papua New Guinea have already made quite significant advances in their local capacity to withstand a disaster (e.g., Thailand and Malaysia).

The most comprehensive and therefore most adopted index-based risk assessment of the world has been worked out by the International Bank for Reconstruction and Development/The World Bank (Dilley et al. 2005). The Bank has over many years successfully tried to establish a generalized risk index: the Global Disaster Risk Index (GDRI) that intends to provide an overall assessment on the risk of mortality from natural disasters for the world in total. Due to the varying quality and quantity of the databases available, the approach aggregated all data accessible to the World Bank into one set of data, and consequently could only provide a very generalized impression. The DRI therefore should not be taken as a source of information on a regional differentiation of the disaster type and its severity and frequency. The Global Disaster Risk Index assessed the distribution of risks worldwide based on two disaster-related outcomes: mortality and economic losses. Both parameters are assessed by combining the regional exposure to earthquakes, volcanoes, landslides, floods, drought, and cyclones with vulnerability data on population distribution and the national gross domestic product. The study presented the first successful approach for an index of the global risk to natural hazards. The calculation was based on grid cells, as such an approach gave a more detailed insight to the subnational and local distribution of the risks than an assessment based on a national scale. The GDRI of the World Bank gives two more sets of information of the global risk distribution: on the total economic losses and the economic loss as a portion of the GDP.

The Natural Disaster Hotspots study identified that East and South Asia, Central America, and large areas of the Mediterranean and the Middle East are at the greatest risk of loss from multiple hazards and indicated that about 3.4 billion people, more than half the world's population, lives in areas where at least one hazard could significantly affect them. Other key findings of the report were:

- About 20 % of the Earth's land surface is exposed to at least one natural hazard.
- 160 countries have more than one quarter of their population in areas of high mortality risk from at least one hazard; more than 90 have more than 10 % of their population in areas of high mortality risk from two or more hazards.
- In 35 countries, more than 1 in 20 residents lives at a relatively high mortality risk from three or more hazards.

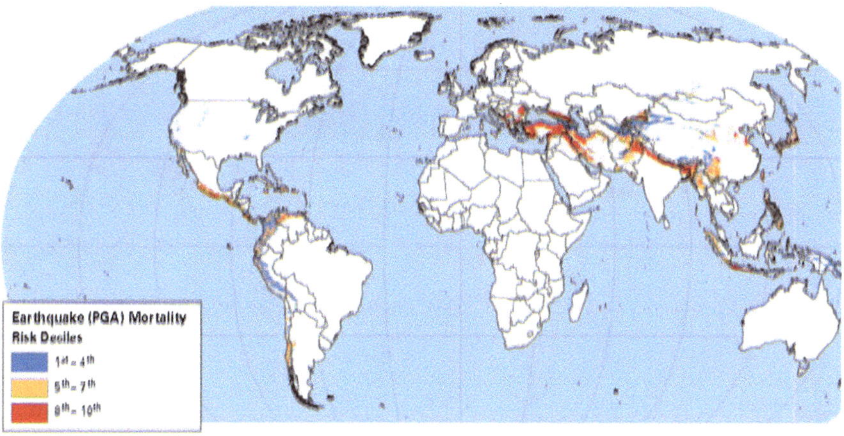

Fig. 3.45 Risk of mortality from earthquakes (*Courtesy* Dilley et al. 2005)

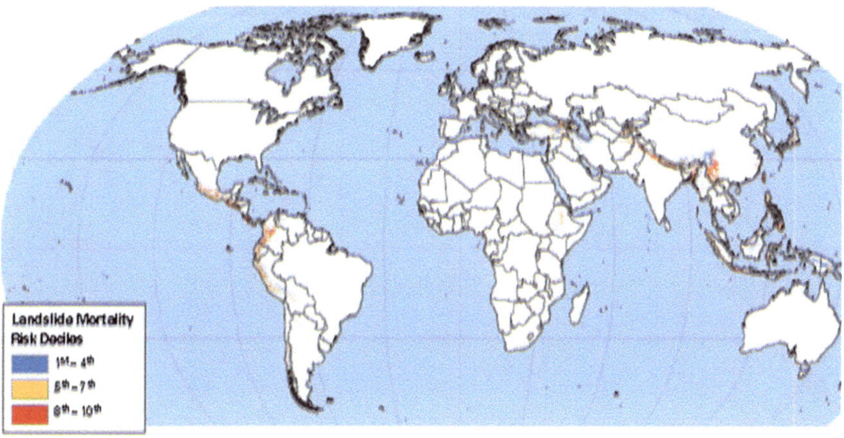

Fig. 3.46 Risk of mortality from landslides (*Courtesy* Dilley et al. 2005)

- Taiwan may be the place on Earth most vulnerable to natural hazards, with 73 % of its land and population exposed to three or more hazards.
- More than 90 % of the populations of Bangladesh, Nepal, the Dominican Republic, Burundi, Haiti, Taiwan, Malawi, El Salvador, and Honduras live in areas at high relative risk of death from two or more hazards.

Although the World Bank (Dilley et al., ibid) approach provided an impressive comparison of the disaster hotspots of the world (Figs. 3.45, 3.46, 3.47, 3.48 and 3.49), it is nevertheless obvious that even such an impressive data collection

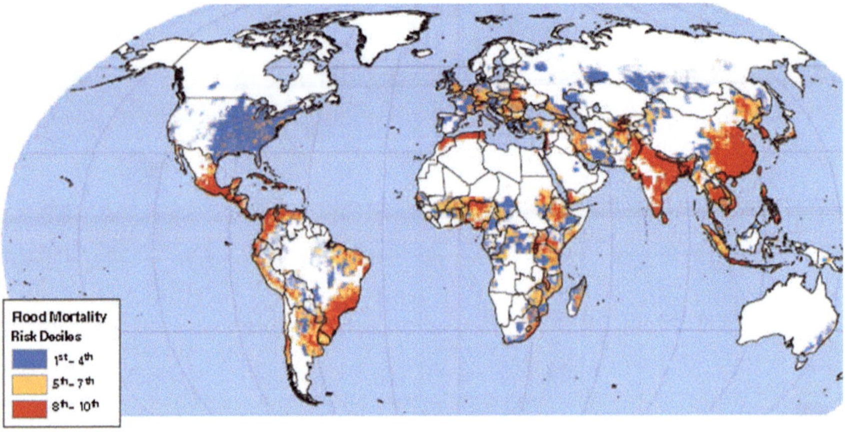

Fig. 3.47 Risk of mortality from floods (*Courtesy* Dilley et al. 2005)

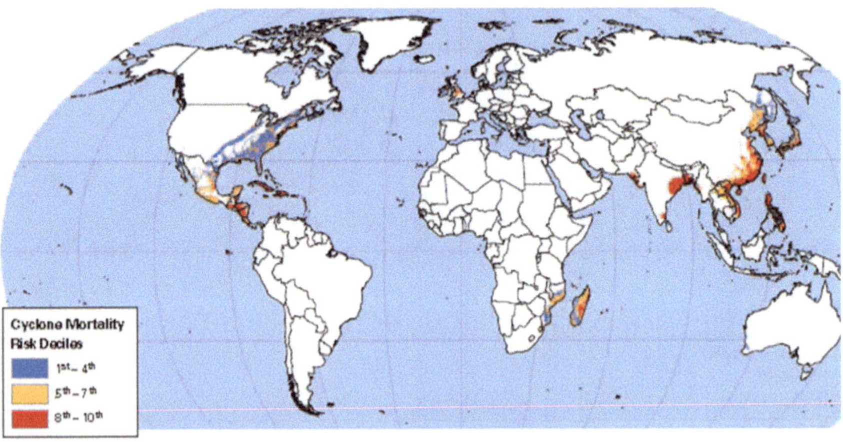

Fig. 3.48 Risk of mortality from cyclones (*Courtesy* Dilley et al. 2005)

cannot cover all parameters that define "risk." For example, the risk distribution from volcanic eruptions has not been incorporated in the study, as the World Bank argues that volcanoes on a world scale are only represented by very tiny spots. The study moreover deliberately left out those areas with low population density or without agricultural importance, a systematic approach that is contradicted by many researchers such as Birkmann (2007) who claimed that many of these areas show a high relative mortality to floods. The World Bank although confessed that the findings should not be overinterpreted as the database is still sparse concerning availability and quality of natural hazards and occurrences as well as on historical economic losses. Therefore the World Bank sees its global hotspot analysis

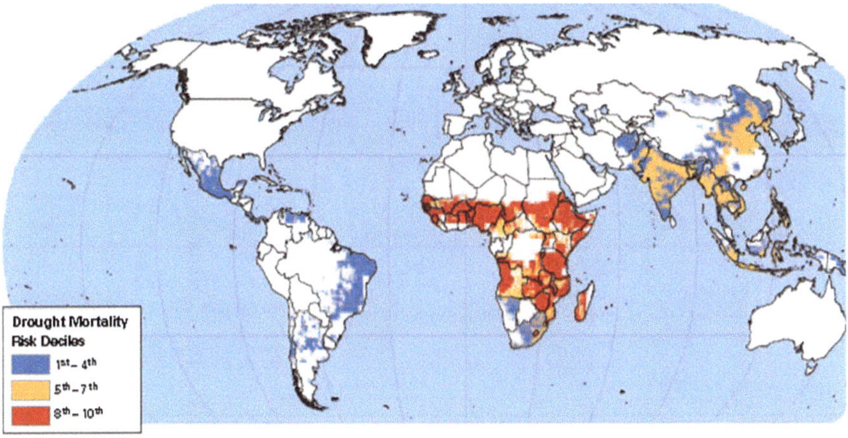

Fig. 3.49 Risk of mortality from droughts (*Courtesy* Dilley et al. 2005)

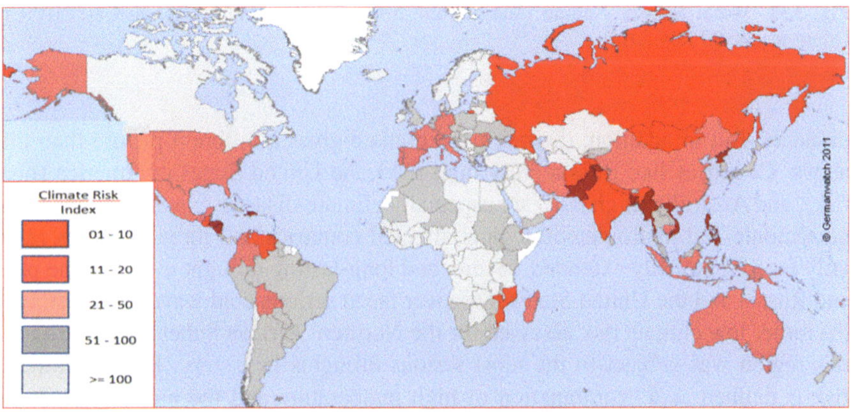

Fig. 3.50 World Climate Risk Distribution (*Courtesy* GermanWatch 2014)

as an instrument for identifying the relative levels of risk rather than an indicator on the absolute risk levels. The fundamental drawback of the study results from the lack of availability of reliable and reproducible indicators of vulnerability. Vulnerability, in the understanding of the concept, cannot simply be determined by past losses of life and economic values. From the many world maps of risk distributions only those from earthquakes, landslides, floods, cyclones, and drought were therefore taken up.

Also index-based is the World Map on Climate Risk (Fig. 3.50) published by the GermanWatch (2014). It is presented here as many of the disasters are climatic in origin. The study mirrors the already-presented risk distributions and revealed that the Asian countries and Central America are at high risk from climate

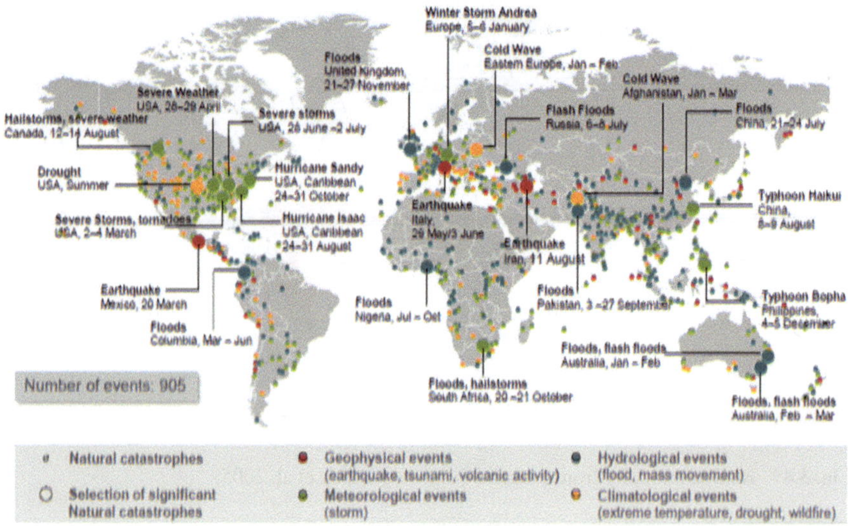

Fig. 3.51 World Map of Natural Catastrophes 2011 (*Courtesy* MunichRe, NatCatSERVICE; online access: 30 July 2014)

disasters. But in addition, the risk from climate gives a different picture than that above. Countries like Russia (very high risk), the United States of America (high risk), and Australia (high risk) are especially climate-disaster exposed compared to geotectonic and hydrometeorological affected countries. All three countries regularly faced regionally extended, severe, and long-lasting drought events in the past, and Russia and the United States moreover faced serious cold waves. Interesting is the rather low climate risk assessed for the Northern African Sahel region although this region was subject to the most serious droughts in history. But as (climate) risk is defined as a "combination of high temperature and the number of people exposed," the region is therefore rated of a rather moderate climate risk.

A further step towards identifying the worldwide distribution of disaster hotspots is to record the natural disasters by number and event type. Such an assessment is regularly carried out by many international organizations. The MunichRe (2013) for more than a decade has published statistics on natural catastrophes every year. When combined with the type of disasters such a map (Fig. 3.51) provides an impressive indication of the world disaster hotspots. The map clearly shows that there are three disaster hotspots. One center is located in the United States, another in Western Europe, and the third in South and East Asia. The regional distribution of this event and type-based assessment fully coincides with the risk-based assessments of UNDP (DRI), WB (GDRI), and the UNU-EHS (World Risk Index).

The broad range of assessment tools in use mainly originated in the different topics the authors intended to cover (scientific, economic, political). Nevertheless each provides valuable information and when combining the different sets of

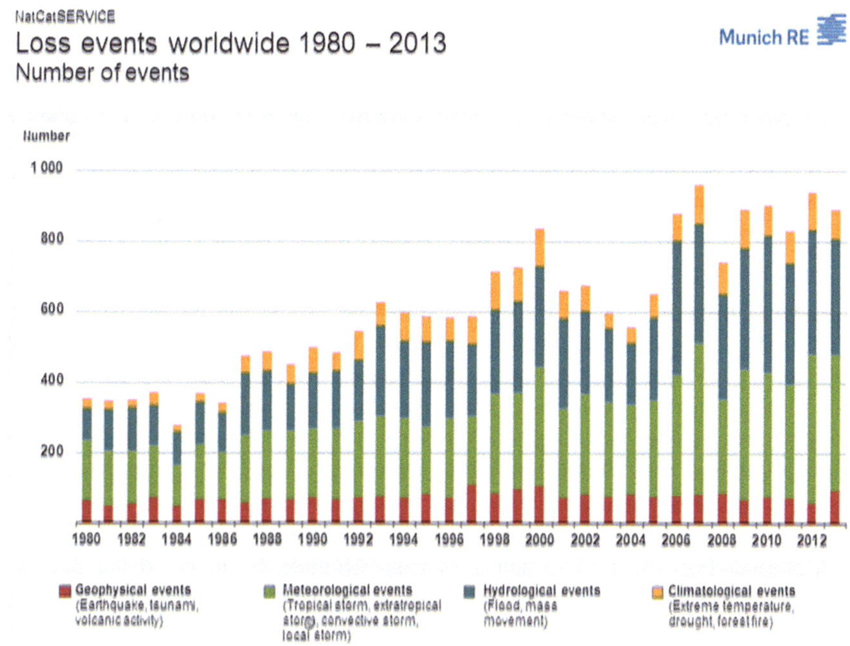

Fig. 3.52 Historic development of natural disaster events from 1980 until 2010 according to type of disaster (*Courtesy* Munich Re, Topics Geo Online, 2013)

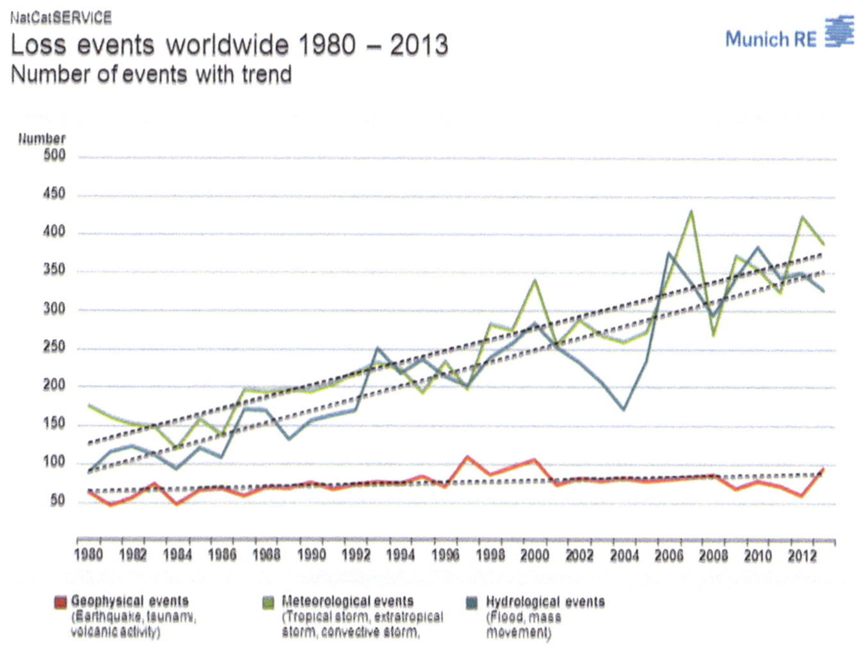

Fig. 3.53 Generalized trend in type of natural disasters from 1980 until 2010 (*Courtesy* Munich Re, Topics, Geo Online, 2013)

information, an ideal insight to the regional distribution of risks from natural disasters is provided.

But disaster hotspots should not be defined only by their number and type of event. Also the development of disasters over the course of time, their frequency, and severity are factors that help to identify the potential future risk. Therefore disaster distribution assessments deserve to be accompanied by a "time" and "impact" component. Only when the data on the historical development of disaster events are combined with severity (impact) does such an assessment become meaningful and open a reliable insight on the risk the people may be exposed to worldwide in future.

A statistical assessment of the disaster events of the last 30 years (Fig. 3.52) published by Munich Re indicates that there is a steady increase in the number of disaster events worldwide, that has more than doubled in this time span. Nevertheless a closer view (Fig. 3.53) reveals that the increase has a different origin. Although geophysical–tectonic disasters more or less occur at a steady rate, disasters of climate and meteorological origin have increased almost threefold. Moreover this increase shows a large year-to-year fluctuation.

Moreover Figs. 3.51, 3.52 and 3.53 from Munich Re do not detail how the people in the different regions are really affected by the disaster. There are many regions that are exposed to frequent events but only experience disasters with

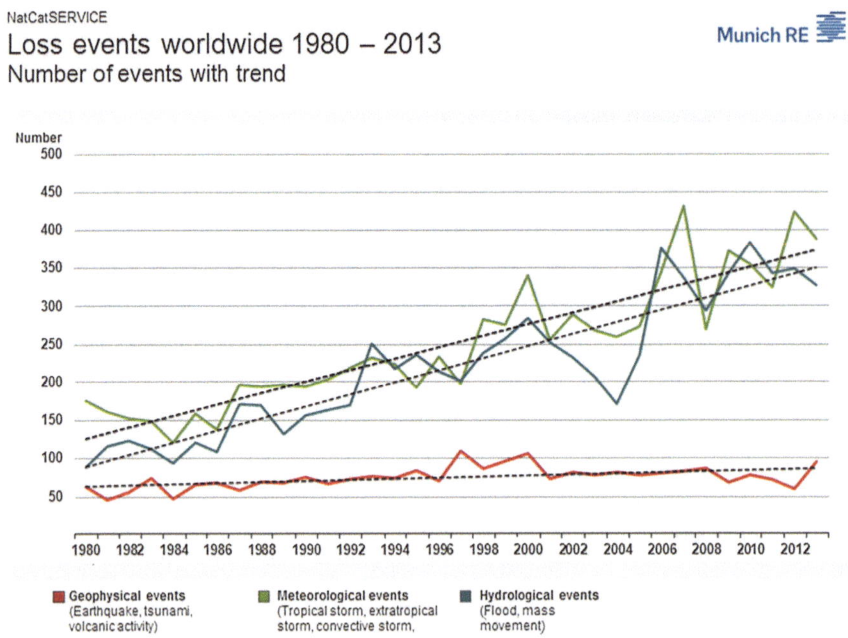

Fig. 3.54 Historic development of overall and insured economic losses "1980 to 2013" (MunichRe 2012)

a low intensity, neither claiming a significant death toll nor causing heavy economic losses, and others that are highly exposed to disasters but face mainly economic losses whereas other countries are hit by rare (extreme) events that cause a high death toll or economic losses. The degree of vulnerability from a natural disaster of an individual or a societal group cannot be realized from such a type of assessment. A further indicator of vulnerability is given by Munich Re by the "Loss Events Worldwide" (Fig. 3.54). The figure clearly indicates that the insured losses and the overall losses are both increasing. Similar to the increase in number of people affected by disasters worldwide the distribution of insured and overall losses is also dominated by single mega events, such as the 1995 Kobe earthquake. Nevertheless the trend in overall losses worldwide clearly mirrors the general increase in economic values accumulated in the developing countries. A very general assumption can be made: the higher the income of a society, the better developed is its capacity to adjust against adverse risk impacts. But that does not say that countries with a low income did not develop effective capacities to withstand adverse impacts or, for instance, societies that are living at the seaside or living in areas exposed to snow avalanches.

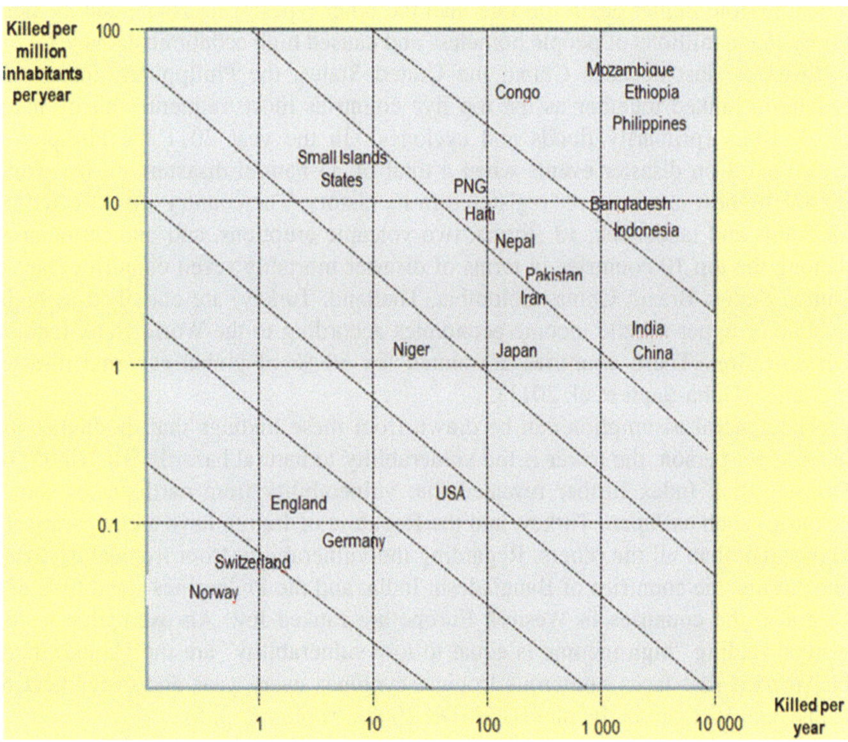

Fig. 3.55 Mortality risk of selected disaster affected countries (based on: UNISDR 2007, 2009a, b, 2013)

The above-given generalized "World Risk Indices" demonstrate that risks from natural hazards are not uniformly distributed over the world. In order to relate the risk distribution to specific countries, UNISDR (2004, 2007, 2009a, b) plotted the relative vulnerability against the hazard exposure for all different risk types. Figure 3.55 sums up losses from droughts, earthquakes, floods, and storms for all countries with the exception of those where risk exposure was rated negligible or where less than 2 % of the total population was exposed. The risk index plotted the absolute figure of persons killed per year and the relative multiple risk, indicated by the people killed per year as a percentage of the total country population (Peduzzi et al. 2009). The plot shows (without surprise) the top countries at risk are in Africa and in eastern Asia. In terms of those killed per year are the most populated countries—China, India, Indonesia, and Bangladesh rank highest—whereas in terms of people killed per million inhabitants per year, the small island states of Vanuatu, Dominica, Mauritius, Antigua, and so on have the highest risk. The plot moreover points out (also not very surprising) that countries such as Germany, France, England, and so on are, although facing natural disasters including floods almost every year, nevertheless rank low, as these disasters (fortunately) do not claim a high casualty ratio. In flood vulnerability Venezuela ranked high, as it has been subject to many flood events in the last decades that once claimed a hundredfold higher death toll rate than the other types of disasters, and as these floods made millions of people homeless and caused high economic damages.

Over the last decades China, the United States, the Philippines, India, and Indonesia ranked together as the top five countries most frequently hit by natural disasters (primarily floods and cyclones). In the year 2011 the Philippines set a record on disaster events when a total of 33 natural disasters hit the country, the highest number ever registered in its history. The country was affected by 18 floods and landslides, 12 storms, two volcanic eruptions, and one earthquake. Among the top 10 countries in terms of disaster mortality seven countries (Japan, United States, Brazil, China, Colombia, Thailand, Turkey) are classified as high-income or upper-middle income economies according to the World Bank Income Classification. These countries accounted for 80 % of global reported disaster mortality (Guha-Sapir et al. 2011).

The general assumption can be drawn from these findings that the higher the income per person, the lower is the vulnerability to natural hazards. The UNISDR Disaster Risk Index further revealed that vulnerability from earthquakes shows countries such as Japan, Turkey, and the Republic of Iran to have a comparatively higher risk than all the others. Regarding the vulnerability from tropical cyclones and storms, the countries of Bangladesh, India, and the Philippines stand first, and here also the countries in Western Europe are ranked low. An exemption to the general finding "high income is equal to low vulnerability" are the United States of America that faces numerous tropical cyclones every year and occur next to China, and Vietnam in the plot (UNISDR 2009a, b, plots 2, 3, 5).

Furthermore, exclusively making statistical evidence a basis for an assessment of a disaster impact can lead to biased assumptions. Much information especially when related to drought disasters is often given in a time span that exceeds one year. Thus it is often difficult to compare drought disaster with another in the same year in the same region. Or a drought followed by a severe dry period one year later but with the victims of the former still in need of external help. For example, if a drought struck a region for a period of, say three years, affected in total 100,000 people, and killed 10,000 people, it is possible that all of the 100,000 are being affected over the entire three years, whereas the 10,000 fatalities are killed in the first year. Statistically the casualty is (normally) attributed to the entire drought time span. Moreover, many drought-stricken countries often do not have the administrative and operational capacities at their disposal necessary to assess the number of victims really affected.

The lack of standardization in drought hazard characterization contributes to the problem of attributing definitive losses. Even if drought information has improved and the methodology applied in CRED-EMDAT has been strengthened over the last years, data still remain inconsistent because of the complexity of droughts, especially in terms of measuring the direct human impact (Below et al. 2007), a situation happening at many refugee camps, for instance, in the sub-Saharan region, even when the international community is rendering assistance. In this context it has to be acknowledged that many countries, especially those who deeply depend on external aid to cover yearly losses (sometimes), tend to exaggerate their death toll in order to keep up tension while raising international aid solidarity.

The GDRI assessment made an important point regarding the timewise and geographical distribution of disaster events. As described above, the assessment did not incorporate, for instance, volcanic eruptions as they only cover very tiny spots and therefore will not be sufficiently represented on a world scale. This evidence clearly indicates that disaster and emergency management cannot be solely based on statistics of time and regional occurrence of natural disasters. The geophysical–tectonic disasters (earthquake and volcano eruptions) mainly occur on short notice and are quite local. Droughts and the other climate-related disaster heat waves, however, are very slow in their onset and are characterized to last over many years and to cover even continents. On the other hand the majority of the hydrometeorological disasters, such as snow avalanches or flash floods are short in onset and duration and normally restricted to morphology. The generating mode of the different disasters—fast and local, or slow and widespread—provides the keys for mitigation and prevention countermeasures. Snow avalanche prevention, for instance, is therefore mainly technical to avoid avalanches reaching the villages whereas heat waves and droughts due to their large areal extension require operational management capabilities helping the victims to survive.

3.4 Database and Data Reliability

Official disaster statistics, such as those preferably used in this book, are normally provided by CRED-EMDAT, UNDP, UNISDR, the World Bank, or other insurance companies such as the Munich Re Insurance Company or the Swiss Re Insurance Company. The most comprehensive database exists with the industrial countries, especially the United States of America, that have collected an enormous databank to serve the needs of the national Natural Disaster Risk Assessment Program "Hazus 99" (FEMA 2014) or the United Nations Global Assessment Report on Disaster Risk Reduction (UNISDR 2013) that every two years comprehensively reviews and analyzes the world disaster risk and the international initiatives of disaster risk management.

The EMDAT database today contains core data on the occurrence and impacts of about 16,000 disaster events in the world dating back to 1900 (Below et al. 2007). Although the quality of the reported data is steadily improving, its data quality still differs strongly. Priority is therefore given to data provided by UN agencies, followed by OFDA, official releases from national governments and the International Federation of Red Cross and Red Crescent Societies. The database provides a functionality to make the information collected from all over the world comparable. The entries are constantly reviewed for redundancy, inconsistencies, and incompleteness. At least two different sources are necessary to confirm figures. The figures are validated according to the priority sources but can be completed by secondary ones. The incoming data are validated monthly and internationally cross-error-checking and are made available for the public every month. Furthermore there is a yearly quality data control.

CRED defines a disaster according to the international definitions of (UNISDR 2004). For a disaster to be entered into the database, at least one of the following criteria must be fulfilled:

- 10 or more people reported killed
- 100 or more people reported affected
- Declaration of a state of emergency
- Call for international assistance

Data on earthquakes are mainly taken from the USGS database, on floods from the Dartmouth Flood Observatory (DFO), on epidemics from the WHO Diseases Outbreak News, on economic losses from the UN-ECLAC and from the global reinsurance companies, whereas data on social aspects are mainly taken from the UN-OCHA ReliefWeb or the UNISDR PreventionWeb.

The EMDAT database provides access to different search possibilities:

- Country profiles
- Natural and technological disaster profiles
- Disaster lists
- Regional maps
- Trends in disaster distribution

Although quite reliable data on casualty and economic loss are easy to obtain, data on the number of affected people very often lack independent proof. Such figures are mainly based on a subjective perception by local risk assessors, often based on a rough estimate of the population settling in the disaster-affected area but not headcount. Therefore still most information given on natural disasters is by the number of casualties. But it should be noted that for every victim about 3000 people are affected by disasters (Nadim et al. 2006a, b).

Another international database that aims to improve evidence-based disaster risk management at the regional, national, and global scales is the Global Risk Identification Program (GRIP) of the UNDP/UNISDR (UNDP 2007) with the support of the ProVention Consortium of World Bank and Columbia University. The program is outlined to strengthen the effectiveness of national and international disaster risk reduction strategies by adding value to, and improving coordination between ongoing international initiatives. It provides a framework where international donor agencies, governments, regional organizations, and research institutes as well as the private sector can share respective knowledge, information, and expertise.

In addition, major insurance companies of the world, especially the Munich Re and the Swiss Re insurance companies have established databanks to suit their specific needs. The Munich Re databank is named NatCat and comprises some 30,000 data records, making it the most comprehensive natural catastrophe loss database in the world. Approximately 1000 events are recorded and analyzed every year.The information is collected by Munich Re to perform risk and trend analyses on individual natural hazard types in various parts of the world. The Munich Re statistics are freely accessible from the NATHAN-Online. The data collection provides information on all major natural disasters since 1980. But it not only comprises data on damages and casualties but also provides basic information on the hazards and risk exposure of the areas at risk. Munich Re regularly publishes a variety of disaster-related brochures, all on the four "Topics Geo Online" (MunichRe 2012, 2013) but also brochures on specific disasters on all natural and technical topics. World famous is the World Map of Hazard Hazards. SwissRe yearly publishes their data collection in the Sigma publication series, a series that like the data collection of Munich Re is freely accessible via the Internet (Swiss Re 2010). Furthermore the Asian Disaster Reduction Center (ADRC) has initiated a new disaster database, called Glide (ADRC 2009) which gives specific features on hazards and disasters by a unique identifier and a number of relevant attributes.

There are several problems in using disaster statistics as they are provided by the different organizations, and often collected with different intentions. The collections are much affected by the monitoring procedure, the way the data are collected, processed, and finally transmitted to the international agencies. Normally local government organizations in the disaster-affected districts collect the data within their jurisdiction and report these to their national data centers. From there the international data centers such as CRED and UNDP are informed. As the different data providers have very different monitoring network densities and equipment, collecting the data at very different intervals and duration, some provide handwritten and visual-based data after a disaster happened, and others provide data online, digitally, and 24 h a day. The different data quality makes the CRED prioritization scheme concentrate on meaningful international data sources as many reporting sources do not cover all disasters due to political limitations or do not cover transboundary effects. For instance, Myanmar did not report any casualties or damage of the 2004 tsunami although neighboring Thailand and Indonesia were seriously hit. Moreover often the data on disaster events are not collected according to the internationally agreed format. The author's own experience revealed that often local administrators in remote districts do not see the relevance of gathering such information and are reluctant to report as they feel that reporting on tragedies and disasters might harm their political standing. Such practice resulted in a very uneven distribution of disaster occurrences in Indonesia (BG/Georisk 2005). It was recognized that those districts properly equipped with, for example, a seismic detection network and are furthermore densely populated and moreover have high media coverage show a much higher frequency of natural disasters than far remote districts although they might have a much higher risk potential. It is a well-known fact that the more people live exposed to a disaster, the more is reported even on smaller disasters. The opposite results from areas that are far remote. There a low population density does not support detailed investigations and thus reports on disaster are often not transferred to the central government.

Moreover disaster information in general is linked to the final type of disaster that struck. For example, landslides that were triggered by earthquakes or by tropical storms are grouped under the triggering event "landslides", although the actual triggering mode was different. Thus far no disaster event statistic is available that also covers the triggering causes. The 2004 tsunami in Banda Aceh is anticipated to have destroyed 80 % of the capital's houses. But that assumption does not take into account that the tsunami was triggered by the third biggest earthquake ever recorded in history. The statistical number of destroyed houses does not indicate how much of the building substance had already been destroyed before the tsunami struck. This problem is difficult to solve, as there are only a few cases known where such an attribution is possible. If the houses in Banda Aceh had been more seismic resistant it is assumed that many of them would have been to withstand the following tsunami.

Although the database is unique worldwide CRED itself acknowledges that reliable assessment of the disaster events cannot be made for the time span before the years 1985–1990. Since then the number of reports on disaster events has at least doubled, a factor that often led to the incorrect assumption that the amount of disasters has increased simultaneously.

The still limited reliability of data often makes assumptions regarding the working out of risk reduction strategies a real challenge. Nevertheless the data record compiled thus far by the many reputed disaster management and research organizations now enables them to draw a realistic picture of the risk situation and to make effective proposals for disaster risk prevention activities, although it has to be confessed that still too often such strategies are based on a transfer of experience from other regions.

References

ADRC (2009): Global Disaster Identifier Number (GLIDE).- Asian Disaster Reduction Center, (ADRC), Bangkok

Apel, H., Thieken, A.H., Merz, B. & Bloschl, G.A. (2006): Probabilistic modeling system for assessing flood risk.- Natural Hazards, Vol.90, p.229-237, Springer, Heidelberg

Bakun, W.H. & Prescott, W.H. (1989): Earthquake Occurrence. -United States Geological Survey (USGS), The Loma Prieta Earthquake, Professional Papers 1550-A through 1550-F, Reston VA

Bakun, W. H. & Lindh, A.G. (1985): The Parkfield, California, earthquake prediction experiment. – Science, Vol. 229, p. 619–624, Washington DC

Barbano, M.S., Rigano, R. & Coppolino, I. (2001): The seismic history of Nicolosi (Catania, Sicily).- in: Glade et al., (eds.): The Use of Historical Data in Natural Hazard Assessment, Vol. 17, p. 55-70, Kluwer Academic Press, Dordrecht

Below, R., Grover-Kopec, E. & Dilley M. (2007): Drought related disasters - A global assessment, Journal of Environment and Development, Vol. 16, p.328-344 Sage Journals

Berckhemer, H. (1990): Grundlagen der Geophysik. – Wissensch.Buchgesellschaft, Darmstadt

BG/Georisk (2005): Mitigation of geohazards in Indonesia – Status report on the GEORISK project. - A contribution to the World Conference on Disaster Reduction, Kobe, Hyogo, Japan, 18-22 January 2005, Badan Geology (BG), Bandung

BGR (2010): Das seismische Ereignis bei Landau vom 15. August 2009 – Abschussbericht Expertengruppe „Seismisches Risiko bei hydrothermaler Geothermie. – Bundesanstalt für Geowissenschaften und Rohstoffe (BGR), Hannover

Bilham, R., Yeats, R. & Zerbini, S. (1989): Space geodesy and the global forecast of earthquakes. - EOS, Transactions American Geophysical Union, Vol. 70, No. 5, Wiley Online Library

Birkmann, J. (2007): Risk and vulnerability indicators at different scales: applicability, usefulness and policy implications . - Environmental Hazards, Vol. 7, Elsevier, Philadelphia PA.

Bluth, G.J.S., Doiron, S.D., Schnetzler, S.C., Krueger, A.J. & Walter, L.S. (1992): Global tracking of the SO2 clouds from the June 1991 Mount Pinatubo eruption. – Geophysical Research Letters, Vol. 19, Wiley Online Library

BMLFUW (2012) Lawinenwinter 1999 und die Katastrophe von Galtür .- Österreichisches Bundesministerium für Land- und Forstwirtschaft, Umwelt und Wasserwirtschaft, Wien (online access: 27.7.2014)

Bondevik, S., Løvholt, F., Harbitz, C., Mangerud, J., Dawson, A. & Svendsen, J.I. (2005): The Storegga Slide Tsunami—comparing field observations with numerical simulations. - Marine and Petroleum Geology, Vol. 22, Elsevier Ltd, Elsevier, Philadelphia PA

Bryant, E. (2001): Tsunami – the underestimated disaster, 2nd edition, Cambridge University Press, Cambridge MD

Bullard, E., Everett, J. E. & Smith, A. G. 1965, The fit of the continents around the Atlantic. In: Blackett, P. M. S., Bullard, E. & Runcorn, S. K. (eds), A Symposium on Continental Drift, Philosophical Transactions of the Royal Society, Vol. 258, p. 41-51, London

Calais, E., Freed, A., Mattioli, G., Amelung, F., Jónsson, S., Jansma, P., Sang-Hoon Hong, Dixon, T.,Prépetit, C. & Momplaisir, R. (2010): Transpressional rupture of an unmappedfault during the 2010 Haiti earthquake. - Letters / Focus, Rosenstiel School of Marine & Atmospheric Science, University of Miami FL (online: 24 Oct. 2010; DOI: 10.1038/NGEO992)

Camassi, R. (2004): Catalogues of historical earthquakes in Italy.- Annals of Geophysics, Vol. 47, No.2/3, Bologna

Carrara, A., Catalano, E., Sorriso Valvo, M., Reali, C., Merenda, L. & Rizzo, V. (1977): Landslide morphometry and typology in two zones, Calabria, Italy. – International Association Engineering Geology, Bulletin, Vol. 16, p. 8- 13, Springer Berlin/Heidelberg

Chui, G. (2009): Seismology: Shaking up earthquake theory. – Nature, Vol. 481, p. 870- 872; San Francisco CA (online 14th Oct. 2009; doi:10.1038 / 461870a)

Costa, J.E. & Schuster R.L. (1988): The formation and failure of natural dams. – Geological Society of America Bulletin, Vol. 100, No. 7, p.1054-1068, Boulder CO

Cruden D.M., Varnes, D. J. (1996): Landslide types and processes; in: Turner A.K.; Shuster R.L. (eds) Landslides: Investigation and Mitigation.- Unites States Transport Research Board (TRB), Special Report, No. 247, pp 36–75, Washington DC

Dale, V.H., Swanson,F.J. & Crisafulli,C.M. (2005): Disturbance, survival, and succession: Understanding ecological responses to the 1980 eruption of Mount St. Helens, in: Dale, V.H.(ed): Ecological Responses to the 1980 Eruption of Mount St. Helens, Springer, New York NY

Dilley, M., Chen, R.S., Deichmann, U., Lerner-Lam, A.L., Arnold, M. with Agwe, J., Buys, P., Kjekstad, O., Lyon, B. & Yetman, G. (2005): Natural disaster hotspots - A Global Risk Analysis - Synthesis Report.- The International Bank for Reconstruction and Development / The World Bank and Columbia University, Washington, DC

Driedger, C., & Scott, W., (2008): Mount Rainier - Living safely with a volcano in your background. United States Geological Survey (USGS), Geological Survey Fact Sheet 2008-3062, Reston VA

Duncan, R.A. & Richards, M.A. (1991): Hotspots, mantle plums, flood basalts and true polar wander. – Review of Geophysics, Vol., 29, p.31-50, Wiley Online Library

EEA (2003): Mapping the impacts of recent natural disasters and technological accidents in Europe. – European Environmental Agency (EEA), Environmental Issue Report, No. 35, Copenhagen (online: http://europa.eu.int)

EERI (2014): Induced seismicity – Earthquake Engineering Research Institute, Oakland CA (online: www.eere.energy.gov)

Erdik, M, & Durukal, E. (2008): Earthquake risk and mitigation in Istanbul. - Natural Hazards, Vol. 44, p. 181-197, Springer Science&Business Media BV, Luxembourg

EU-EWFD (2000): Directive 2000/60/EC of the European Parliament and of the Council establishing a framework for the Community action in the field of water policy" (EU Water Framework Directive-EWFD).- European Commission, Brussels

FEMA (2014): Hazus-MH 2.1 - Multi-hazard Loss Estimation Methodology.- Federal Emergency Management Agency, United States Department of Homeland Security, Washington, DC (online: www.hazus.org)

Fitton, M. (1980): The Benue Trough and Cameroon Line - A migrating rift system in West Africa . –Earth and Planetary Science Letters, Vol. 51, Philadelphia PA

Friess, K.S. (2010): Konzept zur Bewertung lahargefärdeter Regionen zur Anwendung in Entwicklungsländern, basierend auf Erkenntnissen am Vulkan Mt. Rainer (USA). – Bachelor Thesis (un-published), University Goettingen, Faculty of Geosciences, Goettingen

Fritz, H.H., Mohammed, F. & Yoo, J. (2009): Lituya Bay - Landslide impact generated a Mega-Tsunami 50th anniversary . – Pure and Applied Science, Vol. 166, p. 153-166, Birkhäuser Verlag, Basel

Galahaut, K. & Galahaut, V.K. (2010): Effect of the Zipingpu reservoir impoundment on the occurrence of the 2008 Wenchuan earthquake and local seismicity. – Geophysical Journal International, Vol. 183, No. 1, p.277-285. Wiley Online Library (online: doi: 10.1111/j.1365-246X.2010.04715.x)

Geller, R.,J., Jackson, D.D., Kagan, Y.Y. & Mulargia, F. (1997): Earthquakes cannot be predicted . – Science, American Association for the Advancement of Science, Vol. 275, No. 5306, p.1616, Washington DC

Germanwatch(2014): Global Climate Risk Index 2014 - Who Suffers Most from Extreme Weather Events? Weather-Related Loss Events in 2012 and 1993 to 2012. - Germanwatch EV, Bonn (online: www.germanwatch.org/en/cri)

GFZ (2011): Earthquake Map of Germany, Austria and Switzerland (D-A-CH). - Potsdam Helmholtz Research Centre for Geosciences, GeoForschungszentrum, Annual Report 2011, Postdam.

Grove, R.H. (1998): Global Impact of the 1789–93 El Niño". – Nature, Vol, 393, p. 318–319. (online: doi:10.1038/30636)

Grünthal, G. ed. (1998): European Macroseismic scale 1998 (EMS 98).- Cahiers du Centre Europeen de Geodynamique et des Seismologie 15, Centre Europeen de Geodynamique et de Seismologie. Luxembourg, Vol. 99, pp.1998, Helfent-Betrange

GSNRW (1992): Erdbeben bei Roermond am 13. April 1992. - Geological Survey of North Rhine-Westphalia (GDNRW), Krefeld (online: www.gd.nrw.de/zip/l_yroer.pdf)

Guha-Sapir. D., Vos, F., Below, R. with Ponserre, S. (2011): Annual Disaster Statistical Review 2011: The Numbers and Trends. -The OFDA/CRED International Disaster Database, Brussels

Guha-Sapir, D., Below, R. & Hoyois, Ph. (2013): EM-DAT: International Disaster Database (2013) – The OFDA/CRED International Disaster Database, Brussels

Gupta, H. (2002): A review of recent studies of triggered earthquakes by artificial water reservoirs with special emphasis on earthquakes in Koyna, India. – Earth-Science Reviews, Vol. 58, Issues 3–4, p.279–310, Elsevier, Philadelphia PA

Gudmudsson, A. (2008): Magma-chamber geometry, fluid transport, local stresses and rock Behavior during collapse caldera formation.-Development of Volcanology, Vol. 10, Elsevier, Philadelphia PA

Hamberger, M. (2007): Rutschungserkennung mit Klassifikationssystemen am Beispiel Sachseln/Schweiz. – University Erlangen- Nürnberg, Naturwissenschaftliche Fakultäten, Online Ressource (PDF), Erlangen

Hancox, G. T. & Perrin, N. D. (1994): Green Lake landslide: a very large ancient rock slide in Fiordland. - Institute of Geological and Nuclear Sciences Limited (GNS), Science Report No. 93/18, Lower Hutt New Zealand

Hjartarson, et al. (2009): Central volcanoes as indicators for the spreading rate in Iceland.- International Association of Volcanology and Chemistry of the Earth's Interior (IAVCEI), Studies in Volcanology, No. 2

Holzer, T.L. (ed.) (1998): The Loma Prieta, California, Earthquake of October 17, 1989— Liquefaction. – United States Geological Survey (USGS) in cooperation with the National Science Foundation, Professional Paper 1551-B, Reston VA

Hong Yang, Adler, R. & Huffman, G. (2006): Evaluation of the potential of NASA multi-satellite precipitation analysis in global landslide hazard assessment.- American Geophysical Union, Geophysical Research Letters, Vol. 33, No. 22, Wiley Online Library

Hungr, O., Evans,S.G., Bovis, M.J. & Hutchison, J.N. (2001): A review of the classification of landslides of the flow type. - Association of Environmental and Engineering Geologist, American Geological Institute, Vol. 7, No. 3, Alexandria VA

Hutchinson, J. N. (1988): General Report: Morphological and geotechnical parameters of landslides in relation to geology and hydrogeology.- *Proceedings, Fifth International Symposium on* Landslides (Bonnard, C. (ed.), Vol. 1, Balkema, Rotterdam

IASPEI (2002): New manual of seismological observation practice (NMSOP). – International Association of Seismology and Physics of the Earth's Interior (IASPEI), New Manual, Vol. 1. – GeoForschungszentrum (GFZ), Potsdam

ICPR (2010): Our common objective: Living waters in the Rhine catchment. - International Commission for the Protection of the Rhine (ICPR), Koblenz

IFRC (2011a): Public awareness and public education for disaster risk reduction: A guide. - International Federation of Red Cross and Red Crescent Societies (IFRC), Geneva

IFRC (2011b): An Evaluation of the Haiti Earthquake 2010 Meeting Shelter Needs: Issues, Achievements and Constraints. – The International Federation of Red Cross and Red Crescent Societies (IFRC), Geneva (online: www.ifrc.org)

Ilkesik, M. (2002): Istanbul study - disaster prevention/mitigation basic plan in Istanbul Including seismic microzonation in the Republic of Turkey. Metropolitan Municipality Istanbul Earthquake Master Plan. – prepared for the Istanbul Metropolitan Municipality;

Japan International Cooperation Agency (JICA) and the Middle East Technical University, Istanbul, Technical University, Bosporus University and Yıldız Technical University, Istanbul
IPCC (2001): Climate Change 2001: - Third Assessment Report The Scientific Basis, Section 2, "Observed climate variability and change".- Special Report of the Intergovernmental Panel on Climate Change, Cambridge University Press, Cambridge MD
IPCC (2007): Climate Change 2007: Synthesis Report- Contribution of Working Groups I, II and III to the Fourth Assessment Report of the Intergovernmental Panel on Climate Change.- Intergovernmental Panel on Climate Change (IPCC), Cambridge University Press, Cambridge MD
IPCC (2012): Managing the Risks of Extreme Events and Disasters to Advance - Climate Change Adaptation. - Special Report of the Intergovernmental Panel on Climate Change, Cambridge University Press, Cambridge MD
IPCC-SREX (2011): Managing the Risks of Extreme Events and Disasters to Advance – Climate Change Adaptation. - Special Report of the Intergovernmental Panel on Climate Change, Summary for Policy Makers, Cambridge University Press, Cambridge MD
Iverson, R.M., Reid, M.E. & LaHusen, R.G., (1997): Debris-flow mobilization from landslides. – Annual Reviews of Earth and Planetary Sciences, Vol. 25, Palo Alto CA
Jakobsson, S.P. (2007): The Surtsey eruption. 1963-1967. – The Surtsey Research Society, Reykjavik
Jauhari, V.P. (1999): Options Assessment- Large Dams in India - Operation, Monitoring and Decommissioning of Dams.- The International Commission on Large Dams; Report prepared for Thematic Review IV.5 (online: www.dams.org)
Jentsch, V., Krantz, H. & Albeverio, S.(2006): Extreme events in nature and society. - Springer, Heidelberg
Jonas, M., Staeger, T,& Schoenwiese, C.D. (2005): Berechnung der Wahrscheinlichkeiten für das Eintreten von Extremereignissen durch Klimaänderungen – Schwerpunkt Deutschland. - Umweltbundesamt (UBA), Dessau
Kagan, Y.Y. (1997): Are earthquake predictable. - Geophysical Journal International, Vol. 131, No 3, p.505-525, Wiley Online Library
Kagan, Y. Y. & Jackson, D. (1991): The seismic gap theory: Ten years after. - Journal of Geophysical Research, Special Selection, Assessment of Schemes for Earthquake Prediction, Vol. 93, B 13 p.21,419-21,431, Wiley Online Library
Kraaijpoel,D., Dost, B., Sleeman,R. & Goutbeek, F. (2011): Location of induced earthquakes in the Netherlands gas field seismology research. – Royal Netherlands Meteorological Institute, Den Haag
Kowalik, Z., Knight, W., Logan,T. & Whitmore, P. (2005): Numerical Modelling of Global Tsunami – Indonesian Tsunami of 26 December 2004. – Science of Tsunami Hazards, Vol. 23
Kronholm, K. (2004): Spatial Variability of Snow Mechanical Properties with regard to Avalanche Formation. – University of Zurich (ETH), Dissertation, Zurich
Lang, K. (2002): Seismic vulnerability of existing buildings. - Institute of Structural Engineering, Swiss Federal Institute of Technology, Zurich (online: e-collection, ethz.ch)
Lockwood J.P. & Hazlett, R.W. (2010): Volcanoes – Global perspectives. – John Wiley & Sons, Intercorp. Blackwell, Hoboken NJ
McClung, D.M. & Schaerer, P. (1993): The Avalanche Handbook.- The Mountaineers, Seattle, Washington
Mader, C.L. (1974): Numerical simulation of Tsunamis. – Journal of Physical Oceanography, Vol. 4, p.74-82, Boston MA
Matsutomi, H., Shuto,N., Imamura, F. & Takahasi, T. (2001): Field survey of the 1996 Irian Jaya Earthquake Tsunami in Biak Island. - Natural Hazards, Vol. 24. p. 199-212, Springer Link
Mileti, D.S. (1989): Societal Response - The Loma Prieta Earthquake Professional Papers. – United States Geological Survey (USGS), Professional Papers 1553-A through 1553-D, Reston VA

Misereor (2010): Global aber gerecht – Klimawandel bekämpfen, Entwicklung ermöglichen.- Kurzfassung eines Reports des Potsdam –Instituts für Klimafolgenforschung (PIK) und Misereor, Potsdam (online: www.klima-und-gerechtigkeit.de)

Mool, P.K., Wangda, D., Bajracharya, S. R., Joshi, S. P., Kunzang, K. & Gurung, D. R. (2001): Inventory of Glaciers, Glacial Lakes and Glacial Lake Outburst Floods: Monitoring and Early Warning Systems in the Hindu Kush-Himalayan Region – Bhutan. - in: Reynolds, J.M. & Taylor, P.J. : *Mountain Research and Development, International Mountain Society,* Vol. 24, No. 3, p. 272-274, Bern

MunichRe (2005): Topics Geo Edition Wissen - Jahresrückblick Naturkatastrophen 2005. - Munich Reinsurance Company, Munich

MunichRe (2011): World Map of Natural Hazards. – Munich Reinsurance Company, Nathan, 2011-Version. Munich

MunichRe (2012): Topics Geo online – 2012. - Munich Reinsurance Company, Munich

MunichRe (2013): Topics Geo online - 2013. - Munich Reinsurance Company, Munich

Nadim,F., Kjekstad, O., Domaas,U., Rafat, R. & Peduzzi, P. (2006a): Global Landslides Risk Case Study; in: Arnold, M., Chen, R.S., Deichmann, U., Dilley, M., Lerner-Lam, A.L., Pullen, R. & Trohani, Z. (2006): Natural Disaster Hotspots - Case Studies, Disaster Risk Management Series, The World Bank Hazard Management Unit, Washington DC

Nadim, F., Kjekstad, O, Peduzzi,P., Herold,C. & Jaedicke, C. (2006b): Global landslide and avalanches hazard and risk hotspots.- Landslides, Vol. 3, No 2, p. 159-173, Springer Link

Newhall, C. G. & Punongbayan, R.S. (1996): Fire and mud: eruptions and lahars of Mount Pinatubo, Philippines. – Philippines Institutes of Volcanology (PHIVOLCS) and United States Geological Survey (USGS), University of Washington Press, Washington (online: http://pubs.usgs.gov/pinatubo)

Newhall, C.G. & Self, S. (1982): The volcanic explosive index (VEI): An estimate of explosive magnitude for historical volcanism. - Journal of Geophysical Research, Vol. 87, p.1231-1238, Wiley Online Library

Peduzzi,P., Dao, H., Herold,C. with contributions from A. Martin Diaz, F. Mouton, O. Nordbeck, D. Rochette, T. Ton-That, B. Widmer (2002): Global Risk and Vulnerability Index-Trends per Year (GRAVITY) - Phase II: Development, analysis and results. - The Bureau for Crisis Prevention and Recovery of the United Nation Development Program (UNDP), Geneva

Peduzzi, P., Dao, H., and Herold, C.: (2005): Mapping Disastrous Natural Hazards Using Global Datasets. – Natural Hazards, Vol. 35, No. 2, p.265–289, Springer Link

Peduzzi, P., Dao, H., Herold, C. & Mouton, F. (2009): Assessing global exposure and vulnerability towards natural hazards: the Disaster Risk Index. – Natural Hazards Earth Systems Sciences, Vol. 9 p.1149—1159, Copernicus Publications, Goettingen

Pereira, A.S. (2009): The Opportunity of a Disaster: The Economic Impact of the 1755 Lisbon Earthquake. - The Journal of Economic History, Vol. 69, No. 2, p.466-499, The Economic History Association, Tuscon AR

Pichler, H. (2006): Vulkangebiete der Erde.- Elsevier, Spektrum, Heidelberg

Pierson,T.C., Daag, A.S., Delos Reyes, J., Regalado,M.T.M., Solidum,R.U. & Tubianosa, B.S.(1996): Flow and Deposition of Post-eruption hot Lahars on the East Side of Mount Pinatubo, July-October 1991; in: Newhall, C. G. & Punongbayan, R.S.: Fire and mud: eruptions and lahars of Mount Pinatubo, Philippines. – Philippines Institutes of Volcanology (PHIVOLCS) and United States Geological Survey (USGS), University of Washington Press, Washington (online: http://pubs.usgs.gov/pinatubo)

Richter, C.F. (1935): *An instrumental earthquake magnitude scale.- Bulletin of the Seismological Society of America, Vol. 25, No. 1, p.1-32, Albany NY*

Robock, A. (2000): Volcanic eruption and climate. – American Geophysical Union, Reviews of Geophysics, Vol. 38, No. 2, p.193, Washington DC

Ruff, M. (2005): GIS-gestützte Risikoanalyse für Rutschungen und Felsstürze in den Ostalpen (Vorarlberg, Österreich). – Georisikokarte Vorarlberg, Universitätsverlag Karlsruhe, Karlsruhe

Sager, J.W. & Chambers, D.R. (1986): Design and construction of the Spirit Lake outlet tunnel Mt. St. Helens, Washington. – Landslide Dams -Process, Risk and Mitigation; in: Schuster, R.L. (ed): Geotechnical Special Publication, No. 3, American Society of Civil Engineers, New York NY

Scarpa, R. & Tilling, R. I. (eds.) (1996): Monitoring and Mitigation of Volcanic Hazards. – p. 841, Springer Verlag Berlin/Heidelberg

Schmidt, K.M., Roering, J.J., Stock, J.D., Dietrich, W.E., Montgomery, D.R. & Schuab,T. (2001): The variability of root cohesion as an influence on shallow landslide susceptibility in the Oregon Coast Range: Canadian Geotechnical Journal, Vol. 38, Ottawa ON

Schmincke, H-U (2000): Volcanoes. – Springer, Berlin/Heidelberg

Schweizer, J. & Jamieson, J.B. (2003): Snow stability measurements.- Proceedings International Seminar on Snow and Avalanche Test Sites (22-23 Nov. 2001), Grenoble

Segall,P., Mark, P.I., Murray H., & Kenner, S. (2007): Crustal deformation in the San Francisco Bay Area. – Final Technical Report, Earthquake Hazards: External Research, Report 00-HQ-GR-0039, p.14, Stanford University, Stanford CA

Simkin, T., Siebert, L., McClelland, D., Bridge, C., Newhall, C. & Latter, J.H. (1981): Volcanoes of the Earth – A regional directory, gazetteer and chronology of volcanism during the last 10 000 years. – Hutchinson Ross, Stroudsburg

Small, C. & Naumann, T. (2001): Holocene volcanism and the global distribution of human population.- Environmental Hazards, Vol. 3, p.93-109, Taylor Francis Online

Soloviev, S.L., Solovieva, O.N., Chan N. Go, Khen S. Kim, & Shchetnikov, N.A. (2000): Tsunamis in the Mediterranean Sea 2000 B.C.-2000 A.D. - Kluwer Academic Publishers, Dordrecht

Sornette (2002): Predictability of catastrophic events: Material rupture, earthquakes, turbulence, financial crashes, and human birth. - Proceedings of the National Academy of Sciences of the United States of America (online access: 29th July 2014)

Steward, Th. & Bostrom, A. (2002): Extreme event - Decision making. – Workshop report, Center for Policy Research, Rockefeller College of Public Affairs and Policy, University of Albany, Albany NY

Stommel, H, & Stommel, E. (1983): The story of the year without a summer. – Volcano Weather, Seven Seas Press, Newport RI

Swiss Re (2010): Nature and Man-made catastrophes 2010. - Sigma, Nr. 1/2011, Swiss Reinsurance Company, Zurich (online: www.swissre.com/sigma)

Tang, C., Zhu, J., Qi, X. & Ding, J. (2011): Landslides induced by the Wenchuan earthquake and the subsequent strong rainfall event: A case study in the Beichuan area of China.- Engineering Geology, Elsevier B.V. (online: doi:10.1016/j.enggeo.2011.03.013)

Tilling, R.I. (1989): Volcanic hazards and their mitigation: progress and problems. – American Geophysical Union, Reviews of Geophysics, Vol. 27, p.237-269, Wiley Online Library

Tschirky, F., Brabec, B. and Kern, M. (2000): Lawinenunfälle in den Schweizer Alpen - Eine statistische Zusammenstellung mit den Schwerpunkten Verschüttung, Rettungsmethoden und Rettungsgeräte; in: Ammann, W.J. (ed): Durch Lawinen verursachte Unfälle im Gebiet der Schweizer Alpen, Eidgenössisches Institut für Schnee- und Lawinenforschung (SLF), Sonderdruck, Davos

Tyagunov, S., Grünthal, G., Wahström. R. Stempniewski, L. & Zschau, J. (2006): Seismic risk mapping for Germany. – Natural Hazards Earth Systems Sciences, Vol. 6, No. 4, p.573-586, Copernicus Publications, Goettingen (online: doi:10.5194/nhess-6-573-2006)

UNDP (2007): Global Risk Identification Program. – United Nations Development Program, Bureau of Crisis Prevention and Recovery, United Nations, Geneva (online:/content/undp/en/home/ourwork/crisispreventionandrecovery/projects_initiatives/global_risk_identificationprogramme.html)

UNEG (2010): Haiti Earthquake Response, Context Analysis, July 2010.- United Nations Evaluation Group (ALNAP), Secretariat, London

UNFCCC (2012): Climate-related risks that are most relevant to African context. – UNFCCC-expert meeting on a range of approaches to address loss and damages associated with the adverse effects of climate change, including impacts related to extreme weather and slow onset events, The IPCC Special Report on Managing the Risks of Extreme Events and

Disasters to Advance Climate Change Adaptation, Intergovernmental Panel on Climate Change (IPCC), 13th June 2012, Addis Ababa
UNISDR (2004): Living with Risk - A Global Review of Disaster Reduction Initiatives". – United Nation International Strategy of Disaster Reduction (UNISDR), Geneva
UNISDR (2007): Global risk reduction: 2007 Global Review. - United Nations International Strategy for Disaster Risk Reduction (ISDR), ISDR/GP/2007/3, United Nations, Geneva
UNISDR (2009a): Global assessment report on disaster risk reduction – Risk and poverty in a changing climate.- United Nations International Strategy for Disaster Risk Reduction (ISDR), United Nations, Geneva
UNISDR (2009b): 2009 UNISDR terminology on disaster risk reduction. - International Strategy for Disaster Reduction (ISDR), United Nations, Geneva
UNISDR (2010): Statement of Commitment by the Private Sector for Disaster Prevention, Resilience and Risk Reduction. - United Nations International Strategy for Disaster Risk Reduction (ISDR), United Nations, Geneva
UNISDR (2011): Global Assessment Report on Disaster Risk Reduction: Revealing risk, redefining development. – United Nation International Strategy of Disaster Reduction (UNISDR), United Nations, Geneva
UNISDR (2013): Global Assessment Report on Disaster Risk Reduction 2013 - From Shared Risk to Shared Value: the Business Case for Disaster Risk Reduction.- United Nations International Strategy for Disaster Risk Reduction (UNISDR), United Nations, Geneva
UNU-EHS (2012): WeltRisikoBericht 2012. – United Nations University (UNU-EHS), Bonn
USGS (1960): Giant Waves in Lituya Bay, Alaska.- United States Geological Survey (USGS), Professional Paper 354-C, Reston VA
USGS (1981): Landslides from the February 4, 1976, Guatemala earthquake. – United States Geological Survey (USGS), Professional Paper 1204 A, Reston VA
USGS (1993): Seismicity of the United States, 1568-1989 (Revised). – United States Geological Survey (USGS), Professional Paper 1527, Reston VA
USGS (1995a): Earthquake prediction. - United States Geological Survey (USGS), Fact Sheet 096-03, Reston VA (online: pubs.usgs.gov/gip/earthq1/predict.html)
USGS (1995b): Volcanic gas.- United States Geological Survey (USGS), Volcano Hazards Fact Sheets, Open File Report 95-85, Reston VA
USGS (2000): Mount Rainier – Living safely with a volcano in your backyard. – United States Geological Survey (USGS), Fact Sheet 2008-3062, Reston VA
USGS (2001): Landslides triggered by Hurricane Mitch in Guatemala - Inventory and Discussion. - United States Geological Survey (USGS), Instituto Nacional de Sismologia, Vulcanologia, Meteorologia e Hidrologia, Guatemala City, USGS-Open-File Report 01-443, Reston VA
USGS (2005a): An Assessment of Volcanic Threat and Monitoring Capabilities in the United States: Framework for a National Volcano Early Warning System (NVEWS). – United States Geological Survey (USGS), Open-File Report, 2005-1164, Reston VA
USGS (2005b): Mount St. Helens – From the 1980 Eruption to 2000.- United States Geological Survey (USGS), Fact Sheet 036-00, Reston VA (online Version 1.0)
USGS (2005c): Debris-Flow Warning System-Final Report. – National Oceanic and Atmospheric Administration (NOAA), National Weather Service (NWS) and the United States Geological Survey (USGS), Circular 1283, Reston VA
USGS (2008a): The Dynamic Earth - The story of plate tectonics. – United States Geological Survey (USGS); Kious, W.J. & Tilling R.I. (eds.) (online: http://pubs.usgs.gov/gip/dynamic /dymanamic.html)
USGS (2008b): What are volcano hazards. – United States Geological Survey (USGS), Fact Sheet 002-97, revised March 2008, Reston VA
USGS (2008c): The Landslide Handbook—A Guide to Understanding Landslides. - United States Geological Survey (USGS), Circular 1325, Reston VA
USGS (2009a): Surviving a Tsunami – Lessons from Chile, Hawaii, Japan. – United States Geological Survey (USGS), Circular 1187, Reston VA

USGS (2009b): Community Exposure to Lahar Hazards from Mount Rainier, Washington. - United States Geological Survey (USGS), Scientific Investigations Report 2009-5211, Reston VA
USGS (2013): Induced earthquakes. – United States Geological Survey (USGS), Reston VA (online: http://earthquake.usgs.gov/regional/nca/seminars/2013-12-02)
USGS (2014): 2014 Seismic Hazard Maps. - United States Geological Survey (USGS), Earthquake Hazard Program, Open File Report 2014-1091, Reston VA
Vander Linden, P. & Mitchel, J.F.B. (eds.)(2009): ENSEMBLES: Climate change and its impacts- Summary of research and results from the ENSEMBLES project. – Meteorological Office Hadley Centre, Exeter
Varnes, D. J. (1978): Slope movement types and processes. - in: Schuster, R. L. & Krizek, R. J. (eds): Landslide*s: Analysis and Control.* -Transportation and Road Research Board (TRB), *Special Report 176,* National Academy of Science, Washington DC
Verbeek, R.D.M. (1885): Krakatao.- Imprimerie de l'Etat de Batavia. p.495
Voight, B. (1986): How volcanoes work - The Nevada del Ruiz eruption. The San Diego States University, San Diego CA (online: www.geology.sdsu.edu)
Walker, A.B., Redmayne, D.W. & Browitt, C.W.A. (1992): Seismic monitoring of Lake Nyos, Cameroon, following the gas release disaster of August 1986; in: McCall, G.J.H., Laming, D.J.C.,& Scott, S-C. (eds): Geohazards, Springer Link (online: 978-94-009-0381-4)
Wegener, A. (1929): Die Entstehung der Kontinente und Ozeane. - Vierte umgearbeitete Auflage, Braunschweig
WHO (2008): Improving Public Health Responses to extreme weather/ heat-waves – EuroHEAT. – World Health Organization (WHO), Meeting Report 22-23 March 2007, Bonn
Winchester, S. (2003): Krakatao – The day the World exploded, August 27, 1883. – Harper & Collins Publishers
Witham,C.S. (2005): Volcanic disasters and incidents: A new database. - Journal Volcanology and Geothermal Research, Vol. 148, p.191-233, Elsevier B.V., Rotterdam
WMO(2010): Unprecedented sequence of extreme weather events. -World Meteorological Organization (WMO) and United Nations International Strategy for Disaster Risk Reduction (ISDR), Geneva, (online: PreventionWeb, access: 20th July2014)

Chapter 4
Institutional and Organizational Framework for a Disaster Risk Management

4.1 General Aspects

There are a multitude of organizations and mechanisms worldwide targeted at social and human development. The key word for all the different approaches is "sustainability." Although there are many different concepts, strategies, and instruments in practice in development policy, they all have the same aim, "to reach sustainable human development." Berkes et al. (2000) emphasize that for reaching sustainability a holistic approach is necessary that integrates the social and natural components by dealing with their interrelationships. They report on the "Vanua Concept" of the Fiji Islands where in the traditional belief, land, water, and the human being are regarded as an entity. Thus far they state, resource management is purely defined from technical, scientific, and economic viewpoints. They pledge to tear down the arbitrary and artificial delineation between the two systems (social and ecological) and call for a comprehensive incorporation of the system into one systematic approach. Such approaches were practiced for centuries by many traditional societies of the developing world (Ostrom 1990).

Today such problems even have accelerated: globalization of resource markets, depletion of natural resources, financial crisis, environmental degradation, and the changing climate all make human beings on Earth more and more vulnerable. The same holds true for an increase in natural hazard exposure that in many events turns into serious risk to societies. Through disaster risk reduction activities many international institutions mandated with development seek good opportunities to protect livelihoods from shocks by making societies more resilient and more capable of absorbing the impact of, and recovering from, disruptive events. Furthermore, disaster risk reduction creates a multiplier effect that accelerates the achievement of the Millennium Development Goals reflecting the Hyogo Framework of Action and makes disaster risk reduction an integral part of the international efforts to eradicate poverty.

Making a society resilient against the impact of natural disasters requires a conducive legal and operational framework that defines the overall as well as the planning and operational procedures. In the first order, the well-being of a society

© Springer International Publishing Switzerland 2016

U. Ranke, *Natural Disaster Risk Management*,
DOI 10.1007/978-3-319-20675-2_4

is the responsibility of every nation as laid down in the Charter of the United Nations. This principle meanwhile has been incorporated into all basic laws worldwide. In the United States the Natural Hazard Mitigation policy is laid down in the Stafford Act Disaster Relief and Emergency Assistance ACT ("Stafford Act") of 1974 (FEMA 2007). The act is to provide an orderly and continuing federal assistance to state and local governments in carrying out their responsibilities to alleviate the suffering and damage caused by disasters. The act defines the two strategic targets:

- To substantially increase public awareness of risk from natural hazards so that the public demands safer communities in which to live and work
- To significantly reduce the risk of life, injuries, economic costs, and destruction of natural and cultural resources that result from natural hazards

To achieve this and to guide the states and local governments in mitigation planning and implementation 10 principles among others are summarized:

- Proactive measures can reduce disaster costs and impacts.
- Hazard identification and risk assessment are cornerstones of mitigation.
- Risk reduction comprises preventive and corrective measures.
- Risk reduction measures for one natural hazard must be compatible with risk reduction measures for other natural or technical hazards and with the goals of protection of the natural and cultural environment.
- All mitigation is local.
- Those who knowingly choose to settle in hazard exposed areas must accept responsibility for that choice.

According to the Stafford Act, the national government can render financial and technical assistance after a request for a "presidential emergency declaration" by a local movement, when it sees its reaction capacity overstressed. Section 409 of the act deals specifically with the responsibilities of disaster risk assessment and mitigation. According to 409 a hazard mitigation plan is declared the cornerstone of any mitigation measure. In such a plan, the natural conditions and the social vulnerabilities of a hazardous region have to be summarized and evaluated. At a minimum a hazard mitigation plan shall comprise:

- An evaluation of the types of natural hazards
- A description and analysis of the local hazard management programs as well as its local reduction capabilities
- An assessment of the local mitigation objectives, their respective implementation, monitoring, and evaluation of the reduction achievements

The following compilation names the most important organizations and mechanisms targeted at social and human development, those acting according to international agreements and UN mandates making them world leaders in social, economic, and environmental development. Not only organizations are listed, but also the instruments and mechanisms that are prerequisites to implement the political strategies and concepts. Nevertheless such a compilation is somewhat arbitrary and acknowledges that there are quite a number of international, especially nongovernmental, organizations that contribute significantly to human development.

4.2 The United Nations System

With the report "Our Common Future" a new era on human development was initiated. The main outcome of the report was the concept of "sustainable development" that in the aftermath became one of the most successful approaches in international development policy. The concept provided the basis for an international agenda and the attitude towards economic, social, and environmental development. The report was published in October 1987 by the Brundtland Commission (UN 1987), chaired by Gro Harlem Brundtland, the former Norwegian prime minister. The former general of the United Nations, Javier Perez de Cuellar, commissioned the World Commission on Environment and Development (WCED) as a reaction to the first serious signs of heavy deterioration of the human environment and natural resources especially in the developing world. Although people in the developed countries were starting to become more aware of environmental issues stemming from industrialization and growth, developing countries were discouraged because they were not substantially able to reach the higher level of economic growth. As a reaction to this need for growth, developing countries were desperate to use cheap methods in agricultural and industrial production with negative environmental and social impacts and often unethical labor practices. The United Nations saw a growing need to address these environmental challenges and their interrelationship with economic and social conditions.

The Brundtland Reoport "Our Common Future" was the first in history to define the term "sustainable development" as a "Development which meets the needs of current generations without Compromising the ability of future generations to meet their own needs."

The Brundtland Report was the first to officially identify that the majority of the global environmental problems result from both the south's enormous poverty and the north's un-sustainable consumption and production. The report underlined that the complex issue of environmental deterioration should be integrated with human development policy. Moreover it clearly suggested that poverty eradication and environmental conservation will only be solved simultaneously and in a mutual way. The idea of sustainable development designs an attempt to connect environment with development and thus called for a combined strategy that united these two. The report was radical in stating that ecological sustainability cannot be achieved if the problem of poverty is not successfully addressed globally.

It was quite enviable and without doubt one of the objectives, that such a "revolutionary" approach should create an intensive debate. Many argued that the Brundtland Report is a regressive document which strongly supports the traditional belief that growth and affluence are necessary to solve problems related to the environment. Another critical objection was that the report addressed the problems mainly from the viewpoint of the environment, leaving aside the many socioeconomic causes that influence environmental degradation. The most serious objection was that it failed to identify the root causes of the problems and that it favors a continuation of the same developments which were seen by many critics as fundamental to the problems.

Following publication of the Brundtland Report, numerous attempts were made to operationalize sustainable development. The most popular and common attempt is the triangular concept with the three pillars of economy, environment, and society, which was agreed upon at the United Nations summit in Johannesburg in 2005 and later named the P3 concept of "people, planet, and prosperity." And the report paved the way for a multitude of initiatives to introduce the sustainable development concept in national and local policies. The Rio Earth Summit gave this a real boost and afterwards local Agenda 21 documents and action plans were drawn up in a great number of municipalities. Furthermore the United Nations Commission for Sustainable Development started to scrutinize the implementation of the Rio decisions at its annual meetings.

Despite the many objections, the Brundtland Report created a broad and intensive dialogue on human development and natural conservation. The concept has been seen as inclusive and operational enough to make meaningful action in pursuit of sustainable development possible and broadly supported and therefore the international community has continued using it.

4.2.1 United Nations Millennium Development Goals (MDGs)

Eight development goals were internationally agreed upon in September 2000, when 189 world leaders and representatives of more than 20 international aid organizations met at the United Nations after a two-year consultation process, to agree to free more than a billion people from extreme poverty by year 2015. On the occasion of this meeting, called the Millennium Summit, the United Nations Millennium Declaration (UN 2000) presented a roadmap how to achieve these targets, known as the Millennium Development Goals (MDGs).

The goals are:

- Eradicating extreme poverty and hunger
- Achieving universal primary education
- Promoting gender equality and empowering women
- Reducing child mortality rates
- Improving maternal health
- Combating HIV/AIDS, malaria, and other diseases
- Ensuring environmental sustainability
- Developing a global partnership for development

Each of the goals has been given specific targets and dates and indicators for achieving them.

The aim of the MDGs is to encourage development by improving social and economic conditions in the world's poorest countries by addressing issues as poverty eradication, environmental protection, human rights, and protection of vulnerable societies. The MDGs assert that every individual has the right to dignity, freedom, equality, and a basic standard of living that includes freedom from hunger and violence, and encourages tolerance and solidarity. The MDGs were made to operationalize these ideas by setting targets and indicators for poverty reduction in order to achieve the rights set forth in the Declaration on a set 15-year timeline from 2000 to 2015.

To accelerate progress, the G8 Finance Ministers agreed at the Summit to allocate the necessary funds to the World Bank, the International Monetary Fund (IMF), and to the African Development Bank (AfDB). Furthermore the OECD countries abstained from remittance of about US$50 billion outstanding debts owed by the heavily indebted poor countries (HIPC)—the group of the most impoverished countries of the world—when used for national social programs for improving health and education and for alleviating poverty. The MDGSs reiterated the so-called 0.7 % target that the United Nations had already set in 1970 for the OECD countries, to finance development assistance (Official Development Assistance, ODA, of the gross national income, GNI). Although most of the countries (except the northern European countries) failed the 0.7 % target, the total amount of ODA assistance nevertheless summed up to more than US$130 billion. A sum that nevertheless must be put into a relationship of international financial market activities: for example, the investment by the private sector in the developing countries reached more than US$350 billlion at that time and official funding for restructuring East Germany was more than US$130 billion per year, over a time span of more than 20 years.

The United Nations stated that since then the Millennium Development Goals have become the most successful global antipoverty push in history. Significant and substantial progress has been made in meeting many of the eight targets, including halving the number of people living in extreme poverty and the proportion of people without sustainable access to improved sources of drinking water. Remarkable gains have been made in the fight against malaria and tuberculosis. There have been visible improvements in all health areas as well as primary education.

UN Secretary General Ban Ki-moon said, "In more than a decade of experience in working towards the MDGs, we have learned that focused global development efforts can make a difference."

He pointed out that:

- The proportion of people living in extreme poverty has been halved at the global level.
- Over two billion people gained access to improved sources of drinking water.
- Remarkable gains have been made in the fight against malaria and tuberculosis.
- The hunger target is within reach.

But also that there is still much to do inasmuch as:

- Environmental sustainability is under severe threat.
- Progress on maternal deaths reduction is falling short.
- Access to antiretroviral therapy and knowledge about HIV prevention must expand.
- Too many children are still denied primary education.
- There is less aid money overall, with the poorest countries most adversely affected.

For the purpose of "reduction of hazard exposure and achieving a higher level of resilience form natural disasters," the MDG target No. 7 is most relevant. MDG (7) concerns "ensuring the environmental sustainability" and as can be shown in this book, the population of developing countries is at highest risk, especially from climate-related disasters: floods, droughts, and epidemics. Therefore any reduction of vulnerability, risk prevention, and preparedness and increasing the coping capacity on natural disasters at every level of a society, will contribute not only to the MDG 7. As all the MDGs interact and have a high interdependency, risk management will also help to achieve the MDG goals of improving social and economic conditions in the world's poorest countries.

4.2.2 United Nations (UNISDR)

The UN General Assembly adopted the International Strategy for Disaster Reduction in December 1999 and established the UNISDR-secretariat to ensure its implementation. The International Strategy for Disaster Reduction builds upon the experience of the International Decade for Natural Disaster Reduction (1990–1999) (UNIDNDR 2000). Since then UNISDR has been the United Nations central office for disaster risk reduction and thus serves as the focal point in the UN system for the coordination and implementation of international disaster risk reduction activities. The United Nations emphasizes that the national governments have the primary responsibility for protecting their citizens from risks and disaster, and moreover stated that local communities and elements of civil society are identified as the key initiators of disaster prevention actions. Through an early and comprehensive partnership of the decision-making levels with the population at risk the necessary encouragement and support to realize the vision of disaster resilience can be achieved. For regional/subregional and international collaboration is essential, especially with regard to the dissemination of experience and information, scientific and technical applications, continual advocacy, and the coordination of strategies to assist in the development of national capabilities. The United Nations system has been mandated by the international community a special leadership role in global risk and disaster reduction and to serve as a forum for global dialogue.

The vision of UNISDR was given by Kofi Annan, former UN Secretary General: "We must, above all, shift from a culture of reaction to a culture of prevention. Prevention is not only more humane than cure; it is also much cheaper. … Above all, let us not forget that disaster prevention is a moral imperative, no less than reducing the risks of war."

UNISDR aims at enabling all communities of the world, but especially those who lack capacities to work out their own measures making them capable to withstand natural, technological, and environmental hazards. An effective means is seen by UNISDR in the integration of risk prevention into sustainable development. UNISDR has been mandated with a series of goals:

- Increase public awareness on hazards that pose a risk to modern societies.
- Obtain commitment by public authorities to reduce risks to people, their livelihoods, social and economic infrastructure, and environmental resources.
- Increase public participation at all levels of implementation.
- Reduce the economic and social losses of disasters.
- Form a global community dedicated to making risk and disaster prevention a public value.

In order to achieve these goals UNISDR is engaged in the following sectors:

- Stimulate research and application, provide knowledge, convey experience, build capabilities on natural hazards and disasters.
- Allocate necessary resources for reducing or preventing impacts of hazards.
- Extend the operational capacities of the science and technology sector to provide more information to the public decision-making process.
- Develop an interface between the disaster risk management sector and the risk reduction practitioners.
- Link risk prevention with economic competitiveness to enhance economic partnership.
- Carry out and integrate risk assessments in development plans.
- Develop and apply risk reduction strategies and mitigation measures at all societal levels.
- Establish risk monitoring capabilities, and early warning systems as integrated processes.
- Develop and institutionalize public information and educational components for all ages.
- Establish internationally agreed-upon standards/methodologies for the analysis and expression of the socioeconomic impacts of disasters on societies.

UNISDR articulated its principles in a number of major documents in particular, the following.

4.2.3 The Yokohama Strategy

The "Yokohama Strategy for a Safer World" (UNIDNDR 1994) provided the first internationally agreed-upon guideline on how to take action for a worldwide reduction of impacts from natural disasters. The Yokohama Strategy identified major challenges for ensuring systematic action to address disaster risks in the context of sustainable development and in building resilience through enhanced national and local capabilities to manage and reduce risk.

The review stresses the importance of disaster risk reduction being underpinned by a more pro-active approach to informing, motivating, and involving people in all aspects of disaster risk reduction in their own local communities. It also highlights the scarcity of resources allocated specifically from development budgets for the realization of risk reduction objectives, either at the national or the regional level or through international cooperation and financial mechanisms, while noting the significant potential to exploit existing resources and established practices better for more effective disaster risk reduction.

Specific gaps and challenges are identified in the following five main areas:

- Governance: organizational, legal, and policy frameworks
- Risk identification, assessment, monitoring, and early warning
- Knowledge management and education
- Reducing underlying risk factors
- Preparedness for effective response and recovery

4.2.4 Hyogo Framework of Action

The World Conference on Disaster Reduction was held from January 18th to 22nd, 2005 in Kobe, Hyogo, Japan (UNISDR 2005) and unanimously adopted the Hyogo Framework for Action, "Building the Resilience of Nations and Communities to Disasters." The conference was the logical consequence of the UNIDNDR 1990–1999 decade and (eventually) was dated just one month after the disastrous tsunami event of the Indian Ocean. The conference for the first time formulated an international accord for the need for worldwide actions to promote a strategic and systematic approach to reducing vulnerabilities and risks to hazards. It underscored the need for, and identified ways of, building the resilience of nations and communities to disasters. With the Hyogo Framework of Action the international community pointed to the fact that disaster losses are on the rise with serious consequences for the survival, dignity, and livelihood of individuals, particularly the poor. In the past two decades on average more than 200 million people have been affected every year by disasters. Moreover the Action reiterated that disaster risk is becoming a global concern and its impact and actions in one region can have an impact on risks in another. An increase in vulnerability related to changing demographic, technological, and socioeconomic conditions; unplanned urbanization of high-risk zones; environmental degradation; and climate change leads to geological hazards threatening the world's economy and especially the sustainable development of developing countries. The Action Plan is seen as a milestone for the achievement of the Millennium Development Goals.

It further reiterates that the targets envisaged will only be achieved by a systematic integration of disaster prevention and emergency management into national and international policies, plans, and programs. The way to achieve this is seen in a more effective integration of disaster risk management into sustainable development policies, planning, and programming at all levels, with a special emphasis on disaster prevention, mitigation, and preparedness and vulnerability reduction. Moreover institutions, mitigation mechanisms, and capacities at all levels, in particular at the community level, have to be developed and further strengthened. Only by a systematic integration of hazard risk mitigation approaches into design and implementation of emergency preparedness, will response and recovery program sustainable resilience be achieved.

The main objectives of the Hyogo Framework of Action for the decade 2005–2015 are:

- Review the Yokohama Strategy and where found necessary update the guiding framework on disaster reduction for the twenty-first century.
- Share good practices and lessons learned to further disaster reduction within the context of attaining sustainable development, and to identify gaps and challenges.
- Increase awareness of the importance of disaster reduction policies and promote the implementation of those policies.
- Increase the reliability and availability of appropriate disaster-related information to the public and disaster management agencies in all regions, as set out in relevant provisions of the Johannesburg Plan of Implementation.
- Identify activities aimed at ensuring the implementation of relevant provisions of the Johannesburg Plan of Implementation (World Summit on Sustainable Development, Rio de Janeiro 1992) on vulnerability, risk assessment, and disaster management.

With the Action Plan the following outcomes are expected.

- To reach a substantial reduction of disaster losses, in lives and in the social, economic, and environmental assets of communities and countries.

The Action Plan acknowledged that the realization of this outcome will require the full commitment and involvement of all actors concerned, including governments, regional and international organizations, civil society, the private sector, and the scientific community.

4.2.5 United Nations Office of Humanitarian Affairs (UNOCHA)

In deep concern about the suffering of the victims of disasters, the loss in human lives, the flow of refugees, the mass displacement of people, and the material destruction, and in review of the capacity, experience, and coordination arrangements in the United Nations system for humanitarian assistance, the United Nations adopted the resolution (UN 1991) for the strengthening of the coordination of emergency humanitarian assistance of the UN system.

There is a clear relationship among emergency, rehabilitation, and development. In order to ensure a smooth transition from relief and rehabilitation to a sustainable development, emergency assistance should be provided to support recovery and long-term development. The United Nations has been given a central and unique role in providing leadership and coordinating the efforts to support disaster-affected countries. As called for in the International Decade for Natural Disaster Reduction (IDNDR), efforts should be intensified to develop measures for prevention and mitigation of natural disasters and similar emergencies through technical assistance. The United Nations calls for identification of technologies and management strategies for disaster relief and prevention as a prerequisite for an increase in disaster resilience as well as worldwide information dissemination. Moreover, a comprehensive exchange should be initiated of existing and new technical information related to the assessment, prediction, and mitigation of disasters.

The guiding principles are:

- Humanitarian assistance is of cardinal importance for the victims of natural disasters and other emergencies.
- Each state has the responsibility first and foremost to take care of the victims of natural disasters and other emergencies occurring on its territory.
- Humanitarian assistance should be provided in principle on the basis of an appeal by the affected country, and will be given as supplementary aid.
- Humanitarian assistance will be provided in accordance with the principles of humanity, neutrality, and impartiality as laid down in the Charter of the United Nations.

The United Nations sees economic growth and sustainable development as essential elements for a sustainable resilience from natural and other disasters. The international community therefore offers assistance to (especially) developing countries in strengthening their capacity in disaster prevention and mitigation, both at the national and regional levels. The United Nations reiterates disaster relief, prevention, and preparedness as the main elements of disaster and emergency management:

- To assist developing countries to strengthen their capacity to respond to disasters at the national and local levels by improving the capacities to mitigate the effects of natural disasters and to cope efficiently with all emergencies
- To increase awareness of the need for establishing disaster mitigation strategies
- To provide assistance relevant to all sectors of prevention and preparedness
- To increase the capacity of disaster-prone countries to receive and make use of this information as well as their operational capacities for rapid and coordinated response to emergencies
- To develop emergency management procedures within the disaster-prone countries to expedite the rapid procurement and deployment of equipment and relief supplies

By the Office of the United Nations Disaster Relief Coordinator (UNOCHA) and the United Nations Development Program (UNDP) all bodies within the United Nations system (e.g., the International Committee of the Red Cross, the League of Red Cross and Red Crescent Societies, and the International Organization for Migration) are asked to assist UNOCHA in building up a capable unit to analyze the regional natural hazards exposure, disasters, and other emergencies of developing countries. The findings should be disseminated freely to the countries at risk. The international community is urged to provide the necessary support and resources to programs and activities undertaken to further the goals and objectives of the Decade and to assist these countries upon request with the establishment and enhancement of national early-warning systems. Organizations and entities of the United Nations system should continue to respond to requests for emergency assistance within their respective mandates.

The secretary-general should establish under his authority a central emergency revolving fund as a cash-flow mechanism to ensure the rapid and coordinated response of the organizations of the system. This fund should be put into operation with an amount of US$50 million. The fund should be financed by voluntary contributions of the member states. The United Nations should further build up a register of available experts and technical capacities within the different UN organizations.

The UN resolution made the United Nations the only legal authority mandated to pursue contracts with interested governments and intergovernmental and nongovernmental organizations to enable them to have better access to the international emergency and relief capacities. The Office for the Coordination of Humanitarian Affairs was given the mandate including the coordination of humanitarian response, policy development, and humanitarian advocacy. OCHA thus carries out a coordination function primarily through the United Nations Emergency Relief Coordinator (ERC). The ERC's function is seen as most critical in order to ensure comprehensive preparation for, as well as rapid and coherent response to, natural disasters and other emergencies. This ERC was given these responsibilities:

- Processing requests from affected member states for emergency assistance
- Organizing, in consultation with the government of the affected country, a joint interagency needs-assessment mission and preparing a consolidated appeal to be issued by the secretary-general
- Facilitating the access to emergency areas for the provision of emergency assistance
- Managing the central emergency revolving fund and assisting in the mobilization of financial resources
- Serving as a focal point with governments and intergovernmental and nongovernmental organizations to emergency relief operations
- Providing consolidated information, including early warning on emergencies to all interested governments and concerned authorities
- Preparing an annual report on the coordination of humanitarian emergency assistance, including information on the central emergency revolving fund, to be submitted to the General Assembly

At the country level resident coordinators should organize and facilitate the humanitarian assistance of the UN system. The local resident is the central UN authority to advise local authorities in a speedy implementation of the relief operations and should facilitate the transition from relief to development. The resident coordinator should chair the emergency operations under UN auspices and maintain close contact with the national authorities.

4.2.6 Comprehensive Nuclear Test Ban Treaty (CTBT)

Official data of the United Nations Preparatory Commission for the Comprehensive Nuclear Test Ban Treaty Organization (CTBT 2103) stated that more than 2000 nuclear tests have been carried out worldwide since 1945. In the aftermath of World War II, the increased radioactive fallout from atmospheric nuclear tests made the world aware that there is a need for a binding contract that bans all nuclear explosions. One of the first to call for such a treaty was Indian Prime Minister Nehru. His initiative was quickly joined by many other international politicians and scientists and finally after lengthy discussions, the UN General Assembly in 1966 adopted a treaty called "Comprehensive Nuclear Test Ban Treaty" (CTBT). In the treaty the nations agree "not to carry out any nuclear weapon test explosion or any other nuclear explosion, and prohibit and prevent any such nuclear explosion at any place under its jurisdiction or control" either for military or civilian purposes. The ultimate goal is to strengthen the international "nuclear disarmament efforts and the nuclear non-proliferation regime by a total elimination of nuclear weapons and by constraining their development and qualitative improvement." The seat of the organization was decided to be Vienna (Austria). The treaty was ratified as of March 2014 by 162 states, including five of the eight "nuclear-capable states"; another 23 of the so-called Annex 2 states have signed but not ratified (China, Egypt, Iran, Israel, and the United States), and India, North Korea, and Pakistan have not signed it. The CTBT will enter into force 180 days after the treaty has been ratified by all 44 Annex 2 states. Annex 2 states are those who possessed nuclear power reactors or research reactors at the time of the CTBT's negotiations between 1994 and 1996.

As the treaty is not yet in force the United Nations made a strong appeal (CTBT, ibid) and called on all states to "refrain from nuclear weapon or any other test explosions, the development and use of new nuclear weapon technologies and any action that would undermine the objective of the CTBT." However, CTBTO was able to prove the effectiveness of the verification regime on the occasion of the nuclear test explosion, announced by the Democratic People's Republic of Korea on February 12th, 2013. Its international monitoring system (IMS) could successfully detect seismic signals and infrasound signals, providing relevant and useful physical data on the test. Further measurements of radioactive noble gases later in April 2013 also confirmed the sensitivity and specificity of the monitoring network. The verification system comprises a worldwide network of monitoring stations, an International Data Centre (IDC) in Vienna, and onsite inspections. Seismology and other technologies including hydroacustic, infrasound, and the monitoring of radionuclides in the air, are used to monitor for compliance with the treaty. Once the treaty enters into force, onsite inspection (ISO) will provide information where concerns about compliance arise. The monitoring network consists of 337 facilities located all over the globe. As of May 2012, more than 260 facilities have been certified. The IMS, when completed, will consist of 170 primary and auxiliary seismological monitoring stations and 60 infrasound stations to detect very-low-frequency sound waves. Eighty radionuclide stations can detect radioactive particles released from atmospheric, underground, or underwater explosions that are analyzed in 16 laboratories. When conducting onsite inspections quite a number of highly sophisticated detection techniques can be used including visual site observations, passive seismic measurements, and gamma radiation measurements, and the detection of radioactive noble gases such as xenon and argon. The ISO inspection will be carrried out for an initial period of up to 25 days that can be extended up to 60 days to carry out, for instance, resonance seismic measurements, ground-penetrating radar, gravity, and electric and magnetic field mappings. Argon-37 field measurement is a unique technology specially developed for the purpose of ISO. Moreover scientific drilling to obtain radioactive samples from a suspected underground explosion site is also allowed. The data are sent to states that have signed the treaty and are simulataneously transmitted to the IDC in Vienna through a global private data network largely based on satellite communication. The technical and scientific capabilities that have been established will furthermore bring benefits to disaster alert and early warning systems for natural disasters such as tsunamis.

4.2.7 United Nations Framework Convention on Climate Change (UNFCCC)

In 1992 the world community met in Rio de Janeiro (Brazil) at the Rio Earth Summit (United Nations Conference on Environment and Development; UNCED 1992) to agree on methods and instruments to be undertaken by the international community to reduce greenhouse gas emissions worldwide. The UNCED conference was the biggest international conference ever held by the United Nations on the environment. High-level representatives from all nations, 2400 from nongovernmental organizations, and more than 17,000 environment experts and politicians participated in the conference. The most important outcomes were the adoption of Agenda 21 and the Rio Declaration on Environment and Development (outlining the framework for a sustainable development), the "Forest Principles" and the Convention on Biological Diversity (CBD). Following Rio the Commission on Sustainable Development (CSD) was founded to monitor the implementation processes agreed upon. After intensive debates the conference agreed upon the United Nations Framework Convention on Climate Change

(UNFCCC 1992). Since 1992, 195 nations have ratified the convention. Its central aim is laid down in Article No. 2 "stabilization of greenhouse gas concentrations in the atmosphere at a level that would prevent dangerous anthropogenic interference with the climate system."

By the UNFCCC the nations committed themselves voluntarily and significantly to reduce their respective greenhouse gas emissions (GHG). The conference further agreed upon a yearly monitoring of the achievements made in accordance with the convention. The next two conferences (1995; 1996) revealed that the volunteer commitments were not really sufficient to reduce the amount of GHG emissions into the atmosphere as envisaged. Therefore the fourth conference negotiated an agreement (Kyoto Protocol; UNFCCC 1992) that for the first time in human history agreed upon defined and legally binding emission targets (on average 5 % reduction of GHG based on the emission levels in 1990) that are to be reached by the OECD countries. The European Union, however, offered a self-commitment that goes beyond the 5 % reduction target on an average of 8 % until 2012.

The Kyoto Protocol comprises the following GHG:

- Carbon dioxide (CO_2)
- Methane (CH_4)
- Nitrogen (N_2O)
- Fluorocarbons (perfluorinated carbons PFC; hydroflourocarbons H-FCKW)
- Sulphurhexafluorid (SF_6)

The protocol identified two main sectors to reduce GHG: one is the commitment of each member state to reduce emissions by setting up a national GHG emissions reduction strategy and second by fostering renewable energy generation via an emission-reduced agriculture.

4.2.8 The Global Facility for Disaster Reduction and Recovery (World Bank—GFDRR)

The Global Facility for Disaster Reduction and Recovery (GFDRR 2012) is a partnership of the International Strategy for Disaster Reduction (ISDR) to support the implementation of the Hyogo Framework for Action (HFA). Established in 2006, the GFDRR is a partnership of 41 countries and eight international organizations committed to helping developing countries reduce their vulnerability to natural hazards and adapt to climate change. The partnership's mission is to mainstream disaster risk reduction and climate change adaptation in country development strategies by supporting a country-led and managed implementation of the HFA. Under the guidance of a high-level consultative group of the donor countries/organizations, the United Nations and the World Bank, and the government of Sweden, the GFDRR secretariat manages the global operations of this partnership. Its secretariat, located at the World Bank headquarters in Washington DC, acts as the support hub for a decentralized network of disaster risk management experts in the countries. These specialists play a leading role in locally managing the GFDRR program.

GFDRR is a grant-making facility but not a "direct implementer." The grants have increased from US$6 million in 2007 to US$75 million in 2012. Moreover GFDRR also manages special initiatives that focus on particular regions or topics, including a five-year €54 million initiative of the Africa, Caribbean, and Pacific group of states financed by the European Union. GFDRR is responsible for allocating funds entrusted to it in line with priorities set by its donors and partners. In any given country, GFDRR adopts a number of criteria to help in allocating resources, including established vulnerability indicators and past evaluation of impact and donor priorities. Geographic priorities currently include 20 countries financed through a multidonor trust fund, and 11 countries targeted by individual donors. All across the disaster management programs, GFDRR is committed to ensure all projects are gender-neutral regarding hazards and risk analysis, prevention and mitigation actions, monitoring, and evaluation of achieved results. GFDRR is further committed to ensure, wherever possible, that projects include consideration of the particular vulnerabilities of the elderly, children, and the disabled.

The work of the GFDRR is divided into three main business lines:

- Financial support to the secretariat of the ISDR secretariat. This track is designed to enhance global and regional advocacy, partnerships, and knowledge management for mainstreaming disaster reduction.
- Mainstreaming disaster risk reduction and climate change adaptation in country development processes to ensure that risk assessments, risk mitigation, risk transfer, and emergency preparedness are incorporated in all strategic plans and programs. This track is designed to provide technical assistance to enhance investments in risk reduction and risk transfer mechanisms, as well as disaster management planning to low- and middle-income country governments.
- Is aimed at early, post-disaster recovery in low-income countries through its standby recovery financing facility (SRFF). Thus this track is less programmatic than Tracks I and II because it is deployed for post-disaster situations. Nevertheless it is oriented to build national capacity and facilitate knowledge management in the long-term.

Through five pillars GFDRR provides grants to disaster risk reduction measures:

- *Risk Identification Grants*: Supporting partner governments and local experts in the development of detailed national, subnational, or sector-specific risk assessments, including hazard exposure and vulnerability. Risk identification is the foundation to sensitize countries at risk and guide risk reduction, preparedness, and risk-financing work.
- Risk Reduction Grants: Supporting countries at risk by identifying strategies to avoid the creation of new risks (spatial planning; building standards) and by assisting them to address existing risks through the proper development planning and implementation of risk reduction plans and investment programs.

- *Preparedness Grants*: Improving national and institutional capacity to anticipate, prepare for, and respond to disasters. GFDRR works together with the World Meteorological Organization (WMO), for example, to institutionalize alert systems that provide early warning especially to hydrological, meteorological, and flood risk. This comprises the preparation of plans and technical, financial, and manpower resources for rapid response by local or national emergency management agencies.
- *Financial Protection Grants*: Financing governments, businesses, and individuals to cope with the financial and economic consequences of disasters. By this not only the government's budgets are subsidized but an opportunity to strengthen the private sector is also provided.
- *Resilient Recovery Grants*: Assisting countries in implementing reliable and internationally accepted damage, loss, and needs assessment methods. Moreover, GFDRR supports such countries by providing solutions for better post-disaster recovery planning and financing recovery programs.

4.2.9 Intergovernmental Panel on Climate Change (IPCC)

The Intergovernmental Panel on Climate Change (IPCC 1988) is a scientific body under the auspices of the United Nations. It was established in 1988 by the United Nations Environment Program (UNEP) and the World Meteorological Organization to provide a scientific view of the current state of knowledge in climate change and its potential environmental and socioeconomic impacts. The IPCC is an intergovernmental body that is open to all member countries of the United Nations and is hosted on the WMO premises in Geneva. Currently 195 countries are members of the IPCC. Because of its scientific and intergovernmental nature, the IPCC embodies a unique opportunity to provide rigorous and balanced scientific information to decision makers. By endorsing the IPCC reports, governments acknowledge the authority of their scientific content. The work of the organization although policy-relevant, is exclusively policy-neutral. Governments participate in the review process and in delineating the work programs. Hundreds of scientists from all over the world contributed to the work of the IPCC on a voluntary basis. Review is seen as an essential part of the IPCC process, to ensure an objective and complete assessment of current information. The IPCC aims to reflect a range of views and expertise, and its work is guided by a set of principles and procedures.

The IPCC member states meet once a year at plenary level that is in general attended by government officials from member countries, national and international scientific agencies, well-reputed research institutions, and from observer organizations to approve, adopt, and accept the reports. Each IPCC member country has a focal point. The focal points prepare and update the list of national experts to help implement the IPCC work program. The focal points also arrange for the provision of comments on the accuracy and completeness of the scientific and/or technical content and the overall scientific and/or technical balance of drafts of reports.

In addition to the working groups and task force, further task groups and steering groups can be established for a limited duration to consider a specific topic or question. One example is the Task Group on Data and Scenario Support for Impact and Climate Analysis (TGICA).

Hundreds of experts are involved in the preparation of IPCC reports, coordinated by lead authors who were selected by the relevant Working Group/Task Force Bureau from among experts listed by governments and participating organizations, and other experts known through their publications and works. To ensure a high level of quality, transparency, and neutrality, IPCC reports undergo a multistage review process. A team of review editors is introduced, two to four per chapter. Their role is to comment on the accuracy and completeness of the scientific, technical, or socioeconomic content and the overall scientific, technical, or socioeconomic balance of draft reports. None of them is paid by the IPCC.

IPCC regularly publishes reports in different categories:

- Assessment reports
- Special reports
- Technical papers
- Methodology papers

Most important and a matter of a broad and intense international discussion on climate change became the assessment reports.

4.2.9.1 1990 First Assessment Report (FAR)

The report was the first that addressed global climate change issues by an international body (IPCC 1990). It revealed that greenhouse gas emissions resulting from human activities are substantially increasing the atmospheric concentrations of water vapor, CO_2, methane, CFCs, and nitrous oxides. Furthermore it stated that the global mean surface air temperature has increased by 0.3–0.6 °C over the last 100 years. These increases will most probably result in an additional warming of the Earth's surface. The main greenhouse gas, water vapor, will increase in response to global warming and further enhance it. An increase of the global mean temperature during the next century of about 0.3 °C per decade will be greater than over the past 10,000 years and will most probably lead to an average rate of global mean sea-level rise of about 6 cm per decade over the next century.

4.2.9.2 1995 Second Assessment Report (SAR)

The second report (IPCC 1996) confirmed the findings of the first report, this time based on a broader database and more sophisticated scientific assessments. It revealed that carbon dioxide remains the most important contributor to human-induced change of climate. It gave a projection on the future global mean temperature that will lead to a rise in global sea level and confirmed that human activities have the potential to alter the Earth's climate to an extent unprecedented in human history. The findings pointed to long time scales that control both the accumulation of greenhouse gases in the atmosphere and the delayed response of the climate system to those accumulations that will effectively make the climate change irreversible.

4.2.9.3 2001 Third Assessment Report (TAR)

The third assessment report (IPCC 2001) has become one of the central international references on the subject of global warming. An increasing number of observations from all over the world pointed to a warming worldwide and to greenhouse gas emissions related changes in the climate system. It confirmed the global average surface temperature has increased over the twentieth century by about 0.6 °C per year. And if the emissions of greenhouse gases and aerosols continue it will definitively alter the atmosphere and affect the global climate. Overall scientific confidence in the ability of the projections has increased significantly. Although the models cannot yet simulate all aspects of climate and there are particular uncertainties associated with clouds and their interaction with radiation and aerosols, there is nevertheless a high confidence that the models will provide useful projections of future climate. There is strong evidence that most of the warming observed over the last 50 years is attributable to human activities. The TAR estimated an average surface temperature increase by 1.4–5.8 °C over the period 1990–2100, and that the sea level will rise by 0.1–0.9 m over the same period.

4.2.9.4 2003 Fourth Assessment Report (4AR)

The fourth report (IPCC 2007) was released in three principal sections:

- The Physical Science Basis of Climate Change (Working Group I)
- Climate Change Impacts, Adaptation and Vulnerability (Working Group II)
- Mitigation of Climate Change (Working Group III)

It assessed current scientific knowledge of natural and human causes as well as the observed changes in climate over the last decades. It looked at the ability of science to attribute changes to different causes and made projections of future climate change. Simulations were based on various computer climate models. As a result predictions for the twenty-first century were derived: surface air warming was estimated in a "low scenario" to reach 1.8 °C, whereas the estimate for a "high scenario" was 4.0 °C. Based on multiple models (that exclude ice sheet flow) it was estimated that the sea level would rise in a "low scenario" by 18–38 cm and in a "high scenario" by 26–59 cm. The sea ice shields are projected to shrink in both the Arctic and Antarctica. The report furthermore projected that dry regions are to get drier and wet regions to get wetter. Drought-affected areas will become larger and heavy precipitation events are very likely to become more common and will increase flood risk. The water stored in glaciers and snow cover will be reduced over the course of the century. The change in climate will likely exceed the resilience of many ecosystems by a combination of climate change and other stressors. The thus-far effective carbon removal by terrestrial ecosystems is likely to weaken or reverse. This would amplify climate change and lead to more coastal erosion, to an increase in coral bleaching, and a widespread mortality of species unless there is thermal adaptation or acclimatization by corals. The report finally stated that many millions more people will be threatened by floods every year due to sea-level rise up to 2080.

One of the key conclusions of the report was that it is possible to stabilize the greenhouse gas concentrations between 445 ppm and 535 ppm at costs of less than 3 % of global gross domestic product.

4.2.9.5 2014 Fifth Assessment Report (AR5)

The Fifth Assessment Report was like the previous reports based on the findings of the three Working Groups of the Intergovernmental Panel on Climate Change (IPCC). It provides an integrated view of the observed changes in climate and their effects. It stated clearly that the warming of the climate system is unequivocal, as is now evident from observations of increases in global average air and ocean temperatures, widespread melting of snow and ice and rising global average sea level. Again a Synthesis Report (SYR) gives a condensed vision of the findings for the international policy makers.

The findings of WG I are as follows:
The warming of the atmosphere and ocean system can no be longer questioned. There is clear evidence that human influence has been the dominant cause of observed warming since 1950, an assumption that has reached a level of confidence much higher than that in 4AR. It is likely that 1983–2013 was the warmest 30-year period for the past 1400 years. It is virtually certain that the upper ocean warmed from 1971 to 2010. Moreover it can be said with high confidence that the inland ice sheets have been losing large masses in the last two decades and that the northern hemisphere spring snow cover has continued to decrease in extent. There is high confidence that the sea-level rise since the middle of the nineteenth century has been larger than the mean sea-level rise of the previous two millennia. Concentration of greenhouse gases has increased to levels unprecedented on earth in 800,000 years. The total radiative forcing in the earth system is today positive and the most significant driver is the increase in CO_2 atmospheric concentration.

4.2.10 ISO (International Organization for Standardization)

Without standards and norms technical and operational activities cannot be carried out successfully. Standards proved indispensable to define the quality of technical products (ISO 2014). They define production processes and make production and human acting comparable. The more technology and industrialization developed in the last century, the more standardization and norms were required. Standards facilitate trade, spread knowledge, and share technological advances and good management practices. Today almost no activity in households and production is carried out without being underlined by norms and standards. Thus every country on Earth has its national norms, for instance, in Germany the longstanding DIN-Norm, in the United Kingdom the British Standards Institution (BSI), or the United States of America the National Institute of Standards and Technology (NIST). The globalization of markets made it necessary to harmonize the many standards to allow for worldwide marketing. Thus under the auspices of the United Nations an international harmonization of standards was initiated that resulted in the founding of the ISO.

The ISO is a network comprising the national standards institutes of 163 countries and is thus the world's largest developer of voluntary international standards. ISO standards ensure quality, ecology, safety, reliability, compatibility, interoperability, efficiency, and effectiveness. The ISO has over 19,500 standards touching almost all aspects of daily life, including the sectors environment and risk management. The most popular standards covered by the ISO standard families are the ISO 9000 family on quality management, the ISO 14,000 family on environmental management, and the ISO 31,000 family on risk management (ISO 2009).

The ISO 9000 family of standards is related to the fundamentals of quality management systems designed to help organizations ensure that they meet the needs of customers and other stakeholders. Over a million organizations worldwide today make ISO 9001 one of the most widely used management tools in the world today. ISO 9004 is the quality management approach designed for the internal managerial processes of an organization.

The ISO 14,000 family addresses various aspects of environmental management. It provides practical tools for companies and organizations looking to identify and control their environmental impact and constantly improve their environmental performance. ISO 14,001 and its sister 14,004 focus on environmental management and map out a framework that organizations can follow to set up an effective management system. It can be used by any organization regardless of its activity or sector.

A third standard tool that can be applied in disaster risk management is the ISO 31,000 family, in which the risk management standards are described. Risk management is a rapidly developing discipline and there are many and varied views and descriptions of what risk management involves, how it should be conducted, and what it is for. ISO 31,000 provides standards and descriptions on what risk management is. By applying the standard an organization can increase the probability of success, and reduce the probability of failure. Some risks can have both external and internal drivers and therefore can interact in different areas of activity. Such multisectorial risk can be categorized further into types of risk such as strategic, financial, operational, hazard, and so on.

Risk management should be a continuous and developing process that runs throughout the organization's strategy and the implementation of that strategy. It should methodically address all the risks surrounding the organization's activities in the past and in particular in future (IRM 2002). Identification of risk is based on three pillars.

4.2.10.1 Risk Identification

This requires an intimate knowledge of the organization, the market in which it operates; the legal, social, political, and cultural environment in which it exists; as well as the development of a sound understanding of its strategic and operational objectives, including factors critical to its success and the threats and opportunities related to the achievement of these objectives, best displayed by using a table:.

- Name of risk
- Scope of risk
- Type/nature of risk
- Quantification of risk
- Stakeholders
- Risk tolerance
- Risk treatment and control mechanisms
- Potential action for improvement
- Strategy and policy developments

4.2.10.2 Risk Estimation

Next to follow is the estimation of the probability of occurrence and the possible consequence. The assessment can be done either qualitatively or quantitatively (semi-quantitatively) and should be prioritized in terms of threats (downside risks) and opportunities (upside risks) which may be high, medium, or low, and with an estimation of the probability to occur (high, medium, or low). The assessment requires, however, an agreed-upon definition of "threats" and "opportunities." The use of a well-designed structure is necessary to ensure a transparent and objective assessment process.

4.2.10.3 Risk Evaluation

When the risk analysis process has been completed, a further step is to analyze the estimated risks against risk criteria the organization has established. The risk criteria may include associated costs and benefits, legal requirements, socioeconomic and environmental factors, concerns of stakeholders, and so on. Risk evaluation, therefore, is used to make decisions about the significance of risks to the organization and whether each specific risk should be accepted or treated.

4.3 The European Union

The European Union is the world's largest donor of humanitarian aid. Through Directory General ECHO (EU 1992), the European Union provides significant assistance in humanitarian crises. The EU disaster management policy is guided by the principle of solidarity, respect for human dignity, equality, and tolerance. It is based on international law, and on the fundamental principles of impartiality, nondiscrimination, and neutrality. The European Union is well placed to assume a leading role in the pursuit of a disaster reduction strategy within the context of global sustainable development. So far the EU commitment on disaster risk management has been systematically enshrined in only three EU legal documents, nevertheless the legal situation of the European Union allows an almost worldwide coverage of disaster preparedness measures through commission instruments other than ECHO. The commission has acknowledged the challenges imposed by disaster events not only affecting its member countries but especially the developing countries. Every day natural disasters worldwide exhaust the coping capacities of the population bringing these societies deeper into the poverty trap. The European Union therefore sees in a systematic incorporation of disaster preparedness and prevention activities in its members' development policies a prerequisite for increasing natural disaster resilience. The European Union also underpins the avoidance of natural disasters becoming a political imperative in disaster mitigation and requires appropriate disaster preparedness to a more rapid recovery from the effects of disasters. ECHO points out that the following sectors are the cornerstone of its disaster preparedness policy:

- *Preparedness*: Organizational activities that ensure that the systems, procedures, and resources required to confront a natural disaster are available in order to provide timely assistance to those affected, using existing mechanisms wherever possible (e.g., training, awareness raising, establishment of disaster plans, evacuation plans, pre-positioning of stocks, early warning mechanisms, and strengthening indigenous knowledge).
- *Mitigation*: Measures taken before disasters that intend to reduce or eliminate their impact on society and the environment. These measures reduce the physical vulnerability of existing infrastructure or of vulnerable sites that directly endanger the population (e.g., retrofitting of buildings, reinforcing "lifeline" infrastructure).
- *Prevention*: Activities conceived to ensure permanent protection against a disaster. These include engineering, physical protection measures, and legislative measures for the control of land use and codes of construction. These activities reduce the physical vulnerability and/or exposure to risks through infrastructure through improving existing infrastructure and sustainable development practices.

ECHO's disaster preparedness policy is mainly carried out by the Disaster Preparedness Program (EU-DIPECHO 1992). This program finances disaster preparedness and reduction measures in areas of risks from natural catastrophes. In order to optimize the administrative and financing measures, action plans have been established for many of the countries at risk with the focus on areas at high risk of natural disasters and low coping capacities. DIPECHO's main objective is to address disaster prevention and preparedness targeting the most vulnerable populations in the world in the main disaster-prone areas with low coping capacities. DIPECHO's main focus is on preparation, rather than mitigation or prevention. The scope of activities includes:

- Community training/capacity building, (including, e.g., material and services for capacity building, training of disaster brigades, simulation exercises)
- Provision of equipment (including, e.g., equipment for refugees, primary emergency kits, scientific advocacy)
- Small-scale mitigation works for demonstration purposes and awareness raising (including, e.g., reforestation, machines and material, building of emergency shelters, water tanks)
- Early warning systems (including, e.g., radio communications)
- Emergency response planning, hazard mapping
- Public awareness raising, education
- Research and dissemination
- Facilitation of coordination
- Institutional strengthening

4.3.1 Eurocode

The European Commission initiated in 1975 a harmonization and standardization of technical specifications for all kinds of construction within its jurisdiction. Since then the European Committee for Standardization (CEN) published nine building codes, including the EUROCODE 7 and 8 (EU-EN 1998) that are concerned with the geotechnical design and stability of structures for earthquake resistance (European Standard EN 1998-2, CEN/TC259). All national standards organizations (in Germany: DIN-EN 1998-2; in UK: BS-En 1998-2, etc.) were called to give this European Standard the status of a national standard. The Eurocode Standards provide common rules for the design of entire buildings and other technical structures and component products in order to make them seismic resistant and technically safe. The codes serve as reference documents for the following purposes:

- As a framework for drawing up harmonized technical specifications for construction products
- As a basis for specifying contracts for construction works and related engineering services
- As a means to prove compliance of building and civil engineering works with the essential requirements

Eurocode 7 (“Geotechnical Design”) applies to geotechnical aspects of the design of buildings and civil engineering works. In its first part the code covers the general basis for the geotechnical aspects of the design of buildings and civil engineering works, assessment of geotechnical data, use of ground improvement, ground reinforcement, dewatering, fill, and geotechnical design of spread foundations, piles, retaining structures, embankments, and slopes, as well as calculation rules for actions originating from the ground (e.g., earth and ground water pressure). And in its second it covers requirements for the execution, interpretation, and use of results of laboratory tests to assist in the geotechnical design of structures.

Eurocode 8 (“Earthquake”) explains how building and civil engineering structures are to be constructed in seismic-prone regions. It outlines the general requirements and rules for the assessment of seismic risk and seismic risks in combinations with others. It furthermore specifies the different construction elements to increase resistance of buildings, bridges, tanks, pipelines, and tall structures such as towers, masts, or chimneys.

4.3.2 European Water Framework Directive (EWFD)

Water does not stick to national boundaries. This is why EU member states have agreed to establish a framework for the European Union in the field of water policy, the European Water Framework Directive (EU-EWFD 2000). The directive commits the member states to achieve qualitative and quantitative good status of all water bodies within its jurisdiction including marine waters up to one nautical mile from shore by year 2015. Following the successful implementation of the EU Drinking Water Directive and the Urban Waste Water Directive, the European Water Policy addresses the increasing awareness of citizens and other involved parties of their water. The EWFD is a framework in the sense that it prescribes steps to reach common goals rather than adopting the more traditional limit value approach. The directive stipulates that groundwater must have a “good quantitative status” and a “good chemical status” and should be classified as either “good” or “poor”. An assessment of the ecological and chemical status of all surface waters has to follow criteria on:

- Biological quality (fish, benthic invertebrates, aquatic flora)
- Hydromorphological quality (riverbank structure, river continuity, or substrate of the river bed)
- Physicochemical quality (temperature, oxygenation, and nutrient conditions)
- Chemical quality (environmental quality standards, specific pollutants)

With the EWFD the European Union presented a binding directive to be implemented by the national governments with the following key aims:

- Expanding the scope of water protection to all waters, surface waters, and groundwater
- Achieving "good status" for all waters by year 2005
- Water management based on river basins
- "Combined approach" of emission limit values and quality standards
- Getting the prices right
- Getting the citizens involved more closely
- Streamlining legislation

One important aspect of the Water Framework Directive is the introduction of river basin districts leading to a management of entire water bodies based on their natural, geographical, and hydrological units rather than on administrative or political boundaries (the "catchment area"). As many rivers cross national borders, representatives from the respective member states have to cooperate and work together for the management of the basin, the so-called "transboundary basins." The very successful initiatives such as the International Commission of the Protection of the River Rhine (ICPR 2010) or the Elbe River (ICRE) have served as positive examples of this approach, in the case of the Rhine even beyond the EU territory. For each river basin district a river basin management plan has to be established and updated every six years.

A second milestone of the directive is the call for EU-wide coordinated actions to achieve a good status for all water bodies. Key objectives are related to water quality: "good ecological status" and "good chemical status." These objectives are seen as indispensable for integration into every protection initiative at the national level. A set of procedures for identifying these requirements for a given body of water, and establishing particular chemical or hydromorphological standards to achieve it, is provided, together with a system for ensuring that each member state interprets the procedure in a same way. The target "good chemical status" is defined in terms of compliance with all the quality standards established for chemical substances at the European level. The directive also provides a mechanism for renewing these standards and establishing new ones by means of a prioritization mechanism for hazardous chemicals

Moreover the EWFD reiterated that scientific-oriented investigations will not be successful if there is no full integration of the directive into the national legislations and if its implementation is not accompanied by comprehensive public participation. The directive formalizes the rationale for a coordinated application of the respective pollution prevention and reduction measures of science and managers with public participation. But over and above this, implementation also requires a framework for monitoring and controls. The framework therefore comprises a list of priority substances for action at the EU level, prioritized on the basis of risk, and designed the most cost-effective set of measures to achieve load reduction of those substances. The rationale for extensive public participation is at first that the implementation of prevention such as load-reducing measures will have to balance the interests of various groups. The second reason concerns enforceability and public acceptance. The greater the transparency in defining the objectives, during implementation and reporting of standards, the more citizens are invited to participate in environmental protection, the greater the societal acceptance will be.

4.3.3 Natura 2000

With the aim of protecting Europe's most important wildlife areas, the European Commission has set up the NATURA 2000 network of protected areas. It is the most important initiative ever undertaken in the European Union to conserve areas of high importance for threatened species and habitats (NATURA 2000 2000). By protecting these "precious components of biodiversity," NATURA 2000 aims to meet the European Council's goal of halting biodiversity decline within the European Union by 2010 (EU 2003). It represents a major contribution to global nature conservation and a model for international cooperation on sustainable development. NATURA 2000 is the official term for the EU-wide established coherent network that was erected according to the EU-Directive 92/43/EWG and thus consequently follows the 1992 Habitats Directive. Often the directive is abbreviated as Birds Protection Directive. With the establishment of this network the European Union fulfils their obligations under the UN Convention on Biological Diversity.

With the NATURA 2000 network the EU aims at providing an EU-wide transboundary protection of wildlife fauna and flora. It is not a system of strict nature reserves where all human activities are excluded, but rather a network of areas where private use is not restricted although emphasis is given to ensuring that future management is sustainable, both ecologically and economically. Restriction means that new building construction, business, agricultural, or recreational activity within sites is not in general prohibited. But any special development has to follow procedures laid down in the Habitats Directive that asks for an assessment of adverse impacts on the designated sites. The network defines two different areas: Special Areas of Conservation (SAC) and also bird protection areas, called Special Protection Areas (SPAs) both organized by the respective member state under the Habitats Directive. There is no specification given on how much land and water need to be included in NATURA 2000. The directive only states that this depends on the biological richness of the different regions and that the designated areas should be in proportion to the wealth in biodiversity. Up to today the NATURA 2000 network has proposed an area for conservation under the network larger than the size of Germany, equivalent to more than 15 % of the European Union's territory. Thus more than 26,000 areas of highest ecological value between Scandinavia and the Mediterranean are under formal protection against further biological and environmental degradation. The fauna–flora and the bird conservation sites cover nine different geographical regions.

According to NATURA 2000 all EU member states are obliged to ensure full compliance with the legal requirements, regardless of whether they are receiving financial assistance from EU structural funds. In order to achieve comprehensive compliance, the EU Commission informed the member states that noncompliance in presenting lists of NATURA 2000 sites could result in the suspension of payments under certain structural fund programs. With this the commission wanted to prevent its members from irreparable damage to sites before they have been proposed officially for protection under the NATURA 2000 regime. However, it became clear from the very beginning that such an ambitious target would not be achieved on short notice. Therefore the directive allows beginning with activities to increase the ecological situation of areas that are regarded ecologically under stress until 2020. The amelioration process shall be initiated to at least increase the level by one step.

The European Commission has elaborated a number of guidance documents with regard to the management of defined NATURA 2000 sites. A large variety of organizational approaches, experience, and practical examples have been published. The guidance is one of the most important parts in the Habitats Directive as it defines how the sites are to be managed. To assist in the understanding and correct application of the directive, the commission has produced a number of general interpretative and methodological documents on specific provisions of the directive. The guidance stipulates that the member states have:

- To transfer the directive into national law within two years
- To take appropriate conservation measures to maintain and restore the habitats and species for which the site has been designated
- To ensure that no area defined shall be subject to further environmental and biological deterioration
- To avoid damaging activities that could significantly disturb these species or deteriorate the habitats
- To carry out environmental impact assessments on all areas eligible to the directive
- To ensure that any plan or project likely to have a significant effect on a Natura 2000, either individually or in combination with other plans or projects, should undergo an appropriate assessment to determine its implications for the site
- In spite of a negative assessment, a plan or project may nevertheless be allowed, provided there are no alternative solutions and the project is considered to be of overriding public interest.

Regardless of the undisputed success of the directive, there is, however, much criticism from different stakeholders (Ebert 2012):

- Binding definition of the term “environmental protection” is missing.
- This also holds true for regulations that define the “management” of the directive at the national level.
- The funds allocated under the directive are rated not sufficient at all.
- The environmental impact assessments often do not correspond to international standards.

- The criteria and indicators for the assessment of the actual environmental status are often unclearly formulated, leaving much room for interpretation.
- The methodological approach declared compulsory is not suitable to cover the time variance of environmental changes.
- Assessing the mere existence of species as the sole indicator for biological variance does not reflect any changes in the species' quantity.
- Although the member states are called on to identify protection sites the response is lagging a great deal behind the envisaged timeframe.
- Many states had to be forced to action by lawsuits of the European Court of Justice.
- Often states and business lobbyists oppose a straight implementation of the directive with the argument that this will hamper economic and social development in the region.

The environment commissioner of the European Union, Margot Wallström, reacted to the many critics and started an initiative to ensure that NATURA 2000 is better understood. She acknowledged some misconceptions that have provoked unfounded concerns. The EU Commission will respond to these concerns by communicating more effectively, especially with the different stakeholder groups. The European Union is convinced that, through partnership with the member states and the different stakeholder groups, it will be possible to reconcile economic and social goals with their responsibilities towards nature.

4.4 German Regulations of Natural Disaster Management

4.4.1 German National Law to Increase Flood Prevention

According to the German Federal Water Act (Wasserhaushaltsgesetz-WHG; BMUB 2013), flood risk prevention is a task that is in Germany mutually shared between the public and the private sector. The law defines flood-prone areas are to be generally managed in a way that they can either withhold floodwater as well as they can secure damage-less drainage of the floodwater. Moreover the law defines what areas are flood-prone and how they can be safeguarded to enhance their flood-diminishing function. On the other hand, the law requests that every person exposed to flood risk, implement his own measures oriented to decrease the flood risk. This especially concerns flood risks to households, land property, animals, and the living environment, as far as it is technically possible and economically reasonable.

Flood-prone areas are defined by law as areas that are lying along rivers or streams, or between dams and levees, that are known to be at risk from a 100-years flooding or that are defined to have a distinct flood retention function. By act of law, local governments until 2012 had to define such areas within their jurisdiction and to define concrete measures that focused on enhancing ecological sustainability to diminish soil erosion. This also comprises the definition of measures meant to foster damage-less drainage and outline such areas that can function to retain floodwaters (retention areas, flood polders). The law furthermore clarifies that local governments have to inform the public in time and regularly on these decisions and requests to incorporate societal groups in this decision making. In flood-prone areas in general no new buildings are allowed to be constructed unless there are no alternative locations for such settlements, no harm is expected to life and property, and the buildings are not hampering drainage of floodwaters, they do not restrict the retaining function, or their impact can be balanced by other flood-diminishing action in order to avoid harm to the low-lying populations and assets. Local governments are obliged to display their findings and decisions on the areas at risk, on the retention areas, and on the envisaged countermeasures in maps of an appropriate scale and to make this information accessible to the public. The maps have to be updated regularly.

4.4.2 German Advisory Council on Global Change (WBGU)

The rationale for establishing the Scientific Advisory Council to the Federal German Government (WBGU 2014) was given as follows:

> The critical environmental changes are advancing worldwide; include climate change, biodiversity loss, soil degradation and freshwater pollution and scarcity. The continuing spread of non-sustainable lifestyles and the persistence of absolute poverty and a growing global population are fundamental to this environmental deterioration. One consequence of global environmental change is the mounting vulnerability, especially of developing countries to natural disasters, food crises and disease. Thus, environmental degradation has also become a security issue.

Consequently to prepare for the Rio Earth Summit (1992), the German Council on Global Change (WBGU) was installed by the German Federal Government as an independent, scientific advisory body, serving the overall target of advising the federal government on all kinds of global change politics. The council consists of independent experts that are commissioned by the German Ministries of Education and Research (BMBF) and Environment, Nature Conservation, Building and Nuclear Safety (BMUB). The WBGU is free to make its own choice of themes to be addressed. It takes a transdisciplinary approach to assess hazards and risks that derive from changing climate conditions. WBGU identifies critical scenarios for humankind and elaborates precautionary options by which irreversible damage to human societies and natural systems can be avoided. The investigations undertaken enable it to provide guidance for political decision makers. Thus WBGU draws the attention of politicians and the wider public to initiate corrective action. Since 1992 WBGU:

- Analyzes global environment and development problems
- Reviews and evaluates national and international research in the field of global change
- Provides early warning of new issue areas
- Identifies gaps in research to initiate new research options

- Monitors and assesses policies on sustainable development
- Elaborates recommendations for action and research
- Raises public awareness on global change issues

Biannually WBGU publishes flagship reports. Additionally it prepares special reports and policy papers. Meanwhile 14 flagship reports, eight special reports, and seven policy papers have been published covering the full range of global change challenges and human interference with the natural environment.

Of the many reports those that have a significant relation to natural disaster risk management the following ones have been chosen and are summarized below.

4.4.2.1 Flagship Report 1995 "World in Transition: Ways Towards Global Environmental Solutions"

At the first UN Climate Convention in Berlin (WBGU 1996) it became evident that anthropogenic-triggered climate changes made a reshuffling in individual's as well as society's way of thinking and behavior indispensable. This accounts for all areas of global environmental change especially on global trends such as soil degradation, loss of biological diversity, water scarcity, and population growth, that all show little or no sign of improvement. In its 1995 report WBGU considered two aspects to lead to a reduction of environmental degradation and thus can help to sustain living conditions of the poor. WBGU sees an improvement of the societal conditions with a significant increase in environmental awareness, especially in the most disaster-affected regions as a prerequisite. Therefore strong efforts must be taken in environmental education worldwide. A critical element for environmental degradation is located in the steadily increasing world population, despite the fact that the statistical increase is leveling off somewhat. That leveling off, however, should not be taken as an excuse not to address this sector any further. As the persistent trends in poverty migration into the larger conurbations will increase rather than go on as in the past, WBGU pledges for concerted efforts to incorporate global environmental policies including climate change, biodiversity, desertification, and soil degradation into international agreements such as the GATT and WTO treaties. WBGU closed the report with the pledge for the international community to make more efforts to address global environmental problems as they are more evident than ever before.

4.4.2.2 Flagship Report 1998 "World in Transition: Strategies for Managing Global Environmental Risks"

Global risk potentials and their economic, social, and ecological outcomes have already reached global dimensions. This has been driven by a growing global population, particularly in developing countries, and on the other hand by rising human aspirations in the developing countries for an increase in development, as well as in conjunction with specific patterns of resource dependency of the industrialized countries. The report (WBGU 1999) contributed first to a classification of risk and second recommended a number of cross-cutting strategies for international policies. These include worldwide alignment of liability law, creation of environmental liability funds, establishment of a UN risk assessment panel and an implementation of strategies aimed at reducing vulnerability.

The risk classification concept was presented and defined risk according to its severity and frequency, an approach that is also described in more detail in Chap. 7 and can either be "normal," meaning they are of low probability and small severity, leading to low potential of societal conflicts, inequities, and social discrepancy and low economic loss. Or they occur with higher probability ("significantly transcend everyday level") with an uncertain distribution of adverse effects of the populations at risk. Third, risks can occur in the "prohibited area." Such risks will lead to irreversible damage, have ubiquitous spreading, and can lead to severe social and socioeconomic conflicts. The report sees an overall acceptance in societies that such kind of risk should be avoided in general. Although the "normal zone" does not require any corrective action, the "transitional zone" develops as the field of political and social intervention. It is there where disaster risk reduction has to be formulated and implemented accordingly.

4.4.2.3 Flagship Report 2000 "World in Transition: New Structures for Global Environmental Policy"

The 2000 flagship report (WBGU 2000) dealt with the international architecture of environmental agreements seen by WBGU among others as poorly interacting and not interrelated to a degree necessary. Eight years after the Rio Summit, more than 900 bi- or multilateral environmental agreements are in force, but the most urgent environmental problems remain unsolved. The lack of coordination and collaboration provoked the council to set up a vision for a framework of international environmental institutions and organizations to improve coordination called an Earth Alliance to be established with the United Nations.

All activities in the field of natural conservation should be harmonized and restructured under the wings of such new entity (Earth Alliance). This entity is to oversee all activities attributed to three crosscutting sectors: (a) Earth Assessment, (b) Earth Organization, and (c) Earth Funding. The duty of this central entity would be to issue timely warnings of environmental risks. This authority should be deliberately limited in size and have the rights to address the public as needed.

4.4.2.4 Earth Assessment

Knowledge and knowledge assessment are the keys to risk management. Following the example of the Intergovernmental Panel on Climate Change, the WBGU recommended the establishment of an IPCC comparable scientific body to advise and support the environmental activities under the United Nations, for example, combatting desertification and soil degradation, natural disaster risk management, or on biodiversity These tasks should be directed and coordinated by a special Earth Commission to be newly established. The Earth Commission body should provide long-term thinking and also should give motivation for research and political action. It should be composed of not more than 15 leading figures who can command the attention of a global audience, such as the Brundtland Commission has done successfully. The Earth Commission, with its scientific panels, should achieve four accomplishments in particular:

- Elaborating an overall perspective of the Earth system using the existing monitoring systems
- Giving early warning on potentially irreversible environmental damage
- Guiding international environment policy towards nature conservation and prevention based on scientific data and findings
- Annually reporting to the secretary general of the United Nations on the achievements, gaps, and future orientation

4.4.2.5 Earth Organization

The identified lack of coordination and effectiveness of global environment policy calls for a comprehensive reconfiguration of international institutional and organizational structure. In order to improve coordination and cooperation, WBGU pledges a strengthening of the UNEP to become an "International Environmental Organization" to secure a closer linkage among the different secretariats of the international environmental conventions and their scientific panels. The Council proposes among others:

- To proceed in "using protocols to advance the purposes of framework agreements," that define the general goals and implementation strategy, but in order to arrive at a speedier ratification of a convention, the voting procedures have to become more flexible. Therefore the consensus principle of "tacit acceptance" should be used more frequently.
- To modify those decision-making activities that affect human heritage as a whole. In such cases the formal principles "one state, one vote" or in agreements about financial contributions, the common practice of "one dollar, one vote" should be reconsidered in favor of a "one person, one vote."
- To introduce a general scientific appraisal of member states to report on the fulfillment of their commitments in the way of reach at a more transparent international compliance control function.

4.4.2.6 Earth Funding

The increasingly entrenching provision of industrialized nations' contributions for financing global environment policy led the WBGU Council to recommend three measures for a more efficient use of these funds:

- Only when the global earth policy is based on an appropriate and secure financing, will the challenges to master natural damages and climate changing impact be successful. This would require an increase of the 0.7 % of GNP target of financial contributions.
- The system to monitor and review the fund's expenditure of multinational organizations has to be strengthened. Only by a constant evaluation of compliance with the envisaged targets will it be possible to make achievements transparent and reliable.
- The Earth organization should be given overall trusteeship on global common resources such as international airspace, world oceans, outer space, and so on. The use of these resources should be levied from the users. The introduction of such an innovative financial instrument should be used to finance pre-defined mitigation activities.
- Earth Funding needs competition among a variety of individual, innovative financing schemes. The Advisory Council sees in the combination of the various financing instruments a distinct opportunity. It will further encourage global players to assume a more responsible use of global resources and no longer to plan their activities in accord with their own environmental standards. Additionally, global NGOs are in a position to influence the behavior of individuals, groups, and organizations at a point where government standards do not apply effectively and where private initiatives can take over. WBGU therefore rearticulates its call for the creation of institutional framework conditions that can galvanize the private sector and strengthen national noncommercial funds, for example, in combination with a worldwide debt-relief initiative.

4.4.2.7 Flagship Report 2007 "World in Transition: Climate Change as a Security Risk"

On the occasion of the G8 Summit, the Advisory Council submitted its report on "Climate Change as a Security Risk" (WBGU 2007). The core message of the report is that if no resolute counteractions undertaken, climate change will overstretch many societies' adaptive capacities. In some regions of the world, this may result in internal destabilization processes with diffuse conflict structures, interstate conflicts, and an overstretching of the international system. Classic security policy cannot respond adequately, making climate policy and strategies key elements of a preventive security policy. If it fails to do so, climate change will draw ever-deeper lines of division and conflict in international relations.

WBGU analyzed selected regional hotspots in more detail: Southern Africa, the Ganges Delta, and Amazonia. All were found to be especially at risk. Compared with the rest of the world, Africa is already most at risk from destabilization and violent conflict. Millions of people there are refugees from civil war and poverty, and climate change is likely to exacerbate simmering conflicts. A further example is Amazonia, where the collapse of the rainforest would have incalculable economic and social consequences. In addition to the developing countries, however, major ascendant economic regions such as the east coast of China are also at risk: here, storm and flood disasters could in future threaten densely populated cities and industrial regions with severe economic and social impacts.

The flagship report calls on resolute climate policy action within the next 10 to 15 years; in order to avert the socioeconomic disruption and negative implications for international security it recommended among others that the German federal government implement the following measures:

- A reduction in global greenhouse gas emissions of 50 %, against the 1990 baseline by 2050. For the period after 2012, the industrialized countries should commit themselves to a 30 % reduction in GHG emissions by 2020. Germany and the European Union should enter into strategic "decarbonization partnerships" with newly industrializing countries,
- In order to master the security challenges it is essential to promote the participation of the ascendant new world powers, China and India. Germany should undertake the necessary advocacy work within the European Union and work proactively at the international level for the adoption of confidence-building measures,
- The Security Council's mandate must be adapted to meet the security policy implications of climate change. Furthermore, the United Nations Environment Program (UNEP) should be strengthened and upgraded to the status of a UN specialized agency. Finally, the German government should lobby to increase the funds available to finance international crisis prevention and peacebuilding at the UN level.
- Climate change is likely to further destabilize fragile states. The German government is called on to initiate an action plan, "Civilian Crisis Prevention, Conflict Resolution and Post-Conflict Peace-Building." In this context, the European security strategy should also be enhanced with the aim of avoiding military intervention as far as possible. The WBGU also recommends the clear restructuring of military budgets in favor of preventive measures in the field of development cooperation.
- Climate change capacities of most developing countries lack the capability to implement effective adaptation measures. WBGU identifies a particular need for action in relation to water, food security, disaster prevention, and migration policy. In order to master these challenges, the European Union shall increase ODA to 0.56 % of gross national income by 2010 and to 0.7 % by 2015.
- In time of climate change timely warning on extreme events is becoming increasingly important. The German government should therefore continue to

participate in the development of a global early warning system. There is also a need to provide processed data on predicted regional climate impacts, especially for developing countries,

- Environmental migrants currently do not fit into the agreed categories of international refugee and migration law. Under current law, states have no specific obligations in relation to the treatment of environmental migrants, nor are any other legal mechanisms in place for their protection. A multilateral convention for environmental migrants should therefore be established.

4.4.3 German Strategy for Adaptation to Climate Change (DAS)

Acknowledging the fact of a changing climate and in accordance with the Germany's obligations in the frame of the United Nations Framework Convention on Climate Change (UNFCCC), the federal German government announced its central climate policy objective, in order to strengthen the national capacities to limit the increase in global temperature to less than 2 °C above the pre-industrial level. In accordance with Article 4 of the UNFCCC-Convention, the government initiated the "German Strategy for Adaptation to Climate Change" (DAS; BMUB 2011) that formulates a national strategy to master the follow-ups of the changing climate. The legal frame for the adaptation strategy is given by the German Constitution, Article 20a (GG) stipulating the "Protection of Nature and Environment," a national objective.

With the strategy, the government fulfills its obligations within the European Union to formulate an EU-wide harmonized strategy aiming at preventing the expected consequences of climate change. With this strategy, the federal government for the first time adopted a political framework on how to adapt to the consequences of climate change and how to establish a structured mitigation dialogue among all stakeholders. The adaptation strategy pursues an integrated approach to assessing risks and action need by defining adaptation objectives and identifying, developing, and implementing measures for a sustainable development. The internationally heightened debate on climate change issues makes clear that the time horizon of the climate change poses political, scientific, and technical challenges that neither have been experienced in history nor settled by international law thus far.

The adaptation strategy requires these action objectives:

- Identification of the probability of risks and the potential damages
- Creation of awareness with the societal stakeholders
- Incorporation of climate change issues in all public and private sectors
- Capacitation of the stakeholders to take necessary precautions to increase resilience
- Definition of mitigation options and defining responsibilities to implement measures

The adaptation strategy provides a comparative regional risk evaluation for Germany, based on four regional simulation models (ensembles). These ensembles evaluated the corridors of change and led to recommendations on the future risk reduction targets for the different German regions. The following key regions are especially sensitive to climate change:

- Central parts of eastern Germany, the northeast German plain and the southeast German basin and hills could be increasingly affected in future by a reduced water supply.
- The hill country on both sides of the Rhine is expected to see a general increase in precipitation with large consequences for agriculture and forestry, and for flood control. Heat waves in the Rhine rift valley could become more frequent and more intensive, and the risk of flooding could increase.
- The Alpine regions are very sensitive from the point of view of biodiversity. The retreat of the glaciers would have an impact on water resources. The risk of natural dangers, such as rockfalls or mudslides, must be expected to increase.
- The coastal regions could be increasingly at risk from the rise in sea level and changes in the storm climate. However, there is great uncertainty about the probable size of changes in sea level and the storm climate. One aspect of special importance is the potential danger to wetlands and low-lying areas and to regions with high damage potential, such as the port of Hamburg.

Following the issuance of the adaptation strategy, further provisions were made to the Competence Centre on Global Warming and Adaptation (KomPass) at the German Federal Environment Agency (UBA). KomPass will collate and evaluate information and results from the various subject areas and ministries and communicate them via an Internet portal. Second is the establishment of a Climate Service Centre at the Helmholz-Gesellschaft Deutscher Forschungszentren (Hamburg). The center is located at the interface between climate system research and users of the data obtained from scenario and model calculations. The aim is user-oriented acceleration of knowledge dissemination and research processes in the field of climate modeling and scenario development.

With the adaptation strategy the German government contributes to the challenges of an international climate policy. The government stresses the need for a rigorous examination of world development policy strategies, concepts, and programs as to whether they are sufficiently robust in the face of possible effects of climate change and whether they can help to strengthen the capacity of the societies to adapt ("climate check"). This addresses not only the direct impact from climate change but also indirect outcomes, such as those concerning the global security architecture, poverty migration, and sustainable development. With the DAS, Germany has accepted to play an active part in the development of relevant climate risk reduction concepts under the Framework Convention on Climate Change, including the development of suitable mechanisms for financing adaptation measures in developing countries

4.5 The International Risk Governance Council (IRGC)

The IRGC was established in 2003 (IRGC 2003) as a consequence of the heightened level of public concern about the lacks in natural disaster risk management in the late 1990s. Although the council was established by the Swiss Parliament, the IRGC functions as an independent and international body to bridge the natural sciences, technological development, and the socioeconomic sector, the spatial planners, and the risk-affected population. The mission of the council is to act as the catalyst for improvements in the design and implementation of effective risk governance strategies by developing concepts of risk governance. To ensure the objectivity of its governance recommendations, IRGC activities are backed by international scientific expertise from both the public and private sectors (IRGC 2010). The Scientific and Technical Council (S&TC) is the leading scientific authority of the foundation. Its members comprise experts from a range of scientific and organizational backgrounds who overview the scientific quality of IRGC work and assure the quality of IRGC's publications and other deliverables. They provide input and scientific advice to the work program, and advice on selection of affiliates to the IRGC network. The IRGC operates as an independent think-tank with multidisciplinary expertise that can help bridge the gaps among science, technological development, policymakers, and the public. Since 2003 the IRGC has developed a series of fact-based risk governance recommendations for policy makers to help them anticipate and understand emerging risks, as well as the risk governance options, before they become urgent policy priorities.

The main objectives of the IRGC are:

- Identifying potential risk issues at the earliest possible stage
- Understanding the issue and the associated risks as well as the institutions and risk governance structures and processes that are currently in place for assessing and managing the risks
- Identifying governance gaps that appear to hinder the efficacy of the existing risk governance structures and processes
- Making recommendations for overcoming these gaps

Essential to the Risk Governance Framework (IRGC 2009) is to gain a thorough understanding of a risk and to develop options for dealing with it. Therefore the IRGC's risk governance framework distinguishes between analyzing and understanding a risk and deciding what to do about a risk. This distinction reflects the IRGC's strategy for the clear separation of the responsibilities for risk appraisal and management as a means of maximizing the objectivity and transparency of both activities. Those responsible for both should be jointly involved in the other three elements of the framework:

Pre-Assessment

This defines the baseline for how a risk is assessed and managed. It captures the variety of consequences that stakeholders and society may associate with a certain risk and delineates measures and routines that help to improve the resilience of either the individual or certain social groups. The IRGC's pre-assessment approach begins with "early warning" and "framing" the risk in order to provide a structured definition of the problem and how it may be handled. This approach should address the questions:

- What kind of risks and risk reduction opportunities can be addressed?
- What are the dimensions of the risk?
- How are the limits for risk assessment evaluations defined?
- Which indications are already defining the risk?
- Is there a need to act?
- Who are the stakeholders and what are their perceptions of the problem?
- What are the established scientific/analytical tools and methods that can be used to assess the risks?
- What are the current legal/regulatory systems and how do they potentially affect the problem?
- What are the organizational capabilities of the relevant governments, international organizations, businesses, and people involved?

Appraisal

Risk appraisal generates and synthesizes the knowledge base for the decision on whether a risk can possibly be reduced or at least contained. Risk appraisal comprises both scientific risk assessment, the field of the risk's factual, physical, and measurable characteristics including the probability of it happening, and a concern assessment, a systematic analysis of the perceived consequences (benefits and risks) that stakeholders may associate with a hazard. The concern assessment is a particular innovation of the IRGC framework, ensuring that decision makers account for how the risk is viewed when personal perceptions and emotions come into play. Scientific risk assessment deals with these types of questions:

- What are the potential damages or adverse effects?
- What is the probability of occurrence?
- How ubiquitous and persistent could the damage be?
- Is there a clear cause–effect relationship?
- What scientific, technical, and analytical approaches, knowledge, and expertise should be used to better assess these impacts?
- What are the primary and secondary benefits, opportunities, and potential adverse effects?

Characterization and Evaluation

The IRGC deliberately included this element to ensure that scientific evidence-based disaster risk management is combined with a thorough understanding of societal values in order to make the decision whether a risk is "acceptable" (risk reduction is considered unnecessary), "tolerable" (to be pursued because of its benefits), or "intolerable" (to be avoided). This phase involves making a judgment based on such questions as:

- What are the societal, economic, and environmental benefits and risks?
- Are there impacts on quality of life?
- Are there ethical issues to consider?
- Is there a possibility of substitution?
- Can the risks be compared (regionally, socially)?
- Does a choice of a particular risk-reduction technology have an impact on other risks?
- What are the possible technical, financial, and social options for risk reduction?
- Can risk compensation be a matter of mitigation strategy?
- What are the societal values and norms for making judgments about tolerability and acceptability?
- Do any stakeholders have commitments or other reasons for wanting a particular outcome of the risk governance process?

Management

All tolerable and intolerable risks need an appropriate and adequate risk management. According to the IRGC, risk management involves the design and implementation of the actions and remedies to avoid, reduce, transfer, or retain the risks. Risk management includes the selection of appropriate risk-reduction options as well as implementing the selected measures, monitoring their effectiveness, and reviewing the decision if necessary. Based on the range of options the most appropriate risk-reduction measures are to be put into practice. The questions are:

- Who is, or should be responsible for decisions within the context of the risk management?
- Have the decision makers accepted this responsibility?
- What management options could be chosen (technological, regulatory, institutional, educational, compensation, etc.)?
- How are these options evaluated and prioritized?
- Is there an appropriate level of international cooperation and harmonization for global or transboundary risks?
- What are the secondary impacts of particular risk-reduction options?

References

Berkes,F., Folke, C.& Colding, J. (2000): Linking Social and Ecological Systems: Management Practices and Social Mechanisms for Building Resilience. - Cambridge University Press, Cambridge MD

BMUB (2011): The German Adaptation Strategy to Climate Change. – Federal Ministry for the Environment, Nature Conservation, Building and Nuclear Safety, (BMUB), Berlin (online. www.bmub.bund.de)

BMUB (2013): Wasserhaushaltsgesetz (WHG) - Gesetz zur Ordnung des Wasserhaushalts vom 31. Juli 2009 (BGBl. I S. 2585), Bundesministeriums für Umwelt, Naturschutz, Bau und Reaktorsicherheit (BMUB), Berlin (online. www.bmub.bund.de)

CTBT (2013): Final declaration and measures to promote the entry into force of the Comprehensive Nuclear-Test-Ban Treaty. - Conference on Facilitating the Entry into Force of the Comprehensive Nuclear-Test-Ban Treaty, United Nations, New York NY

Ebert, W. (2012): Natura 2000 - bis hierher und wie weiter? Ehrgeiziges Naturschutzprojekt mit durchwachsener Bilanz. –Naturschutzbund (Nabu), Naturmagazin Berlin-Brandenburg, Vol. 4, Berlin

EU (1992): European Commission's Humanitarian Aid and Civil Protection Department (ECHO).- European Commission, Brussels (online: http://ec.europa.eu/echo)

EU (2003): Memo on commission strategy to protect Europe's most important wildlife areas – frequently asked questions about NATURA 2000. - ec.europa.eu/nat2000/2003_memo_natura.pdf

EU-DIPECHO (1992): DIPECHO program.- Disaster Preparedness Program DIPECHO.- European Commission, Brussels (online: http://europa.eu.int/comm/echo/index_en.html)

EU-EN (1998): EUROCODE 8 – Design of structures for earthquake resistance.- European Committee for Standardisation, European Commission, Brussels (online: http://ec.europa.eu/enterprise/sectors/construction/index_en.html)

EU-EWFD (2000): Directive 2000/60/EC of the European Parliament and of the Council establishing a framework for the Community action in the field of water policy" (EU Water Framework Directive-EWFD).- European Commission, Brussels

FEMA (2007): Robert T. Stafford Disaster Relief and Emergency Assistance Act, as amended, and Related Authorities.- Federal Emergency Management Agency (FEMA), Publication 592, Washington DC

GFDRR (2012): Strategy for the Global Facility for Disaster Reduction and Recovery (GFDRR) – Managing Disaster Risks for a resilient Future. - The International Monetary Fund / The World Bank, Washington DC (online: http://www.gfdrr.org/gfdrr/node/1301)

ICPR (2010): Our common objective: Living waters in the Rhine catchment. - International Commission for the Protection of the Rhine (ICPR), Koblenz (online: www.icpr.org).

IPCC (1988): Intergovernmental Panel on Climate Change (IPCC).- United Nations General Assembly Resolution 43/53 of 6th Dec. 1988, New York NY

IPCC (1990): Climate Change : The IPCC Scientific Assessment 1990: Report prepared for the Intergovernmental Panel on Climate Change.- Intergovernmental Panel on Climate Change (IPCC), Cambridge University Press, Cambridge-New York NY

IPCC (1996): IPCC Second Assessment - Climate Change 1995: A Report of the Intergovernmental Panel on Climate Change (SAR).- Intergovernmental Panel on Climate Change (IPCC), Cambridge University Press, Cambridge-New York NY

IPCC (2001): Climate Change 2001: - Third Assessment Report The Scientific Basis, Section 2, "Observed climate variability and change".- Special Report of the Intergovernmental Panel on Climate Change, Cambridge University Press Cambridge MD

IPCC (2007): Climate Change 2007: Synthesis Report- Contribution of Working Groups I, II and III to the Fourth Assessment Report of the Intergovernmental Panel on Climate Change.- Intergovernmental Panel on Climate Change (IPCC), Cambridge University Press, Cambridge MD

IRGC (2003): International Risk Governance Council (IGRC), Lausanne (online: www.igrc.org).

IRGC (2009): Risk Governance: Coping with Uncertainty in a Complex World. – prepared for the International Risk Governance Council (IGRC), Natural Hazards, Vol. 48, No. 2, p. 313–314, Springer-online

IRGC (2010): What is Risk Governance. - International Risk Governance Council (IGRC), Lausanne.

IRM (2002): A risk management standard.- Institute of Risk Management; The Association of Insurance and Risk Managers (AIRMIC) and The Public Risk Management Association, Alexandria, VA
ISO 31000 (2009) Risk management – Principles and Guidelines. - International Organization for Standardization, Geneva (online: www.iso.org)
ISO (2014): International Organization for Standardization (online: www.iso.org.)
NATURA 2000 (2000): Council Directive on the Conservation of Natural Habitats and of Wild Fauna and Flora. – European Commission, EU-Directive 92/43/EWG, Brussels
Ostrom, E. (1990): Governing the Commons - The evolution of institutions for collective action. - Cambridge University Press, Cambridge MD
UN (1987): Brundtland Report - Report of the World Commission on Environment and Development "Our Common Future".- United Nations General Assembly, World Commission on Environment and Development, New York NY
UN (1991): Strengthening of the coordination of humanitarian emergency assistance of the United Nations.- United Nations General Assembly, Office for the Coordination of Humanitarian Affairs (OCHA), Resolution A/RES/46/182, New York NY
UN (2000): Millennium Development Goal. – United Nations General Assembly, United Nations Millennium Declaration, New York NY
UNCED (1992): Earth Summit – Agenda 21.- United Nations, UN Conference on Environment and Development (UNCED), Rio de Janeiro
UNFCCC (1992): United Nations Framework Convention on Climate Change (UNFCCC).- United Nations, United Nations Conference on Environment and Development (UNCED), New York NY
UNFCCC (1997): The Kyoto Protocol.- United Nations, United Nations Framework Convention on Climate Change, New York NY
UNIDNDR (1994): Yokohama Strategy for a Safer World: Guidelines for Natural Disaster Prevention, Preparedness and Mitigation and its Plan of Action - A safer World in the 21st Century: Disaster and Risk Reduction, Yokohama (23–27th May, 1994), International Decade of Natural Disaster Reduction (IDNDR), United Nations, Geneva
UNIDNDR (2000): The International Decade for Natural Disaster Reduction (IDNDR), 1990–1999 - a report on the activities of the UK National Coordination Committee for the International Decade of Natural Disaster Reduction (IDNDR), Geneva
UNISDR (2005): Hyogo Framework for Action 2005–2015: Building the Resilience of Nations and Communities to Disasters. - Extract from the Final Report of the World Conference on Disaster Reduction (A/CONF.206/6), p. 25, Geneva
WBGU (1996): Welt im Wandel: Herausforderung für die deutsche Wissenschaft .- Jahresgutachten 1996, Wissenschaftlicher Beirat der Bundesregierung Globale Umweltveränderungen, Springer-Verlag Berlin Heidelberg, New York NY
WBGU (1999): Welt im Wandel – Strategien zur Bewältigung globaler Umweltrisiken. – Wissenschaftlicher Beirat der Bundesregierung Globale Umweltverän-derungen, Hauptgutachten 1998, Springer, Heidelberg
WBGU (2000): Welt im Wandel: Neue Strukturen globaler Umweltpolitik.- Jahresgutachten 2000, Wissenschaftlicher Beirat der Bundesregierung Globale Umweltveränderungen, Springer-Verlag Berlin Heidelberg, New York
WBGU (2007): Welt im Wandel: Sicherheitsrisiko Klimawandel.- Wissenschaftlicher Beirat der Bundesregierung Globale Umweltveränderungen, Hauptgutachten 2007, Springer Heidelberg
WBGU (2014): Homepage „Wissenschaftlicher Beirat der Bundesregierung Globale Umweltveränderungen" (WBGU).- Minister für Bildung und Forschung (BMBF) / Minister für Umwelt, Naturschutz und Reaktorsicherheit (BMU), Berlin

Chapter 5
Hazard Assessment

5.1 General Aspects of Data Monitoring and Measuring

Natural disasters strike the people in general randomly, infrequently, and without prior warning. But frequency and impact of the natural disasters are not at all equally distributed over the world. There is a clear geographical and time-related preference that results in some regions of the globe being more often and more heavily affected than others. World Bank reports have recorded more than 3 million deaths from natural hazards recorded since 1970, making an average death toll of about 82,000 a year (Kreimer and Munasinghe 1991; Dilley et al. 2005). The emergency disaster database of CRED (Guha-Sapir et al. 2013) listed over 8800 events that claimed the lives of about 2.2 million people. In the same period economic losses worth US$1500 billion were caused by disasters. But mortality and economic losses are not equally distributed over the globe. There are regions that are more prone to fatal disasters and areas that are more exposed to economic losses: in Asia more people die from natural disasters whereas in Northern America more assets are subject to destruction. All disaster records clearly reveal that disasters predominantly affect the poor in the developing countries, a fact that has been highlighted in many publications especially by the UNDP Report (2004, 2013; World Bank 2013).

There is a variety of definitions of "hazard" (see Annex A). Very generally, hazards are: "natural conditions that can cause damages on property, claim death and injury or destroys the economic and ecological base of human life."

From a geological standpoint the Earth is a "product" of geological processes that define the distribution of land and sea, of mountains and plains, rivers, and lakes, of meteorological processes that constitute our atmosphere. All of these processes are steadily changing the face of our Earth and are thus a phenomenon of which human mankind is a part. But these changes are not in every case running smoothly, gently, and peacefully, rather sometimes are very sudden, eruptive, and incidental. A volcano, for example, that has long been producing lava and ashes in the middle of the Saharan desert can be seen from the standpoint of a geologist or volcanologist a fascinating event, but does not constitute a hazard, as long as the

© Springer International Publishing Switzerland 2016

U. Ranke, *Natural Disaster Risk Management*,
DOI 10.1007/978-3-319-20675-2_5

gases from the eruption are not changing the world's climate. Also seawater that for centuries regularly floods marshland is a natural process that helps sustain the local ecology, but also is not a hazard, as long as the marshland is envisaged to be used for agriculture. Winds that are drifting Saharan desert sands westward to the Atlantic Ocean are a phenomenon but do not constitute a hazard as long as the winds are not eroding fertile soils thus deteriorating the local population's living source.

But not only nature can be hazardous to humankind; human activity also often interferes with the natural system that may again interfere with the human environment. Human-induced hazards originate from technological or industrial accidents, dangerous procedures, infrastructure failures, or certain human activities, which may cause the loss of life or injury, property damage, social and economic disruption, or environmental degradation (e.g., industrial pollution, nuclear activities and radioactivity, toxic wastes, dam failures, transport, industrial or technological, all result from modifications of natural processes in the Earth's system caused by human activities which accelerate and aggravate the damage potential (e.g., land degradation, landslides, forest fires). Human-made hazards also often interfere with the climate and have already led the sea-level increase and warmed the global temperature.

Natural processes or phenomena are hazards when they have the potential to become a physical event (natural disaster) that can harass human life. In this regard it should be acknowledged that the term "natural disaster" is in the strict sense of the word not correct, as the term comprises two different viewpoints: one is "natural" that defines all processes that occur on the planet and thus are subject to natural sciences, and "disaster" that describes a paradigm defining adverse interference with humankind and the living environment. Nevertheless the term has overall acceptance. Consequently the term "natural disaster management" should be extended to "natural disaster risk management" as in light of the above, nature cannot be managed: only the impact from such processes to the living environment can be managed and sometimes even changed for the better.

Natural hazards occur all over the world, although the regional distribution of the different hazard types follows a distinct relationship to geological and climate patterns (see also Sect. 3.3). Hazards can be single, sequential, or combined in their origin and effects. Each hazard is characterized by its location, area affected, magnitude and intensity, duration, probability, and frequency as well as the speed of onset (Fig. 5.1).

The graph clearly demonstrates that the different indicators describing a hazard vary a great deal and that there is no "single" relationship between them. There "fast" hazards ("lead time to onset") such as a snow avalanche or tsunami that on the other hand have very different areas of impact ("regional distribution"). A snow avalanche in the Alpine region is normally restricted to a certain valley whereas tsunamis may be extended over very large areas, like the 2004 tsunami that affected the entire Indian Ocean region. Drought and heat waves share slow to very slow onset speeds (sometimes decades) and both are regionally largely extended and their probability is (comparatively) low. Earthquake and volcano

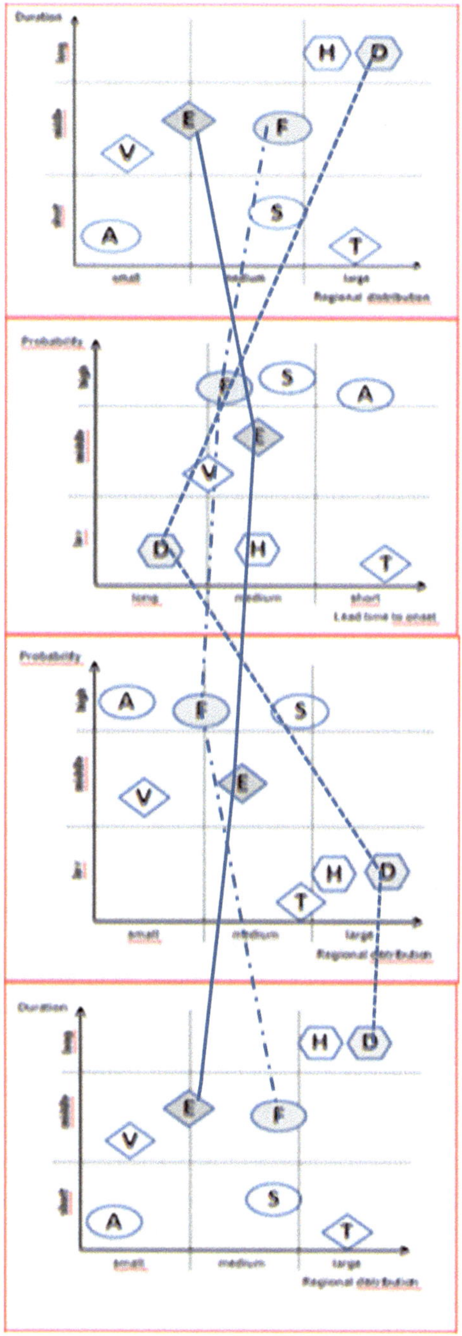

Fig. 5.1 Comparison of different hazard types by probability, regional distribution, lead time, and duration (*A* snow avalanche, *E* earthquake, *D* drought, *H* hurricane, *S* storm, *T* tsunami, and *V* volcanic eruption; Own graph)

eruption are in general local events that, however, experience a high probability but moderate lead times.

There is no hazard type that has the highest probability, the largest regional distribution, the strongest impact, the longest duration, or the shortest lead time. On the other hand there is no hazard type that has the lowest probability, the shortest duration, the smallest regional distribution, or the longest lead time. The very different character of the hazard types makes working out a generalized risk-reduction strategy difficult, not to say impossible. However, there is in the large difference in hazard types the clue for hazard-specific risk-reduction measures.

5.2 International Classification Scales of Hazard Intensity

Disaster risk mitigation requires objective means to compare different disaster events at different locations either globally or within a country, for example, on volcanic eruption, earthquakes, flooding, tsunamis, forest fires, heat, wind, and storm surges, hail, avalanches, and so many others. There are quite a number of scales in use that enable scientists and risk managers to better qualify each event according to its severity and frequency. The main criteria are in general physical parameters that were instrumentally recorded or assessed by visual inspection. The criteria mainly comprise casualty or injury of disaster victims, and actual or possible damage to buildings or the natural environment. The indicators chosen are in general independent of the site of damage occurrence, are reproducible, and can be traced back into geological history. They are simple, neutral, and indicative in order to also serve the needs of the nonspecialist.

All major disaster management agencies worldwide such as PHIVOLCS, UN-OCHA, the European Union, or the Swiss-BUWAL (see below) have set up classification schemes of natural hazards to better address the hazards. All such schemes have in common that they intend to standardize and harmonize disaster impact data in order to make them understandable, transferable, and comparable worldwide. Therefore it is not very important which of the many classification schemes is used; important is that for one assessment the same scheme is in use and the risk assessors are well trained.

In the following, some examples of commonly used alert scales are reproduced from different published sources. The choice gives an overview of how different nations are addressing the alert level definitions. Many of the great industrial and technical oriented countries often set a high value on the geoscientific background and administrative operational aspects of warning, whereas many developing countries (e.g., the Philippines) often provide additional information for the public at risk on how to assess the individual risk and how to initiate mitigation measures.

Using the different magnitudes in identifying the severity and frequency of natural hazards is daily practice in disaster hazard assessment, although it may lead to significant problems. The information on the severity and frequency of disaster events is generally based on scientific measurements, given in ordinal numbers from zero to ten or more. Such standardization is meaningful, if the momentum of just one hazard type, for example, an earthquake, is compared to another within a country or between different countries. An M5 earthquake in Indonesia has released the same energy as an M5 earthquake in the United States. But when different kinds of hazards are compared, for example, in order to assess the overall hazard exposure of a certain area, different hazard scales are to be combined and compared. But how can a volcano eruption with a severity of say, VEI 4 be compared with a flood level measured say, 5 m above normal. There is a methodological dichotomy to be solved. Generally this bias is bridged by attributing the different quantitatively generated figures, a qualitative description, often ranging from nil to very high.

A problem arises from the fact that an arbitrarily described hazard exposure ("low–moderate–high") for a certain region for hazard management purposes has to be transferred back into quantitative indicators. For example a "high" ranked flood level, can in the following serve as the baseline for a flood threshold indicator. For this the formerly arbitrarily given ranking "high", subsequently will be transferred into an ordinal number, say 7 m above normal. A 7-m flood level will then be the threshold value that starts a flood warning or evacuation.

5.2.1 Earthquake Magnitude and Intensity Scale

Richter Scale and Mercalli Scale

Information for the public and scientific purposes on earthquake occurrence is in general answering the questions of what happened where, by what severity and causes, and what kind of impact. The basic scientific indicator for ranking the strength of such events is the Richter scale already described in detail in Sect. 5.2. The Richter scale measures the energy of an earthquake by its ground acceleration, however, the Mercalli scale is a measure of the earthquake's intensity. However the Mercalli intensity is not exclusively determined by the earthquake's magnitude, rather by geological cum sedimentological features as well as by a building's substance. For engineering purposes it is therefore indispensable that a correlation between magnitude of the quake and intensity of damage is assessed, based upon several factors including the depth of the earthquake, the site-specific geological and sedimentological features of the location, building structure stability, as well as factors such as population density and the like.

5.2.2 Volcano Explosivity Index

The Volcanic Explosivity Index (VEI) was established in 1982 by the US Geological Survey to provide an assessment tool that ranks the severity of volcanic eruption (Newhall and Self 1982). The index is based on the records of over

8000 historic and prehistoric eruptions and provides a good knowledge of the frequencies of highly explosive, moderately explosive, and nonexplosive eruptions. It is a composite estimate of the magnitude of past explosive eruptions. It is based on the volume of volcanic material (ash, volcanic bombs, lapilli, etc.) ejected, the height of the eruption cloud, how long the eruption lasts, the occurrence of pyroclastic flows, the height of the ash column, and some specific qualitative observation indicators using terms ranging from "gentle" to "mega-colossal". The VEI only refers to andesitic/dacitic lava and does not comprise basaltic lava as it has, according to lava composition, no explosivity potential. Also the density of volcanic material and the gas content (vesicularity) are not considered. Moreover the VEI cannot give an indication of how strong the ejection was, as there is no way to measure such a value reproducibly. Each factor is given in a numerical qualifier on an open-ended scale with the largest volcanoes in history given magnitude 8. A value of 8 represents a "mega" explosive eruption that can eject more than 1000 km^3 of tephra and have a cloud column (plume height) of over 50 km. The scale is logarithmic, with each interval on the scale representing a tenfold increase in observed ejected material, with the exception of between VEI 0, VEI 1, and VEI 2.

The authors defined the index as a simple scheme for estimating explosive magnitude, although knowing its limitations, as it provides a combination of quantitative or semi-quantitative assessments as a basis for comparing explosive eruptions. The index has a simple numerical index of increasing magnitude (1–8), as eruptions greater than 8 could not be reconstructed from history. The volume of tephra ejected increases by a factor of 10 for each VEI-interval; except the step from VEI 1 to VEI 2 where the tephra ejection volume increases by 100. The VEI furthermore provides a generalized assessment of an expected climate impact of the eruptions as well as the number of eruptions since Holocene times. For comparison, recent volcanic eruptions are added, such as the ones of Mt. St. Helens or Krakatoa.

The problem establishing the index scale was that records of eruptions in historic times often are highly incomplete and lack quantifiable data: therefore many of the historic data had to be estimated, especially where lava temperatures are concerned. Only the largest eruptions of Mt. Tambora (1815), Indonesia, had a complete record, but reliable information is not dated before 1960. The compilation of the magnitude of historic volcanism was mainly based on professional experience and expertise for each eruption and an analogous reporting system did not exist in those times. In this regard the VEI differs much from other scales such as the Richter scale that is based on instrumental records.

Table 5.1 summarizes the VEI-scale and is extended by the factor "dispersion", defined as the "outer limit where the volcanic deposit thickness decreases to 1 % of its maximum thickness" (Walker 1973; cited in Cas and Wright 1988).

All data on volcano eruption events are documented in the largest volcano databank of the world: the Global Volcanisms Program of the Smithsonian Institute (GVP). This databank is the worldwide unique documentation of current and past

Table 5.1 Volcanic explosive index (summarized from Newhall and Self 1982; Walker 1973)

	Ejecta volume	Classification	Description	Plume	Frequency	Stratospheric injection	Dispersal	Example
1	>10,000 m^3	Strombolian	Gentle	<1 km	Daily	None	>0.05 km^2	Nyiragonqo, Raoul lid
2	>1 mio m^3	Strombolian	Explosive	1–5 km	Weekly	None	>2.5 km^2	Unzen Galeras
3	>10 mio m^3	Pelean	Severe	3–15 km	Few months	Possible	>5 km^2	Nevada del Ruiz soufriere
4	>0.1 km^3	Pelean/Plinian	Cataclysmic	10–25 km	>1 year	Definite	>200 km^2	Mayan Mt. Pelee
5	>1 km^3	Plinian	Paroxysmal	20–35 km	>10 years	Significant	>500 km^2	Vesuvius Mt. St. Helens
6	>10 km^3	Plinian	Colossal	>30 km	>100 years	Substantial	>1000 km^2	Krakatao Pinatubo
7	>100 km^3	Ultra-Plinian	Super-colossal	>40 km	>10,000 years	Substantial	>5000 km^2	Thera Tambora
8	>1000 km^3	Super-volcanic	Mega-colossal	>50 km	>10,000 years	Substantial	>10,000 km^2	Yellowstone Toba

activity for all volcanoes on the planet active during the last 10,000 years. The GVP‘s mission is to document, understand, and disseminate information about global volcanic activity. The databank that was established in 1968 documents current eruptive activities and regularly publishes on volcano activities in the “Weekly Volcanic Activity Report” and in the *Bulletin of the Global Volcanism Network*. The “Weekly Volcanic Activity Report” is a cooperative project between the Smithsonian's Global Volcanism Program and the US Geological Survey's Volcano Hazards Program updated every Wednesday. The reports do not intend to provide a comprehensive list of all of Earth's volcanoes erupting during the week, but rather provides comprehensive reporting on recent eruptions on a longer time horizon. The *Bulletin of the Global Volcanism Network*, however, carefully reviews the different reports on various volcanoes and is published monthly. The GVP website presents more than 7000 reports on volcanic activity, provides access to the baseline data and eruptive histories of Holocene volcanoes, and is openly accessible.

5.2.3 Examples of Volcanic Alert Levels

Volcano Alert Level (United States of America)

By the Stafford Act, the law on natural disaster emergency management of the United States of America, the US Geological Survey (USGS) is mandated to monitor volcanic activity in the United States and give a warning according to an

Table 5.2 Volcano alert levels of the United States of America (USGS 2006)

Level	Criteria
Normal	Volcano is in a noneruptive state. This level allows for periods of increased steaming, seismic events, deformation, thermal anomalies, or detectable levels of degassing as long as such activity is within the range of typical noneruptive phenomena based on its monitored history. Or the volcano has returned from a higher level of activity to a noneruptive background state
Advisory	Volcano is exhibiting signs of elevated unrest, increased steaming, seismic events, deformation, thermal anomalies, or detectable levels of degassing above known background level. But a progression towards eruption is not certain. After a change from a higher level or volcanic activity has decreased significantly, but requires still requires close monitoring for possible renewed increase
Watch	Volcano is exhibiting heightened or escalating unrest with increased potential of eruption, or minor eruption is underway but poses limited hazards. This level is used for both heightened precursory unrest and for minor eruptive activity because both states require continuous monitoring but no immediate hazardous effects are expected. When changing from advisory to watch this implies an increased potential for an eruption. When changing from warning to watch, the volcano is still showing signs of heightened activity that may lead to renewed highly hazardous activity
Warning	This level implies that a major or highly hazardous eruption is underway, is confirmed, imminent, or at least suspected. When the activity cannot be confirmed by visual evidence, ground-based monitoring data may indicate the probable onset, duration, size, intensity, or explosivity of the eruption as well as the potential impact on landscape and atmosphere

international standardized procedure. The USGS assigns an alert level based upon a dense monitoring network and interpretation of changing phenomena. Four alert levels are defined (Table 5.2) reflecting the status at the volcano and the expected or ongoing hazards and function as guidelines for scientists to assess the level of hazardous volcanic activity as well as for public officials and the public to consider when deciding what actions they need to take.

Volcano alert notices are accompanied by explanatory text to give a fuller explanation of the observed phenomena and to clarify hazard implications to affected groups. Updates that describe the ongoing activity are issued on a regular basis, at increasing frequency at higher activity levels.

Volcano Alert Level (Taal Volcano, Philippines)

The following volcano alert levels have been established by the Philippine Institute of Volcanology (PHIVOLCS 2014a, b) for the purpose of volcano warning of Mt. Taal, one of the most active volcanoes in the Philippines. The warning scale (Table 5.3) offers additional information for the public settling the area on the potential development of the unrest. Similar alert levels have been set up for many other volcanoes in the Philippines, for instance, of the volcano Hibok-Hibok on Mindanao.

Table 5.3 Volcano alert levels of the Mt. Taal Volcano, the Philippines (PHIVOLCS 2014a, b)

Alert level	Criteria	Interpretation
No alert ("normal")	Seismic background noise Volcano quiet	No eruption in foreseeable future
1 ("abnormal")	Low level of seismicity Fumarolic Other activity	Magmatic, tectonic of hydrothermal disturbances No eruption imminent
2 ("alarming")	Low to moderate level of seismicity, persistence of local but unfelt earthquakes. Ground deformation measurements above baseline levels. Increased water and/or ground probe hole temperatures, moderate bubbling at Crater Lake	(a) Probable magmatic intrusion, could eventually lead to an eruption (b) If trend shows decline volcano may soon go back to level 1
3 ("critical")	Relatively high unrest manifested by seismic swarms including increasing occurrence of low-frequency earthquakes and/or harmonic tremor (some events felt). Sudden or increasing changes in temperature or bubbling activity or radon gas emissions of Crater Lake pH. Bulging of the edifice and fissuring may accompany seismicity	(a) If unrest is increasing, eruption may be possible within days to weeks (b) If trend is decreasing, volcano may soon go to level 2
4 ("eruption imminent")	Intense unrest, continuing seismic swarms, including harmonic tremor and/or low-frequency earthquakes that are usually felt, profuse steaming along existing and perhaps new vents and fissures	Hazardous explosive eruption is possible within days
5 ("eruption")	Base surges accompanied by eruption columns or lava fountaining or lava flows	Hazardous eruption in progress Extreme hazards to communities west of the volcano and ashfalls on downwind sectors

5.2.4 *Volcano Alert Levels for International Air Traffic (Aviation Color Code)*

Volcanic eruptions pose a steady threat to aviation safety when clouds of volcanic ash disperse into the air and cross the flight paths of jet aircraft. Numerous incidents of aircraft flying into volcanic-ash clouds have experienced the life-threatening potential of this type of hazard. In the aftermath of such incidents the International Civil Aviation Organization (ICAO 2004) has established the International Airways Volcano Watch System (IAVW) with nine Volcano Advisory Centers worldwide that control air traffic routes of occurrences of explosive

Table 5.4 ICAO Volcanic alert color code

ICAO color code	Status of Volcano's activity
GREEN	Volcano is in normal noneruptive state. Or, after a change from a higher level: Volcanic activity considered to have ceased, and volcano reverted to its normal, noneruptive state
YELLOW	Volcano is experiencing signs of elevated unrest above known background levels. Or, after a change from a higher level: Volcanic activity has decreased significantly but continues to be closely monitored for possible renewed increase
ORANGE	Volcano is exhibiting heightened unrest with increased likelihood of eruption. Or: Volcanic eruption is underway with no or minor ash emission
RED	Eruption is forecast to be imminent with significant emission of ash into the atmosphere likely. Or: Eruption is underway with significant emission of ash into the atmosphere

(*Courtesy* ICAO-online: http://www.icao.int/NACC/Documents/eDOCS/ATM/ATMVolcanicAshContingencyPlanEN.pdf)

volcano eruptions and the whereabouts of airborne ash clouds globally. The ICAO furthermore institutionalized an internationally binding color code for quick reference that describes the actual volcano hazard alert levels (Table 5.4).

5.2.5 *Beaufort Wind Scale (United Kingdom, Royal Navy)*

Traveling the sea, wind is a major factor defining the maritime business. However, since weather observations were first made there was for a long time no standard scale that enabled the sailors to compare weather observations at different locations. In 1805 Sir Francis Beaufort succeeded in standardizing a wind scale. Based on previous work of others he defined what is, for instance, a "storm", a "gale", or a "breeze". The scale he invented (Table 5.5), was made up by increasing classes from zero to twelve, mainly making reference to qualitative wind conditions that in those days affected the sail canvas and the ship's maneuverability. Today windspeed is instrumentally measured by anemometers; nevertheless the classes are harmonized to the Beaufort Scale. Windspeed is affected by a number of physical factors and climatic and geographical situations. The main constituents are the wind direction and local weather conditions. Windspeed is mostly linked to the air pressure gradient, which describes the difference in air pressure inside and outside a low or high pressure location.

5.2.6 *The Saffir–Simpson Hurricane Wind Scale*

The wind scale is a 1–5 categorization based on the hurricane's intensity. The scale, originally developed by Saffir and Simpson (1971) has proven an excellent tool for comparing the various intensities of hurricanes. Mirroring the utility of the

Table 5.5 Beaufort scale (Wikipedia; online access, 20.7.2014)

Force	Knots	WMO classification	On water	On land
0	Less than 1	Calm	Sea surface smooth and mirrorlike	Calm, smoke rises vertically
1	1–3	Light air	Scaly ripples, no foam crests	Smoke drift indicates wind direction, still wind vanes
2	4–6	Light breeze	Small wavelets, crests glassy, no breaking	Wind felt on face, leaves rustle, vanes begin to move
3	7–10	Gentle breeze	Large wavelets, crests begin to break, scattered whitecaps	Leaves and small twigs constantly moving, light flags extended
4	11–16	Moderate breeze	Small waves becoming longer, numerous whitecaps	Dust, leaves, and loose paper lifted, small tree branches move
5	17–21	Fresh breeze	Moderate waves taking longer form, many whitecaps, some spray	Small trees in leaf begin to sway
6	22–27	Strong breeze	Larger waves, whitecaps common, more spray	Larger tree branches moving, whistling in wires
7	28–33	Near gale	Sea heaps up, waves, white foam streaks off breakers	Whole trees moving, resistance felt walking against wind
8	34–40	Gale	Moderately high waves of greater length, edges of crests begin to break into spindrift, foam blown in streaks	Twigs breaking off trees, generally impedes progress
9	41–47	Strong gale	High waves, sea begins to roll, dense streaks of foam, spray may reduce visibility	Slight structural damage occurs, slate blows off roofs
10	48–55	Storm	Very high waves with overhanging crests, sea white with densely blown foam, heavy rolling, lowered visibility	Seldom experienced on land, trees broken or uprooted, considerable structural damage
11	56–63	Violent storm	Exceptionally high waves, foam patches cover sea, visibility more reduced	
12	64 and more	Hurricane	Air filled with foam, waves over 45 feet, sea completely white with driving spray, visibility greatly reduced	

Richter scale on earthquakes, they devised a 1–5 categories scale, based on windspeed that today is the internationally agreed-upon sole measure to classify tropical hurricanes in the Western Hemisphere east of the International Date Line. The other areas use different scales to categorize their tropical storms. In the Pacific Ocean they are called "Taifoons" and in the Indian Ocean "Cyclones." The scale is based on the maximum sustained surface windspeed that prevails at least for 1 min, measured at 10 m over ground. Earlier versions of this scale, known as the Saffir–Simpson Hurricane Scale, incorporated central barometric pressure and storm surge as components of the categories (Table 5.6). The central pressure was used until the end of 1980 because accurate windspeed intensity measurements from aircraft were not routinely available for hurricanes. Storm surge was also quantified by category in the earliest published versions of the scale dating back to 1972 but since then has been abolished.

To be classified as a hurricane Category (1), a tropical cyclone must have at least a maximum sustained windspeed of 119 km/h, whereas in the highest Category (5) windspeeds exceed 251 km/h. Combined with information on the

Table 5.6 Saffir–Simpson hurricane wind scale

Category	Sustained winds	Types of damage due to hurricane winds
1	119–153 km/h	Very dangerous winds will produce some damage. Well-constructed frame homes could have damage to roof, shingles, vinyl siding, and gutters. Large branches of trees will snap and shallowly rooted trees may be toppled. Extensive damage to power lines and poles likely will result in power outages that could last a few to several days
2	154–177 km/h	Extremely dangerous winds will cause extensive damage. Well-constructed frame homes could sustain major roof and siding damage. Many shallowly rooted trees will be snapped or uprooted and block numerous roads. Near-total power loss is expected with outages that could last from several days to weeks
3	178–208 km/h	Devastating damage will occur. Well-built framed homes may incur major damage or removal of roof decking and gable ends. Many trees will be snapped or uprooted, blocking numerous roads. Electricity and water will be unavailable for several days to weeks after the storm passes.
4	209–251 km/h	Catastrophic damage will occur. Well-built framed homes can sustain severe damage with loss of most of the roof structure and/or some exterior walls. Most trees will be snapped or uprooted and power poles downed. Fallen trees and power poles will isolate residential areas. Power outages will last weeks to possibly months. Most of the area will be uninhabitable for weeks or months
5	252 km/h or higher	Catastrophic damage will occur. A high percentage of framed homes will be destroyed, with total roof failure and wall collapse. Fallen trees and power poles will isolate residential areas. Power outages will last for weeks to possibly months. Most of the area will be uninhabitable for weeks or months

Courtesy NOAA, US National Weather Service (2014)

estimated time of landfall, the general pathway the hurricane will follow, and on the rainfall volume, the scale provides reliable indications of the potential damage the hurricane will cause. The Saffir–Simpson Hurricane Wind Scale (SSHWS) underwent a minor modification in 2012 in order to resolve some problems associated with conversions among the various units used for windspeed assessments. The scale provides, in addition, examples of the type of damage and impacts associated with winds of the indicated intensity. In general, damage rises by about a factor of four for every category increase. Historical examples provided in each of the categories correspond with the observed or estimated maximum windspeeds from the hurricane experienced at the location indicated. The scale does not address the potential for other hurricane-related impacts, such as storm surge, rainfall-induced floods, and tornadoes. It should also be noted that these wind-caused damage general descriptions are to some degree dependent upon the local building codes in effect and how well and how long they have been enforced.

5.2.7 Tsunami Watches and Warnings

Australia

The Joint Australian Tsunami Warning Centre (JATWC 2014) operates 24 h a day to monitor, verify, and warn of any tsunami threat to the coastline of Australia. Table 5.7 shows the different stages of the tsunami warning chain. The first stage of the process is a "Tsunami Watch," when seismologists have determined that there is the potential for an undersea earthquake that may cause a tsunami threat. If sea-level observations and further data confirm the tsunami threat, or if any potential first point of impact is less than 90 min away a "Tsunami Warning" is issued. Once separate Tsunami Watch

Table 5.7 Tsunami watch and alert system of Australia

Level	Criteria
Watch	Seismologists or oceanographers have determined the potential for an undersea earthquake that may cause a tsunami threat, or if a tsunami is (although unconfirmed) identified by sea-level observations and any potential first point of impact on Australia is more than 90 min away
Warning	Seismologists or oceanographers have determined a tsunami threat, or if a tsunami is (although unconfirmed) identified by sea-level observations and any potential first point of impact on Australia is less than 90 min away
Warning "summary"	Issuance list of all watches, warnings, and cancellations that are in effect for the current tsunami event
Watch "cancellation"	Issued when the main threat has passed or if a tsunami didn't eventuate
Warning "cancellation"	Issued when the main threat has passed or if a tsunami didn't eventuate
Event "summary"	Issued after the event

Courtesy (JATWC 2014)

or Tsunami Warning have been issued a "Tsunami Warning Summary" will also be issued listing all the watches, warnings, and cancellations that are in effect for the current tsunami event. When the main threat has passed or if a tsunami didn't eventuate, "Tsunami Watch Cancellation" or a "Tsunami Warning Cancellation" is issued. Finally a "Tsunami Event Summary" is issued after the event.

All relevant emergency management organizations, regional and local authorities, as well as the media and the public in general are automatically informed in the case of a tsunami event, according to the national "Tsunami Warning Distribution List." The information is disseminated through the organization's website and media information channels and furthermore provides a complementary graphic showing the coastal areas currently under threat. The graph distinguishes between "green" not tsunami threat, "yellow" as an indication that there might be a tsunami threat, and "red" as a regional/local warning that a tsunami is imminent and poses a risk to the identified coastal region.

Tsunami Alert Scheme (Philippines)

Tsunami alert levels established by the Philippines additionally provide information to the public on how to react as citizens of affected places (Table 5.8).

Table 5.8 Philippines tsunami alert scheme

Alert level	Threat	Recommendation for affected places
0	A large earthquake has been generated but either 1. There is no tsunami generated by this event or 2. A tsunami was generated but will not reach the Philippines	No evacuation needed The bulletin is issued for information purposes only
1 ("ready")	There is potential threat to the Philippines	No evacuation order is in effect but coastal communities that may be affected are advised to be on alert for possible evacuation
2 ("watch/ observe")	There will be a minor sea level disturbance (A confirmed tsunami is expected to arrive at the Philippines with wave heights of less than 1 m above the expected ocean tides.)	The public concerned is advised to be on alert for unusual waves People are advised to stay away from the beach and not to go to the coast to watch the tsunami. People whose houses are located very near to the shoreline are advised to move farther inland
3 ("go")	Destructive tsunami is generated with life-threatening wave heights (A confirmed tsunami is expected to arrive to the Philippines with wave heights greater than 1 m above the expected ocean tides.)	Coastal communities that may be affected are strongly advised to immediately evacuate to higher ground Owners of boats in harbors, estuaries, or shallow coastal waters of the affected provinces should secure their boats and move away from the waterfront. Boats already at sea are advised to stay offshore in deep waters until further notified

Courtesy PHIVOLCS-online; access: 20.7.2014

5.2.8 European Snow Avalanche Danger Scale

In April 1993 the avalanche warning services of the Alpine countries agreed upon a uniform, five-part European Avalanche Danger Scale (Table 5.9; EU-EAWS 2014). Until then the individual countries had used various scales with differing numbers of danger levels (e.g., seven levels in Switzerland, eight levels in France) and a variety of definitions of the individual dangers. Since the adoption of a uniform scale, user groups in all countries can refer to the same warning levels, which is of great benefit for all people living in the areas as well as to all snow sport recreationists. The new scale provides a uniform scale with equivalent formulations that furthermore pays tribute to the sensitivities of the different languages in the region. It also explains the individual danger levels by describing the relevant avalanche-triggering probabilities.

The European Avalanche Danger Scale contains five ascending danger levels: low–moderate–considerable–high–very high. These danger levels are described by reference to the snowpack stability and the avalanche-triggering probability, as well as the geographical extent of the avalanche-prone locations and the avalanche size and activity. The snowpack stability forms the basis of all statements concerning the avalanche danger because it directly controls the probability of an avalanche being released. In general a strong and well-bonded snowpack is associated with low avalanche danger. A similar rule applies in reverse: low snowpack stability, in other words, a weak bonding of the snow layers, indicates elevated avalanche danger. The avalanche-triggering probability is a measure of likelihood that depends directly on the snowpack stability. It indicates both the conditions that exist in the absence of external influences (for natural avalanches) and the probability of avalanches being released by additional loads as originating from snow sport participants, etc.).

Fujita Tornado Intensity Scale

The Fujita Scale is a scale for rating tornado intensity, based primarily on the damage tornadoes inflict on human-built structures and vegetation. Originally the scale was developed by Fujita (1971) to connect the Beaufort Scale and the Mach-Number scale (Table 5.10).

The Fujita scale (NOAA), was introduced in 1971 to differentiate tornado path, windspeed, and intensity based on visual inspection. Moreover, although the scale gave a general description for the type of damage a tornado could cause, it did not taken into consideration how the stability of a construction and other natural-related factors might influence the damage. This methodological bias was addressed with the Modified Fujita Scale introduced in 1992. Nevertheless the scale was again modified in 2007 and since then replaced by the Enhanced Fujita Scale (EF-Scale). It accounts for five different levels of damage that occur related to different building types and topographical situations. The enhanced scale now provides a refined damage indicator and thus sets a tornado damage standard. It also is thought to provide a much better estimate for windspeeds, and sets no upper limit on the windspeeds for the strongest level (EF5).

Table 5.9 European snow avalanche danger scale

Danger level	Snow pack stability	Avalanche trigger probability	Consequences for infrastructure	Consequences for persons outside secured zones
Low	The snowpack is generally well bonded and stable	Triggering is generally possible only with high additional loads on very few steep extreme slopes. Only sluffs and small natural avalanches are possible	No danger	Generally safe conditions
Moderate	The snowpack is only moderately well bonded on some steep slopes; otherwise it is generally well bonded	Triggering is particularly possible with high additional loads, mainly on the steep slopes indicated in the bulletin. Large-sized natural avalanches not expected	Low danger of natural avalanches	Mostly favorable conditions. Careful route selection, especially on steep slopes of indicated aspects and altitude zones
Considerable	The snowpack is moderately to weakly bonded on many steep slopes	Triggering is possible, sometimes even with low additional loads mainly on the steep slopes indicated in the bulletin. In certain conditions, a few medium and occasionally large-sized natural avalanches are possible.	Isolated exposed sectors are endangered. Some safety measures recommended on those places.	Partially unfavorable conditions. Experience in the assessment of avalanche danger is required. Steep slopes of indicated aspects and altitude zones should be avoided if possible
High	The snowpack is weakly bonded on most steep slopes.	Triggering is probable even with low additional loads on many steep slopes In certain conditions, frequent medium and also increasingly large-sized natural 6 avalanches are expected	Many exposed sectors are endangered. Safety measures recommended in those places.	Unfavorable conditions. Extensive experience in the assessment of avalanche danger is required. Remain in moderately steep terrain/heed avalanche run-out zones
Extreme	The snowpack is generally weakly bonded and largely	Numerous large natural avalanches are expected, even on moderately steep terrain	Acute danger. Comprehensive safety measures required	Highly unfavorable conditions. Avoid open terrain

Courtesy EU-EAWS, online access: 20.7.2014

Table 5.10 Enhanced Tornado Fujita scale introduced in 2007

Danger level	Intensity	Windspeed (km/hr)	Probability of occurrence (%)	Damages
EF 0	Gale	64–117	38	Some light damage to chimneys; branches broken off trees; shallow-rooted trees pushed over; sign boards damaged
EF 1	Weak	118–180	35	Moderate damage, the lower limit is the beginning of hurricane windspeed; roofs begin peeling off; moving cars pushed off the roads; attached garages may be destroyed
EF 2	Strong	181–253	20	Significant damage. Roofs torn off frame houses; mobile homes demolished; boxcars overturned; large trees snapped or uprooted; highrise windows broken and blown in
EF 3	Severe	254–332	5	Severe damage. Roofs and some walls torn off well-constructed houses; trains overturned; most trees in forest uprooted; heavy cars lifted off the ground and thrown
EF 4	Devastating	333–418	1	Devastating damage. Well-constructed houses leveled; structures with weak foundations blown away some distance; cars thrown and large missiles generated
EF 5	Incredible	419–512	<1	Incredible damage. Strong frame houses lifted off foundations and carried considerable distances to disintegrate; automobile-sized missiles fly through the air in excess of 100 m (110 yard); trees debarked; steel reinforced concrete structures badly damaged

Courtesy Fujita (1971)

Table 5.11 Indicators of rainfall Intensity

Time	Heavy precipitation (mm)	Very heavy precipitation (mm)	Examples
5 min	5.0		16.3 mm Berlin-Tegel (25.08.2006)
10 min	7.1		126 mm in 8 min Füssen (25.05.1920)
20 min	10.0		
1 h	17.1	25	108.3 mm Berlin-Tegel, (25.08.2006)
6 h		35	
1 day	10.0	20	European definition (http://eca.knmi.nl)

Courtesy German weather service (DWD-online access: 20.7.2014)

Rainfall Intensity

As are the other natural hazards, rainfall intensity is also ranked according to an internationally harmonized scale. The ranking in general follows the principle of combining rainfall intensity and time. All meteorological surveys worldwide exchange their data online under the auspices of the World Meteorological Organization (WMO). The most important aspect in assessing rainfall intensity is to define the threshold values for "heavy rainfall" and "very heavy rainfall". Table 5.11 summarizes indicators given on such threshold values based on information by the German Weather Service (DWD). Next to the definition of the threshold values, the ranking "very heavy rain" is also defined by statistical assessment. For this the rank is defined by using percentiles of a normal rainfall intensity distribution, for example, a rainfall that oversteps the 1, 5, or the 95 % percentile. Another threshold value defining "heavy rain" is the 24-h rainfall that exceeds 20 % of the quarterly amount. In practice a combination of percentile values, descriptive statistics, and the frequency distribution of a particular threshold value allows an overview of the regional rainfall distribution and provides an easy classification. The main problem of rainfall intensity assessment is that, for instance, a monitoring station that is just hit by a heavy rainfall normally represents a larger region, where in this case the rainfall was less. Rainfall pattern is very much influenced by the local geomorphology and the microclimatic situation thus making a regionalization of the particular value difficult (Baumgartner and Liebscher 1997). This difficulty is the reason that up to now an internationally agreed-upon definition was not found.

5.3 Hazard Assessment

5.3.1 Single Hazard Assessment

The next step is to go from data on a particular hazard to a regional hazard distribution. The aim of hazard assessment is to identify and map the different hazard types according to their regional distribution, severities, and frequencies on the Earth's surface. In regional disaster prevention hazard maps play a fundamental role, as only by this instrument can areas of equal levels of hazard be identified: aggregated to low, middle, or high severity ranks. Based on local data, as well as on assumptions drawn from equal hazard exposures or from a similar geological condition of a well-known region, quite reliable hazard maps can be drawn. Knowledge of regional hazard distribution is indispensable for spatial planners to design hazard prevention countermeasures. The conceptual approach of a hazard map (and this holds true also for disaster distribution maps) therefore has to be chosen according to the specific topic that is to be worked out. Nevertheless, the more reproducible data are available and the denser the data and time coverage is, the more reliable assumptions on the hazard distribution can be drawn on the potentially affected populations and respective economic assets.

Hazard assessment serves two different aims:

- To assess the hazard potential of a certain area. Therefore detailed information on a small map scale is required.
- To assess the general hazard exposure of an entire region or a country. Consequently the map scale has to be much larger and the information density is less detailed and more generalized.

Assessing the regional distribution of natural hazards deserves at first to define what type of hazard should be analyzed. This decision automatically defines the map scale to be chosen (Fig. 5.2). Looking at climate change related hazards

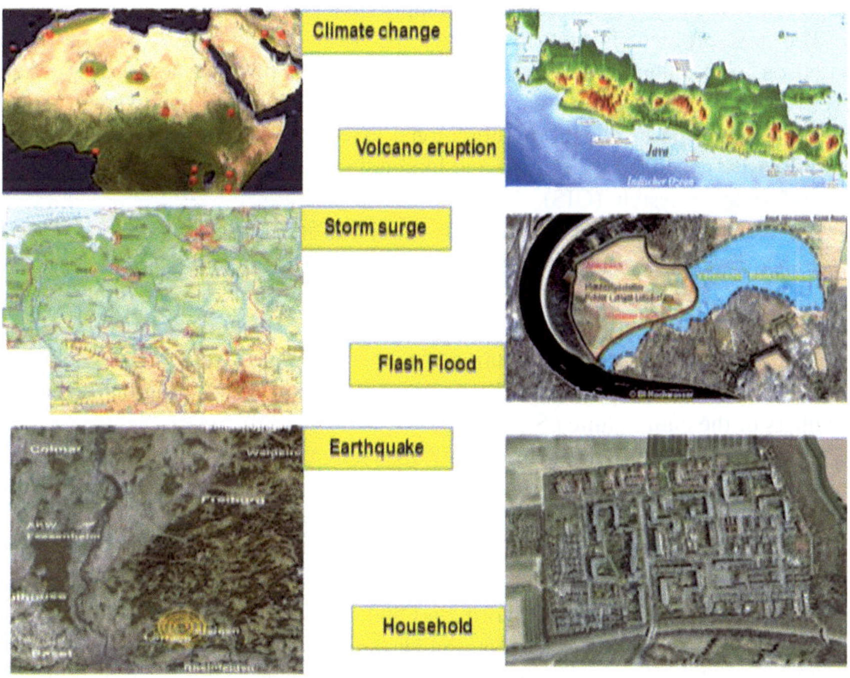

Fig. 5.2 The type of hazard defines the map scale (Own graph)

(drought, heat wave) deserves a map scale that allows for a very general assessment of the hazardous potential thus covering continents and therefore has a transnational scope. Assessing storm surge hazards, however, has a more regional focus, covering an entire country or a part of it. Earthquake or volcanic hazards are in contrast rather local phenomena that deserve information densities in map scales that allow for in-depth local assessments. An assessment of the hazard potential at household level, however, must be based on data at very small scales collected from the communities or equally exposed households.

A problem in regional hazard distribution maps is how to distinguish between local hazards that are identified to have low severity and those that have a high impact potential. The World Map of Natural Hazards (MunichRe 2011) therefore only displays, for instance, earthquake hazards when they have a probable maximum intensity of more than 20 % to be expected in a period of 50 years. The Global Seismic Hazard Assessment Program (GSHAP) of the United Nations (Giardini 1999) on earthquake events only provides those earthquakes with peak ground acceleration that have a 10 % chance of exceedance in 50 years and a return period of 475 years. The seismic hazards were calculated probabilistically based on available earthquake catalogues and databases combined with earthquake source characteristics and an evaluation of the regional ground motion patterns. The GSHAP program was thus able to clarify the long-standing opinion that only earthquakes with magnitudes higher than 6.5 are to be ranked as potentially hazardous is no longer valid. Even small- to medium-sized earthquake hazards can have disastrous impacts when the geological conditions amplify the ground shaking (liquefaction).

The next problem is how to identify the areal distribution of a hazard type based on local values. There are different approaches possible to attach a certain event type to its reasonable area of impact. Hazards that are related to small scales, such as household accidents or subregional assessments such as traffic accidents can be easily geocoded by latitude and longitude and then entered into a geographical information system (GIS).

For example, for assessing hazards from traffic accidents, in Germany the insurance business breaks the data on losses down to the postal code numbers. MunichRe (2013) was able to demonstrate that the probable maximum losses when attributed to the specific policyholder (household) can be reduced by 1–2 % compared to a postal code attribution. Another example is the US HAZUS-MH Hazard Assessment (FEMA 2002) that based its hazard distribution on the county numbers of the entire United States.

However, displaying hazard distribution on a transboundary or even world scale deserves quite a different approach. Such event types often cover more than one country or district and thus have to be much more generalized and aggregated than at a local level. The technical means for displaying such hazard types can be achieved by using different color codes such as the example given for Germany on the national distribution risk of storm and hail (Fig. 5.3) or for a largely extended country such as Indonesia (e.g., "Natural Hazard Distribution; Fig. 5.4).

Alternatively a country affected by a disaster can be defined by a circle the diameter of which can indicate the severity of the impact, as it has been given

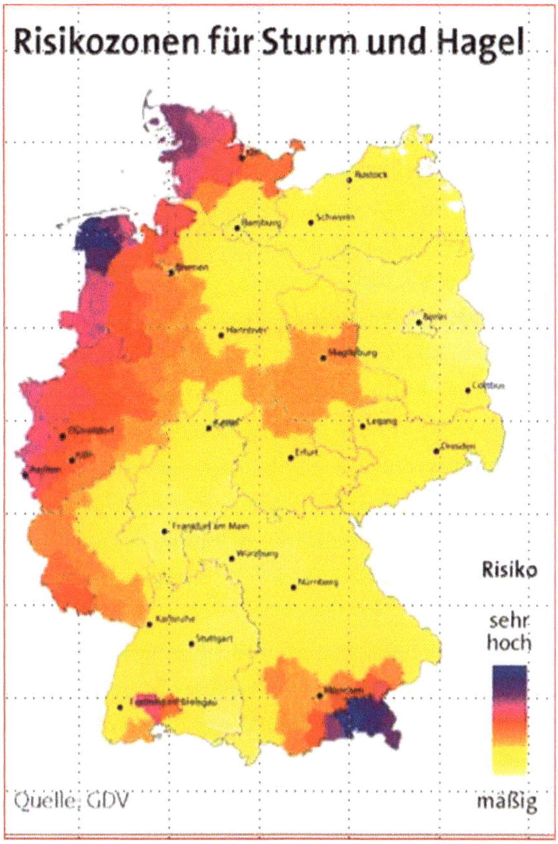

Fig. 5.3 Regional distribution of Germany on risk from storm surges and hailstorms (*Courtesy*: Gesamtverband der Deutschen Versicherungswirtschaft, Berlin)

by the MunichRe (2012) (Fig. 5.5) on the aggregated losses of the continents for the time span 1980–2010 as compared to the volume of insured losses in 2011. Although such a display has the big advantage to provide an easily understandable comparison—here a significant shift of insured losses from America to Asia—it nevertheless does not provide a deeper insight to the specific conditions at risk.

The broader the scope of work or area of interest, the larger is the map scale. Whereas local hazards are based on investigation at the household, street, or city level, transnational hazards such as droughts or floods deserve information collected from very large areas. Large areas cannot be assessed by area-related monitoring. For such kind of hazard assessment remote-sensing techniques are best applied. Remote sensing refers to processes of monitoring the Earth's surface by optical sensors mounted either on aircraft or on satellites. The technique is especially applicable to natural hazards assessment because nearly all geologic, hydrologic, and atmospheric phenomena are recurring events or processes that leave evidence of their previous occurrence. Moreover as natural hazards are highly dependent on the morphology at the location, morphology can play a significant role in hazard assessment. Therefore the type of hazard and its specific areal

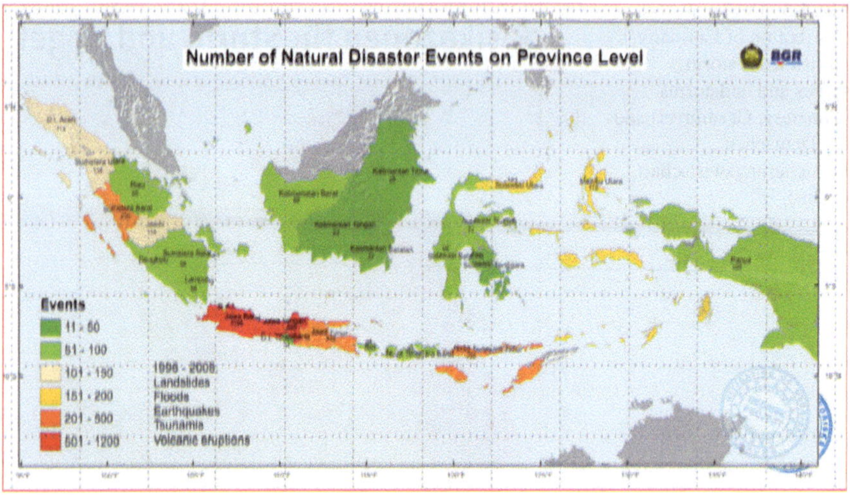

Fig. 5.4 Regional distribution of disaster events of Indonesia at provincial level (*Courtesy* Badan Geology/Gitews 2005, Indonesia)

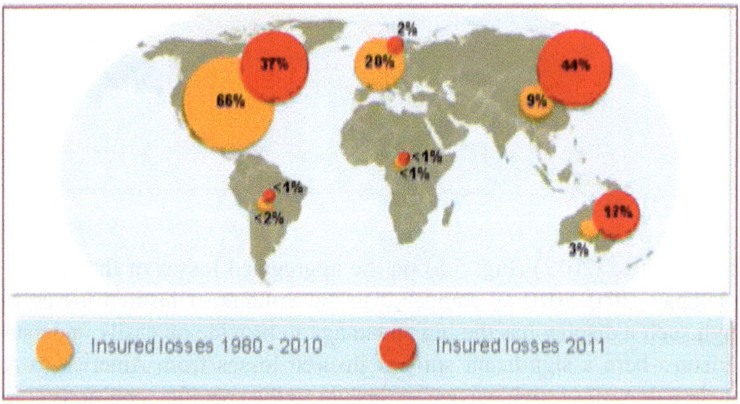

Source: MunichRe, NatCatSERVICE, as at January 2012

Fig. 5.5 Generalized overview of disaster impact severity indicated by *circles*: comparison of insured losses per continent (*Courtesy* MunichRe 2012)

distribution is to be overlain by a digital elevation model (DEM). Natural hazards including volcano eruptions, earthquakes, floods, or landslides in general follow distinct geological patterns. It is therefore the central task of any natural hazard assessment to identify those patterns carefully. Volcanoes are in general easily located from geographical maps and are visible in nature, whereas earthquakes (in general) follow plate tectonic and fault zones and are thus in general to be located by earthquake monitoring. Floods generally occur along river catchment areas and droughts are highly related to zones with no or little rainfall. But it is not only the geological setup alone that makes a natural phenomenon hazardous, rather than the combination of geology with morphology, vegetation, local climate conditions, and so on. Hazard assessment is thus to identify these different root causes and to put them in a meaningful cause–effect relationship.

Figure 5.6 gives an example of a landslide hazard assessment of a river catchment area north of the city of Ende (Flores Island, Indonesia; BG/Georisk 2004). In early 2003 there was a minor landslide that killed 23 people after several days of heavy rainfall. For hazard assessment the trigger elements of that particular landslide event were analyzed: geology, morphology, prevailing vegetation, slope inclination and orientation, and rainfall pattern, all entered into a GIS system to identify those areas made up of the same parametric given in the figure as red-marked areas. Thus it was possible to identify those areas along the river that are composed of the same hazard elements that generated the landslide in 2003 and that have a similar hazard potential.

Even more local scales are in use when it comes to the assessment of the hazards from (local) flooding, hailstorms, landslides, or snow avalanches. Figure 5.7 gives an example of a detailed map of the snow avalanche hazard of the village Hinterrhein (Switzerland).

Other than the "classical" geological- and tectonic-based natural disasters, snow avalanche disasters in most cases result from human interference with nature.

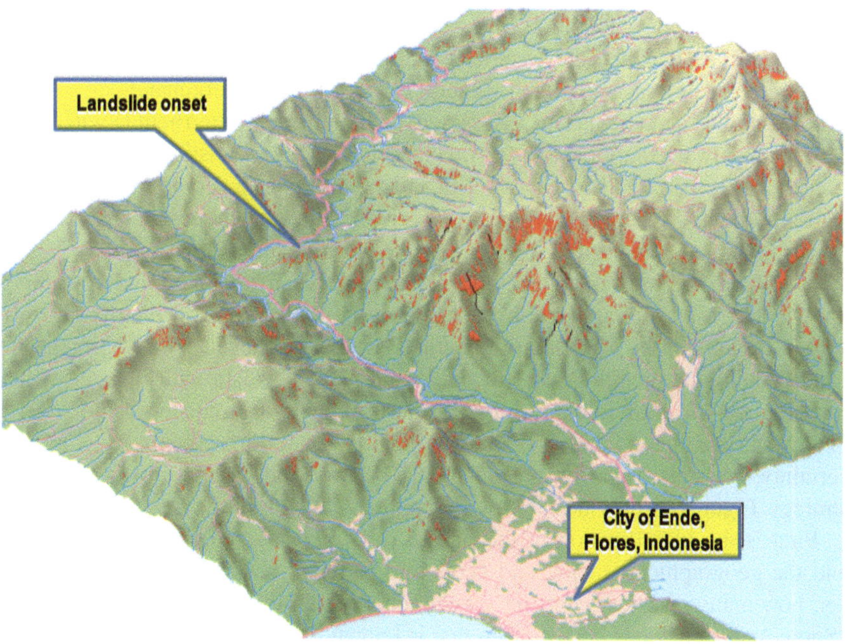

Fig. 5.6 Landslide hazard map of the village of Ende, Flores Island, Indonesia (*Courtesy* BG/Georisk 2004)

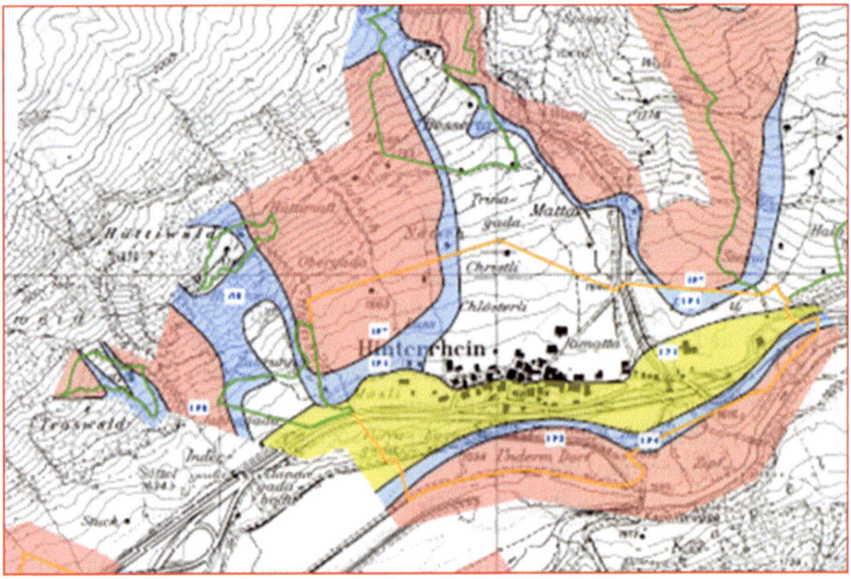

Fig. 5.7 Snow avalanche hazard map; village Hinterrhein, Switzerland (*Courtesy* Amt für Wald, Gemeinde Hinterrhein, Mittelbünden/Moesano, Switzerland)

A snow avalanche hazard results from the physical interrelationship of four critical variables: the slope inclination, the snow pack, the weather, and human activity. Nevertheless even when the weather conditions and the snow quality can change rapidly, the hazardous situation in general can be assessed beforehand.

The snow avalanche hazard assessment of the village Hinterrhein (Switzerland) carried out by the national snow avalanche by the Swiss Institute for Snow and Avalanche Research (SLF) gave a clear indication that the people in former times (most probably based on long-time experience) located their village where the lowest to almost a negligible snow avalanche hazard occurs.

Based on the multitude of evidence, experience in the past and transfer of knowledge from similar hazard-prone areas, the natural hazard distribution of a certain region can be assessed in general at desk study level. By combining published geological, geophysical, volcanological, or other data with disaster events, for example, of earthquake events, location, magnitude, frequency, and severity of impact, it is theoretically possible to identify the hazard potential at a level of certainty, very often already sufficient to work out the baseline for a prevention strategy. Figure 5.8 provides a stepwise approach to work out a hazard assessment.

First is the identification of the hazard to be investigated, where it is located, and the geographical, social, and economic patterns that define the area, moreover, how many disaster events have been recorded. In the example the hazard type "Snow Avalanche" has been chosen that is known to pose a high threat to a village in the Swiss Alps, a village that is highly dependent on winter sports recreationists. When a snow avalanche is assumed to pose the only threat to the village, this

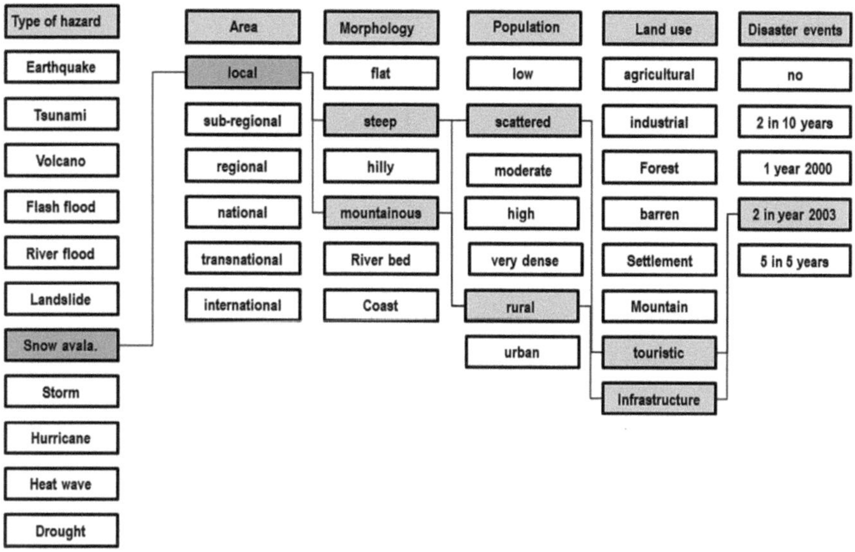

Fig. 5.8 Hazard assessment matrix for desk study purpose example: "snow avalanche" (Own graph)

automatically leads to the assumption that there is no threat in the summertime. For the desk study analyses it is normally sufficient to resort to the information that is easily accessible (mainly topography, geography, geology, land use, population census, social and economic data, etc.). For the Swiss Alp example it can be stated that the area at risk is small, comprises a steep to mountainous morphology, and has a small and very scattered population distribution. The area is highly tourist-oriented and experienced two snow avalanches in the year 2003.

The collection of hazard indicators as given in Fig. 5.8 just serves as an example. The list of indicators should be worked out according to the data available and felt necessary. But not necessarily will the assessment be better the more information there is listed: the indicators should be chosen according to their hazard-related specification and whether they are representative for area under investigation. There are indicators that are qualitative in nature and others can only be quantified. The challenge is how to combine qualitative and quantitative indicators in order to arrive at a harmonized understanding of the risk from snow avalanche hazard to a certain area. In hazard assessment practice it turned out that data on the regional hazard distribution, number of previous events, the amount of damage and casualties, the morphology, land use practice and precipitation, the population distribution, and the critical infrastructure are best to be listed. To follow the above-given example on a snow avalanche, the data collected can moreover be organized in the form of a multifunctional diagram, here for example: a pentagon (Fig. 5.9). Such a display helps to better assess and value the area at risk from a snow avalanche hazard and is especially recommended when different areas have

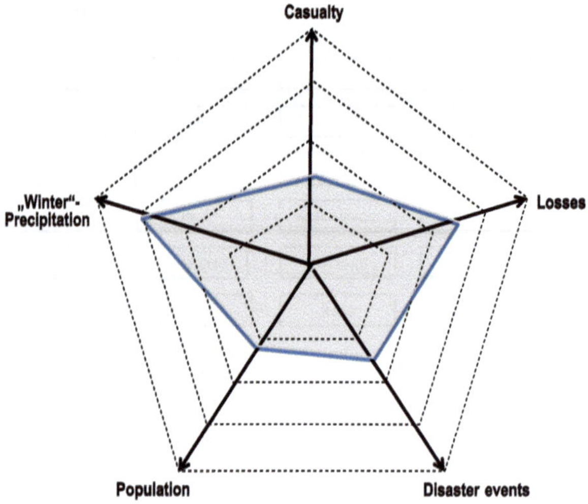

Fig. 5.9 Hazard assessment pentagon "fictitious" (Own graph)

to be compared. A diagram such as a pentagon, sechseck, or other enables the risk assessor to either understand the relation of the different sets of indicators to each other or simultaneously assess their respective level of severity.

When all these data are entered into an area-specific topographical map, a first hazard assessment of the area can be drawn. However, no risk assessment can be derived based on such a set of hazard indicators. Such a desk study will rather lay the base for in-depth local hazard investigations and can perfectly serve to concentrate on those areas that are prominently exposed to a certain hazard. Nevertheless the following steps to a "reliable" and "reproducible" hazard assessment will not differ very much.

Another tool for an easy, fast, and generalized hazard assessment is a so-called "decision tree" (also called "decision support system"; Fig. 5.10). The principle for such a rapid appraisal procedure is given in the example of a tsunami generated by a volcano eruption. The volcano eruption is supposed to occur with either a low to negligible VEI (0–2) or is expected to have a higher VEI (2–5). The volcano might occur submarine or on land, near the coast. Whereas near coast volcano eruptions even with a VEI higher than 2 are not expected to generate a tsunami, submarine volcano eruptions experience much different generation potentials. Underwater volcano eruptions with a VEI of 2–5 can either originate from an older volcano mountain slope or may produce its own caldera. If the slope is quite steep and is to occur near the coast (<100 km) the tsunami generation potential is ranked high, the same as a newly generated volcano that is to be supposed to form a caldera. All other combinations (small caldera diameter, location more than 100 km away from the coast) are supposed to be moderately or non-tsunamogenic.

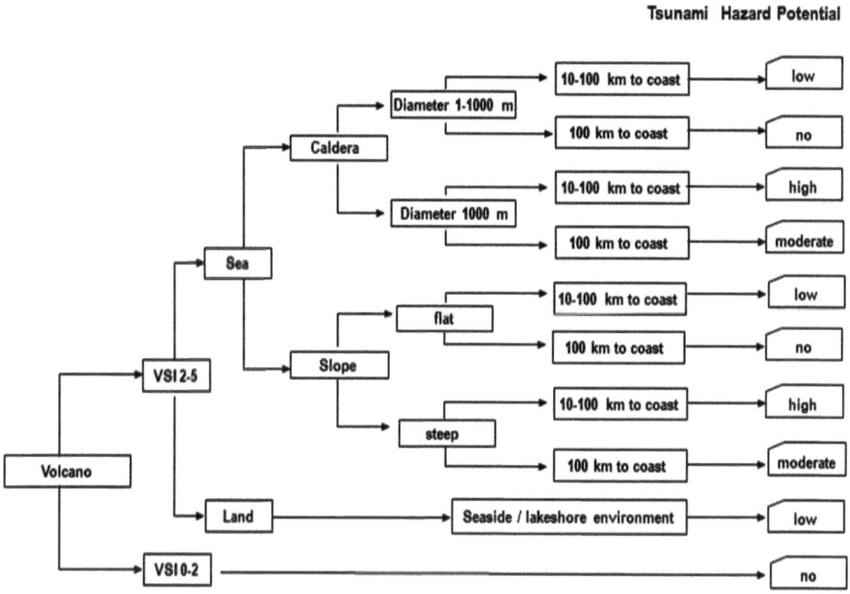

Fig. 5.10 "Decision support system" for identifying the tsunami hazard potential from a volcano eruption "fictitious" (Own graph)

5.3.2 The Hazard Cascade

A regional earthquake occurring along a known tectonic fault or a very local volcano eruption is normally not the kind of event that really puts many people at risk. The risk from an earthquake rather is that ground motion lets construction walls collapse and kills people living in that house. Or, for example, a local and not life-threatening volcanic eruption produces a large volume of ashes that is disposed at the volcano flanks. The eruption, when accompanied by heavy rainfall (very often the case) often triggers large landslides or lahars that at the foot of the volcano can flood the houses of citizens settling there. So the lahar that destroyed the house was a secondary effect (if not tertiary or more). Such a succession of different types of natural disasters is called a "disaster cascade" as displayed in Fig. 5.11.

Natural hazards are mainly a result of plate tectonics or climate factors. Such factors can directly trigger ground motions, volcanic eruptions, a variation in rainfall patterns, or many other hazardous event types. Indirectly they can cause secondary hazards, such as landslides that then can develop into mudslides when simultaneously heavy rainfall sets in. Or a tsunami is generated from an earthquake when the oceanic plate is suddenly lifted up. Hazards caused by other hazards are referred to as concatenated hazards or cascading hazards. From Fig. 5.12 it becomes clear that these multihazard relationships can be very

Fig. 5.11 Volcano-lahar multihazard cause-effect relationship (Own graph)

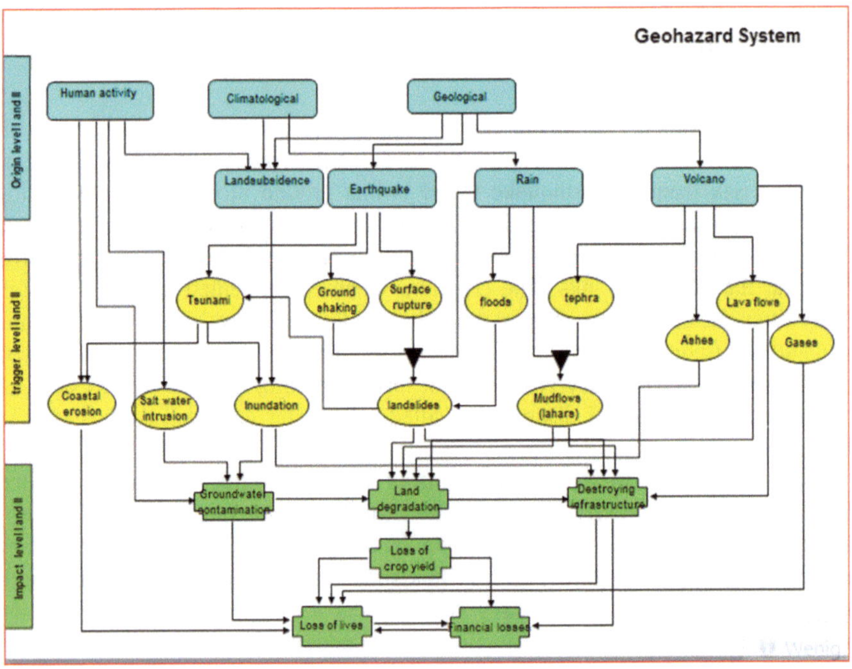

Fig. 5.12 Generalized cause-effect relationship of natural hazards (Own graph)

complex to assess. At the end natural disasters either of tectonic or climatic origin all affect society in general or the individual in particular. There are quite a number of different hazard cascades to be drawn, that can generally be distinguished to be of tectonic origin:

- Plate tectonics—ground motion—earthquake—collapse of house—victims
- Plate tectonics—ground motion—earthquake at seafloor—oceanic plate uplift—tsunami—lowland inundation—destruction of houses—victims
- Plate tectonic—up rise of magma—volcanic eruption—ash deposition—rain—lahar—flood—damage—victims

Or of climatic/meteorological origin:

- Combustion of carbon—increase in CO_2 in atmosphere—change of climate—melting of ice caps—increase in sea level—inundation of lowlands—flood-damages—victims
- Climate change—modification of regional climate regime ("ENSO")—increase in extreme weather—heavy rains—flooding—damages—victims

As most of the hazard information will be available in the form of highly scientific, technical, or sociologically based information, it is thus difficult to understand by nonscientists. The maps and their verbal explanations have to be "translated" for user-application purposes (VanWesten 2004). In such a way "worked over" hazard maps (single and multihazard) can be a significant contribution in the decision-making process, best by having the scientists explain their findings.

In this context it becomes obvious that many of today's disaster statistics do not indicate what type of natural hazards really were the cause and what hazard type (secondarily) contributed to the disaster. Normally disasters are attributed to the final type of hazard (lahar/landslide) and not to the lahar-triggering volcano eruption. The damages caused or the death toll claimed consequently are identified as lahar/landslide generated, making a realistic disaster assessment quite difficult. The 2004 tsunami caused damage of two thirds of the buildings of the Aceh capital. But the damage figures did not distinguish between the tsunami and the foregoing earthquake event, although it was the third biggest in history. Thus far our disaster assessment lacks reasonable operation procedures to identify such a complex means–end relationship. To demonstrate that this discussion is not purely scientific: if the earthquake resiliency of the many Acehnese houses had been better, many houses would (most probably) have been able to withstand the deadly tsunami flood.

5.3.3 *Multiple Hazard Distribution*

Multihazard assessment is the basic means for regional and spatial planning. Single-hazard assessment although it provides valuable information for a preventive risk management, is still mostly in use to assess specific scientific and technical problems related to natural hazards often carried out by natural scientists and risk assessors. With multiple hazard assessment the focus shifts from a single event to a complex relationship between different types of hazards in a certain region. No longer is a particular problem in the focus but the "risk exposure" of a region instead. Consequently multihazard assessment has to take the different hazard types and their interdependencies into account thus requiring a higher level of generalization of the data compared to a single-hazard assessment. By applying multiple hazard assessment the natural scientists, engineers, and technicians are entering the social dimension of risk management, or at least are contributing to it. In any case multiple hazard assessment demands from natural scientists to address targets that (often) are seen to lie beyond scientific responsibility; but it is this final orientation natural scientists have to understand. The society at risk clearly calls for the provision of reliable and robust data in order to arrive at sustainable solutions to increase social, economic, and ecological resiliency.

The combination of overlay or dependency of different natural hazards to finally sum up to the hazardous situation a region is exposed to is also called "cascading" or "concatenating" hazards. Such hazardous situations can best be assessed by a multihazard assessment, "an instrument that synthesizes information derived from several sources. The data is agglomerated in a single assessment to allow for a composite picture of the hazard types, their areal distribution and severity of impact" (VanWesten ibid). The multiple hazard distribution is displayed in overlay maps, a map type that has already proven an excellent tool in risk management, as it provides a "regional and agglomerated assessment of hazard on an equitable basis."

Multiple hazard maps deserve, as do single-hazard maps, a pre-emptive decision to the type of map that should be established:

- What kind of questions shall be addressed?
- What kind of data are available?
- What kind of data are required?
- What groups of society or spatial planners should be addressed?
- What kind of data and in what form do the users need the information?
- What kind of disaster prevention activities should be triggered with the information?

The assessment of the hazards is normally based on a set of data (statistical, historical, heuristic or scenario-based), that has been collected in the past on the area of interest. Therefore a hazard type (e.g., landslide run-out, area of volcano ash deposition, flood levels, earthquake peak ground acceleration distribution, etc.) is listed, ranked, and then displayed in a map according to the specific areal distribution. Often such a database proves insufficient for reliable and meaningful assumptions. Then data from regions that are made up of a similar geological, morphological, and hazard setting are transferred to supplement the database (see Fig. 5.6). Such a data "transfer" is justified as long as the data transfer is taken into account in the final assumption.

In Fig. 5.13 an example of a "synoptic" generalized hazard distribution for the Asia-Pacific region is given (UN-OCHA 2011). The map shows the areal distribution at risk from earthquake damage (yellow to dark brown), tropical storms (light to dark blue), and volcanic eruptions (triangles), based on internationally established risk scales. The earthquake intensity is shown according to the Modified Mercalli Scale (MMI) indicating zones where there is a 20 % probability that the intensity level will be exceeded in 50 years. For this assumption the 12° MMI data of MunichRe (NATHAN) were condensed to five in order to increase understanding. The tropical storm risk was taken from the World Map of Natural Hazards and indicates tropical storm intensity based on (condensed) five windspeeds of the Saffir–Simpson hurricane scale. The zones indicate where there is a 10 % probability of a storm of this intensity to strike in the next 10 years. Triangles indicate the hazard exposure distribution of volcanoes that have shown activity in 11,500 years (Holocene). The information the synoptic map revealed is

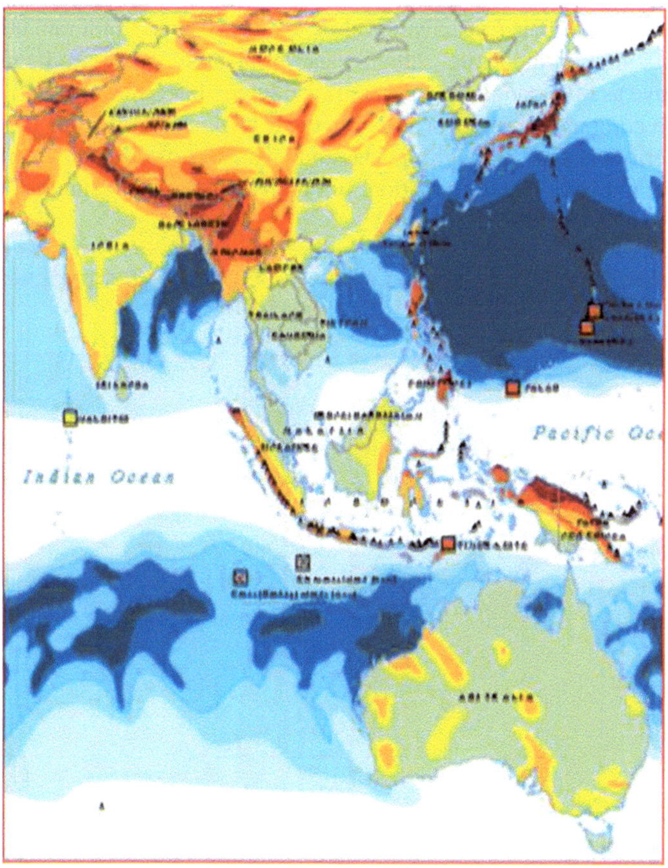

Fig. 5.13 Synoptic map of risks from natural hazards in Asia-Pacific (*Courtesy* UN-OCHA 2011)

that countries such as Bangladesh and Myanmar are also in future highly exposed to earthquake damage and tropical storms, whereas both countries do not show a risk for hazards from volcanic activity. The Philippines, however, will most probably be exposed to all three hazard types. In contrast, the western part of Indonesia shows high exposure to volcanic activity and earthquake hazards and its eastern parts will mostly face (just) an increased level of earthquake damage. Nevertheless it should be considered that such type of synoptic maps only give a very generalized picture of the risk or hazard distribution of the area under investigation. Moreover they do not provide information on how the hazard types may interact, if they occur simultaneously.

A methodology to derive at such a synoptic (synthesized) distribution of risks or hazards is, for example, given for a "peak ground acceleration hazard" cum "landslide hazard" based on an analogue-grid-cell approach for a certain region in Fig. 5.14. First the distribution of each hazard ("earthquake"; "landslide") is laid down in a specific map based on an arbitrary chosen grid. To be able to compare severity and regional distribution of the different hazard types they have to be classified according to their respective level of severity. The use of a ranking between "very low", "low", "moderate", "high", and "very high" proved to be meaningful and effective as often tangible (measured) and intangible (arbitrarily assessed) data have to be combined. Moreover it did turn out to be impractical to introduce too many severity classes, although the classification is arbitrarily and should be chosen according to the specific situation. In Figure 5.14 the following hazard classes

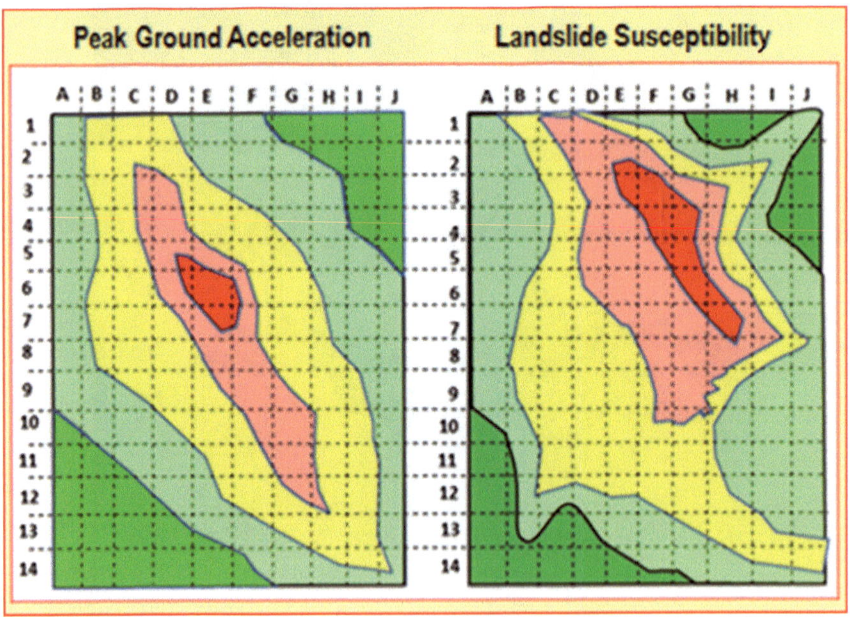

Fig. 5.14 Multiple hazard assessment based on peak ground acceleration and landslide susceptibility ("fictitious"): *dark green* 1; *lush green* 2; *yellow* 3; *pink* 4; *red* 5 (Own graph)

have been identified: very low (green) = 1; low (light green) = 2; moderate (yellow) = 3; high (light red) = 4; very high (dark red) = 5. In order to arrive at an overall hazard distribution of a certain region all hazards are added up (overlain) to one synoptic multiple hazard distribution map. The example given displays the sum of severity values at a grid cell base (peak ground acceleration plus landslide susceptibility). Such a methodological approach opens the chance to aggregate as many different hazards as required, although the different hazard types are not identified according to their specific geological, tectonic, or climatic interdependencies, for example, an earthquake that triggers a landslide.

The analogue grid cell approach is only given here in order to explain the operational procedure. Normally such an aggregation is carried out by a GIS system. A GIS has (when supported by a digital elevation model) the further advantage that can display the multiple hazard distribution according to its real morphological, geological, and land use settings.

In a next step the value of each grid cell is then added to arrive at a synoptic multihazard assessment as an arbitrary estimation of the hazard exposure level of a certain area from both landslide susceptibility and peak ground acceleration (Fig. 5.15).

A generalized understanding of a region's hazard exposure is often given by the average hazard exposure (50 %). To arrive at this, the percentages of the different hazard levels are added and then the "mean" hazard value as representative for the entire region displayed (Fig. 5.16).

Finally, multiple hazards cannot be assessed properly without taking the human factor into consideration. When a flood, although being a predominantly meteorological and (somehow climate-related) phenomenon, is entering a human activity altered flood plain (*versiegelte landschaft*), the above-normal increased water level

Fig. 5.15 Grid-based synoptic multihazard assessment (Own graph)

	A	B	C	D	E	F	G	H	I	J
1	4	6	7	7	5	4	3	2	2	2
2	4	5	6	7	7	7	6	5	3	2
3	4	5	5	7	7	7	6	4	3	2
4	4	5	6	7	7	7	7	5	3	2
5	4	5	6	8	8	7	8	5	4	3
6	4	5	6	8	9	8	8	6	4	4
7	4	6	6	6	7	8	7	7	5	4
8	4	5	6	6	7	8	7	7	5	4
9	4	5	5	6	7	8	8	6	5	4
10	3	4	5	6	6	7	7	5	5	4
11	2	3	5	5	6	7	7	5	5	4
12	2	3	5	4	5	6	7	6	5	4
13	2	3	2	2	4	4	5	6	6	4
14	2	2	2	2	2	3	4	5	5	5

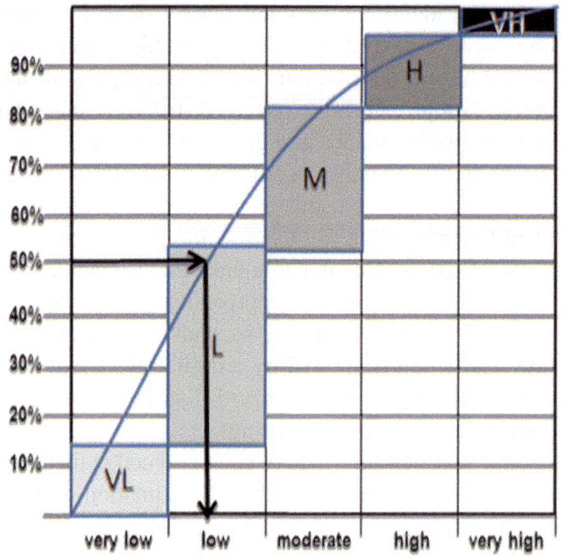

Fig. 5.16 Assessment of the "average hazard" level (50 %) of an area (Own graph)

can result in a hazard that in turn can develop into risk to the people. Even in developing countries there are almost no culturally unaltered flood plains any more. There is a multitude of human interference that changes the former natural conditions of an area, making it hazardous. Even when we acknowledge that the root causes for floods are essentially uncontrollable and cannot be prevented entirely, it is human activity that finally creates the hazard. Therefore hazard assessments should not be restricted to the natural conditions that contribute to the hazard but also to analyze where and how human activity is exacerbating the hazard. For instance a flood-prone broad valley that normally gives ample space to spread heavy rainfall becomes a hazard when the valley is densely populated, with settlements that deteriorate the soil–water uptake capacity leading to an increased likelihood of adverse flood impact. This especially holds true for rivers that are very long or have a large catchment area. The hazard situation gets even more severe when the river crosses countries with different risk reduction capacities. For instance, the Danube River crosses on its 2800 km way to the sea 10 different countries. All of them can be affected, if in the case of technical measures upstream that increase the runoff.

In order to give all river neighbors an equal chance to cope with such flooding, all major rivers riparian worldwide have founded alliances to organize their waterways. Famous for such an alliance is the International Commission for the Protection of the River Rhine (ICPR 2010) that since 1950 coordinates and concerts activities on the hydrological and ecological protection of the river and particularly on flood protection and prevention of its five neighboring countries. Such a form of harmonized flood risk management is an integral part of the so-called Integrated Water Resources Management (IWRM). Other examples to manage shared water bodies and thus to solve inherent conflicts from the use of the water in times of droughts or the release of floodwaters to a low-lying neighbor is the bilateral Ganges Joint River commission of India and Bangladesh established in 1972. Or the Nile Basin Initiative (NBI 2014), that was once anchored in a vision to achieve socioeconomic development through the equitable and sustainable utilization from the common Nile Basin water resources. The NBI was founded to provide a convenient and permanent forum that, based on a legal and institutional framework, helps to ensure the equitable and sustainable utilization of the waters. Although the Nile basin carries water to serve the needs of about 800 million Nile riparian by 2025, it still lacks political harmony and standardization to settle conflicts from the many national interests. The flood-related hydrological conditions along the Danube river and the ones prevailing along the Nile river actually form the "real" hazard of both areas.

5.4 Example of a Multiple Hazard Assessment (Ende, Flores)

Based on the author's own research activities in the Flores City of Ende an assessment of the general hazard exposure of the area was carried out in 2003 by the Badan Geology-GEORISK Project (BD/GEORISK 2005). The district of Ende was subject to a series of natural disasters in the last 50 years, among them an earthquake in 1961, and a volcanic eruption in 1969 when Mount Iya, a volcano just 5 km from the city erupted, producing many flows of lava and several large lahars. The eruption claimed three lives and destroyed almost 200 houses. In 1988 heavy rainfall triggered massive landslides and mudflows that claimed 48 lives. Another earthquake in 1992 killed 25 people and let many buildings collapse. In spring 2003 a massive landslide was triggered after days of heavy rain along the upper parts of the river reaching the sea just next to the city. The landslide most seriously hit the small village of Detumbawa north of Ende where 27 people were killed. In the aftermath the Ende District Administration addressed the GEORISK Project for a rapid and general assessment of the risk exposure level and for assistance in working out and disseminating community-based mitigation and prevention measures. To meet the demand comprehensive field surveys have been carried out, evaluating the morphological, geological, geophysical, and volcanological base parameters and combining the findings with the land use pattern, rainfall data, and population distribution (Fig. 5.17).

The specific hazard-related occurrences were mapped as shown in Fig. 5.18:

- Volcanic hazard
- Landslide hazard
- Liquefaction hazard
- Tsunami hazard.

Finally all four hazard types identified were combined according to their spatial distribution to a synoptic multihazard assessment. The hazard assessment (hazard specification) was based on a Bayesian expert knowledge assessment by which the hazard level was identified from experience of the Indonesian geoscientists and from perceptions of the people of Ende. The findings were entered into an assessment matrix.

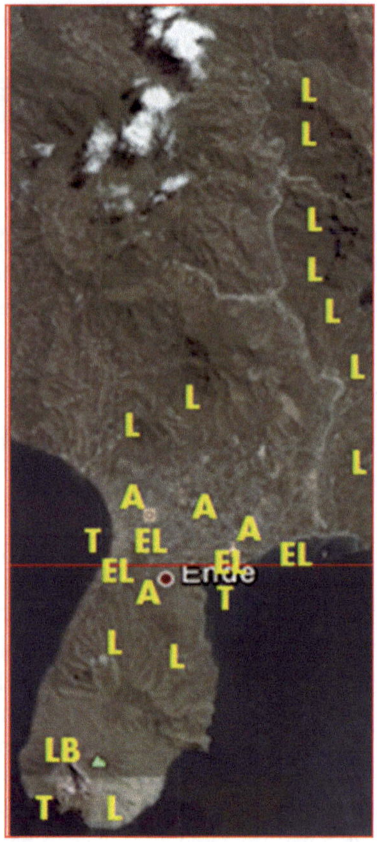

Fig. 5.17 Identified natural disaster occurrences in Ende District (Flores, Indonesia) *L* landslide; *A* volcanic ashes; *LB* volcanic bombs and lapilli; *T* tsunami; *EL* earthquake/liquefaction. (Google Earth, city of Ende, Flores, Indonesia; access March 2014)

Multiple hazard assessment matrix (*Courtesy* Badan Geologi /GEORISK project)

Type of hazard	Condition of hazard	Hazard exposure level		
		1	2	3
Earthquake/Liquifaction	Potential	Moderate	High	Very high
Tsunami	Run-up height	< 3 m	3–6 m	>6 m
Landslide	Susceptibility	low	Moderate	High
Volcanic ash fall	Thickness	< 5 cm	5–10 cm	>10 cm

One area along the eastern shoreline was specifically investigated for its hazard exposure level as it hosts a number of critical infrastructure elements (airport, jetty, central full tanks, power station) of the Ende capital. The assessment revealed that there the hazard level is ranked 8 out of 9.

The volcanic hazard exclusively derives from the Mt. Iya volcano that already erupted several times claiming casualties and damages. But from volcanological surveys around the crater it became evident that significant fractures occurred around the vent, indicating that if Mt. Iya erupted it would (most probably) be directed southwest of the peninsula right into the sea. The eruption would be

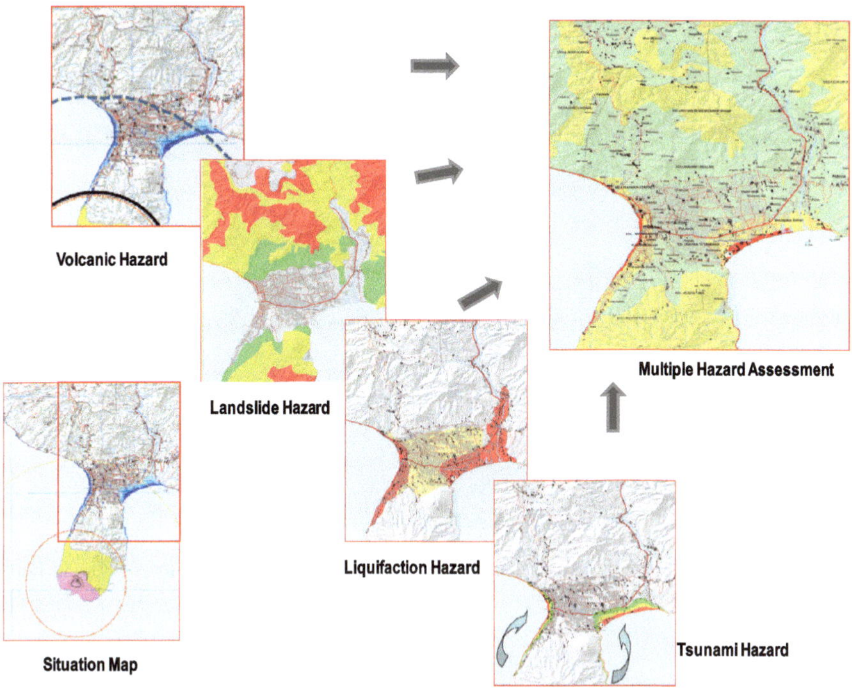

Fig. 5.18 Multihazard assessment of Ende district (*Courtesy* BG/GEORISK 2004)

similar to the lateral blast of Mt. St. Helens (USA). An assessment of the volume of volcanic material that may be thrown out from the flank might be enough to generate a tsunami (BD/GEORISK ibid). The volume of lava flows and the amount of erupted bombs and lapilli in the past were altogether quite small and were not believed to pose a considerable risk to the city of Ende. Moreover the volcanic ashes in the past did not pose a threat to the Ende citizens that would not overstep their coping capacity. Thus the volcanic hazard was ranked low to moderate.

The landslide hazard assessment revealed that the northern parts of the city of Ende are exposed from low to moderate, whereas people settling along the river towards the north were seen to be exposed to higher levels. As with the landslide event in Detumbawa in April 2003, a landslide also destroyed the only bridge over a small tributary river just several hundred meters north of the small village Aemuta, blocking any rescue and relief assistance to the people living upstream in case of a disaster (Fig. 5.19).

The parameters defining the Detumbawa event revealed, when entered into a digital terrain model, a general overview about the landslide hazard exposure of the entire region (see Fig. 5.6). The red-marked spots are concentrated at the eastern flank of the valley. But when superimposed on the locations of villages, roads, bridges, and irrigation canals, it turned out the "only" 10 spots pose a real risk for disasters. Such an assessment helps disaster risk managers to concentrate on the most critical hotspots.

Fig. 5.19 Bridge destroyed from a landslide/debris flow N' of Aemuta, Flores (Google Earth, City of Ende, access: March 2014)

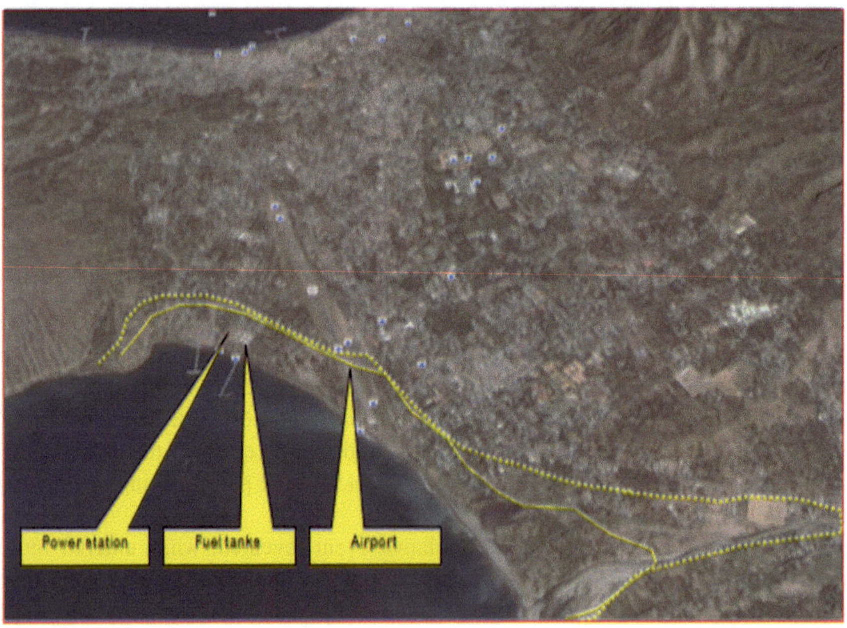

Fig. 5.20 Critical infrastructure locations direct at the shoreline; city of Ende Flores, Indonesia; Google Earth, access, March 2014)

The liquefaction hazard is generated from soils becoming liquid when exposed to physical energy. The energy is in general brought into the system by earthquakes. Thus earthquake distribution and unconsolidated fine–medium grained soil favor liquefacation pose the hazard. In Ende the soils that are conducive to liquefaction are distributed along both shorelines. They are of volcanic origin or were deposited at the river mouth east of Ende. But as the soils were not very thick and their areal distribution is restricted to the broader shorelines, the liquefaction hazard of the city of Ende was rated low to moderate.

When combining all four hazard types into one single multiple hazard assessment it turned out that the city of Ende is primarily exposed along its east coast to a tsunami hazard. There the risk for disaster is quite significant, as many critical infrastructure elements (airport, fuel tanks, power station) are located just there (Fig. 5.20). As the airstrip is placed at heights between 5 and 15 m above sea level, a tsunami experiencing a run-up height of more than 6 m already poses a risk to the airport. And also, because the city of Ende lies more than 2000 km east of Java, any rescue and relief operation can only be effective when the airport is not risk exposed. A similar situation also defines the hazard levels of the fuel depository and the power station. Recommendations by the GEORISK Projekt were made to the district administration to level-up the airstrip pavement by at least 2 m and to protect the tanks and the power station by 6 m high earthen dams.

References

Baumgartner, A., Liebscher, H.-J. mit Beiträgen von Benecke, P., Brechtel, H. u.a. (1996): Allgemeine Hydrologie - Quantitative Hydrologie, 2. Auflage, Band 1, in Lehrbuch der Hydrologie, p. 694 Seiten, Gebrüder Bornträger, Berlin, Stuttgart

Baumgartner, A. & Liebscher, H.J. (1997): Lehrbuch der Hydrologie - Allgemeine Hydrologie, 2. Auflage, Band 1, Borntraeger Berlin-Stuttgart

BD/Georisk (2005): Mitigation of geohazards in Indonesia – Status report on the GEORISK project. - A contribution to the World Conference on Disaster Reduction, Kobe, Hyogo, Japan, 18–22 January 2005, Badan Geology, Bandung

BG/Georisk (2004): Project report on Natural disaster occurrence – Georisk Project.- German Technical Cooperation with Badan Geologi Indonesia, Bandung

Cas, R.A.F. & Wright, J.V. (1988): Volcanic suscessions. A geological approach to processes, products and sucessions. - Chapman and Hall, London

Dilley, M., Chen, R.S., Deichmann, U., Lerner-Lam, A.L., Arnold, M. with Agwe, J., Buys, P., Kjekstad, O., Lyon, B. & Yetman, G. (2005): Natural disaster hotspots - A Global Risk Analysis-Synthesis Report.- The International Bank for Reconstruction and Development / The World Bank and Columbia University, Washington, DC

DWD (2014): Rainfall intensity scale.- German Weather Service (DWD), Offenbach (online: www.dwd.de)

EU-EAWS (2014): European Avalanche Danger Scale (EADS).- European Avalanche Warning Services (EAWS) (online: www.avalanches.org)

FEMA (2002): Building Support for Mitigation Planning – How to Guide. - Federal Emergency Management Agency (FEMA), Publication 386, Vol 1–9, Washington DC

Fujita, T.T. (1971): Enhanced Fujita Scale (EF-Scale) - Proposed Characterization of Tornadoes and Hurricanes by Area and Intensity. - National Oceanic and Atmospheric Administration (NOAA), Short Message Research Paper, No. 91, Washington DC

Giardini, D. (1999): The Gobal Seismic Hazard Assessment Program (GSHAP - 1992–1999). - Annali di Geophysica, Vol. 42, No. 6, Rome

Guha-Sapir, D., Below, R. & Hoyois, Ph. (2013): EM-DAT: International Disaster Database (2013) – The OFDA/CRED International Disaster Database, Brussels

GVP (2014): Smithsonian's Global Volcanism Program and the US Geological Survey's Volcano Hazards Program.- (online: www.volcano.si.edu/).

ICAO (2004): Handbook on the International Airways Volcano Watch (IAVW) – Operational Procedures and Contact List. - International Civil Aviation Organization (ICAO), 2nd edition

ICPR (2010): Our common objective: Living waters in the Rhine catchment. - International Commission for the Protection of the Rhine (ICPR), Koblenz

JATWC (2014): Australian Tsunami Warning System (ATWS).- Bureau of Meteorology/ Geoscience Australia (GA), Melbourne/Canberra (online: www.bom.gov.au/tsunami)

Kreimer, A. & Munasinghe, M. (1991): Managing the Environment and Natural Disasters.- The International Bank for Reconstruction and Development/The World Bank, Washington DC

Manton, M.J., Della-Marta,P.M., Haylock,M.R., Hennessy,K.J., Nicholls,N., L.E. Chambers, D.A. Page,M., Pahalad,J., Plummer,N., Salinger,M.J., Suppiah, R., Tran,V.L., Trewin, B.,Tibig,I.J. & Yee. D. (2001), Trends in extreme daily rainfall and temperature in southeast Asia and the South Pacific: 1916–1998. – International Journal of Climatology, Vol.21, p.269-284, Wiley Online Library

MunichRe (2011): World map of naturals hazards. – Munich Reinsurance Company, Nathan, 2011-Version. Munich

MunichRe (2012): Topics Geo online – 2012. - Munich Reinsurance Company, Munich

MunichRe (2013): Topics Geo online - 2013. - Munich Reinsurance Company, Munich

NBI (2014): Understanding the Nile Basin. - Nile Basin Initiative (NBI), Entebbe, (online: www.nilebasin.org)

Newhall, C.G. & Self, S. (1982): The volcanic explosive index (VEI): An estimate of explosive magnitude for historical volcanism. - Journal of Geophysical Research, Vol. 87, p.1231–1238, Wiley Online Library

NOAA (2014): Saffir-Simpson Hurricane Wind Scale. – NOAA, National Weather Service National Centers for Environmental Prediction, National Hurricane Center, Miami, FL (online: www.nhc.noaa.gov/aboutsshws.php)

PHIVOLCS (2014a): Taal Volcano Alert Signal.- Philippine Institute of Volcanology and Seismology (PHIVOLCS), *Manila (online:* www.phivolcs.dost.gov.ph)

PHIVOLCS (2014b): Philippines Tsunami Alert Scheme. - Philippine Institute of Volcanology and Seismology (PHIVOLCS) in collaboration with the DOST-Advanced Science and Technology Institute, Manila (online: www.tsunami-alarm-system.com/

Richter, C.F. (1935): An instrumental earthquake magnitude scale.- Bulletin of the Seismological Society of America, Vol. 25, No. 1, p. 1–32, Albany CA

Saffir, H. & Simpson, B. (1971): Saffir-Simpson Hurricane Wind Scale.- National Weather Service (NOAA), National Hurricane Center, Miami FL (online: www.nws.noaa.gov)

UNDP (2004): Reducing disaster risk – A challenge for development. - United Nations Development Program (UNDP), Bureau of Crisis Prevention and Recovery, United Nations, Geneva

UNDP (2013): Human Development Report 2013 - The Rise of the South: Human Progress in a diverse World.- United Nations Development Program (UNDP), United Nations, Geneva

UN/OCHA (2011): Natural Hazard Risk in Asia-Pacific (Risk Map). - Regional Office for Asia and the Pacific (ROAP), Bangkok

USGS (2006): U.S. Geological Survey's Alert-Notification System for Volcanic Activity. United States Geological Survey (USGS), Fact Sheet 2006–3139 (online: volcanoes, usgs.gov/activity/alertsystem)

VanWesten, C.J. (2004): Remote Sensing and GIS for Natural Hazards Assessment and Disaster Risk Management. - Faculty of Geo-Information Science and Earth Observation (International Training Center (ITC), University of Twente,-Enschede

Walker, G.P.L. (1973): Explosive volcanic eruptions.- A new classification scheme.- Geol. Rundschau, Vol. 62, p.431–446

World Bank (2013): World Development Report (WDR) 2014. - International Bank for Reconstruction and Development/The World Bank, Washington DC (online: www.worldbank.org)

Chapter 6
Vulnerability Assessment

6.1 General Aspects

In 2010 there were three earthquakes all of magnitude about 8 on the Richter scale that caused totally different impacts:

- In Chile the earthquake claimed the life of 500 people, affected 3 million, and caused a loss of US$30 billion.
- In China the quake killed 2200 people, affected 30 million people and caused a loss of US$6 billion.
- In Haiti 220,000 people were killed and 4 million affected and the economic loss amounted up to US$8 billion.

This example clearly shows that the impact can vary from earthquake to earthquake, thus raising the questions of why one country is more vulnerable to a disaster than others and what really makes people vulnerable. And other questions immediately follow:

- Why are particular populations vulnerable?
- How they are vulnerable?
- What group of a society really is vulnerable?

These questions lead to the central question: "What is vulnerability"?

The above-given example clearly points out that vulnerability is related to social processes and is generally based on the societal, economic, and political fragility of a country exposed to various hazards. In other words, although disasters are biophysical in nature and their occurrence (locations), their respective impact (severity) is the result of a lack of resilience. The prevailing socioeconomic conditions of a country (Chile, China, and Haiti) are in general the dominant factors favoring the adverse impacts thus creating the risk. The factors defining vulnerability have to be identified to lay the base for designing disaster risk mitigation options. In this regard it is necessary to stress that vulnerability assessment is not

© Springer International Publishing Switzerland 2016

U. Ranke, *Natural Disaster Risk Management*,
DOI 10.1007/978-3-319-20675-2_6

and end in itself, but an opportunity to learn from experience in order to identify future risk exposure.

There are 29 different definitions of the term "vulnerability" collected by Thywissen (2006) that cover a wide range from technical assessment to impact on a specific vulnerable group to the behavioral reaction of the global society. Today's generally accepted definition is given by UNISDR (2004) and stresses the factors that make a society vulnerable: "The conditions determined by physical, social economic and environmental factors or processes, which increase the susceptibility of a community to the impact of hazards."

This definition comprises the physical root causes of a disaster as well as the outcome for the population at risk, although other definitions such as that of (UNDRO 1991) put the focus more on the damage that potentially may derive from a disaster. Or in course of the climate change debate the focus shifted more to the sensitivity of the global systems to cope with assumed impacts (Klein 2003).

Although there is still some confusion in the use of the term "vulnerability," the many definitions in use can be viewed essentially as complementary. Moreover the intensified debate on vulnerability terminology will help to master the complexity of the concept especially in its relation to the state of development. Still much research on vulnerability merely describe it based on tangible outcomes affected by stochastic forces of nature, such as aspects of undernourishment, poverty, or lack of shelter, and still often ignore the social impact or the adjustment capacities (Kasperson et al. 1988).

The vulnerability definition has no universally acceptance up to now, yet it can generally be characterized as a function of societal exposure to natural hazards and of the susceptibility of an individual or a societal group to damage or harm, as well as its capacity to absorb or cope with the effects of these threats. In this regard vulnerability is a systemic approach that focuses on coping or adaptive capacities. But as vulnerability is a multidimensional construct it is not easy to capture in one single figure. In this context vulnerability is not an "outcome" but rather a "state or condition of being" as a result of social inequities, for example, limited access to resources, lacking of mitigation opportunities as they were derived from historical patterns of social domination, and marginalization (Eakin and Luers 2006). When we accept that vulnerability comprises more than the tangible manifestation in loss and harm, then lack of capacity is essentially equivalent to vulnerability.

As a risk is always closely related with the uncertainty of its occurrence, risk can be best described by the term "probability". What is the probability that I will be injured by a bicycle accident in my home town, or how probable will another 9.2 earthquake offshore Fukushima be? But in this regard another discussion, closely related to this, complicates the matter: the difference between "probability" and "frequency". Kaplan and Garrick stated (1981, p. 17) that "probability" is, although being a numerical measure, describing the degree of belief, or of the state of confidence and knowledge, that such an event will occur, whereas "frequency" is the (tangible) measurable and objective outcome of an experiment or of a trial.

"Frequency" thus describes the hard facts, whereas "probability" is soft and subjective in this sense actually "non-measurable." The way the probability scale uses to compare different kinds of uncertainty needs a calibration for which the frequency will give the standard reference (Kaplan and Garrick ibid., p. 18) "showing the intimate connection between frequency and probability" and continue saying: "when 'frequency' is the science of handling statistical data then 'probability', as a subject, we might say is the science of handling the lack of data." They summarize, "when one has no or insufficient data at hand, than probability is the only thing that can be used."

6.2 Methods and Instruments

Following this argument, vulnerability assessment has to be based on analytical methods and approaches that next to the assessment of the physical manifestations of a disaster, also analyzes the social entitlements and mitigation capabilities. This would shift the vulnerability paradigm away from food security, sustainable livelihood, and poverty assessment to a human-oriented conception in which the people, their activities, their institutions, and their coping capacities will become the center of analysis (UN-AFD 2004; Benson and Twigg 2007). In its strategic sense, emergency management is not just about understanding hazard causation; it is about understanding the relationships of environmental, political, social, and economic forces that influence and shape the frequency, nature, and location of emergencies (Buckle 1998).

Next to the hazard assessment the vulnerability assessment addresses the social, economic, and ecological aspects of the risk-exposed population. Therefore the categories for vulnerability assessment mainly have to answer the following questions:

- Who are the vulnerable groups?
- What factors make them vulnerable?
- What are the social and economic outcomes of the disaster exposure?
- What are the needs of the vulnerable groups to increase their level of resilience?
- What are the capacities and abilities the vulnerable groups need to increase their level of resilience?

These questions (Benson and Twigg, ibid) consequently lead to address vulnerability by using indicators. As the disasters are by their frequency, severity, their probability, and socioeconomic impact difficult to be objectively measured, resort has been taken to a set of indicators to "measure" the major key elements of vulnerability. For practical application purposes only a limited number of indicators can be used, that have to be aggregated and that are relative in nature. Using indicators to "measure" vulnerability provides a methodological approach that make at least a generalized vision and impression of the risk exposure of a population possible, and that furthermore provide a means for comparing risks of different regions and evaluating the effectiveness of risk mitigation activities. The main requirements indicators have to fulfill to be used as standard means are that they are transparent, robust, representative, replicable, comparable, and easy to understand. Consequently it has to be accepted that the methodologies have their limitations, reflecting the complexity of what is to be measured and what can be achieved. Therefore it has in many cases to be accepted that a lack or scarcity of data may make it necessary to accept assumptions, analogies, and tradeoffs from other regions (Cardona et al. 2003).

Although the discussion has since long incorporated all kinds of vulnerability, still most people understand "vulnerability" as only related to economic losses, the loss of lives, or the number of injuries. This perception is quite understandable as such elements are easy to assess and communicate. They can be counted, proved on money terms, or identified on the mere number of casualties. But such elements only describe a minor fraction of the lives of a society. The reason that social aspects of vulnerability are very often neglected is based on the fact that such elements are very difficult to assess. And moreover such an assessment will undoubtedly give very different results in different societies, thus making a comparison on a national basis inhomogeneous, not to say impossible. But such a comparison is indispensable for a nationwide risk reduction strategy.

Thus any vulnerability assessment has to identify the different elements:

- Direct
- Indirect
- Tangible
- Intangible

Figure 6.1 visualizes the complex relationship between direct and indirect, tangible and intangible impact from disasters as well as the relationship among these groups. The graph clearly shows that there is much more in disaster management to take into consideration than just to count victims and losses.

The relationship between the different elements at risk (Fig. 6.2) can be explained by an example of the Fukushima-Daichii Nuclear Power Plant catastrophe.

A	Direct–tangible	Economic losses, respectively, the costs for reconstruction of the Fukushima-Daichii Nuclear Power Plant as well as costs for reconstruction of the damaged regional infrastructure
B	Direct–intangible	Death toll from earthquake and tsunami as well as from the nuclear power plant explosion and radioactive fallout
C	Indirect–tangible	Economic losses from power failure, interruption of industrial and agricultural production as well as gross social product
D	Indirect–intangible	Mistrust in the socioeconomic and cultural leadership of national and local governments

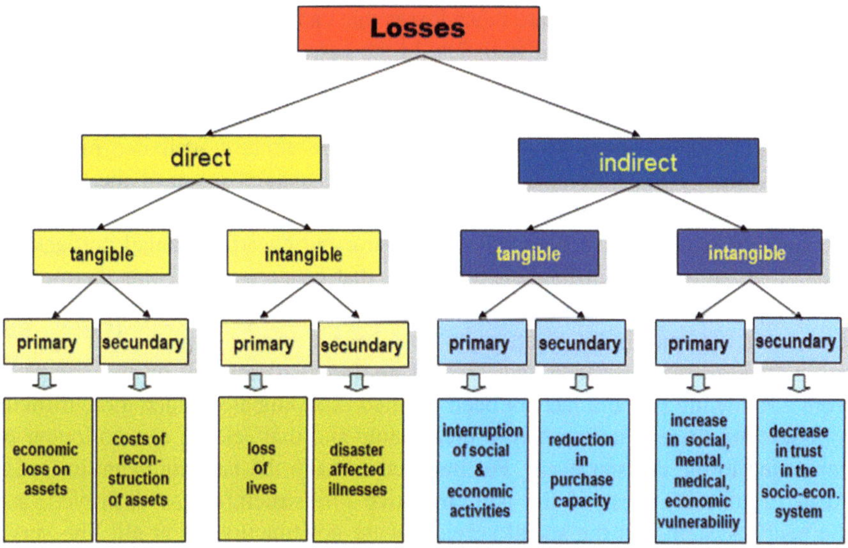

Fig. 6.1 Direct and indirect social and economic vulnerability (Own graph)

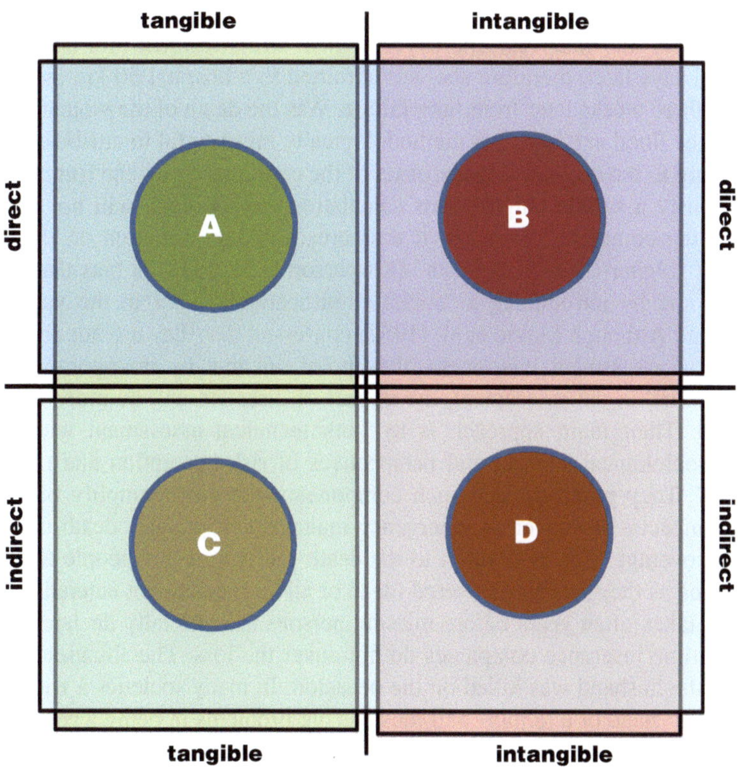

Fig. 6.2 Relationship between different elements at risk (direct and indirect, tangible and intangible; Own graph)

Direct and tangible losses are those losses that can be measured, calculated, and valued in terms of money and statistics. In the case of an earthquake, such losses comprise the number of houses destroyed and their reconstruction costs, as well as the costs for the rehabilitation of the technical and material infrastructure. Indirect and tangible losses comprise losses that cannot be directly attributed to the disaster anymore. For instance, the earthquake has damaged much of the industrial assets and machinery, so that industrial production cannot start again; or most of the labor workers have been killed, so that experienced personnel are lacking. Intangible but directly attributable are losses that concern the victims themselves. As a rule, although it was for a long time not seen as ethical to economically value a person's life quite a lot of attempts have been carried out in the last years to value the statistical life (see Sect. 7.2). Thus far the death toll of an earthquake from a following tsunami has not been counted economically. Much more difficult to assess are losses that are hardly measurable and that seldom can no longer be connected to the disaster event. For instance, as with the tsunami event in Banda Aceh (Indonesia) most of the administrative infrastructure was destroyed, and the health system and the social services were not functioning at all. The survivors often died or suffered from the missing services. Imagine that the tsunami destroyed many local health service centers in a far remote area. Among the victims was an elderly woman who survived the flood without any injury, but lost all her family and relatives. The psychological stress resulted in a heart attack a couple of days later, therefore she was admitted to a hospital 50 km away, where she died three weeks later from lung failure. Was the death of the woman a follow-up from the flood and how it is methodologically meaningful to attribute the lethal lung failure to the original disaster onset of the earthquake plus the tsunami.

Obviously a simple tangible loss calculation methodology will not be able to estimate the complete losses and it thus creates a bias situation on the adverse effects of "higher-order of impacts" (Kasperson et al. 1988), a bias that can only be overcome by introducing a "social amplification" factor in the vulnerability assessment. Although Slovic et al. (1982) confessed that they are not able to offer a complete assessment instrument, they point out that by incorporating "social amplification" in the assessment, the impact of disasters will be more reasonably evaluated. Their main approach is to "link technical assessment with psychological, sociological and cultural perspectives of risk perception and risk-related behavior." They point out that such components can either amplify or attenuate the felt impacts. Moreover, in emergency management practice death is the basis for loss coverage. But in addition to the death toll, very often people are missing and as long as they are not recovered (dead or alive) they are not entered in the statistics. It takes often years before missing persons are officially declared dead. In the meantime insurance companies do not cover the loss. The situation even gets worse if the husband was killed on the occasion. In many societies a widow is not accepted as "head of a family" and thus has big problems in being a partner to settle financial and social formalities. In many countries, moreover, the risk insurance operational practice is the so-called 72-h rule, meaning that all incidents occurring 72 h after an event are not attributed to the risk insurance liability. On one hand this deadline makes sense in regard to how else can live insurance companies handle the many kinds of illnesses following an event. But the multitude of dead-full crisis and conflicts results very often in serious and long-term traumatization of the victims who are consequently not liable for insurance. Thus far there is no methodological instrument at hand that combines the two aspects of risk management in a meaningful and operationally practical manner.

The normally applied instrument of valuing direct tangible losses is by using reconstruction costs. If insured, such costs are covered by the insurance company either totally or partially. But what if the house owner was not able to pay the risk premium? Then the loss would be fully on him. Secondly, reconstruction can mean to repair just the damages based on the value of the house at the time of the disaster. But what if it were an old house with a very low value; the repair cost coverage would be low, although it might not be livable anymore, a situation that exists quite often in developing countries.

The following indicators provide possible descriptions of the factors that make up "vulnerability": natural disaster risk exposure, socioeconomic weaknesses, level of societal resilience, and so on:

Social sector

- High population density in settlements (large housing complexes)
- Settling high-risk areas (flood-prone lowlands, landslide-prone slopes)
- Low-quality houses (not seismic structured, bad electricity, no sanitation)
- Lack of mobility (no transport facilities to evacuate from a disaster), limited access to resources (no financial means, poor technical knowledge), exclusion from development decisions (no participation)
- Risk perception differs strongly
- Lack of education
- Poverty

Institutional sector

- Poor risk and emergency management
- No hazard/vulnerability/risk assessment
- Low interest in problems of socially and economically marginalized groups
- Urban planning not reflecting natural hazards
- Unplanned urbanization
- Lack of proper risk planning
- Lack or poor implementation of identified mitigation measures
- No or low level of enforcement of law and order (failed states)

Physical/technical sector

- Poor construction quality (bad bricks, no cement, no tightening of roofs to walls)
- Infrastructure not or badly maintained
- Critical infrastructure exposed to high risk

Economic sector
- High import and low export quota
- Dependency on monocrop agriculture
- No diversification of industrial and agricultural production
- No added-value production
- Subsistence economy
- Bad ratio of indebtedness to export earnings
- Dependency on foreign aid (budget, relief, welfare)
- Sale of fertile land to international investors (landgrabbing)

Environmental sector
- Erosion of fertile land (degradation)
- Deforestation
- Flooding or drought
- Pollution of soils and groundwater bodies
- Destruction of mangrove forests
- Local change in climate (hot summers, dry winters)

This meta-category structure helps to identify the specific dimensions that define people's exposure to disasters that makes them vulnerable. As the central aim of vulnerability assessment is to provide information on how to diminish problems, it is better to avoid future vulnerability in total, and serious information can only be derived from many impacts and events of the past. For this quite a long timeline of disaster information is required; unfortunately often such timelines are missing or scarce. Furthermore, acquiring the necessary information is in general easier to be achieved from individuals and well-organized communities that have robust and reliable information at their disposal, whereas societies settling in remote areas often lack this kind of information system. The dilemma is that those who are well organized and living at a state-of-the-art disaster resilience normally are not those strongly exposed to natural disasters, and those living in marginalized environments suffer more. In order to cover the information gap making specific assumptions and drawing analogies to assess at least the dimension of hazardous threats is very often unavoidable.

There are at least three different levels of interventions to identify vulnerability:

- The local level, where the information is in general quite easily structured as the problems have a distinct cause–effect relationship (e.g., local flood after heavy rainfall).
- At the national level (e.g., drought in one of the Sahel zone countries) different socioeconomic and political root causes interact.
- At the regional level a strong participation of local stakeholders as well as the national level is required, as there, many "local" mitigation activities need national backing for implementation.

In order to design appropriate countermeasures at the local level, participation of the people at risk is indispensable, as only they can provide information according to long-time experience and, moreover, give the identified countermeasures political legitimization. At the national level an involvement of the many stakeholders, affected population, decision makers, as well as disaster risk managers will open the chance to combine insights on the local problems and experience of disaster management with the authority to implement the findings. The larger the area at risk (regional/national level) the more it will be necessary to draw knowledge from best-practice examples based on expert opinion, analogous or historical studies, and/or modeling.

As we have seen, the impact of natural disasters is not randomly distributed over the globe. There are distinct patterns to be recognized, where the disaster impact is superimposed by the social status of the exposed populations, their "membership in a devalued social category," their social organization, and their way of interaction (Bolin and Stanford 1998). There is great disparity to be seen between the rich and poor leading to an increased isolation of the poorest in settlements on metropolitan fringes, in inner cities, in remote rural areas, or in areas that are historically known to be unsuitable for any settlement. Moreover, all development statistics prove that this trend will increase in the future. The impact is that poor households recover more slowly, likely have higher mortality rates, and many never fully regain pre-impact levels, increasing their vulnerability to future hazards (Morrow 1999; Wisner et al. 2004). In the beginning, emergency management concentrated mainly on material and economic (tangible) indicators, and in the course of time it was extended to include human resources (labor, income, education), family and social resources (medical care, access to water, energy), and political resources (political participation, self-determination), making risk assessment today a complex task that deserves a working together of natural scientists, sociologists, medical doctors, and even psychologists (traumatization) and others. The "social order and its everyday relation to the habitat and the larger historical circumstances" define the vulnerability. The resiliency of a household not only depends on the social status (income, education, etc.) but is also clearly defined by its integration into the community. Only when the affected populations formally participate in the prevention and awareness-raising efforts, will they be able to develop mechanisms that enable them to master their living circumstances. Thus they would no longer be exposed to the decisions by the governing political structures that too often ("only") promote their particular interests (Cannon 2008; Morrow ibid) by neglecting the social and economic processes to govern the impact.

Cannon (ibid) further stated that there are "sets of unequal access to opportunities and unequal exposures to risks as a consequence of the socio-economic system." This social paradigm not only is true for societal minorities or, for example, women or ethnic minorities, but also holds true for many least-developed countries as the Haitian earthquake disaster clearly has demonstrated.

Emergency management therefore has to recognize the importance of designing prevention policies to suit the specific needs of the risk-exposed societies. It has to focus on the socioeconomic and cultural patterns of vulnerability, not on the geographical alone. The first step for such an emergency management is to work out a community vulnerability inventory, sometimes referred to as a community hazard and risk assessment profile by registering the population density, its age, sex, and income distribution as well as the percentage of elder, younger, and disabled people, who all are likely to need special attention in an emergency. Although it is unrealistic to collect such vulnerability data for every individual household, planners should make use of communal databases, or from areas under detailed investigation to extrapolate the percentage of vulnerable groups. A community vulnerability inventory can comprise (among others) these indicators:

- Population distribution
- Ethnic, cultural, and religious minorities
- Population concentration: <15, >65
- Ill, disabled
- Women, children
- Poor/large households
- Residents rendered homeless
- Material infrastructure (communication, traffic)
- Access to resources (water, power, sanitation)
- Access to medical care, schools, hospitals
- Distribution of income, labor, insurance, political participation

By many societies and based on tradition, experience, and ethnic belief, disasters are often seen as a token of external origin. But when carefully evaluated, it turns out that in many cases disasters are a result of an interaction of natural conditions that are superimposed and/or juxtaposed to human activity. According to the risk definition given in Chap. 7, risk is seen as the social, economic, and ecological impact of a natural hazard with serious adverse effects, and having in mind that disasters are often a result of a cascading nature of disasters, a more complicated picture develops. For example, a dam that is constructed in the upper part of a valley to manage the numerous floods there increases the water flooding rate and the water level that reaches the low-lying riparian, making that region more vulnerable than it was before the risk management measures were undertaken. Thus not only rainfall and the shape of the catchment area define the flood risk but also very often interference of the human being.

6.3 Vulnerability Indicators

The basic intention of a vulnerability assessment is get a measure that enables the risk assessors to compare the threat to which a society or a social group in different regions or countries is exposed. In order to make the vulnerabilities comparable it is necessary to develop sets of specific hazard-related indicators that rank the levels of vulnerability in relation to the likelihood of future disasters. Based on such indicators local vulnerability maps can be worked out that help to identify where high-risk groups are concentrated and, where found necessary, mitigation countermeasures can be designed and implemented accordingly. As often there are not enough data available on a specific area at risk, data from other analogous regions and generalized assessments can be incorporated to put the assessment on a more reliable base. But vulnerability maps should not only identify the areas at risk but also identify community self-help potentials and resources, such as shelters, community health services, capacity of water and energy, local volunteer service groups or neighborhood response networks, and so on. Combining vulnerability indicators with the hazard distributions is best achieved by using a GIS system. Such GIS-based information provides effective crisis-response tools. Emergency managers are provided with indispensable baseline data to design appropriate disaster prevention measures as well as provide the populations at risk with information to improve their disaster resilience capacity. The information compiled in the vulnerability maps is a strategic tool for awareness-raising initiatives, disaster mitigation programs, evacuation plans, humanitarian relief distribution, and other response services (FEMA 2012). And they moreover allow working out the disaster prevention infrastructure in accordance with locally adoptable technologies. As mitigation is costly and—as all over the world the financial means for prevention are generally too small—such prevention measures can thus be prioritized and implemented accordingly.

In 2006 the United Nations carried out a worldwide assessment in order to identify the populations and localities that are at risk, as well as to understand the different causes leading to human vulnerability. The Gravity Project (Peduzzi et al. 2002, 2009), as a part of the United Nations International Decade of Natural Disaster Reduction, aimed to highlight the populations with the highest needs in risk reduction to mitigate hazardous exposure and to decrease possible future casualties. This assessment was mainly based on a collection of information and datasets from open accessible data of leading disaster risk management organizations including UNOCHA, UNISDR, UNEP/GRID, or the USGS. The main difficulty in comparing human vulnerability of different societies consists of differentiating whether populations are affected because of a high frequency and/or magnitude of events, or because of a high vulnerability from the different situations in which they live. The geoscientific-related aspects of the Gravity Project were derived from comprehensive databanks such as the Global Seismic Hazard Assessment Program, the Smithsonian Institution's Global Volcanism Program, or many others.

The project combined extensive geographical, geological, and hydrometeorological datasets in order to estimate the physical disaster exposure (frequency, magnitude, and population exposed), as well as data on the socioeconomic elements. The advantage of the Gravity Project was that it successfully combined data of purely geoscientific origin with those of social and economic origin. The main limitation while mixing scientific and socioeconomic parameters lies in the difference of regional expansion and in the time scale of the datasets. Social data are normally clearly to be defined in number and value (population density, income distribution, unemployment rate, and so on) whereas geoscientific-related data are often not tangibly accessible. Moreover, socioeconomic factors can change within in short time span, whereas geological phenomena in general take centuries to result in visible changes. Hazard occurrences are not confined to national boundaries or sovereignty delineations. Another difficulty that concerns geoscientific as well as socioeconomic data is quality. The state of science and statistics is highly different all over the world. For instance, much of Africa had never been properly assessed for its volcanic and earthquake exposure. But social data are in many developing countries also highly difficult to assess, although the United Nations Development Programme (UNDP) has been able to classify the living conditions of all countries for more than 30 years by the Human Development Index (HDI) or the Human Social Index (HSI) and others. Moreover, it is complex to compare, for instance, seismic-structured buildings or well-maintained flood protection dams that were installed by prosperous nations enacted under comprehensive building codes with impacts from disasters in countries that lack all of such legal, technical, and financial means. An aspect relating to neighboring countries, for instance, regards a water catchment area. How does one compare the hazard exposure when the upper riparian does not comply with agreed-upon flood protection measures and thus the lower riparian territory is flooded regularly. The most critical step in every vulnerability assessment is when data of very different scales and measures are categorized in classes from 0 to 10 or 0 to 100, and so on. In the next step these newly created classes are then compared exclusively on these harmonizations (Table 6.1) that mean, for instance, data on peak ground acceleration measured by instruments are then treated as equal to data based on personal perception, such as on the felt earthquake impact based on

Table 6.1 Factors used by the Gravity Project to compare socioeconomic status of countries under investigation

Economy	Gross domestic product at purchasing power parity
	Total dept. service in percentage of the exports of goods and services
	Inflation rate and annual food price index
	Unemployment rate as percentage of total labor force
Land use	Percentage of arable land
	Percentage of urban population
	Percentage of woodland and forests
	Percentage of irrigated land
Demography	Annual population growth rate
	Growth rate of urban population
	Population density as percentage of total population density
	Age–dependency ratio
Health and sanitation	Average calorie supply per capita
	Percentage of people with access to adequate sanitation
	Percentage of people with access to safe water
	Number of physicians per 1000 inhabitants
	Number of hospital beds
	Life expectancy at birth
Politics	Transparency index (corruption)
Infrastructure	Number of radios per 1000 inhabitants

Courtesy Peduzzi et al. (2002)

Table 6.2 Example of vulnerability indicators for country "A" (fictitious)

Sector	Very high vulnerability	High vulnerability	Moderate vulnerability	Median vulnerability	Low vulnerability	Very low vulnerability
Population	>1 million	1 million–500,000	500,000–250,000	250,000–100,000	100,000–50,000	<50,000
Urban planning	No planning at all	Poor planning, no disaster risk management	Disaster risk management only generally mentioned in planning documents	Disaster risk management mentioned in detail in planning documents	Disaster risk management is an integral part of planning process but prevention measures not implemented	Sustainable planning that integrates disaster risk management and prevention measures are implemented
Stability of buildings	Adobe	Brick	Concrete	Reinforced concrete–steel framing	Steel-structured construction	Seismic structure
Agricultural sector	High dependency on international food aid	Monocrop, most food must be imported	Food supply based on 2–3 crops	Agriculture partly diversified	Diversified agriculture	Highly diversified and value- added production lines
Natural disasters (earthquake)	>9 m	8–9	7–8	6–7	5–6	<5
Economic sector	Single product, no added-value production (gold)	Single product, added-value options	Added-value production, still high import dependency	Diversified product, market highly volatile	Multiple products, fair market opportunities	Added-value production, high export earnings

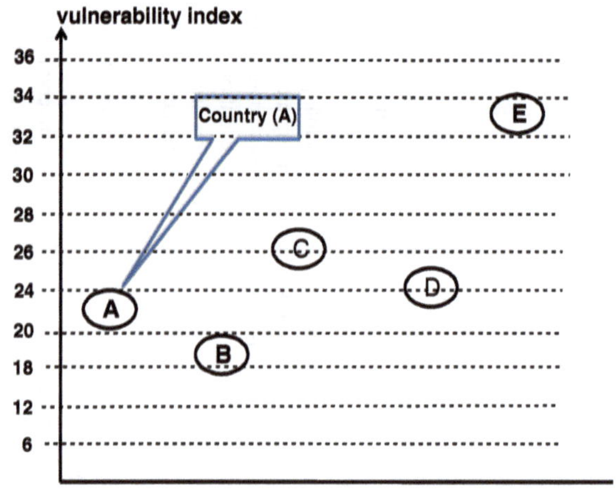

Fig. 6.3 Comparison of vulnerability of different countries (A) to (E) based on indicators given in Table 6.2

the empirically developed Mercalli scale. The Gravity Project made use of all such kinds of datasets and successfully arrived at a single standard for natural hazards.

Methodologically the vulnerability categories are entered into a table attributing the different values (severity) in a harmonized scale. The scale can be set arbitrarily (see also: PAHO/WHO 1998; Dwyer et al. 2004) in order to harmonize the validity of the categories as given in Table 6.2.

For example, the vulnerability of a country (A) is assessed in relation to other countries based on the above given "harmonized" vulnerability indicators (based on Table 6.2): very high = 6, to very low = 1). Summing up the different vulnerability levels for country (A) to (E) it turns out that the vulnerability index ranges between 19 and 33. Country (B) has the lowest vulnerability (19), whereas (A) is low to moderate and (E) is the highest (33). With such an operation it is possible to compare vulnerabilities of different regions, as long as the investigation is based on the same set of indicators (Fig. 6.3). The method does not provide a measure of the "objective" vulnerability of a certain region rather than the relationship of the vulnerabilities of different regions.

6.4 Vulnerability Assessment

According to the term, "the past is the key to the future," such a vulnerability assessment can also be used to identify the level of resilience a society intends to achieve. Therefore vulnerability should not only be restricted to information on the present state of risk exposure but also as a measure of potential consequences (casualties, physical damage, cultural deterioration, etc.) such events may cause (Table 6.3).

The basic means to assess future vulnerability are to identify the specific hazard exposure of people and livelihoods by either evaluating the level of exposure

Table 6.3 Assessment of potential consequences from natural disaster (fictitious)

		Consequences				
		In-significant	Minor	Moderate	High	Very high
Likelihood	Most probably	Moderate	High	High	Extreme	Extreme
	Likely	Moderate	Moderate	High	Extreme	Extreme
	Seldom	Low	Moderate	High	High	Extreme
	Rare	Low	Low	Moderate	High	High
	Unlikely	Low	Low	Moderate	High	High

deterministically or probabilistically. The basis for the evaluation is in any case experience and expertise, experience that has been encountered by the risk-exposed population or knowledge-based by the expert's judgment (Bayesian assessment). In reality both ways are combined and result in the most usable information gain. Following this argument, vulnerability assessment has to be based on analytical methods and approaches that next to the assessment of the physical manifestations of disasters, also analyzes the social entitlements and mitigation capabilities. This shifts the vulnerability paradigm away from food security, sustainable livelihood, and poverty assessment to a human-oriented conception in which the people, their activities, their institutions, and their coping capacities become the center of analysis (UN-AFD 2004; Benson and Twigg 2007). In its strategic sense, "emergency management is not just about understanding hazard causation; it is about understanding the relationships of environmental, political, social and economic forces that influence shape the frequency, nature and location of emergencies" (Buckle 1998).

The method that is today mostly in use to define the dependency of "frequency" and "probability" is given in the following graphs, derived from log-normal distribution of probability versus frequency (see also: WBGU 1999). The example in Fig. 6.4 describes the basic elements of a vulnerability assessment procedure. It deliberately refrains from any mathematical/stochastic algorithms and only wants to present the main operational steps. The figure gives the weekly flood-level record of a particular year. The river normally experiences a low-water level of about 1 m, that can rise up to 10 m in times of flood. The number of water levels recorded is summed up and then categorized accordingly. The graph gives a three-modal water-level distribution with flood peaks in February, August, and December. As the water level rises up to, for instance, the 5-m level, it has also passed the 1 to 4-m levels. These water levels are also entered into the calculation.

When entered into a cumulative curve, the probability of exceedance of a mean flood level (50 %) can be calculated, as well as every other flood level (Fig. 6.5). For instance the 10 % exceedance percentile that means a flood level of 10 m will be exceeded only in 10 % of all floods, or the 90th percentile with water levels higher than 1.2 m. If the level of the hundred-year flood is entered into the graph, it becomes obvious from which water level on flood risk reduction measures have to be initiated.

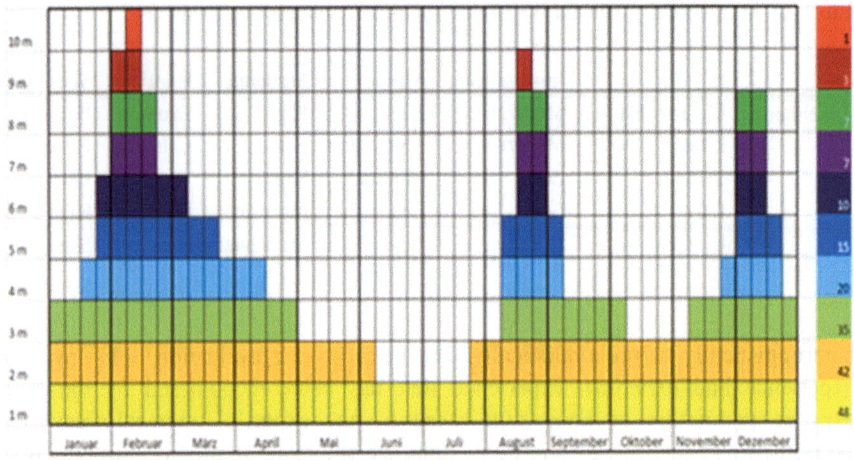

Fig. 6.4 Water level record of a river over a one-year period (fictitious)

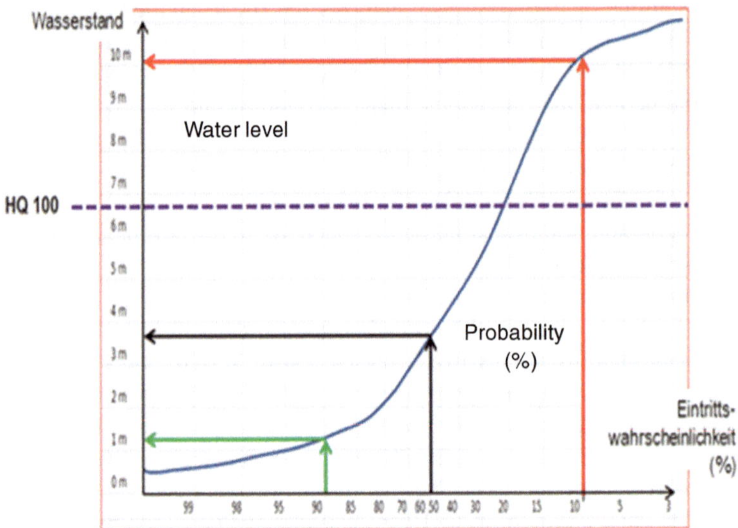

Fig. 6.5 Cumulative curve on probability of exceedance of flood levels (fictitious)

As the vulnerability is according to what has been mentioned before composed of two different elements:

- Risk exposure
- Risk preparedness

a method is required to enable us to distill the indicators describing these two subjects and enable us to arrive at just one figure. As large countries such as India or the United States are known to be exposed to a multitude of natural disasters and both countries are at places populated by hundreds of thousands of inhabitants and

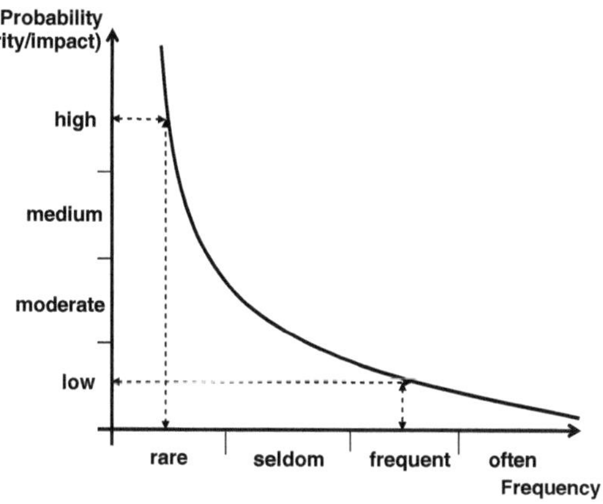

Fig. 6.6 Idealized mathematical relationship between probability and frequency (Own graph)

other regions only have a small population, an indicator has to be developed that can work out the difference between such different regions.

The mathematical dependence between probability and frequency is based on the general idea: Catastrophic events that are very strong, occur only rarely; while those that are of a low in intensity occur quite often. From this assumption, the graph in Fig. 6.6 can be derived.

Probability is one of the governing factors in risk assessment as it is "an equally weighted parameter that helps us to quantify and prioritize mitigation actions" (FEMA 2004, 2012). Probability is seen by many risk managers as a solely technical parameter, based on tangible counting of disaster events and its direct losses. Several probability calculation tools are in use, especially by the insurance sector mostly applying Monte Carlo simulations or other stochastic probability calculation methods. But to an individual affected by a hundred-year flood it is difficult to communicate that he was victim to an event with just a probability of 0.001 % per year. To overcome this misunderstanding, it turned out that probability is better expressed by qualitative expressions, using words that describe the chance of an event occurring.

Possible descriptions of likelihood of risk are:

- Certain (>99 % probability) = one or more occurrences per year.
- Likely (50–90 % probability) = one occurring every 1 or 2 years.
- Possible (5–49 % probability) = one occurring every 2–20 years.
- Unlikely (2–5 % probability) = one occurring every 20–50 years.
- Rare (1–2 % probability) = one occurring every 50–100 years.
- Extremely rare (<1 % probability) = one occurring every 100 or more years.

Accordingly possible verbal descriptions of risk consequences are:

- Catastrophic = Large number of fatalities and severe injuries, extended and large number of people requiring hospitalization.
- Major = Extensive injuries, significant hospitalization; fatalities.

- Moderate = Medical treatment needed but no fatalities; some hospitalization.
- Minor = Small number of injuries but no fatalities; first-aid treatment required.

The methodological concept behind this type is the functional relationship that low-grade events occur quite often, and strong and severe events occur rather seldom. Such curves are known from all kinds of probability distribution functions, for instance, the health sector (increase in illnesses vs. age), damage from car accidents, or crop yield versus years, and has found broad application also for the assessment of natural disasters: earthquakes, floods, landslides, and so on. For example, the many hailstorms in Germany over the last decades were monitored by German risk insurers. They plotted the diameters of the hail particles and compared them with the frequency of their occurrence (Fig. 6.7). It turned out that hail of 1–2 cm in diameter occurred very often, whereas hail the size of 5–7 cm has been very rare. Nevertheless in summer 2013 there was a hailstorm in Germany that locally produced hail particles of 8 cm in diameter; the biggest particle size ever recorded in Germany leading to damages of €1 billion within in 1 h. From such type of plot, risk insurers calculate their hailstorm portfolio. Superimposing the hail's size-induced damages (severity) with frequency an insurance company may decide that it might be economically unreasonable to allocate financial means to cover losses, for instance, for hail damage smaller than 5 cm in diameter, but to raise the premium for larger hail-size damage.

In order to arrive at a probability distribution of floodwater levels, the recorded flood levels are categorized according to their severity levels and juxtaposed to the respective frequencies. The indication on severity and frequency can be chosen arbitrarily (Table 6.4; Fig. 6.8). A further advantage of such type of probability distribution is it opens the chance to extrapolate the distribution function to assess, for instance, flood levels that have not been experienced thus far.

The probability distribution function for the vulnerability from earthquake is broadly used. From the geological record, say 30 earthquake events were taken with magnitudes ranging from 4.1 up to 7.8 on the Richter scale over a time span

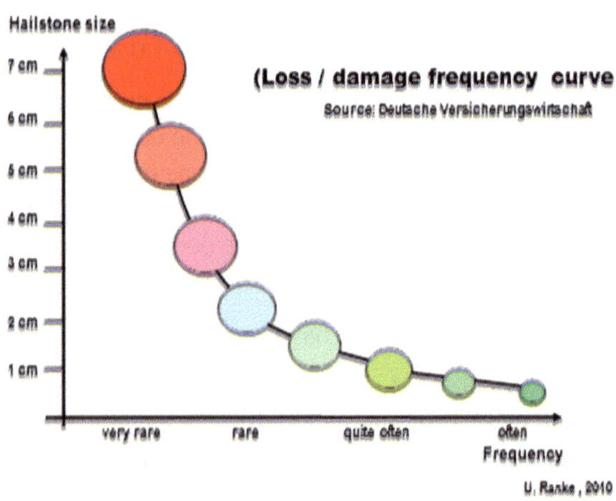

Fig. 6.7 Hail-size distribution as an example of a loss damage frequency curve (*Courtesy* Gesamtverband der Deutschen Versicherungswirtschaft, Berlin)

Table 6.4 Flood levels categorized according to severity and frequency (fictitious)

Water level (m)	Severity level	Frequency	
>9	Very high	<10 %	Very rare
9	High	10–25 %	Rare
8			
7	Increased	25–40	Seldom
6			
5	Moderate	40–55	Occasionally
4			
3	Low	55–70	Often
2			
1	Very low	70–85	Very often
<1	Nil	>85	

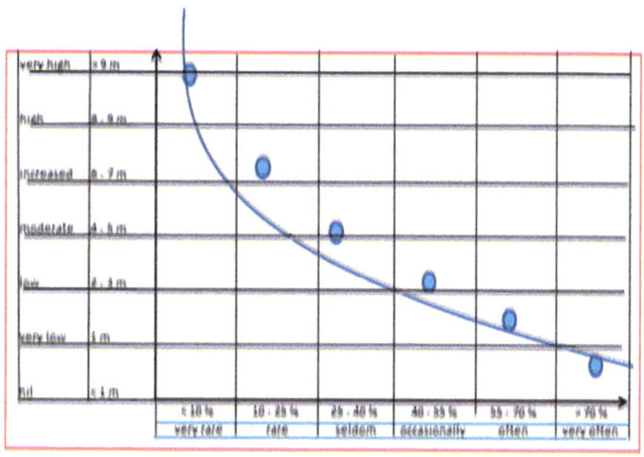

Fig. 6.8 Probability of flood-level distribution (fictitious)

of 80 years in a certain area; the vulnerability is assessed by a vulnerability curve (Fig. 6.9). From the record it can be revealed that 16 earthquakes with a magnitude 4.0–5.0 have occurred, and earthquakes with magnitudes >7.0 were very seldom (3). The magnitudes (4.1–7.8) were grouped into 4–5 (low), 5–6 (middle), and 7–8 (high) Richter scale classes and were then plotted against their frequency (1–3, 3–6, 6–10, >10 times of occurrence). The boundaries were set arbitrarily. The vulnerability curve indicates that earthquakes with low magnitudes in the past have occurred and will in future be expected to occur about 10 times in a 80-year interval, whereas earthquakes of magnitudes >7.0 on the Richter scale should occur (only) four times in the same time span. The vulnerability curve thus provides a quite realistic vision of the future vulnerability from earthquakes for the area under investigation, stating that earthquakes higher than M7.0 will occur very rarely (time scale has to be defined arbitrarily) whereas earthquakes of a lower severity (M < 4.5) may occur very often.

The instrument of the vulnerability curve can also be applied for very specific vulnerability assessments, for instance, on earthquake-exposed buildings. An example for this is the assessment of the vulnerability for residential buildings in

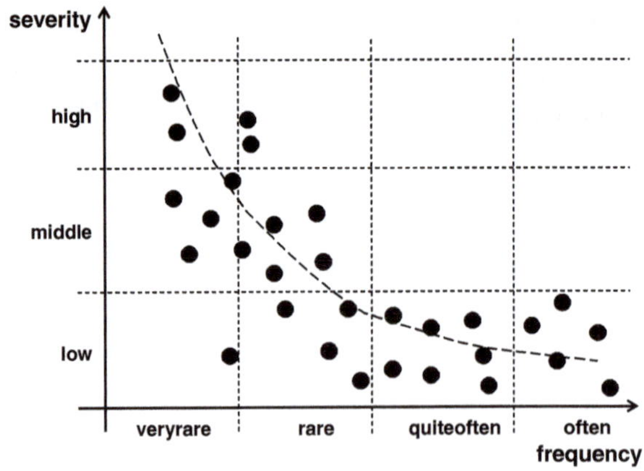

Fig. 6.9 Probability estimate of possible earthquake occurrence from historical magnitude and frequency distribution (fictitious)

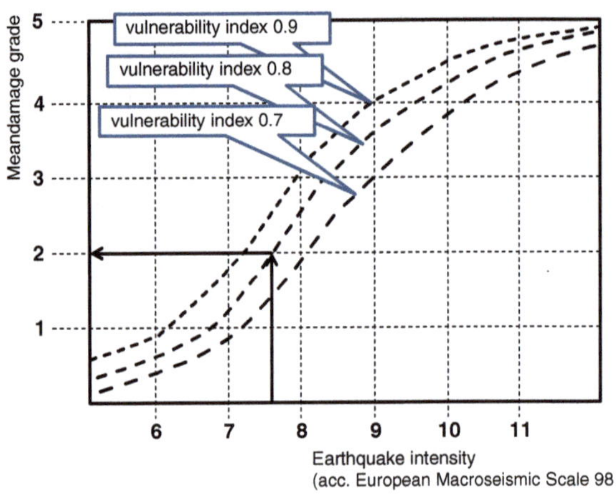

Fig. 6.10 Mean damage grade based on vulnerability curves from relating to vulnerability indices of the Manjil earthquake 1990 (*Courtesy* Omnidar et al. 2011)

Iran (Fig. 6.10) to provide a basis for estimating the damage features by future earthquakes (Omvidar et al. 2011). For such a vulnerability assessment, all structural damages from the Manjil earthquake (Iran 1990) have been assessed, using a set of indicators that comprehensively describe the disposition of a building for earthquake events such as adobe walls, unreinforced brick walls with wooden beams, unreinforced steel frame without bracing, steel frame with bracing, reinforced concrete frames or reinforced concrete shear walls, and many more. When the findings are plotted against the seismicity's experience, a set of vulnerability curves is derived indicating what damage ratios are to be expected if in the area under investigation an earthquake, say of magnitude 7.8 struck buildings of a vulnerability index, say 0.8 (=damage ratio of 2). Such an assessment can then be made the basis for designing and implementing seismic mitigation measures in order to improve local earthquake resilience.

Following this methodological concept Kammel (2012) made a much generalized assessment of the vulnerability of central Java (Indonesia) based on published disaster event data. He calculated the specific vulnerability distribution of earthquakes, landslides, floods, and tsunamis and then combined the respective vulnerability curves in one plot, deriving the overall vulnerability from the said hazard types of that particular region (Fig. 6.11). Although the earthquake vulnerability shows a broad range of frequency compared to severity, the other vulnerabilities (landslides, floods, tsunamis) show different curves according to their disaster event characteristics. In the next step all curves are aggregated into one graph revealing a corridor of vulnerabilities for the country "A". The average vulnerability for the country is given by the red curve, and the upper and lower limits of the vulnerability corridor are marked by the pink curves. By arbitrarily dividing the vulnerability corridor into meaningful sections, for example, 30, 60 %, higher or lower than the average, the potential vulnerability for country "A" can be assessed. The disaster vulnerability curves pinpoint the assumption that country "A" has a potential vulnerability on average for a disaster to occur with a <60 % probability ("very rare") with a high severity, and there is a >60 % probability ("quite often") that a disaster of a severity "low to medium" will occur.

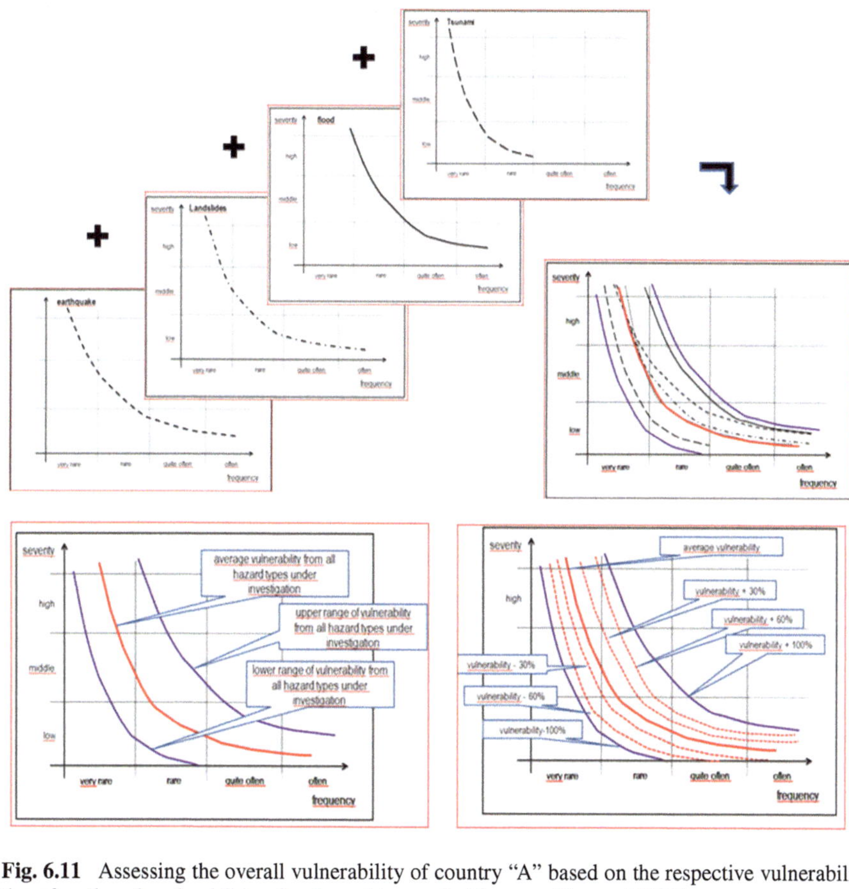

Fig. 6.11 Assessing the overall vulnerability of country "A" based on the respective vulnerabilities of earthquakes, landslides, floods, and tsunamis (*Courtesy* Kammel 2012)

References

Benson, C. & Twigg, J. (2007): Tools for disaster risk reduction - Guidance Notes for Development Organisations. - International Federation of Red Cross and Red Crescent Societies; the ProVention Consortium Secretariat, Geneva

Bolin, R. & Stanford, L. (1998): The Northridge earthquake: community-based approaches to unmet recovery needs.- Disasters, Vol. 22, No. 1, p. 21-38, Wiley Online Library

Buckle, P. (1998): Re-defining community and vulnerability in the context of emergency management. –Queensland State Emergency Recovery Unit, Department of Human Services, Australian Journal of Emergency Management, Mount Macedon, Victoria

Cannon, T. (2008): Vulnerability, "innocent" disasters and the imperative of cultural understanding.- Disaster Prevention and Management, Vol. 17, No. 3, p.350-357, Emerald Group Publishing Limited (online: www.emeraldinsight.com/0965-3562.htm)

Cardona, O. (2003): The need for rethinking the concept of vulnerability and risk from a holistic perspective: A necessary review and criticism for effective risk management; in: Bankoff, G., Frerks, D. & Hilhorst: D. (eds): Mapping Vulnerability: Disasters, Development and People.- Earthscan Publisher, London

Dwyer, A., Zoppou, C., Nielsen, O., Day, S. & Roberts, S. (2004): Quantifying social vulnerability: A methodology for identifying those at risk to natural hazards.- Geoscience Australia Record 2004/14, Canberra

Eakin, H. & Luers, A.L. (2006): Assessing the vulnerability of social-environmental systems. – Annual Revue Environmental Resources, Vol. 31, p.365–394, Palo Alto CA

FEMA (2004): Using HAZUS-MH for Risk Assessment - HAZUS-MH Risk Assessment and User Group Series Using; How to Guide. - Federal Emergency Management Agency (FEMA), Publication 433, Washington DC

FEMA (2012): Emergency Response Plan- Resources for Protective Actions for Life Safety.- Federal Emergency Management Agency (FEMA), Ready Campaign, Washington DC (online: www.fema.org; last update 12/19/2012)

Kammel, L. (2012): Methodischer Ansatz zum Vergleich der Risikobewertung verschiedener Naturgefahrentypen am Beispiel von Zentraljava, Indonesien. - Bachelor Thesis (unpublished), Faculty of Geosciences University Goettingen

Kaplan, S. & Garrick, B.J. (1981): Risk Analysis, Vol. 1, No 1, Wiley Online Library

Kasperson, R.E., Renn, O., Slovic, P., Brown, H.S., Emel, J., Goble, R., Kasperson, J.X. & Ratick S. (1988): The social amplification of Risk: A Conceptual Framework. – Risk Analysis, Vol. 8, No. 2, Wiley Online Library

Klein, R.J.T., Nicholls, R.J. & Thomalla, F. (2003): Resilience to natural hazards: how useful is this concept?.- *Global Environmental Change, Part B: Environmental Hazards,* Vol. 5, No.(1–2, p.35–45 (online: doi:10.1016/j.hazards.2004.02.001)

Morrow, B.H. (1999): Identifying and mapping community vulnerability. - Disasters, Vol.23 Issue 1, p. 1-18, Wiley Online Library

Omvidar, B., Gatmiri, B. & Derakhshan, S. (2011): Experimental vulnerability curves for the residential buildings in Iran. – Natural Hazards, Springer Link (online: DOI 10.1007/s11069-011-0019-y)

PAHO/WHO (1998): Natural disaster mitigation in drinking water and sewerage systems – Guidelines for vulnerability analysis.- Pan American Health Organization – World Health Organization, Disaster Mitigation Series, Washington, DC

Peduzzi,P., Dao, H., Herold,C. with contributions from A. Martin Diaz, F. Mouton, O. Nordbeck, D. Rochette, T. Ton-That, B. Widmer (2002): Global Risk and Vulnerability Index-Trends per Year (GRAVITY) - Phase II: Development, analysis and results. - The Bureau for Crisis Prevention and Recovery of the United Nation Development Program, Geneva

Peduzzi, P., Dao, H., Herold, C. & Mouton, F. (2009): Assessing global exposure and vulnerability towards natural hazards: the Disaster Risk Index. – Natural Hazards Earth Systems Sciences, Vol. 9 p.1149—1159, Copernicus Publications, Goettingen

Slovic, P., Fischhoff, B. & Lichtenstein, S. (1982): Why Study Risk Perception?. – Risk Analysis, Vol. 2, Issue 2, p.83-93, Wiley Online Library

Thywissen, K. (2006): Core terminology of disaster reduction: A comparative glossary; in: Birkmann, J.: Measuring vulnerability to natural hazards - Towards disaster resilient societies . - United Nations University Press, Bonn

UN-AFD (2004): United Nations Development Program's Adaptation Policy Framework for Climate Change: Developing strategies, policies and measures. - United Nations Development Program (UNDP), Cambridge University Press, Cambridge

UNDRO (1991): Mitigating natural disasters phenomena, effects and options. – A manual for policy makers and planners. – Office of the United Nations Disaster Relief Co-ordinator, United Nations, Geneva

UNISDR (2004): Living with Risk - A Global Review of Disaster Reduction Initiatives". – United Nation International Strategy of Disaster Reduction (UNISDR), Geneva

WBGU (1999): Welt im Wandel – Strategien zur Bewältigung globaler Umweltrisiken. – Wissenschaftlicher Beirat der Bundesregierung Globale Umweltverän-derungen, Jahresgutachten 1998, Springer, Heidelberg

Wisner, B., Blaikie, P., Cannon, T. & Davis, I. (2004): At Risk - Natural hazards, people's vulnerability and disasters. - Routledge, London - New York NY

Chapter 7
Risk Assessment

7.1 General Aspects—Definition and Explanations

The international discussion on risk has mostly originated from the assessment of risk in the fields of medical services, the chemical industry, and especially after the nuclear catastrophes of Three Mile Island (United States) in 1979 and Chernobyl (former Soviet Union). The discussion is not detailed here, but authors including Paul Slovic, Stanley Kaplan, Ulrich Beck, Ortwin Renn, and many others are the main representatives who have given a multitude of explanations and examples of risk assessment. Today in every nation worldwide are institutionalized organizations, such as the Wissenschaftliche Beirat der Bundesregierung Globale Umweltveränderungen (WBGU) in Germany, the Swedish Nuclear Fuel and Waste Management Corporation, the Swiss Nagra, the British Civil Contingencies Secretariat (SCC), or the US Federal Emergency Management Agency/Department of Homeland Security (FEMA). These organizations deal with all kinds of technical and man-made but also natural disasters and all have contributed substantially to the subject.

Basically natural hazards are defined by the cause–effect relationship between the natural situation described by its physical exposure that may interact in a way to become a threat to people and their living environment/livelihood. As an example, a volcano in the Saharan Desert is just a volcano that can be described geologically, volcanologically, geomorphologically, and so on, but it does not pose a risk as (almost) no people are living there. But a volcano with a great potential to erupt near a megacity, for example, Mt. Popocatepetl, poses a considerable threat to the people in Mexico City. But volcanoes not only pose a risk to the people by their mere physical appearance; such a volcano achieves hazardous potential when people interact with the volcano, for instance, by settling on the steep flanks. This definition leads to the assumption that as long as no one is settling near the volcano, no one is making his living from it, nor has a cultural heritage there, such a hazardous situation will not become a risk.

Tobin and Montz (1997) therefore pledge to broaden the context of hazard and risk by emphasizing that it is not only the action of nature on the human being, but

© Springer International Publishing Switzerland 2016

U. Ranke, *Natural Disaster Risk Management*,
DOI 10.1007/978-3-319-20675-2_7

also society's interaction with nature that defines a risk. For them, hazards "represent the potential interaction between humans and natural events. It represents the potential or likelihood of an event (it is not the event itself)" that may become a disaster. Renn (2003) defines "risk of natural disasters to be at the crossroad between natural events (partially altered and amplified by human action such as the emission of greenhouse gases), economic, social and technological developments and policy driven actions." And for coping with risk they pledge "a holistic approach to hazard identification, risk assessment and risk management." In 2009 Aven and Renn stated, "Risk is more than just multiplying the losses with the probability of occurrence, it rather deals with uncertainty about the occurrence and the consequences."

They give the following example of this: everyone who crosses a busy street unconsciously assesses her risk of a traffic accident and automatically analyzes ways and means to manage the situation based on her experience. But there are, for instance, hazard exposures that pose a threat that is not to be assessed right from the beginning in all its consequences.

Purdy (2010) even goes a step further by extending the former risk definition. He points out that there is a "shift in from the possibilities of an event to the possibilities of an effect," thus "making the risk management a process of optimizing resilience that makes risk management an inseparable aspect of disaster management." For him the process of risk management comprises "a steady communication and consultation process internally as well as externally to gain input as well as to trigger impact on decision making by monitoring and review."

Another definition points to the fact that a natural hazard automatically and implicitly comprises a "component of damage probability that is quantitatively assessable" (Felgentreff 2008). Moreover, hazard assessment is predominantly based on natural sciences and technical indicators; the assessment of the population's vulnerability is exclusively derived from socioeconomic factors.

The broad discussion on risk can be summarized as following the general definition as given by the United Nations International Strategy for Disaster Reduction (UNISDR 2004): "Risk is the product of the exposure of a society to a hazardous natural or mankind generated hazard situation, that may result in (physical, psychological, social, financial, etc.) threats to the individual or the society and its/their living environment, juxtaposed to the capacity of the society or the individual to cope with the impact."

The general relationship of hazard and risk is given in the mathematical equation:

$$\text{Risk} = \text{hazard} \times \text{vulnerablity} : \text{coping capacity}.$$

There is a variety of applications on the risk formula in use, a simplified equation of risk:

$$\text{risk} = \text{hazard} \times \text{population} \times \text{vulnerability}$$

When the population is replaced by physical exposure:

$$\text{risk} = \text{physical exposure} \times \text{vulnerability}$$

The following formula used by insurance companies generally constitutes the two components:

$$\text{risk} = \text{frequency} \times \text{potential maximum economic loss}$$

There have been multiple attempts to define the term "disaster" and to distinguish it from an emergency or a catastrophe (Quarantelli 1989). Researchers emphasize the need to understand and conceptualize the term, however, they agree that a disaster is an event caused by human or natural forces and resulting in an enormous loss of life and property. All definitions (UNDRO 1979; Coburn et al. 1991, p. 49) point to the same basic features of risk assessment: "In general, risk is always the outcome of a potential threat and thus is derived from three basic components:"

- A component that describes how likely a potential hazard may occur (frequency × severity = hazard exposure)
- A component that describes the consequences that may occur from a disaster (e.g., annual casualties versus average population at risk = vulnerability)
- A component that describes the ability of a society to withstand the disaster impact based on technical capacity and experience (preparedness = coping capacity)

The formula proposed by UNDRO (1979) for modeling risk combines the above three components by multiplying them with each other. Multiplying was introduced because if the hazard is nil, then the risk is also nil: 0 (hazard) × population (4) × vulnerability (5) = 0 (risk). The risk is also nil if nobody lives in an area exposed to hazard (population = 0); the same situation applies if the population is invulnerable (vulnerability = 0, induce a risk = 0).

The hazard exposure of a society to a certain risk is a clear indicator whether such a society is at risk. The critical relationship between hazard and risk can be explained by the "tightrope walker" example given by the British Columbia Ministry for Public Safety (Canada) that explains:

> The risk to a tightrope walker in about 100 m height is falling-off and getting killed. When considered that the wire is installed only one meter above the ground, the chance of falling off the rope still remains the same as 100 m above the ground, but the risk to die is nil. Thus risk does not mean (solely) the statistical chance or probability to get killed. Risk is rather a total concept of likelihood of occurrence of a hazard and the severity of possible impacts. Moreover the likelihood for a risk of injury even exist to the crowd below the tightrope walkery. The severity of impact to the tightrope walker and the crowd can be mitigated by a safety net, the chance of falling can be reduced by special training and the extent of injury the crowed can be diminished by emergency medical response capability: all components that are summarized as coping capacity.

A risk definition that points in the same direction was given by UNDRO (1979) and Burton et al. (1993, p. 34) that both gave strong emphasis to the importance of the time factor in risk definition: "The term risk refers to the expected losses from a particular hazard to a specified element at risk in a particular future time period. Loss may be estimated in terms of human lives, or buildings destroyed or in financial terms."

Although there are quite a number of different definitions of risk, all of them have in common that risk is something harmful to people, the living environment, or the ecology. We do not add another risk definition, but for sake of a better understanding, the book underlines what Kaplan (1997) said about risk. Risk can indirectly better be understood when we ask three questions about the specific hazard to which a population at risk might be exposed:

- What can happen?
- How likely is it that an event may occur?
- What are the consequences for those affected?

Kaplan stated that when posing the above questions, a risk definition becomes more flexible and will provide a much better starting point for further risk management considerations. From these three questions it immediately becomes clear that risks are highly complex and very dynamic in nature and that it will therefore not be possible to generalize the disaster impact beforehand. Following Kaplan, the starting point for a risk mitigation concept lies in the combination of the answers to the three questions. He therefore pledges that according to the complex nature of risk a specific assessment of each and every hazard situation is required and has to consider its specific geological characteristics. But also the level of resilience that defines the society's potential to withstand disaster is indispensable.

Kaplan raised the following question: when a risk is "subjective" from the standpoint of the observer, is then the risk this person might be exposed to really a risk or does he or she (only) feel that risk ("perceived risk"). When the risks are "perceived" does this mean there also exist "absolute risks" or "objective risks?" Objective in this regard would mean that such risks have to be risks to which the entire society is exposed. As this discussion will only lead to a very philosophical discussion, Kaplan suggested an easier approach to the matter, by introducing another term, "probability," making the story much clearer and more manageable.

Over the centuries risk-exposed societies all over the world have developed a multitude of instruments, methods, and capabilities that have all—based on real disaster experience—led to effective disaster mitigation. For example, people living on the North Sea coast on tiny Warden Island have broad experience to protect themselves against storm surges. Such surges are normally effective when northern winter storms press seawater into the German bight and additionally the full moon causes an exceptional rise of the tides. Similarly many people living along the rivers Rhine and Mosel have tiled the ground floors up to the highest water levels experienced, retrofitted their windows and doors to be flood-proof, and keep electrical appliances in the upper floor during flood season.

But figures on death tolls and demolished houses normally only rarely represent the real extent of a disaster. Even information on the economic losses does not fully describe the disaster's extent. A figure, given on this by the UNDP (2004a, b, c), states that for every single man killed in a disaster, about 3000 worldwide are exposed to it. Furthermore disaster information normally does not comprise information on the injured, the displaced, and the homeless. Although this phenomenon is not yet fully understood and has not been investigated in depth, a group of leading American risk managers and sociologists (among them Stan Kaplan and Paul Slovic but also the German Ortwin Renn) identified that social disparities are the highly forming factors that "govern risk perception" (Kasperson et al. 1988; Slovic 1987, 2000; Renn 1989). Consequently they introduced a new term in risk management, "the social amplification of risk." According to their investigations "social structures and processes in risk experience resulting in repercussions on individual and group perceptions and the effects of these responses on community, society and economy compose a general phenomenon."

Gender shapes the social worlds and is thus a primary factor of social organization before, during, and after a natural event occurs. Not gender but rather gender inequality puts girls and women at risk (Enarson 2000). Therefore there are distinct differences in emergency preparedness, coping strategies such as voluntary self-help initiatives, emergency communication, stress, fears, and post-traumatic stress. Mortality patterns are also strongly gender related as well as the exposure to injury or to become homeless. Experience from the Indian Ocean tsunami of 2004, earthquakes in China, and cyclones in Asia revealed that women disproportionally died from disaster of such kinds than men, although men are killed, for instance, more from disasters such as lightning, or technical and man-made disasters.

Women are very proactive responders. According to their traditional role in the family they care much more for household security and family sustainability and thus are recognized by the International Decade for Natural Disaster Reduction (UN-ISDR) as "keys to prevention" actors. Masculinity norms, however, orient on immediate, "heroic" (risk anticipating) action during the search and rescue period, debris removal, and reconstruction. But restoring the social aspects of the system is often left to women.

Most disaster assessments that focus on natural disasters thus far defined the severity of natural events preferably according to their impact on the individual and on economic losses, as such data (numbers of victims, amount of economic losses) are easily assessed and are very often highly impressive. But many studies of the last years, especially the long data record series of CRED-EMDAT, reveal that the amount of people killed is steadily decreasing whereas that of economic losses is rising (see Fig. 2.3). The number of people who are affected by natural disasters as opposed to the death toll is rising significantly. There is a clear relationship to be seen between the increase of risk-exposed people and the population increase in the big conurbations. Numerous reasons for this can be found, but the most important is that since the mid-1990s more and more people are migrating into the big cities so that already today more than 50 % of the world's population lives in megacities, such as Shanghai, Tokyo, Jakarta, and the like. And there is no sign that this trend will change in the decades to come. As most of the big megacities are located on coastal plains or flat-lying extended river valleys, the exposure of the population to natural disasters is rising. Therefore this change in the basic disaster assessment parameters (death toll, economic loss) to describe the severity of a disaster impact, will in future lead to inappropriate figures and judgments. Thus in future, the impact of a disaster should be based mostly on a description of the people and their living conditions exposed to a risk in order to get a more realistic picture of their risk exposure and in order to make such assessments more comparable internationally. Vulnerability has to become the important indicator, as it describes the potential risks an individual or a societal group is exposed to as well as defining their future living conditions. Vulnerability should therefore be taken as the cornerstone of a modern natural disaster risk management as only by such an assessment can the potential of the population risk for self-help be defined as well as the duties of the public sector to set up a legal and technical framework suitable to increase social resilience.

7.2 Valuing Statistical Life (VSL)

In the last 10 years a variety of studies has been carried out by economists worldwide to calculate a monetary value for the loss of life or an injury, an approach that is called "valuing statistical life" (VSL). The calculations aim at identifying an economic value of the statistical life of an individual at risk. The calculations, however, vary significantly mainly due to risk exposure, prevailing socioeconomic situations, technical standards of prevention, and the type of living environment. The calculations clearly reflect that different societies make different choices over risks and have different perceptions of the values of life-saving initiatives. The risk variables used in several of the empirical studies were in general based on job-related accident and mortality data.

Different methodological approaches have been used to analyze this strategic financial sector, all derived from economic and actuarial science. One is the "willingness-to-pay" approach that was used by many authors, for example, by Lindhjem et al. (2011). Their investigation revealed a VSL mean value of about US$7 million with a variance of US$190 billion maximum and US$5000 minimum values and a median of US$2.3 million, based on a US dollar of the year 2005. By screening out the upper and lower 2.5 %, the values range between US$5–6 million, a value that comes close to the values given by the US Environmental Protection Agency (EPA 2010), that estimated the VSL at about US$8 million. Another Swedish assessment carried out by Carlsson et al. (2010), also based on the "willingness-to-pay" approach, came to the conclusion that the value of statistical life ranges between US$2 to 3 million.

Other methodological approaches were found by Viscusi and Aldy (2002) in the literature. All of them had in common "an occupation-specific measure" reflecting an average of several years of observations for fatalities, on relatively rare events. They stated that an ideal measure of the risk to be killed or injured on job would reflect both the "worker's perception of such risk and the firm's perception of the risk." Because these variables consider both, the worker's choice is for a certain job (income) and his risk exposure perception, as well as the company's interest to produce at as low cost and secure a safety level as possible. If the individual's and the company's risk perception arrive at an objective measure of fatality risk, then such objective risk data could be used to identify the range of the risk premium. Viscusi and Aldy (ibid) suggest that

> Refining VSL for the specific employment characteristics of the affected population at risk remains an important priority for the research community and the government agencies conducting these economic analyses. Improving the application of VSLs can result in more sustainable government interventions to address market failures related to environmental, health, and safety mortality risks.

The study reported that in the United States and many other highly industrialized countries the value of statistical life for a prime-aged worker falls within a range of US$4–9 million, with an average value is about US$7 million, when converted into year 2000 US dollars. The VSL for developing country labor markets, however, shows significantly smaller values of statistical life. In South Korea, the VSL of the industry sector was estimated at approximately below US$1 million. In the United States individuals with jobs with very minor risks have values of statistical life ranging from US$12–22 million, a value that drops between US$10–18 million when the risk increases tenfold. People working in jobs with very high risks result in much smaller values of statistical lives (far below US$10 million). White collar employees and members of US labor unions benefit from higher risk premiums than others. Moreover, it was found in the literature that the value of a statistical life decreases with age (Fig. 7.1).

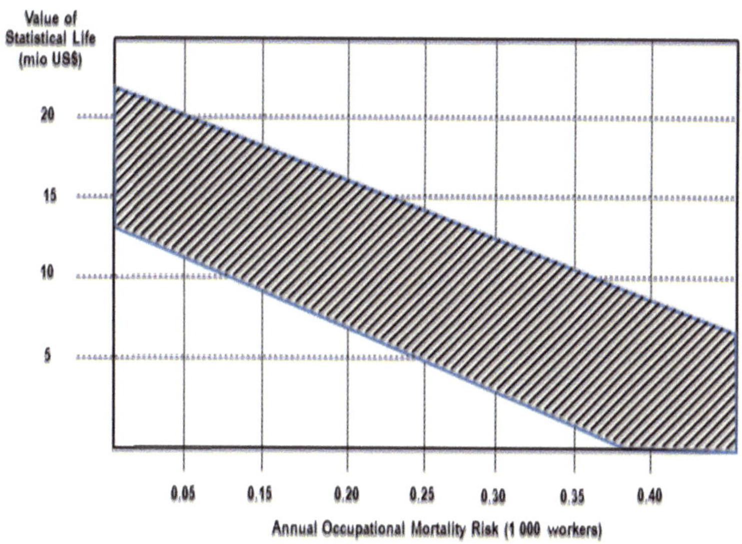

Fig. 7.1 Range of statistical life assessments (Based on Viscusi and Aldy 2002)

7.3 Deterministic Risk Assessment

The aim of risk assessment is to define the likelihood that a certain disaster will threaten a society. Such estimations are carried out based on natural evidence recorded from a single disastrous event or on many disasters over a longer time span (historic dataset). There are two different methods in practice to assess risk. One is based on single point data defining an individual risk that can be called a worst-case estimate of exposed factors from which the worst-case scenario is derived. On the other hand, risks are assessed by true probabilistic estimates (see below). There is no a priori presumption that risk assessment can or should only be estimated by a deterministic or probabilistic analysis (Power and McCarry 1996), but over the time the probabilistic assessment method has proven to have many significant analytical advantages, mainly greater insight into the magnitude of expected effects.

For the sake of completeness and to give easy access to quantitative risk assessment, the deterministic method for risk assessment is briefly described here, although knowing that the "dynamics of nature systems reduce the validity of using simple deterministic approaches" (Power and McCarry, ibid).

Traditionally most assessments have focused on estimating single effects and exposure values to determine the risk using a "quotient method" to determine frequency and severity over a time period. The likelihood is assessed quantitatively by using deterministic models and simulations. Next to quantitative data (tangible data such as the amount of casualties, the number of destroyed houses) also qualitative data (such as a statement on perceived risk) are entered into the model when no numerical data are available (intangible data such as socioeconomic data, data on societal risk perception, risk experience, etc.). But it should be taken into consideration that even such quantitative data have to be transformed into qualitative data by numerically classifying a descriptive scale, although knowing that such transformation is arbitrary and depends strongly on the experience of the risk assessor. In order to address that problem the assessors must indicate the level of confidence in their estimates. The chosen level of confidence will be presented, along with final results, to indicate to decision makers where some low or very low confidence assessments may necessitate caution in interpretation.

The deterministic risk assessment generally correlates the dependency of two random variables or sets of data to indicate a predictive relationship of causal factors that, for example, describe the hazard/risk exposure of a natural, social, and/or economic origin. The causal factors and their relationship can be described by numbers, physical characteristics, or by statements. Predictive are relationships that follow an entirely known rule (law, equation, or fixed procedure) so that the state of each component and of the entire system can be foreseen "at any time for any time in the past and future." Such a system is called deterministic (Kirchsteiger 1999). The aim of such a correlation is to investigate the factors of dependency or behavior on the basis of its components' changes with time. The results of the analysis can be expressed in qualitative statements and quantitative numbers. The application of a deterministic approach in risk assessment is based on the assumption that in general, a lack of adequate and reliable data as well as the behavior of a system can be better predicted using more basic components. The risk data are processed by a point estimate method that allows for an estimation of the levels of risk and to compare between different risk situations based on similarities. Deterministic risk assessment is used when the risk or hazard distribution and the probable consequences show a very direct, easily identified relationship. But it has to be acknowledged that such a direct exposure–risk relationship in reality is very rare.

However, statistical dependence is not sufficient to demonstrate the presence of such a causal relationship, "Correlation does not imply causation" (Wikipedia, access: 19.11.2013). An example of earthquake risk may illustrate this dependency. A correlation of the average annual number of earthquakes and their magnitude does not say anything about the pattern of earthquake occurrence in the past being repeated in the future. Statistical models usually assume independent events and a constant trend, but in reality such assumptions merely provide a first approximation. To reach a higher degree of reliability requires a larger dataset (data volume and time span).

For undertaking deterministic risk analysis it is necessary to classify the data according to severity categories for both likelihood (severity) and consequences (impact) of the identified hazards and to assess the severity category of each particular hazard (Table 7.1).

Following the assumptions in Table 7.2 a deterministic risk matrix can be established according to the following approach. The cells of the matrix express the likelihood/consequence category to which each individual hazard has been assigned. The shading in the figure indicates an increasing risk. High assessments such as i 4–i 5 and F 4–F 5 (black squares) are estimated to have a very

Table 7.1 Example of definition of severity categories of an identified hazard (fictitious)

Likelihood (frequency)	Qualitative definition	Quantitative definition (occurrence per year)	Severity category
	Likely once in next year	0.2–2	F 1
	Possible but not likely	0.02–0.2	F 2
	Unlikely	0.002–0.02	F 3
	Very unlikely	0.0002–0.002	F 4
	Remote	0.00002–0.0002	F 5
Likelihood (impact)	Qualitative definition		Severity category
	Catastrophic, multiple fatalities		i 5
	Major single fatality, multiple injuries		i 4
	Very serious, permanently disabling injury		i 3
	Serious injury, full recovery		i 2
	Minor losses, short time injury, short absence from work		i 1

Table 7.2 Deterministic risk matrix to outline the likelihood/consequence relationship to which an individual hazard can be assigned (fictitious)

	Consequences "impact"				
Likelihood "frequency"	i 1	i 2	i3	i 4	i 5
F 1					
F 2					
F 3					
F 4					
F 5					

severe risk. Dark grey squares have a moderate to severe risk; light gray squares have low risk assessments, and the white squares are considered to have no risk. A scheme of this kind can be worked out for any kind of risk assessment and can be organized arbitrarily according to the situation under investigation. Such schemes have proven an effective tool for obtaining a simple overview of a range of hazards with regard to onsite risks, and for prioritizing them for improvement actions.

For undertaking deterministic risk analysis it is necessary to classify the data according to "severity categories" for both likelihood and consequences of identified hazards and to assess the severity category of each particular hazard in terms of these. Using this approach requires the use of a quantitative definition for the risk factors, although they are in general expressed (only) by qualitative definition. To ensure consistency it might be advisable to incorporate benchmarks or semi-quantitative data in the analysis. In schemes of this type, the assessment team, usually comprising geoscientists, risk assessors, safety engineers, and operation managers will first identify all hazards, and then assign a severity category to each of these for both likelihood and consequences. Figure 7.2 shows another form to assess deterministically the risk, here from earthquakes for Germany, California (United States), Indonesia, and Japan. The assumption is fictitious and not based on real earthquake data. First the hazard exposure is identified by a cross-plot of severity versus frequency. It reveals that in Germany the earthquake hazard exposure is very low compared to California and Japan. The risk factor of vulnerability was assessed by combining the factors of average annual death versus average of earthquake-exposed people. Here Germany and California show comparably low vulnerabilities whereas Indonesia and Japan are highly vulnerable. The third risk factor, coping capacity, reveals that "technical resilience" and "disaster experience" are greatest in California and Japan whereas in Germany, for instance, the technical expertise is high but only little experience of earthquake mitigation exists. Calculating the risk according to the risk formula given above,

$$\text{Risk} = \text{hazard} \times \text{vulnerablity} : \text{coping capacity}$$

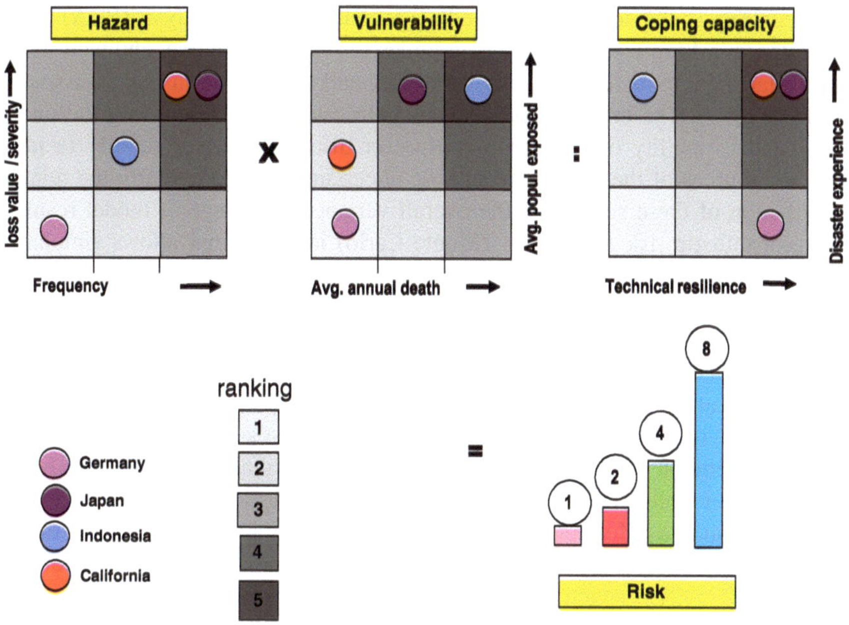

Fig. 7.2 Example of deterministic risk assessment (earthquake risk of Germany, California, Indonesia, Japan (fictitious)

reveals that Indonesia and Japan have the highest risk from earthquakes. Germany has almost no risk and California although it faces a multitude of earthquakes every year, has a comparable lower risk, mainly due to its well-developed risk mitigation capacity.

7.4 Probabilistic Risk Assessment (Computer-Based Hazard and Risk Assessment)

Emergency management practice in the last decades showed that probabilistic distribution functions are becoming increasingly important for risk assessment. Risk managers, regulators, and decision makers realized that physical sciences and engineering are best capable to provide reliable and independent data to assess the uncertainty of risk, although the mathematics involved are quite complex and it often proved difficult to make the findings understandable to the users. Moreover it should be taken into consideration that risk assessment is an approach in which often qualitatively acquired data are transferred into quantitative expressions of uncertainty and the limits of the method are based in the mathematics of probability.

In contrary to the deterministic model a probabilistic risk analysis (PRA) using, for instance, a Monte Carlo analysis provides for viable, effective, and well-established statistical tools for analyzing variability and uncertainty in risk assessments (USEPA 1997). The probabilistic analysis is to estimate quantitatively the uncertainty and variability of a hazard exposure or disaster risk; secondarily to identify the "sources of the variability and the uncertainty and to quantify the relative contribution of these sources to the overall variance and range of model results." The probabilistic risk assessment (Monte Carlo) is a tool that allows simulating the risk variables, making it a dynamic process. One of the important features of PRA is that very often the data input are qualitative in nature and are then transformed into quantitative variables. Probabilistic PRA analysis is best applied when it is necessary to rank hazard or risk exposures, to identify exposure pathways and locations that are risk exposed as well as populations at risk. Moreover PRA can provide helpful indications to identify the most suitable and effective and best implementable risk mitigation measures or remedial actions as a scientifically based tool for political decision making. Although the use of probabilistic risk simulations such as Monte Carlo is increasing considerably, there is still a debate on how strong the data input may influence the simulation output. Bukowski et al. (1995, pp. 215–219) stated that "The choice of input distribution in the simulations likely has a greater effect on the resultant risk distribution than does the inclusion or exclusion of variable correlations."

Probabilistic risk assessment is based on the following two pillars.

1. Bayesian Analysis

The general model for a quantitative assessment of risk uncertainty is called Bayesian analysis. The "Bayesian or subjective view is that the probability of an event is the degree of belief that a person (a disaster expert but also an individual at risk) has, given some state of knowledge, that the event will occur" (USEPA 1997). The subjective view on probability is in general based on scientific knowledge, expert judgment, and past experience, but also on intuition that these subjective probabilities can later be combined with new data to reach an updated output. In hazard exposure and risk assessment, representative and complete datasets are rarely available and inferences in these situations are inherent. The decision as to the appropriateness of following the Bayesian approach is based on the quality of data and the risk manager's subjective experience. The application of a quantitative numerical statistical assessment of uncertainty requires first the establishment of an appropriate likelihood function for the observed data to be formulated upon the general understanding of the hazard or risk situation. Combined with a Monte Carlo simulation Bayesian/Monte Carlo simulation offers an effective numerical technique for carrying out uncertainty analysis by generating a representative sample of uncertainty distribution just from the probabilistic distribution function of the input variables. Through a mathematical model the input variables are then processed with the help of computer software. Such IT-aided models are today available (in general) by open source software packages (see page 248 ff). The selection of input distributions for the uncertainty analysis can be taken from the literature, from similar risk exposure situations, or experimental studies. In practice the final selection of the form of the data input and the processing parameters significantly depend on the experience of the risk assessor.

Complex systems such as the ones that prevail in natural disaster distribution require hundreds of input parameters and it is rarely obvious which of the parameters are of more and which are of lesser importance. Disaster risk assessment experience, however, revealed that in reality only a relatively small subset of input parameters really influence the distribution. The choice of input parameters (parameterization) is in general the analyst's objective, based on her personal experience with the subject. The expert's opinion then has to be valued according the parameters:

- *Uncertainty*: Describes the variability/range of input/output parameter.
- *Sensitivity*: Describes the rate of change in the output when the input parameter changes.
- *Importance*: Describes the degree to which the output changes in response to the uncertainty.

The threat a population at risk is exposed to cannot be assessed solely by the number of disaster events. Therefore a realistic risk assessment requires next the number of events and also an understanding about the negative impacts of a disaster. The "public interest normally focusses on disasters which are dramatic in their impact, that claim a high death toll and as it was to be seen in the World Trade Center attack that have a specific psychological evidence" (Slovic 2002). This behavior attributes disasters very much with a subjective-related component. Disaster risk assessment thus becomes quite a challenge as it demands an objective-related viewpoint. Disaster events and their impacts when assessed impartially generally follow a probability distribution function: hazards with a high intensity rate occur very seldom, whereas hazards with a low intensity occur very often (high frequency). On the contrary the relationship between hazard consequences (impact) and intensity shows the opposite distribution. Hazards with a high intensity and a high frequency result in high consequences (impact) whereas those that are seldom and have a low intensity result in low consequences.

2. Regression Analysis

Regression analysis is in use in a multitude of different applications in economic, social, and also natural science. The analysis is best applied when the interaction between the input parameters is minimal and direct and the functional relationship between input and output is independent and linear. Furthermore linear regression analysis can also be used to provide a prediction model, indicating the probability of a future event, for instance, the probable flood level based on historic flood-level time series analysis (Fig. 7.3).

Regression analysis in natural hazard and disaster assessment often lacks an equally distributed database. In general there are many reliable and interpretable data available in one or two regions of the area of interest, but for a large part of the area such a robust database is lacking. Then the data gap has to be filled by analogue data transferred from regions that have a similar or almost similar hazard/disaster occurrence distribution. Such data are validated as to whether they are

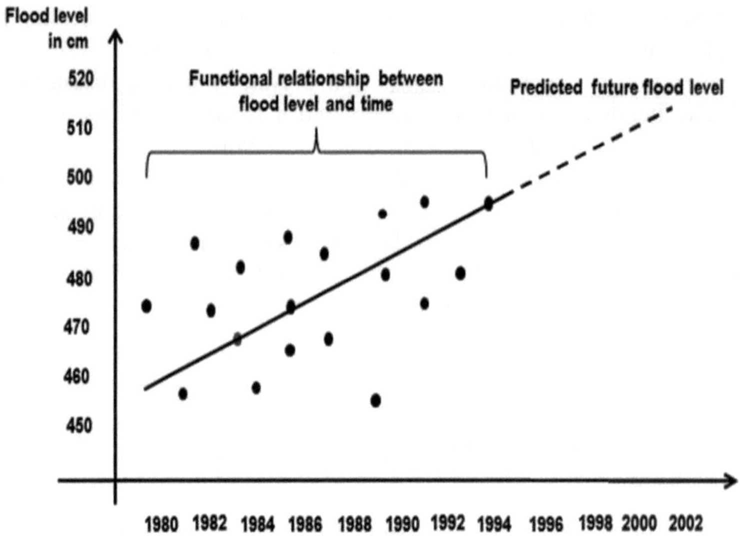

Fig. 7.3 Linear regression analysis to predict probable future flood levels based on historic flood-level time series analysis (fictitious)

applicable by undergoing an assessment method called the ordinary least squares method (OSL). If the data are usable for a generalized assessment the OSL method proves reliable with a sufficient closeness.

In general probabilistic risk assessment is based either on the data that have been recorded or taken up from analogue cases. Thus the type of data can be of a physical nature (exact measurement), of a statistical survey, or on a qualitative assumption such as an emotional feeling of a disaster impact. In order to make these very different types of datasets applicable for assessment, they have to be harmonized. For such a harmonization mostly the statistical relationship of frequency and severity of disaster events—already described in Chap. 6—is in use. The relationship is based on the assumption that severe disaster events quite rarely occur, whereas disasters with low impact occur quite often. Probability and severity are then ranked according to occurrence in respect to a certain pre-defined area (country, region, or a specific locality).

Figure 7.4 shows the principle behind working out a risk assessment matrix. Such a matrix provides the technical base for development decision making. It enables the political decision makers to value and compare disaster prevention measures to reach a sustainable risk reduction strategy. The figure shows, according to the data input, that the general risk of the area under investigation is medium to be increased. For the purpose of comparing risks of different regions, such an assessment has again to be generalized. Thus, for instance, maps such as the World Map on Natural Hazards (Munich Re 2011) only display risks that have a 50 % exceedance of probability.

The figure shows on the left side the general probability function of natural disasters derived from the disaster events that occurred in a certain area.

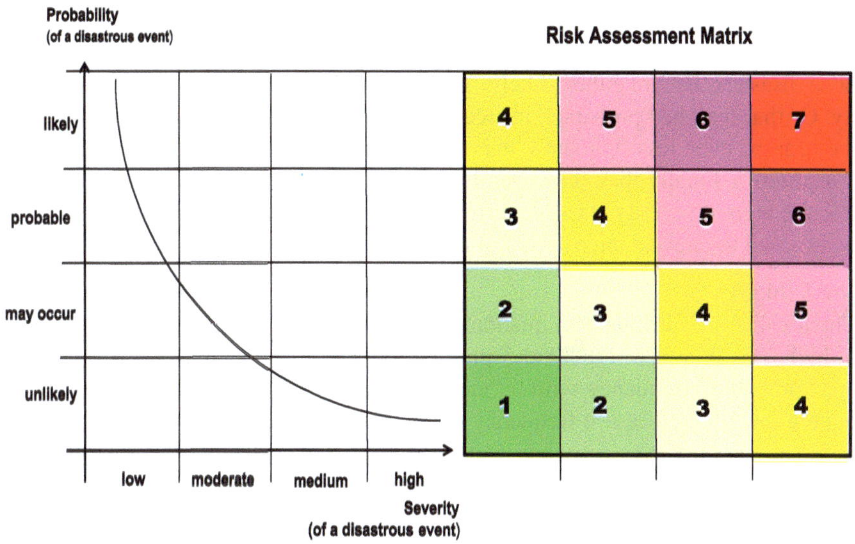

Fig. 7.4 Risk assessment matrix (Own graph)

The ranking of severity and probability was based on the available or assumed data and was attributed to different quadrangles. On the right side of the graph the ranking was given, arbitrarily, specific values from 1 to 9. These values serve as the bases for any further assessment procedure.

In quadrangle:

No. 1 (low)
The probability of a disaster to occur is very "unlikely" and its severity is "low". That means quadrangle 1 comprises a hazard potential that only poses a low and rather unlikely threat to the society.

No. 2 (moderate)
The probability of a disaster to occur exists ("may occur") although its severity is ranked still "low". Or it is quite "unlikely" to occur, but if does, it may result in a "moderate" impact. Societies exposed to quadrangle 2 threats may be harmed to a certain extent (injury, property damage, financial loss), all at a level that does not seriously interrupt the society's resilience.

No. 3 (medium)
The probability of a disaster event is ranked quite high ("probable"). But when it occurs it is anticipated to have a "low" impact ("severity"). Or it "may occur" but if it happens, the severity is ranked higher than low ("moderate"). Moreover the risk can, although "unlikely" in its probability, be of an increased ("medium") impact. For societies exposed to quadrangle 3 threats this means that such events may cause serious harm (injury, rather high economic losses, strong damage to the infrastructure as well as to livelihood, agricultural, and industrial production).

No. 4 (increased)
The probability and severity of quadrangle 4 risks is most widespread. They comprise disasters that are either "likely" to occur, but may result only in a "low" severity. Or that may be "probable" in frequency, with a "moderate" severity. Moreover they can be even less frequent ("may occur") but with an already increased impact ("medium"). Finally they may even be "unlikely" to occur, but simultaneously fall into the highest severity class identified. Societies at threat of quadrangle 4 risk may be exposed to unusual interruption of their societal functioning.

No. 5 (high)
Threats that are attributed to quadrangle 5 describe exposures that enter the level of high risk. Whether of a "likely" probability with a "moderate" severity, an average "probable" frequency with a "medium" severity level, or even disasters with quite a low ("may occur") frequency combined with a "high" severity, all of them pose serious threats to a society (severe injury, extended financial and economic losses) as well as interruption of the societal functioning that exceeds the self-help capacities of the populations at risk.

No. 6 (very high)
Threats that are ranked very high either substantiate by combinations of the highest level of probability ("likely") with an impact that is ranked "medium", or by a somewhat lower probability ("probable") with a "high" severity level. Quadrangle 6 risk comprises a large number of death casualties and a large number of injured and homeless people, as well as widespread, extensive, and lasting interruption of human safety. Societies that are exposed to such kind of risk are no longer capable of managing the disaster impact.

No. 7 (extreme)
Threats that have a "likely" probability to occur and that are combined with "high" severity of impact define quadrangle 7 risks. Disasters of this ranking have a destructive impact on the area, causing a high death toll and a great number of casualties, making thousands of people homeless, result in extensive, widespread, and long-lasting economic losses, severe damage to the infrastructure, and cut victims off from life-supporting assistance. Societies that are exposed to such kinds of risk are totally dependent on external assistance for the immediate crisis response as well as for the following rehabilitation measures.

7.5 Probabilistic Risk Assessment in Practice "HAZUS-MH"

Making people more resilient from natural disasters needs to develop a regional understanding of the "general" risk exposure. Therefore it is necessary to begin with data concerning one specific hazard (location, severity, frequency) to come to an assessment of the hazard potential of an entire area.

Basically data on hazards exist symbolized as either:

- Point data
- Line data
- Polygons

The most suitable operational tool for working on a regional hazard assessment is a geographical information system. There are several hazard assessment tools available based on GIS-software that have all proven their usefulness. Nevertheless the most appropriate and best elaborated instrument for hazard analysis is the US "HAZUS-MH Risk Assessment Guide." The assessment guide has been worked out by the Federal Emergency Management Agency (FEMA 2004, 2014) and exclusively concerns the United States of America territory. According to FEMA "HAZUS-MH" risk assessment includes the largest compilation of geo-referenced data made available by the federal US government and provides a nationally applicable standardized methodology for estimating potential losses from earthquakes, floods, and hurricanes. Furthermore HAZUS-MH provides software tools to assess mitigation and recovery activities as well as preparedness and response. The software was used for a long time by states and local government planners and emergency managers to perform economic loss scenarios for the said natural hazards. But meanwhile the number of HAZUS-MH users increased and more and more private companies and homeowners now make use of it to assess their particular risk. HAZUS-MH uses GIS to illustrate graphically the limits of identified high–moderate–low risk locations and thus can visualize the spatial relationships between populations and the economic assets or natural/cultural resources at risk. Emergency managers can moreover use HAZUS-MH to determine the most beneficial mitigation approaches, giving the program a crucial function in the pre-disaster planning process.

HAZUS-MH analyzes potential loss estimates on:

- Physical damage to residential and commercial buildings, schools, critical facilities, and infrastructure
- Economic loss, including lost jobs, business interruptions, repair and reconstruction costs
- Social impacts, including estimates of shelter requirements, displaced households, and population exposed to scenario floods, earthquakes, and hurricanes

HAZUS-MH is available free of charge online by the FEMA Map Service Center (MSC) Web Store (www.fema.gov/fima/planning_toc4.shtm). The program (MH 2.1) is compatible with Windows 7 and Windows XP and with ArcGIS 10.0. Menu options in the user interface of the three modules in HAZUS-MH (earthquake, flood, and hurricane) have been adjusted to provide greater consistency.

The software is accompanied by a voluminous guide that provides the methodological steps for estimating potential losses. As the tool has many software options to be integrated, HAZUS-MH software can be applied as a method to guiding principles for hazard and risk assessment for natural disasters to occur at other places than the United States of America.

One of the first tasks in hazard assessment is to define the level of hazard analysis envisaged to be achieved at the local, regional, national, or even transnational level. Accordingly, the level of data aggregation is different. A small-scale aggregation provides information in detail, but has only limited evidence for a larger area, whereas for a more regional or national scale the many detailed pieces of information have to be generalized. But they then cannot be used to identify the hazard potential of a particular location. For example, the general hazard assessment of the United States reveals that earthquakes occur most frequently and most severely in California, floods mostly in the riverine areas of the Mid-West, and hurricanes mostly strike the Gulf and the Atlantic coasts. But such a very general assessment neither provides information on the many earthquakes that are also known in the Mid-West, nor on the floods generating landslides along the Rocky Mountains or the Appalachian mountain chain.

The following methodological approach is mainly based on the FEMA 386 publication family (FEMA 2002a, b, 386-1 to 386-9) and FEMA 433 (FEMA 2004). The program presumes to begin with the identification of the level of data aggregation. At first all types of hazards are listed (geotectonic, hydrometeorological, climate, human, technical) and then, as shown in Table 7.3, the hazard types are indicated that are potentially relevant for the area of investigation.

Next, as shown in Table 7.4, the specific characteristics of the identified hazards are listed (modified from FEMA 386 on the Austin, Texas region). Note that the more detailed information is available, the better will be the assessment. Nevertheless the level of detail is dependent on the scope of the assessment (local, regional, national, transnational).

The next step (Table 7.5) goes from the hazard assessment to a risk assessment. The likelihood of occurrence of a potential hazard, its severity, and frequency can in general be assessed from past evidence. Such risk profiling is a key element of HAZUS-MH. Based on the "characteristics of the identified hazards" their respective area of occurrence (geographical coordinates), number of events (frequency per year), duration, the statistical probability, and their magnitude, respectively, intensity, the hazard exposure can be elaborated in more detail. When entered in the GIS system, the areal distribution of the newly collected hazards combined

Table 7.3 Relevant hazard types of the area of interest

Dam failure	No
Drought	No
Earthquake	Yes
Heat wave	No
Flood (coastal)	No
Flood (riverine)	Yes
Hail storm	Yes
Wind/storm	No
Fire	Yes
Hurricane	Yes

Courtesy FEMA (2002, 2004)

Table 7.4 Characteristics of the identified hazards

Hazard	Year	No. of events	Impact	Available data
Earthquake	1850–2005	160	154 deaths (1907), 66 deaths (1943), 45 deaths (1967), 34 deaths (1973), 23 deaths (1985)	USGS, Texas A&M, Texas Dept. of Geology, Nat. Earthquake Reduct. Prog.
Flood (riverine)	1987–2002	64	35 deaths (1915), 215 deaths (1921)	USGS, Flood Insurance Rate Map
Hailstorm	1993–2000	223	No data available	NOAA Hail Storm Map
Fire	1990–2001	151	Average 4 deaths/year	Austin City Fire Brigade
Hurricane	1897–2002	50	1 windstorm caused one death (year?), 3 windstorms in 1967 caused 3 deaths and 17 injuries	American Society of Civil Engineers, NOAA Hurricane Track map, historic data

Courtesy FEMA 386, Worksheet 2-3 on Austin, Texas

Table 7.5 Prioritization of identified hazards

Hazard	Frequency	Duration	Severity	Intensity	Ranking	Qualifier
Snow avalanche	2	2	1	1	6	$6 \times 1 = 6$
Dam failure	1	1	1	1	$4 \times 1 = 4$	$4 \times 1 = 4$
Flood (riverine)	4	3	4	4	$15 \times 3 = 45$	$45 \times 4 = 180$
Earthquake	4	4	5	5	$18 \times 4 = 112$	$112 \times 4 = 448$
Landslide	4	3	4	3	$14 \times 3 = 42$	$42 \times 3 = 166$
Volcano	4	2	2	2	$10 \times 2 = 20$	$20 \times 2 = 40$
Liquefaction	3	2	5	4	$14 \times 4 = 56$	$56 \times 4 = 224$

Ranking: no = 1; low = 2; moderate = 3; high = 4; very high = 5

with historical data, provide the best available database for the ongoing assessments. HAZUS-MH furthermore provides a software routine that allows integrating data from various sources that experience an analogous geological setting, or hazard assessment models from other organizations or scientists, if the collected and historical dataset is rated insufficient. Such kind of data may concern: topography, surface roughness, vegetation coverage, and change in land use patterns, seismic structuring, earthquake intensity distribution, all-hazard insurance coverage, vectorial diseases, coastal salt water intrusion, and many others.

The program continues by a qualitative prioritization of the identified hazards. For this all the hazards are rated according to their frequency, duration, severity, and intensity ranked from no, low, moderate, and high to very high. The software also offers the inclusion of other factors if this is felt necessary. Moreover it is also possible to enter a qualifier for each factor that weighs the factor, for example, the building quality in areas that experience high storm surges (reinforced masonry vs. wooden construction).

Table 7.6 Ranking on hazard based on evidence and experience (qualifier)

Hazard	Ranking (max. 500)
Snow avalanche	No (<100)
Dam Failure	No (<100)
Flood (riverine)	Moderate (100–200)
Earthquake	Very high (>400)
Landslide	Moderate (100–200)
Volcano	No (<100)
Liquefaction	High (>200)

Table 7.5 is modified from FEMA 386 Worksheet 2-3.

Following this procedure a ranking ("qualifier") on the hazards can be given (Table 7.6).

As the intention of a hazard assessment is to anticipate the potential risk of a community it is necessary to assess what groups of society and what kind of technical and economic assets might be exposed to what type of hazard. Note: HAZUS-MH does not provide an assessment of human losses, deaths, or injuries (see Sect. 7.2).

HAZUS-MH as a next step provides for a geographical hazard assessment based on the population distribution and the locations of hazard-exposed technical assets. In this regard it should be recognized that in developing countries more than 60 % of the assets at risk are with the private sector, whereas in the developed countries the ratio is almost opposite. Nevertheless it is a well-proven fact that even in developing countries a set of basic and essential facilities (critical infrastructure) is indispensable to uphold societal life. Therefore every risk assessment has to compound the risk exposure of such facilities in order to ensure efficient maintenance of life after a disaster strikes. Therefore HAZUS-MH included the population distribution as well as the location, type, and values of public assets including building stocks, streets, bridges, traffic lifelines, hospitals, schools, waterworks, power generation facilities, and the like (see Sect. 8.10). The characteristics of the above assets can be displayed in special vulnerability maps identifying the location of hospitals, fire brigades, and so on in the form of point features; streets, bridges, and pipelines as line features; and industrial areas, living quarters, airports, harbors, and such in the form of polygons. Table 7.7 gives an example for a categorization of assets according to their (arbitrarily) assessed level of hazard exposure. The percentage of hazard exposure is based on the total amount of the particular asset (=100 %).

The table provides useful indications on how strong and what kind of assets are exposed to what level of vulnerability. But as regional planners have to make the entire city area risk resilient they have to know the spatial distribution of each of the particular hazards. From the data in the table (point, line, polygon features) specific hazard maps can be created. But also very important for special planning is to create a synoptic hazard map that comprises all hazards in one map. Such synoptic hazard maps should also include indicators for social and economic

Table 7.7 Categorization of assets according to their "arbitrarily assessed" level of hazard exposure

	Inventory	No hazard	Low hazard (%)	Moderate hazard (%)	High hazard (%)	Very High hazard (%)
Population Assets	Residential	30	30	15	15	10
	Public	25	20	20	15	20
	Commercial	15	35	30	15	5
	Critical	50	25	5	10	20
	Industrial	10	5	5	15	65
	Main traffic routes	20	40	25	10	5
Lifelines	Bridges	70	20	6	3	1
	Waterways	5	15	25	50	10
	Pipelines	40	0	10	25	5
	Sewage system	35	5	10	25	25
	Power lines	30	50	10	5	5
	Waterworks	0	20	0	0	80
Utilities	Sewage treatment	0	0	0	0	100
	Power generation	55	0	0	45	0
	Airport	0	0	100	0	0
	Harbor	0	0	0	0	100
	Telecommunication	0	30	30	40	0
	Dams	5	50	25	15	5
	Hospitals	30	35	25	10	10

Courtesy FEMA (2002)

features, such as income distribution; migration of high-income or low-income groups to hazard-exposed living quarters; or the age, gender, and race distribution of the population and their anticipated changes. They can moreover give insight to the quality of the building stock, multiple-storied buildings, rented house occupancy, and many other issues according to the task that is to be done.

The loss estimation in the HAZUS-MH Risk Assessment Program provides a special risk assessment tool (RAT) that helps to assess the loss and damage not only regarding the costs for reconstruction of the structural/building substance (tangible losses) but also to assess the functional downtimes due to the damage. Still there is no accepted tool in disaster management at hand to assess the loss due to casualties. The question is how to assess the impact on human life from a disaster. Thus far casualties are just counted by head and the number of injured or homeless. But such an assessment does not really cover the social and financial damage that occurs. In Indonesia there were villages where almost no women were living after the tsunami, leaving behind a great number of orphaned children. Their life began with a significant traumatization, still not possible to be quantified seriously. On the other hand, international assistance helped to reconstruct almost all the houses destroyed by the tsunami in the city of Banda Aceh, that today provide a much higher standard of living than before. But how to quantify such an impact? Furthermore risk assessment comes to answer the question on the value of life, for example, if a father was killed and can no longer feed his family (see Sect. 7.2).

HAZUS-MH points to the fact that estimating the losses on building structures can only be made by (arbitrarily) generalizing the different loss data and can thus not define the potential loss of a particular building (that would require an assessment by itself). Furthermore loss estimation cannot be carried out without a simplification and generalization of historic loss data.

To obtain the economic loss two different methodological approaches are in practice:

- *The deterministic loss scenario*: This correlates the potential losses based on geological, physical, or technical evidence of real previous disaster impacts. Such an approach only describes the worst-case scenarios in general and does not take the frequency distribution of a specific disaster type into account.
- *The probabilistic loss scenario*: HAZUS-MH propagates a probabilistic loss scenario to obtain a more realistic assessment of potential future losses. Probabilistic assessment is generally based on statistical and historical evidence on the likelihood that a certain event will occur with a specific frequency and severity and with a potential loss. The losses are given in annualized losses (average loss over a certain time span) or on an expected average return period. For this the economic losses or the return period and other risk assessment factors are correlated versus time to develop a vulnerability curve (see Chap. 6).

The next step in the program is the estimation of losses. For example, HAZUS-MH therefore comprises, for a loss prediction on earthquake, data input next to the location of the epicenter, the event's frequency and severity, also an input of the geological environment of the earthquake hazard area: distribution of soil type, liquefaction potential, morphology-induced landslide probability, as well as water depth in order to assess the tsunami potential. All such data can be entered into the software in order to arrive at a more realistic loss estimation scenario. The additional data have to be ranked (arbitrarily) between 1 and 5 (very low–moderate–high–very high). Concerning flood loss estimation the program provides, moreover, a software tool to enter additional data on the morphology (digital elevation model), the water catchment area, flood plain boundaries, the upstream and downstream limits of the area of interest, as well as technical installation flood mitigation (FEMA 2008).

When all available and necessary data are entered, the HAZUS-MH program offers to run probabilistic scenarios either:

- Pre-defined from overall data on the United States territory
- Individually defined according to the specific site of investigation

Next come risk assessments carried out with the risk assessment tool. The tool compiles the findings on the natural hazard distributions, the data on the technical and material inventory, and on the population distribution as well as on the loss estimation results obtained. The RAT runs the assessment automatically based on the above-given datasets. The software program then gives the final risk assessment for the areas of interest in the form of quantitative loss estimation for earthquakes, floods, or on the risk exposure, for instance, landslide susceptibility or on wildfire hazard. HAZUS-MH, however, recommends not simply taking the data from the program, but reviewing the results according to their reasonableness.

A multitude of software packages are offered on the Internet and are accessible, free, and open source, among them:

RStudio

The software integrates the comprehensive state-of-the-art statistical package with a superb user interface, available both as a desktop application and as a browser-based server application. The Web version of RStudio offers a great opportunity for research advisors and IT departments for customized applications without much hand coding.

SciGraphica

The software is a scientific application for data analysis and technical graphics. It has many similarities with Sigmaplot and the very popular commercial application, Microcal Origin. The main features are:

- Easy plot functions and manipulation of data in worksheets.
- Several worksheets can be used interactively at the same time.
- The plots are fully configurable.
- Output in PostScript quality.
- Software has a native XML file format.
- Data manipulation and fitting features are in a roadmap.
- Programming language available for Windows, Linux, NextStep, and Mac.
- Provides a wide variety of statistical (linear and nonlinear) modeling, classical statistical tests, time-series analysis, classification, and clustering.
- Provides graphical techniques and is highly extensible.

Gnu Regression, Econometrics, and Time-Series Library (gretl; Free Software Foundation).

It is a cross-platform software package for econometric analysis, written in the C programming language. The main features are:

- Easy intuitive interface (now in English, French, Italian, Spanish, and other).
- A wide variety of least-squares based estimators, including two-stage least squares and nonlinear least squares.
- Single commands to launch software applications such as Vector Autoregressions, etc.
- Output models as LaTeX files, in tabular or equation format.
- Integrated scripting language: enter commands either via Gnu or via script.
- Command loop structure for Monte Carlo simulations and iterative estimation procedures.
- GUI controller for fine-tuning Gnuplot graphs.
- Links to further data analysis software packages possible.

OpenStat
A widely applicable statistics package for Windows and Linux, with a detailed manual in PDF format.

Vista
A statistical software program with broad graphical user interface/desktop for Windows, Mac, and Linux. The advantage of this software package lays in its broad data visualization capacities.

ADE-4(2004)
This is statistical software for multivariate analysis and graphical visualization and includes Freeware for Windows and Mac applications.

Dataplot (2002)
This is software mainly for running nonlinear model related statistical analysis with good data visualization. The software is applicable for Windows, UNIX, and Linux.

7.6 Risk Assessment (Disaster Risk Index)

The complex nature of risk-defining elements and their manifold interaction and even superimpositions, for a long time made it very difficult to come up with a reliable analytical instrument that condensed all the different elements of disaster information into one single set of data. Such assessments are required to enable risk assessors to quantify standardized and harmonized information, in order to compare the disaster risk exposure of one location with another, at the national scale as well as internationally. The most appropriate instrument allowing for such national as well as global risk quantification was seen by many scientists in the development of a Natural Disaster Risk Index (NDRI). The index must be based on the most reliable available data on the natural hazard impact on a society. Such an index has to specify the level of exposure and vulnerability of a certain area by integrating the different risk elements into one comprehensive picture. However, for methodological reasons it has to be stated that such an index can only be based on past data and for a projection of the identified risk pattern into the future much care should be taken. The main limitation to establish such an index is that the database is (still) very heterogeneous and it turned out that it was in many cases not comparable. From many countries, for example, the United States, the Hazus 99 assessment provided a complete set of data but countries such as Bangladesh mostly lack such an inventory. The challenge in risk indexing is to find out what kind of data can be used for a worldwide comparison. There have been several attempts but they were all very limited in their conclusions. Only after a considerable volume of information has been collected by CRED-EMDAT, are the risk researchers today better disposed to generate harmonized datasets, enabling them to generate a disaster risk assessment at a global scale.

Such information can help to create a better understanding of the root causes of natural disasters to occur and the potential mitigation, prevention, and preparedness measures, including prioritization of resources, targeting of more localized and detailed risk assessments, implementation of risk-based disaster management and emergency response strategies, and development of long-term land-use plans and multihazard risk management strategies.

"There is a great need for a global index to compare country's risk exposure" (Peduzzi et al. 2005, 2009). The reason that such a comparison was missing for so long is the complex nature of the indicators defining hazards, vulnerability, and risk. The difficulty derives from the fact that such a method has to be able to compare different countries (large and small), very differently exposed to very different kinds of disaster: floods, droughts, or earthquakes and volcano eruptions. One of the biggest difficulties occurs when trying to compare human vulnerability from natural hazards with that from socioeconomic parameters, especially because of the different time scales: social and economic changes occur within months or years, whereas geological phenomena can take centuries or decades, as well as climate changes. Other difficulties are inherent to global scale. How does one compare the situation of earthquakes in South America with the problem of drought in Africa? Not only the number of people affected is very different, but also the percentage of occurrence varies greatly for each continent. Disasters may occur very slowly (like climate change) or set in very rapidly. It may occur largely extended like a drought or flood that can destroy the country's harvest, or very locally, like an earthquake that strikes a small settlement or an industrial estate, killing a large number of people or making them homeless or jobless. Moreover, one hazard may trigger another one, so that at the end the people are victims of a totally different disaster type (hazard cascade; see Sect. 5.3). For instance a drought is often followed by a flood and the deteriorated living conditions make such a society much more vulnerable from the following flood. Or the situation of the tsunami-stricken capital Banda Aceh (Indonesia) where definitively the third biggest earthquake damaged large parts of the building structure that then was swept away by the tsunami. The problem with disaster risk comparison is that many disasters are not attributed to the root causes, rather than to the final type of disaster: a volcanic eruption that triggers a lahar that at the foothill turns into mudflow that then damages a house is statistically attributed to mudflow. Mathematically it is thus far not possible to incorporate the hazard cascade into the calculation. Such an assessment can therefore not be carried out without extensive "simplification and normalization" (Peduzzi et al. 2009).

7.6.1 Disaster Risk Index (DRI–UNDP)

The Disaster Risk Index (DRI–UNDP) was the first functioning instrument to monitor the evolution of risk (UNDP 2004a, b, c; Peduzzi, et al. 2002, 2005, 2009). The DRI was a major step towards understanding the causal relationship between social and economic development and disaster. With the DRI the UNDP aimed towards assessing outcomes of disasters rather than their root causes. In order to present a generally applicable method the UNDP-DRI defined a disaster not exclusively as an impact from a natural event, but as related to triggering agents, mainly human activity, and from the interaction between these two spheres.

The methodological approach of the DRI was based on identifying the relevant disaster impact hazards and socioeconomic variables for specific geographic locations. This methodological approach made the selection of locally adoptable indicators on disaster risk variables an arduous and time-consuming task. But it should be noted that such tasks are often handed over to experts to discuss what could result in outcomes of socially debatable appropriateness, particularly in the realm of policy implications.

The UNDP Disaster Risk Index distinguishes several classes of natural risks at the national level based on a model developed by UNEP-GRID and CRED-EMDAT, the Global Resource Information Database (Peduzzi et al. 2002). As a primary goal and intention, the DRI seeks to explain the correlation of a country's development status and its vulnerability to natural hazards. It thus combines physical exposure to hazards (population density) and vulnerability (socioeconomic situation) to calculate the mortality risk for a certain hazard type. In addition to physical exposure and mortality, data on the purchasing power parity per capita, the percentage of arable land, urban growth, forest coverage, as well as the percentage of cropland to total land coverage were calculated. The Disaster Risk Index shows impressively that the highest risk on mortality from natural disasters occurs in South and Southeast Asia, from Iran down to Papua New Guinea. Another mortality hotspot is eastern and central Africa, whereas along the western Latin American coast, although high, the threats are comparably lower (see Fig. 3.43).

Several years before, the United Nation Development Program (Peduzzi et al. 2005) had initiated the first statistical evidence of the links between hazard distribution and social vulnerability. The study carried out a multicriteria analysis for identifying a country's risk exposure, The World Vulnerability and Risk Report—Global Risk And Vulnerability Index Trend per Year (GRAVITY). GRAVITY consists of several reports assessing the different aspects of international risk distribution.

The DRI calculated worldwide the average risk of death per country in disasters associated with the four natural disasters: earthquakes, tropical cyclones, floods, and droughts, based on internationally available data from 1980 to 2000. In the DRI, risk refers exclusively to the risk of loss of life and excludes other facets of risk because of a lack of datasets available at the global scale. The DRI further enables us to assess the socioeconomic and environmental variables that are correlated with risk of death in respect to a certain area that is at risk. In order to calculate the risk of life, variables such as the number of people living in an area, the economic value of the infrastructure, and economic activities were combined with the frequency of hazard events. All of these data are expressed in the DRI variable "physical exposure", as the average number of people exposed to a hazard event in a given year. In the DRI, physical exposure is expressed both in absolute terms (the number of people exposed in a country) and in relative terms (the number exposed per million people).

But physical exposure is insufficient to explain risk, as countries with similar levels of physical exposure to a given hazard can experience widely differing levels of risk. This factor is indicated by "vulnerability." In the DRI, vulnerability refers to the different variables that make people able to absorb the impact and recover from a hazard event. These may be economic, social, technical, or ecological. The DRI, therefore, calculated the relative vulnerability of a country to a given hazard by dividing the number of people killed by the number exposed. A total of 26 variables were selected for each hazard type. The Global Map on Disaster Risk summarizes UNDP findings on the distribution of risk classes worldwide. The map outlines that South and Southeast Asia from Afghanistan to China as well as countries in parts of central Africa experience the highest risk of life on Earth.

7.6.2 Global Disaster Risk Analysis (World Bank)

A similar approach as the UNDP-Disaster Risk Index has been developed by the Global Disaster Risk Analysis of the World Bank and Columbia University (Dilley et al. 2005; Arnold et al. 2005). Like the DRI, the Global Disaster Risk Analysis expresses risk as a function of hazard exposure and social vulnerability. The analysis was targeted to identify the disaster risk potential of geographical areas of the world (hotspots) by a synoptic view of six major natural hazards: earthquakes, volcanoes, landslides, floods, drought, and cyclones. The analysis was based on publicly accessible data mainly from CRED-EMDAT, UNEP/GRID, GSHAP, or the World Bank and succeeded in quantifying the risk from the identified natural disasters. It furthermore has given much emphasis to disasters that originated from multiple hazard impacts. The analysis was based on the calculation of two fundamental types of risks: the risk of mortality and the risk of economic losses, by combining the following three components that contribute to the overall risk of natural hazards:

- Probability of occurrence of the six hazard types
- Physical elements exposed to these hazards (public and private)
- Vulnerability of the elements exposed to specific hazards indicated by a gridded population and their respective gross domestic product (GDP)

The analysis was carried out in respect to pre-defined grid cells. Using grid cells instead of entire countries made it possible to calculate the relative risks at subnational scales. Moreover the calculation was restricted to the time span of the last 25 years, as the database beyond that date proved to be insufficiently reliable in quality and data density.

Although the World Bank analysis provides an impressive and useful tool to assess the world's disaster risk distribution it is, however, unsuitable to identify the absolute risk level of a specific hazard of a certain country or region. For such an analysis the database is still not adequate. It was, however, adequate for identifying areas that are at a relatively higher single- or multiple-hazard risk. It could identify those areas that are at higher risk, for example, of a flood loss than others and at higher risk of earthquake damage than others, or at higher risk of both. Moreover, the World Bank analysis provides the ability to assess in general terms the level of risk exposure and the potential magnitude of losses to people and their assets in these areas.

In Fig. 7.5 the relative risk exposure for selected countries has been summarized based on the World Bank data (Dilley et al., ibid). The table clearly shows that the level of risk of a country is not just a matter of risk exposure, but rather derives from a complex interaction between the exposure and the prevailing socio-economic conditions. Although countries such as China, due to their large territory and their extremely high population density automatically arrived at a very high risk level, countries such as Japan and the Philippines, although lying in the same risk area of Southeast Asia, however, show distinctly lower risk exposures. Comparing the United States and Bolivia give an even more different picture. Both countries share the same risk pattern although they are highly different in size, population, and economic potential.

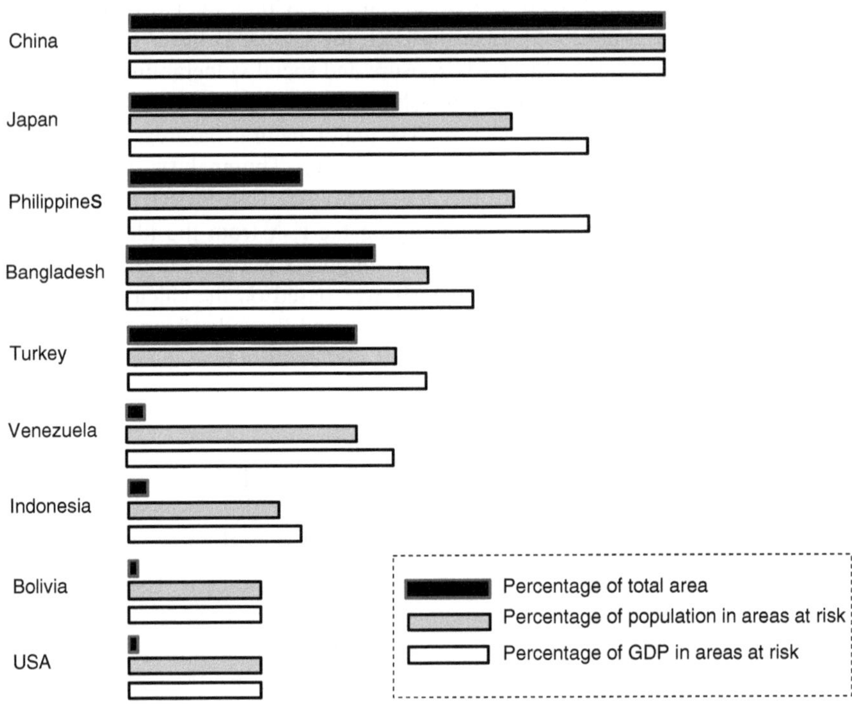

Fig. 7.5 Comparison of risk levels of selected countries (Based on: Dilley et al. 2005; Table 7.2)

7.6.3 *Indicators of Disaster Risk and Risk Management (Inter-American Development Bank)*

The Inter-American Development Bank (Cardona 2005) has developed a risk assessment tool, the "Indicators of Disaster Risk," that mainly focused on vulnerability (see also ECLAC 2003). The tool was developed for an application in the Latin America context for the time span from 1980 to 2000 for twelve Latin American countries. Based on historic loss intensity patterns, the probable maximum future losses were quantified for disaster return periods of 50, 100, and 500 years. The methodological approach of identifying the economic and social factors that affect risk exposure of a population in a mathematical-probabilistic form makes application to other regions of the world possible. The assessment tool comprises a system of risk-defining criteria that was based on four different indicators, representing the main elements of vulnerability.

Disaster Deficit Index (DDI)

The DDI is based on historic geoscientific evidence to forecast potential disaster losses, combined with data on existing and probable future values of public and private assets. By this combination the risk assessors are enabled to identify the expected vulnerability at a country level. The assessment tool is based on the use of "relative indicators" defining the relationship between the "demand for contingent resources to cover the losses caused by the 'Probable Maximum Loss' (PML) and the public sector's economic resilience that describes the available financial resources to restore damages ('Economic Resilience')" (Cardona, ibid). The potential losses are calculated probabilistically based on the indicators "frequency" and "severity" that have been already described in Chaps. 6 and 7 and "economic resilience", that is denominated by the volume of the internal and external financial resources and capabilities to recover, on the government side as well as by the private sector. To give an example of the aspect "resilience" this is incorporated into the calculation by, for instance, the:

- Number of insured assets
- Availability of disaster reserve funds
- Volume of the government's budget provisions for disaster remediating
- External and internal creditability

In many developing countries also the amount of aid donations received in the past and that may be expected in the future from private and international donor agencies can be used as an indicator for a country's economic resilience.

Local Disaster Index (LDI)

This indicator is composed of three subindicators: number of deaths, number of affected people, and the direct costs of rebuilding private and public structures and agricultural losses from geotectonic and hydrometeorological disasters. In order to reveal the LDI value, the ordinal numbers of the three indicators were added and normalized according to the area under investigation to which the indicator corresponds.

Prevalent Vulnerability Index (PVI)

This indicator measures a country's predominant vulnerability conditions by adding up the physical exposure, the direct and indirect impact (fragility) and the lack of social coping capacity (resilience). The composite nature of this indicator index makes it necessary to sum up the indicators to derive at the PVI.

Exposure combined from data on:

- Annual population growth rate
- Urban annual growth rate
- Percentage of population that lives on less than US$1/day
- National capital stock in US$ million per square kilometer
- Import/export as percentage of GDP
- Gross domestic investment as percentage of GDP
- Arable land as percentage of total land area

Fragility combined from data on:

- Human poverty index (HPI)
- Percentage of people depending on income of working population
- GINI coefficient
- Unemployment rate
- Annual food price increase
- Percentage of agriculture on GDP
- Debt service as percentage of total GDP
- Percentage of (human) soil degradation

Resilience combined from data on ("inverse relationship"):

- Human Development Index (HDI)
- Gender-Related Development Index (GDI)
- Environmental Sustainability Index (ESI)
- Social expenditure on health, education, and so on as percentage of GDP
- Infrastructure issuance coverage as percentage of GDP
- Number of TV sets per 1000 people
- Number of hospital beds per 1000 people

Risk Management Index (RMI)

The RMI describes the risk management capability of a society to withstand the adverse nature of disasters by benchmarking the risk management capacities according to pre-defined achievement levels. The RMI was constructed to quantify the different aspects of public risk reduction policies by summing up all components and dividing by four:

- Risk identification (RI) as a measure of individual risk perception and the official risk assessment
- Risk reduction (RR) quantifying the implemented mitigation and prevention measures

- Disaster management (DM) as a quantification of response and recovery activities
- Financial protection capability (FP) measuring the state's capacity to institutionalize risk transfer

With this set of indicators the Inter-American Development Bank (IDB) was able to identify the root causes of vulnerability and to quantify the risk of countries under investigation. For the economic losses that a particular country may suffer from natural disaster impact it can provide an estimation of the extreme impact, for example, from a hurricane, floods, tsunami, or earthquake event during a given exposure time. It moreover quantifies the financial and social resources that are needed to restore the pre-disaster situation. The tool furthermore contributes to an international, cross-border comparison of risk based on the same set of criteria. It assesses the risk level by defining the probability of exceeding a pre-determined level of economic, social, and environmental consequences at a certain place and time, called "convolution." This term describes the interaction between hazard and vulnerability (see Chap. 6) and refers to the fact that if no one lives in an area that is subject to natural hazards, or if a society is settling an area not exposed to hazards it will not be at risk and vice versa.

The assessment of the World Bank revealed that in respect to the DDI Index all countries investigated with the exception of Costa Rica, are lacking an adequate prevention and mitigation policy to cope with natural disasters. Regarding the time span of the next 500 years, Colombia is at the most critical situation. The calculation revealed probable losses of US$20 billion in the next 500 years are to be expected. When the next 50, respectively, the next 100 years are regarded, the maximum possible losses are expected to occur in Mexico with about US$4, respectively, 7 billion.

7.6.4 Global Climate Risk Index (Germanwatch)

Another approach to quantifying the impacts of natural disasters is the yearly Global Climate Risk Index 2014 (CRI) presented by Germanwatch (2014), with special regard to the impact of extreme weather events. Similar to the former reports, the seventh edition of the CRI also ranks the countries according to their number of fatalities and economic losses. The data are provided mainly by Munich Re NatCatSERVICE (MunichRe 2012, 2013) and the International Monetary Fund (IMF) on economy and population distribution. Both are worldwide databases and are among the most reliable and complete on this matter. The result is an average ranking of countries in four indicators, of the absolute and a much stronger weighing of the relative impacts:

- Number of deaths
- Number of deaths per 100,000 inhabitants
- Sum of losses in US$ in purchasing power parity (PPP)
- Losses per unit of gross domestic product (GDP)

As in previous years, the 2012 Global Climate Risk Index analyzed to what extent countries have been affected by the impacts of weather-related loss events: storms, floods, heat waves, and so on. Geological and tectonic disaster types such as earthquakes, volcanic eruptions, or tsunamis were not incorporated, although there was quite a reliable database available, but for the purpose of weather-related disasters, they were rated not relevant. The index represents one of different steps possible in the overall "puzzle of climate-related impacts and associated vulnerabilities" and as it is based on past data, the findings should not be used for a linear projection of future climate impacts. Furthermore the data only reflect the direct losses and fatalities of extreme weather events; for instance, disasters such as heat waves that often lead to much stronger indirect impacts (e.g., through droughts and food scarcity) were not incorporated. Also, the total number of affected people (in addition to the fatal casualties) was not included, since the compatibility of such data is very limited.

The presented analyses comprise all loss events that have caused substantial damage to property or persons. It is important to note that the index originates from an event-related examination that does not allow for an assessment of continuous changes of climate parameters. Thus far Germanwatch (2014) is methodologically not able to quantify effects such as the "long-term decline in precipitation that as it co-occurred in some African countries as a consequence of climate change. Such parameters nevertheless often substantially influence important development factors like agricultural outputs and the availability of drinking water." The Climate Risk Index does not provide an all-encompassing analysis of the risks, but should be seen as an analysis regarding countries´ exposure and vulnerability to climate-related risks. Each country´s index score derived from a country's average ranking in all four analyses, according to the following weighting: death toll 1/6, deaths per inhabitants 1/3, absolute losses 1/6, and losses per GDP 1/3. The quantification revealed that the most affected countries in 2010 were Pakistan, Guatemala, Colombia, and Russia (see Fig. 3.50). For the period from 1991 to 2010, Bangladesh, Myanmar, and Honduras rank highest. The long series of climate risk assessments revealed that in most cases, countries that were already afflicted in the past will most probably be endangered in future, although Germanwatch refrained from giving a deterministic projection of the future.

The methodological approach of identifying the risk through relative and not absolute values is deliberately chosen, as this allows for analyzing country-specific data on damage in relation to real socioeconomic conditions. Germanwatch reiterates that, for example, that US$1 billion in damage poses much smaller economic consequences for a rich country than for one of the world's poorest countries.

7.6.5 WorldRiskIndex (German Alliance Development Works)

Another assessment of risk at the global level was carried out in 2013 for the third time by the German Alliance Development Works (2013), publishing the

World Risk Index, which provides a methodology enabling the risk manager to assess the risk from natural disasters on a worldwide scale. In order to compare the risk worldwide, a standardized and internationally accepted methodology has been developed that enabled the risk assessors to compare findings of one country to another. Such an approach cannot be successfully addressed when it is not clear, where, what kind of risk exists to whom, and at what frequency and severity. The World Risk Index has been calculated for 173 countries based on 28 indicators that were grouped into four components, describing the level of risk exposure as well as the societal capacity of a country to withstand the impact:

- Hazard exposure
- Hazard vulnerability/susceptibility
- Coping capacities
- Adaptive capacities

The data used for the assessment were mainly called up from the Internet which ensures transparency and verifiability. In order to be aggregated into indices, the indicators were transformed in dimensionless rank levels accordingly. The transformation into percentage values and classification was carried out with the help of the quintile method integrated into the ArcGIS10 software packet. Five classes were calculated and translated into a qualitative classification: very high–high–medium low–very low. The risk calculated by the index is determined by the extent to which communities are exposed to natural hazards such as droughts, storms, or earthquakes, but also by their degree of vulnerability. The latter is dependent on social factors such as the public infrastructure, medical services, the prevailing nutritional situation, governance, the level of education, the availability of insurance that might help deal with economic losses in an event, as well as the condition of the environment.

Acknowledging the fact that the database for such a global assessment is very heterogeneous, the authors see the presented index more as a communicating instrument than a technically sophisticated assessment instrument. The aim of the index is to sensitize the public as well as political decision makers towards the important topic of disaster risks. The index wants to draw attention to the people and countries at risk, as the disaster coping capacity is basically defined by the national socioeconomic resilience. The broad press coverage confirmed the necessity for such an assessment instrument and emphasized its need to set a base for political decision making (IPCC 2012).

Figure 7.6 displays the general methodological approach. The risk assessment is basically divided into the factors "people exposed" and "vulnerability," that account for 33 %, respectively, 66 % of the total risk. The factor "vulnerability" is composed of four components that all are set to account for 100 %. Within "vulnerability" the components "exposure", "susceptibility", and "coping capacities" are furthermore divided, for instance, into the basic features, "sanitation" and "access to water" (both weighted by 50 %). Only the component "adaptive capacity" is not further divided. The four components are taken to contribute

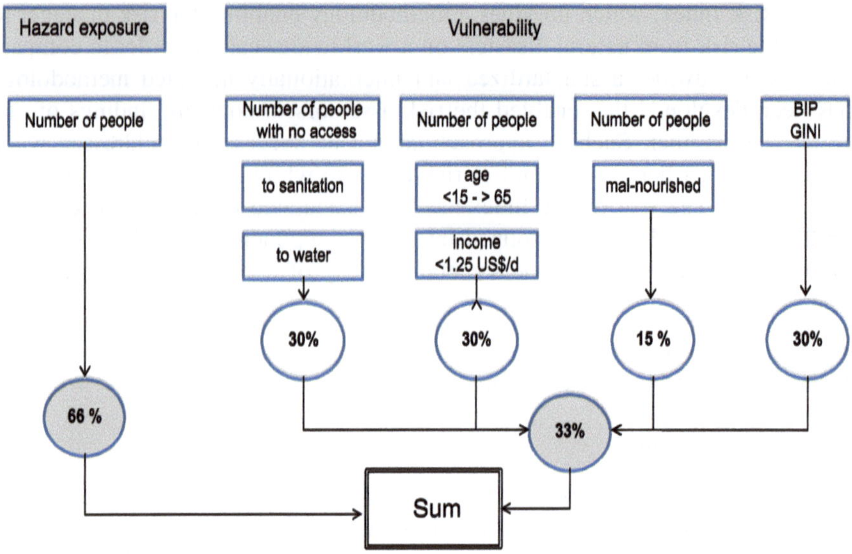

Fig. 7.6 Methodological approach for a risk assessment combining "people exposure" and "vulnerability" (Own graph)

by 28 %, respectively, 14 % ("adaptive capacity"). By such a weight-attributed assessment, the risk index wants to value the importance of the different components. The World Risk Index proved to be a successful attempt to standardize risk assessment worldwide. The approach, however, does not say that there are no other options for a universal risk index. Nevertheless it should be noted that the basic risk-defining components will not change. If a different methodological approach is chosen, it will most probably only alter the weight factors attributed to the different factors.

According to the report, global hotspots for disaster risk are concentrated in Oceania, Southeast Asia, and the southern Sahel, and especially in Central America and the Caribbean. Within the Pacific, the island state of Vanuatu has the highest disaster risk, whereas Malta and Qatar face the lowest risk worldwide. In these regions a high risk exposure is exaggerated by a highly vulnerable society. For example, the small island states have a very high risk exposure owing to their proximity to the sea that make them particularly exposed to cyclones, flooding, and sea-level rise and are facing a low coping capacity. On the other hand the Netherlands shows a similar risk to flood exposure, what makes the country rank 11 among the states most at risk worldwide. But due to a very high standard of social, economic, ecological, and institutional development, the Netherlands is able to reduce its risk enormously; its risk ranking is much lower (rank 51).

7.6.6 Earthquake Disaster Risk Index (EDRI)

Other than the risk indices described before that focus on the general risk distribution in order to provide a measure for comparing exposure from different risks, the Earthquake Disaster Risk Index (EDRI) of Stanford University, California (Davidson and Shah 1997) was developed especially to assess the risk from earthquakes. Such a single-risk-oriented assessment was therefore called a "topic-related risk index." Generally speaking, the EDRI assessment works on the assumption that by a comparison of earthquake risk-determining indicators it will be possible to identify the risk at different locations. The EDRI was calculated for 10 highly earthquake risk-exposed cities of the world from San Francisco to Tokyo. The assessment was based on five sets of indicators, earthquake hazard, earthquake hazard exposure, earthquake vulnerability, external political and economic context, and emergency and recovery capability (Table 7.8). The choice of indicators used and their attribution to different factors was based on a broad range of data from different sources. Although ERDI provides an effective tool for earthquake risk assessment it should be analyzed whether the method can serve the needs of the disaster risk assessment on earthquakes in other regions of the world or on other types of disasters.

The main objective was to arrive at one single number and to attribute the most probable value for each indicator, a process called "weighing values." The datasets available from many open accessible sources were graphically, mathematically, and statistically combined and weighted according to the severity of the recorded disasters. Figure 7.7 shows how the different indicators are weighted. The assessment moreover was accompanied by a subjective assessment routine to counterweight the technical figures by assigning values by the risk assessor based on expert knowledge. Then the values for each indicator were calculated to work out an index figure for all five indicators that at the end were all combined in the EDRI.

The EDRI on earthquakes of 10 international cities was then compared to the earthquake hazard factors known from geologic evidence. The EDRI for the city of Tokyo, for instance, revealed a high risk index and also a high hazard exposure factor. In contrast, Santiago de Chile has a low EDRI and also experiences a low earthquake hazard, whereas the city of San Francisco experiences a high hazard exposure although its EDRI is comparatively low (due to its strongly developed institutional and operational disaster risk management capacity).

7.6.7 Natural Disasters Risk Index (Maplecroft)

The Natural Disasters Risk Index (NDRI; Fig. 7.8) was released by global risks advisory firm Maplecroft and their findings are summarized here (Maplecroft 2010). The index has been developed to enable businesses and insurance companies to

Table 7.8 Factors used for assessing the earthquake disaster risk index

Main factors	Factors	Indicators
Hazard	Ground shaking	MMI: 50-year return period
		MMI: 500-year return period
		% urbanized area with soft soils
	Collateral hazards	% urbanized area with high liquefaction susceptibility
		% buildings constructed of wood
		Population density
		Tsunami potential indicator
Exposure	Physical infrastructure exposure	Population (number)
		GDP per capita
		Number of houses
		Urbanized land area
	Population exposure	Population
	Economy exposure	GDP per capita
Vulnerability	Physical infrastructure vulnerability	Seismic code indicator
		City wealth indicator
		City age indicator
		Population density indicator
		City development speed indicator
	Population vulnerability	% of pop. age <4/>65
External context	Economic external context	Economic context indicator
	Political external context	Political country context indicator
		Political world context indicator
Emergency response and recovery planning	Planning	Planning indicator
	Resources	GDP per capita
		10 years average of real growth in GDP
		Housing vacancy rate
		Number of hospitals per 100,000 people
		Number of physicians per 100,000 people
	Mobility and access	Extreme weather indicator
		Population density
		City layout indicator

Courtesy Davidson and Shah (1997)

identify risks to international assets. It is calculated by measuring the human impact of natural disasters in terms of deaths per annum and per million of population, plus the frequency of events over the last 30 years. The methodology has been refined to reflect the likelihood of an event occurring and covers disasters including earthquakes, volcanic eruptions, tsunamis, storms, flooding, drought, landslides, extreme temperatures, and epidemics. The NDRI includes the Geophysical Disasters Index, the Hydrometeorological Disasters Index, and the Natural Disasters Economic

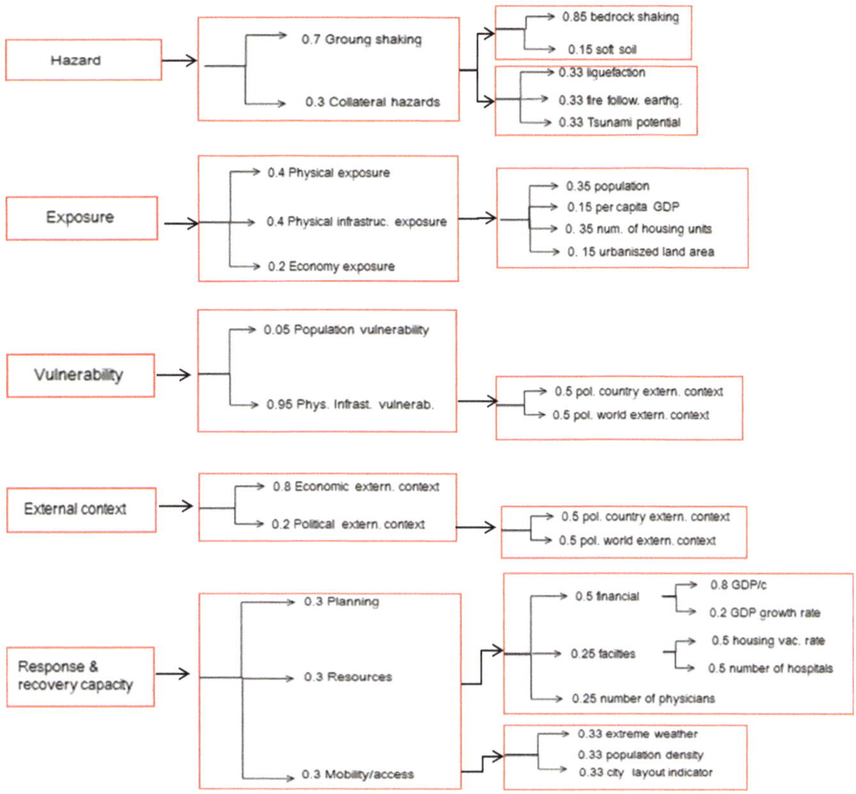

Fig. 7.7 Weight of risk factors according to the severity of the disasters recorded: EDRI (*Courtesy* Davidson and Shah 1997)

Losses Index. Data sources for the NDRI include EMDAT, the International Disaster database, Centre for Research on the Epidemiology of Disasters, USAID's Office of Foreign Disaster Assistance, the World Bank, and IMF. Maplecroft assumes that due to climate change, extreme hydrometeorological events are to increase posing a risk to business and property. This will also lead to interruptions or a complete breakdown of infrastructure such as transport and power and communication networks; plus there can be devastating impacts on local workforces. This makes it essential that businesses plan for the possibility of a disruptive natural disaster. Poverty is an important factor in these countries exposing large parts of the societies to the impact of natural disasters. In combination with a generally poor infrastructure, plus dense overcrowding in high-risk areas such as flood plains, river banks, steep slopes, and reclaimed land, this will in future result in even higher casualty figures. The index ranked countries including Bangladesh, Indonesia, Iran, India and China at the highest risk category.

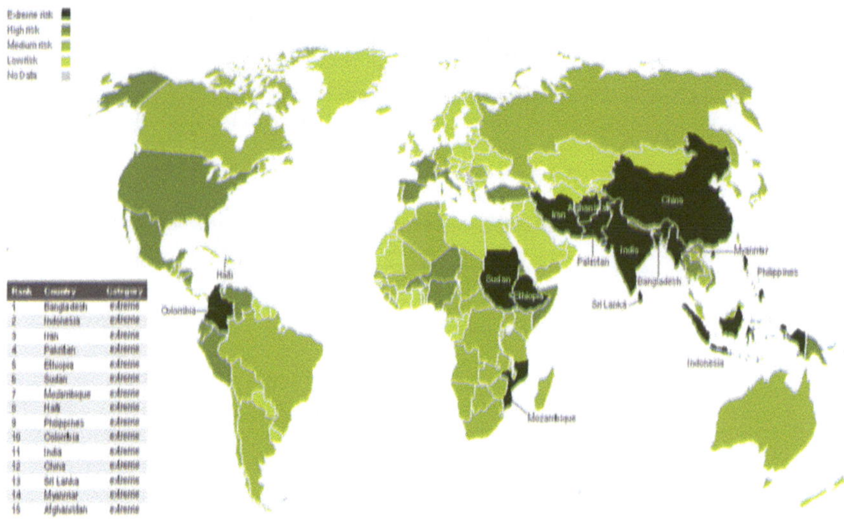

Fig. 7.8 Natural disaster risk index 2010 (*Courtesy* Maplecroft, London)

7.7 Final Remarks on Disaster Risk Indexes

The many disaster risk indexes that have been developed, some of which are described above, all have in common that they intend to provide a means to understand the generating risk factors as well as to provide a worldwide applicable measure to compare the levels of risk at different locations. But the large variety of concepts and methodological approaches presented contradicts this intention to have an index available that serves all needs and purposes all over the world. The number of risk-identifying methods and tools available, all of them developed by reputable organizations and generally based on a multitude of data, did not fulfill the target of an overall applicability. There is still not that "one" risk index that can be used for all countries worldwide and that covers all the different disaster risk situations known. But such a world index is actually needed.

Figure 7.9 very generally depicts the bias in the findings of different organizations. As examples (far from being representative), the findings of UNU-EHS, the World Bank, Germanwatch, and the Maplecroft Natural Disaster Risk Index are displayed for a selected choice of countries. The country rankings (0–130) follow the given rankings as they have been identified by the four organizations. In general the higher the ranking, the less is the risk exposure. It turns out that the findings of UNU-EHS differ a great deal from the findings of the other organizations. For example, UNU-EHS ranks countries such as the United States, Bolivia, Turkey, and Venezuela low risk, whereas most of them were assigned to the moderate and even the high-risk sectors by all the others. Japan is ranked by the organizations to be from highest to lowest risk. It is not the intention here to question the respective findings, but as long as findings differ that strongly a world unique risk index is still pending.

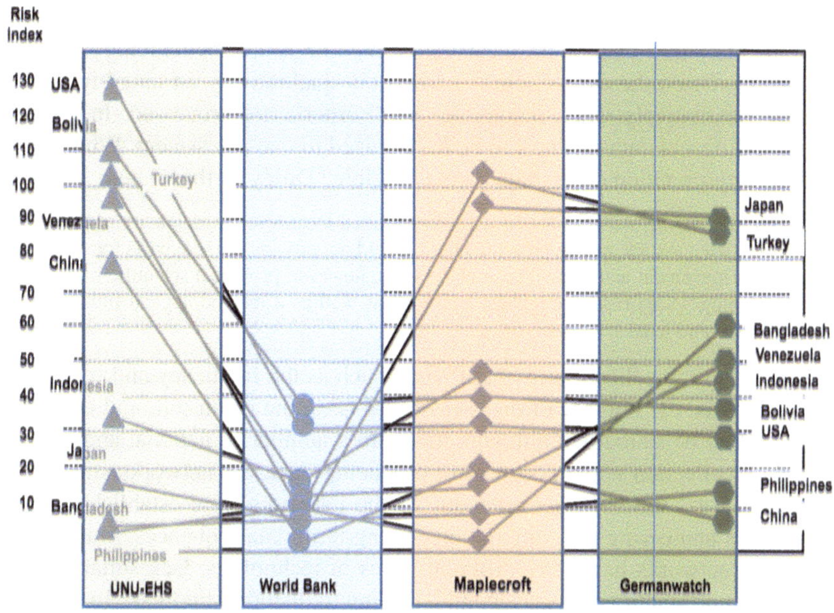

Fig. 7.9 Comparison of world risk indices of different organizations for selected countries (Based on UNU-EHS, Worldbank, Maplecroft, Germanwatch)

The main reason for these differences is seen in the way the basic parameters were chosen to measure hazard distribution and vulnerability. The main obstacle seen by all organizations begins with the first analytical step: who is collecting, which kind of data, and originally for what purpose. Are the data collected at the same time intervals (calendar years, fiscal years, or according to agricultural production cycles)? The vulnerability data were collected unsystematically therefore a countrywide representative coverage is not achievable and thus often assessed by rule of thumb. This holds true for remote areas as well as for the densely settled fringe areas of the big conurbations. There is a huge difference in data reliability between industrialized countries such as the United States that hold a tremendous amount of datasets on all technical, economic, and social issues and many developing countries such as Indonesia, that can hardly keep up their statistical record.

Sometimes the data collection method differs from country to country and the statistical methods to process the raw data still are not harmonized to a degree that makes a worldwide comparison possible.

For example, the World Bank (Dilley et al. 2005) defines their disaster hotspots by dividing the country into grid cells and bases the population density on deciles to arrive at more reasonable results and furthermore allow weighing each hazard individually. Other risk assessment data processing schemes take the data as representative for an entire country. In fact, more information comes from the densely settled areas where the administration is well functioning, thus exaggerating the weight of such areas. Moreover it is important to recognize that "Complex systems are often defined by multiple variables (physical, social, cultural, economic, and environmental) that cannot be measured using the same methodology" (Dilley et al., ibid.). Often the assessment is carried out qualitatively on tangibly defined issues or qualitatively when intangible "soft" criteria are processed. In the next step these different datasets have to be harmonized to one assessment. What finally makes an interpretation of the data barely possible? Dilley further stated,

> As a global analysis was conducted with very limited local-level participation and based on incomplete data, the results presented here should not provide the sole basis for designing risk management activities. The analysis does, however, provide a scientific basis for understanding where risks are highest and why, as well as a methodological framework for regional- and local scale analysis.

Concerning the collection of "hard" facts such as the frequency and severity of natural hazards, the number of disaster occurrences, the death toll, and economic losses, it must be stated that although there are a number of national and international data collection networks steadily producing data, the networks still do not cover the Earth evenly. Moreover many of the networks were established in the last 50 years hampering the compilation of long-term data sequences. Many of the monitoring networks have very different states of technology, lack suitable accuracy, or are not functioning permanently, leading to distorted findings. To give two examples, in the United States according to information of the USGS (2005) the current monitoring level of active volcanoes revealed that only 3 of the 18 very high threat volcanoes are monitored at the highest (level 4); or in Indonesia on occasion of the eruption of the volcano Mt. Leroboleng in year 2004. In order to at least document the seismicity after the eruption, a seismometer had to be brought in by motorcycle over a distance of more than 80 km.

There is growing recognition of the need for better data and information on hazard distribution and disaster-associated impacts both at local and regional levels, on national and international levels, as well as on historical and current data (Subbiah et al. 2008). The first direction to improve the underlying databases is to follow each possible opportunity to reach a higher level of data reliability on disaster risks and economic losses. A range of new global-scale datasets is currently under development with the intention to get broader and more comprehensive regional datasets for areas at risk and on the subnational distribution pattern of poverty and hunger. Still the data monitoring is hampered by often unclear and inappropriate definitions. The characterization of what water level defines a flood, how many dry weeks constitute a drought, or whether a ground motion is derived from a new earthquake or should be counted an aftershock.

Another direction to improve data reliability is to institutionalize investigation networks of long-term key processes, for example, (Dilley et al., ibid.) on the:

- Potential effect of changes in hazard frequency due to human-induced climatic change coupled with trends in human development patterns.
- Degree changes in tropical storm frequency, intensity, and tracks interacting with continued coastal development (both urban and rural) may lead to an increase in risk in these regions.

- Question of whether agricultural areas that are already under pressure from urbanization and other land-use changes are likely to become more or less susceptible to drought, severe weather, or floods.
- Can new types of disaster hotspots occur if the changing land-use patterns interact with the changing of drought, landslide, and deforestation distribution?

Although many aspects of these questions have already been addressed in general, the full range of hazard probability and the evolving vulnerability of the risk-exposed people have not been fully explored to the extent necessary. Moreover an integrated and interdisciplinary focus on data acquisition as well as in data processing is required to better understand the nonlinear relations of the hazard vulnerability context and the social and economic complexity and dynamics of societies. The task ahead is to follow the direction already started by the many national, regional, international, public, and private sector and scientific institutions, but it is recommended that the many and often duplicating activities of the different organizations should be harmonized, best under the auspices of the UN ISDR secretary and CRED-EMDAT, in order to come up with an overall agreed-upon and formally binding risk assessment tool. This would lead to achieving a broader and deeper insight to the complex system of hazard, vulnerability, and risk and would result in more effective risk management by the different stakeholders involved in risk reduction or adaptation decision making.

References

Arnold, M., Dilley, M., Deichmann, U., Chen, R.S. & Lerner-Lam, A.L. (2005) : Natural Disaster Hotspots: A Global Risk Analysis. – The World Bank, Disaster Management Series, No. 5, Washington DC

Aven, T. & Renn, O. (2009): The Role of Quantitative Risk Assessments for Characterizing Risk and Uncertainty and Delineating Appropriate Risk Management Options, with Special Emphasis on Terrorism Risk. - Risk Analysis, Vol. 29.4, p.587, Wiley Online Library

Bukowski, J., Korn, L. & Wartenberg, D. (1995): Correlated inputs in quantitative risk Assessment - The effects of distributional shape. – Risk Analysis, Vol. 15, No 2, Wiley Online Library

Burton, I., Kates, R.W. & White, G.F. (1993): The environment as hazard (2nd edition).- The Guilford Press London/New York NY

Cardona, O. (2005): Indicators of disaster risk and risk management: Program für Latin America and the Caribbean- Summary Report.- Inter-American Development Bank (IBD), Sustainable Development Department, The World Bank, Washington, DC

Carlsson, F., Daruvala, D. & Jaldell, H. (2010): Value of statistical life and cause of accident: A choice experiment. - Risk Analysis, Vol. 30, No.9, p.975-986, Wiley Online Library

Coburn, A.W., Spence, R.J.S. & Pomonis, A. (1991): Vulnerability and risk assessment. – United Nations Development Program/United Nations Disaster Reduction Organization (UNDP/UNDRO), Disaster Management Training Program, Geneva

Davidson, R.A. & Shah, H.C. (1997): An urban earthquake disaster risk index.- John A. Blume Earthquake Engineering Center, Stanford University CA

Dilley, M., Chen, R.S., Deichmann, U., Lerner-Lam, A.L., Arnold, M. with Agwe, J., Buys, P., Kjekstad, O., Lyon, B. & Yetman, G. (2005): Natural disaster hotspots - A Global Risk Analysis - Synthesis Report.- The International Bank for Reconstruction and Development / The World Bank and Columbia University, Washington, DC

ECLAC (2003): Handbook on estimating the socio-economic and environmental effects of disasters. - United Nations Economic Commission for Latin America and the Caribbean (ECLAC), and the International Bank for Reconstruction and Development / The World Bank, Washington DC

Enarson, E. (2000): Gender and Natural Disasters - InFus Programme on Crisis Response and Reconstruction,. – International Labor Organization, Recover and Reconstruction Department (ILO), Working Paper No. 1, Geneva

EPA (2010): Frequently Asked Questions on Mortality Risk Valuation.- National Center for Environmental Economics, U.S. Environmental Protection Agency (EPA), Washington DC

Felgentreff, C. & Glade, Th. (2008): Naturrisiken – Sozialkatastrophen in: Felgentreff, C. & Glade, T (eds.): Naturrisiken und Sozialkatastrophen, Spectrum, Heidelberg

FEMA (2002a): Building Support for Mitigation Planning – How to Guide. - Federal Emergency Management Agency (FEMA), Publication 386, Vol 1-9, Washington DC

FEMA (2002b): Earthquake Hazard Mitigation Handbook. – FEMA Hazard Mitigation Handbook Series, update June 2002, Federal Emergency Management Agency (FEMA), Washington DC

FEMA (2004): Using HAZUS-MH for Risk Assessment - HAZUS-MH Risk Assessment and User Group Series Using; How to Guide. - Federal Emergency Management Agency (FEMA), Publication 433, Washington DC

FEMA (2008): Flood Insurance Manual – National Flood Insurance Program. – Federal Emergency Management Agency (FEMA), Washington DC (online: www.fema.gov/.../flood-insurance-manual-effective-may-1-2008)

FEMA (2014): Hazus-MH 2.1 - Multi-hazard Loss Estimation Methodology.- Federal Emergency Management Agency, United States Department of Homeland Security, Washington, DC (online: www.hazus.org)

German Alliance Development Works (2013): WeltRisikoBericht 2012. - German Alliance Help, Bonn

Germanwatch (2014): Global Climate Risk Index 2014 – Who suffers most from extreme weather events? Weather-related loss events in 2012 and 1993 to 2012. – Germanwatch e.V., Bonn (online: www.germanwatch.org/en/cri)

IPCC (2012): Managing the Risks of Extreme Events and Disasters to Advance - Climate Change Adaptation. - Special Report of the Intergovernmental Panel on Climate Change, Cambridge University Press, Cambridge MD

Kaplan, S. (1997): The words of risk analysis. – Risk Analysis, Vol. 17, Issue 4, p. 407-417.- Wiley Online Library

Kasperson, R.E., Renn, O., Slovic, P., Brown, H.S., Emel, J., Goble, R., Kasperson, J.X. & Ratick S. (1988): The social amplification of Risk: A Conceptual Framework. – Risk Analysis, Vol. 8, No. 2, Wiley Online Library

Kirchsteiger, C. (1999): On the use of probabilistic and deterministic methods in risk analysis. - Journal of Loss Prevention in the Process Industries, Vol. 12, p.399– 419, Elsevier, Philadelphia PA

Lindhjem, H., Navrud, S., Braathen, N.A. & Biausque, V. (2011): Valuing mortality risk reductions from environmental, transport and health polices: A global meta-analysis of stated preferences studies.- Risk Analysis, Vol. 31, No.9, p.1381-1407, Wiley Online Library

Maplecroft (2010): Natural Disaster Risk Index 2010. – Maplecroft United Kingdom, Bath (online: http://maplecroft.com/)

MunichRe (2011): World Map of Natural Hazards. – Munich Reinsurance Company, Nathan, 2011-Version. Munich

MunichRe (2012): Topics Geo online – 2012. - Munich Reinsurance Company, Munich

MunichRe (2013): Topics Geo online - 2013. - Munich Reinsurance Company, Munich

Peduzzi,P., Dao, H., Herold,C. with contributions from A. Martin Diaz, F. Mouton, O. Nordbeck, D. Rochette, T. Ton-That, B. Widmer (2002): Global Risk and Vulnerability Index-Trends per Year (GRAVITY) - Phase II: Development, analysis and results. - The Bureau for Crisis Prevention and Recovery of the United Nation Development Programme, Geneva

Peduzzi, P., Dao, H. & Herold, C. (2005): Mapping Disastrous Natural Hazards Using Global Datasets. – Natural Hazards, Vol. 35, No. 2, p.265–289, Springer Link

Peduzzi, P., Dao, H., Herold, C. & Mouton, F. (2009): Assessing global exposure and vulnerability towards natural hazards: the Disaster Risk Index. – Natural Hazards Earth Systems Sciences, Vol. 9 p.1149—1159, Copernicus Publications, Goettingen

Power,M. & McCarry, L.S. (1996): Probabilistic risk assessment: Betting on its future. - Human and Ecological Risk Assessment, Vol. 2, No. 1 p.30-34, Taylor Francis Online

Purdy, G. (2010): ISO 31000:2009 — Setting a New Standard for Risk Management. - Risk Analysis, Vol. 30, No. 6, p.881–886, Wiley Online Library

Quarantelli, E.L. (1989): Conceptualizing disasters from a sociological perspective. – International Journal of Mass Emergencies and Disasters, Vol. 7, No. 3, Mattoon IL

Renn, O. (1989): Risikowahrnehmung und Risikobewertung in der Gesellschaft – in: Hosemann, G. (ed): Risiko in der Industriegesellschaft, Analyse, Vorsorge and Akzeptanz. - Erlanger Forschungen, Vol. 19, p.176-192, Erlangen

Renn, O. (2003): Social Amplification of Risk in Participation: Two Case Studies; p.374-401: in: Pidgeon, N., Kasperson R.E. & Slovic, P. (eds.): The Social Amplification of Risk. -Cambridge University Press, Cambridge MD

Slovic, P. (1987): The perception of risk. – Science, Vol. 236, p.280-285, Washington DC

Slovic, P. (2000): The Perception of Risk.- Earthscan Publications, p.473, London

Slovic, P. (2002): Terrorism as hazard: A new species of trouble.- Risk Analysis, Vol. 22, Issue 3, p.425-426, Wiley Online Library

Subbiah, A.R., Bildan, L. & Narasimhan, R. (2008): Background paper on assessment of the economics of early warning systems for disaster risk reduction. Report submitted to the World Bank Group, Global Facility for Disaster Reduction and Recovery (GFDRR / Contract 7148513). - Regional Integrated Multi-Hazard Early Warning System, Asian Disaster Preparedness Center, Bangkok

Tobin, G.A. & Montz, B.E. (eds)(1997): Natural Hazards – Explanation and Integration. - The Guilford Press, London/New York NY

UNDP (2004a): International patterns of risk. – in: Reducing disaster risk - A challenge for development, Chapter 2; United Nations Development Programme (UNDP), Bureau for Crisis Prevention and Recovery, United Nations, Geneva

UNDP (2004b): International patterns of risk. – in: Reducing disaster risk - A challenge for development, Chapter 2; United Nations Development Programme (UNDP), Bureau for Crisis Prevention and Recovery, United Nations, Geneva

UNDP (2004c): Reducing disaster risk – A challenge for development. - United Nations Development Programme (UNDP), Bureau of Crisis Prevention and Recovery, United Nations, Geneva

UNDRO (1979): Natural disaster and vulnerability analysis. - Office of the United Nations Disaster Relief Co-ordinator (UNDRO), Report of Expert Group (9-12 July, 1979), United Nations, Geneva

UNISDR (2004): Living with Risk - A Global Review of Disaster Reduction Initiatives". – United Nation International Strategy of Disaster Reduction (ISDR), United Nations, Geneva

USEPA (1997): Guiding principles for Monte Carlo Analysis. – Risk Assessment Forum, United States Environmental Protection Agency EPA) (online: www.EPA/630/R-97/001)

USGS (2005): An Assessment of Volcanic Threat and Monitoring Capabilities in the United States: Framework for a National Volcano Early Warning System (NVEWS). – United States Geological Survey (USGS), Open-File Report, 2005-1164, Reston VA

Viscusi, W.K. & Aldy, J.E. (2002): The Value of a Statistical Life: A Critical Review of Market Estimates troughout the World.- Harvard Law School, John M. Olin Center for Law, Economics and Business, Discussion Paper Series, No.392, Cambridge MD

Chapter 8
Integrated Disaster Risk Management

8.1 General Aspects: The Concept of Risk Governance

The term "governance" comprises a multitude of different definitions; all of them have proven their usefulness, although there is much difference in understanding. The political definition as it is used in our daily life is that governance means running the country. For this a government has set up necessary structures and processes to develop national decision making and an according implementation. Good governance aims at an early and comprehensive societal participation, but nevertheless even a high level of participation of the society will not result in outcomes that satisfy the expectations, needs, and interests of all the different societal groups. Thus risk governance is a systemic approach to enable development processes associated with social, ecological, natural, and technological risks. The approach is based on the principles of participation of all parts of a society to reach a consensus on how to achieve disaster reduction targets and simultaneously fulfill the requirements of sustainability. "An effective disaster risk management that is convergent with other public and private policies seeks to reduce risk exposure and vulnerability by filling gaps in order to avoid or reduce human casualty and economic losses caused by disasters" (Aven and Renn 2010). In order to achieve the envisaged targets two groups of stakeholders exist: the political and implementing institutions mandated with risk management and the other the population at risk. An effective risk governance invites both groups jointly to formulate management strategies that serve the needs of the society in total. The role of the population in this regard is to participate in the decision-making process by contributing personal experience and express expectations and fears. The population at risk is on the one hand benefiting but there are also groups in the society that might be negatively affected by the decisions. The process of bringing these different expectations into a meaningful equilibrium is called "vertical governance". In this regard risk governance is following the principles of good governance when the governance procedures and structures encompass criteria such as transparency, accountability, and participation. As described by the International Risk Governance Council (IRGC 2009):

© Springer International Publishing Switzerland 2016

U. Ranke, *Natural Disaster Risk Management*,
DOI 10.1007/978-3-319-20675-2_8

Governance refers to the actions, processes, traditions and institutions by which authority is exercised and decisions are taken and implemented. Governance is a permanent and important part of life and the willingness and capacity to take and accept risk is crucial for achieving economic development and introducing new technologies. Many risks, and in particular those arising from emerging technologies, are accompanied by potential benefits and opportunities. Better risk governance implies enabling societies to benefit from change while minimizing the negative consequences of the associated risks.

But a government not only acts on the society but is also connected with other governments to bring national interests into the international interconnectedness, as many of the risk types are systemic and are not confined to national borders. Therefore such kind of risk governance requires linking national and international actors to implement a multistakeholder approach within governments, business, science, and civil societies. The integration of national bodies into the international risk governance architecture can either concern nations that share the same targets, interests, and opinions as well as those who do not. Such governance is called "horizontal governance".

In disaster risk management "reality," very often vertical and horizontal governance approaches are closely linked. For example, a flood in the upper catchment area of a large river may call for a local mitigation concept that might deteriorate the flood mitigation capacity of the low-lying neighbor country. Furthermore risk governance should not be seen as synonymous with providing information on what the government does, but on how risk management is organized throughout the country for the sake of society. Thus the term "risk governance" can be described as built up by the three pillars of risk assessment, risk management, and risk communication assembled under one roof, a "risk trinity," as can be seen from Fig. 8.1.

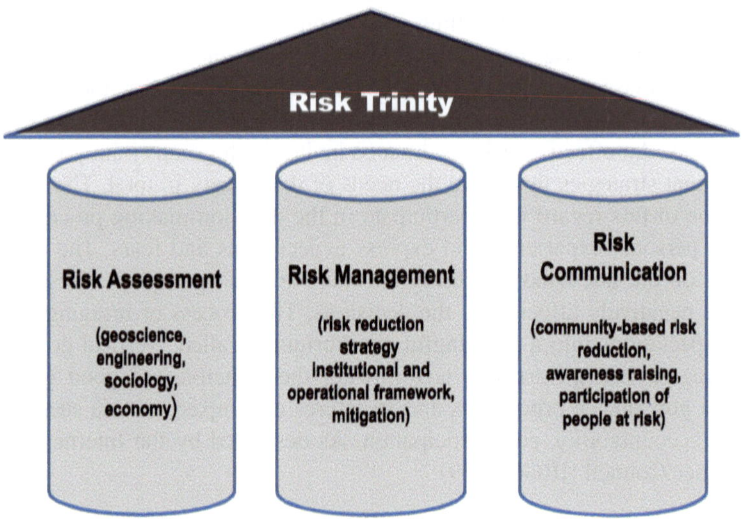

Fig. 8.1 Trinity concept of risk (Based on Böschen et al. 2002)

The "trinity" concept as given by the Office of Technology Assessment at the German Bundestag (TAB), although affirming the traditional form of allocating responsibilities between technically oriented risk assessment and welfare-oriented participation, strongly pledges not to be used as individual standalone pillars. The advantage of the three viewpoint experiences should be used to make the boundaries permeable and profit from interconnecting the inputs. The boundaries result from the different methodological approaches and reflect the classical tasks of scientists and decision makers. TAB points out that the more these methodological boundaries become transitional this will help to overcome the traditional boundaries between technology and sociology for the sake of a prevention-oriented disaster management culture.

The classical methodological approach of risk assessment is, without doubt, not at all free from methodological problems, but as long as there are no convincing alternatives available, science and technology are still given the highest priority in risk assessment (Böschen et al. 2002). The debate runs along the following lines: sociologists strongly pledge that as natural science does not know the term "risk" all evaluations of a risk exposure have to be named "risk analysis". A real "risk assessment" only fulfills its requirements when the social, cultural, and economic framework is taken into consideration. Risk management even goes a step further and starts when the operational, administrative, and legal base for implementing risk prevention is provided. "Risk management depends, however, not only on scientific input. It rather rests on three components: systematic knowledge, legally prescribed procedures and social values" (Aven and Renn 2009).

To give an example for an effective and reliable risk governance initiative: every morning and every evening the North German Radio broadcasts information on the expected high tide level at the North German Sea coast. The text of the message is: "Along the North German Sea Coast the evening high tide level is expected to set in three to four decimeter higher than the mean high tidal sea level".

To make the early warning on sea level at the coast effective, the German government has enacted a special institution:

- The Federal Maritime and Hydrographic Agency (BSH) was mandated to support maritime shipping and the maritime industry and to promote sustainable use of the oceans. In general such a function is handed over to state-owned bodies; but there are also models where such a mandate was given to the private sector. As maritime bodies all over the world are internationally linked, the BSH is the formal representative of the German government in this sector.
- To set up the institutional framework composed of the scientific cum operational capacity as well as the allocation of a sustainable financial budget.
- Establishing and maintaining a pool of experience and expertise in order to fulfill the state's obligation to deliver information on a level of competence that copes with international standards.
- Enacting a national law of broadcasting that obliges the public-financed news agencies to fulfill a "societal" duty on information dissemination. Thus the news, for instance, on an expected sea level are regularly broadcast to the public.

Although the BSH example is tailored to the specific demand of the people living along the coast, the example perfectly highlights how risk assessment, risk management, and risk communication can be organizationally integrated to serve the needs of the public. A central aspect in tide-level forecast is the role of natural scientists and maritime engineers to define, for instance, "the mean sea level" as a regulatory standard. The tasks are daily worked out on certified quality standards so that they can be exchanged via the worldwide communication networks.

There are moreover quite a number of similar institutions mandated with, for example, snow avalanche or extreme weather forecasts.

The governing factor of risk assessment and communication differs with the national culture, political traditions, and social norms. They directly influence the mechanisms and institutions for elaborating knowledge and expertise to be fed into the policy arenas. Policy analysts have developed a classification of governmental styles that address these aspects and mechanisms (IRGC ibid; Aven and Renn ibid):

- The *adversarial approach* is characterized by an open forum in which different stakeholders compete for social and political influence. For this the actors need scientific evidence to support their position. Policy makers pay attention to formal proofs of evidence. Risk management, communication, and stakeholder involvement are mandatory.
- In the *fiduciary approach*, the decision-making process is confined to a group of patrons. The public can provide input but is not allowed to be part of the negotiation or policy formulation process. The system relies on producing faith in the competence and the fairness of the patrons involved in the decision-making process. Scientific and technical advisors are selected according to expertise, sometimes according to national prestige or personal affiliations.
- The *consensual approach* is based on social groups and scientists working together to reach a pre-defined goal. In general such negotiations take place in closed circles of influential actors. Controversies are not made public and conflicts are reconciled on a one-to-one basis. Stakeholder participation is only required to the extent that further insights from the affected groups are needed.
- The *corporatist approach* is similar to the consensual approach, but is far more formalized. Well-known experts are invited to join a group of carefully selected policy makers representing the major forces in society. Often the groups represented within the club are asked to organize their risk management and communication to enhance the credibility of the decision-making process.

8.2 Institutional Framework for Risk-Based Planning

According to the international consensus, disasters whether natural or man-made have to be reduced in order to increase the living conditions of societies. This understanding is the rationale for social and economic development planning and was first formulated by the Brundtland Commission in 1987. The commission introduced the term "sustainable living" in regard to resources, ownership, access to basic needs, and livelihood security. The intention of the vision is that people's capacities to generate and maintain their means of living in order to enhance their well-being and that of future generations should be strengthened wherever possible and wherever necessary. The discussions following the recognition of climate change have especially made such a reorientation necessary. According to DFID (Twigg 2002), sustained individual and societal livelihood comprise a pentagon of human, technical, financial, natural, physical, and social assets that enable them to cope with the adverse impacts of disasters. Thus livelihood resilience describes the capacity of a population to adapt and adjust to actual or potential impacts from natural disasters as well as from the changing climate. After the Second World War this approach has widely been used under the term "civil defense" but in the last decades efforts to cope with threats to local communities are called "emergency management" or "disaster planning" (Quarantelli 1995).

A sustainable disaster resilience will not be achieved without a national strategy that describes short-term responses to periodic stress, as well as long-term perspectives in response to anticipated future challenges. As a result of these outcomes, the Hyogo Framework of Action initiated a multitude of documents and research papers on the current status of disaster risk management and called for an integration of disaster risk management into national strategies for sustainable development. Meanwhile that initiative has been adopted in nearly all countries of the world and disaster risk reduction has been taken up by governmental bureaucracies. All over the world the countries have meanwhile integrated natural/man-made disaster risk reduction in their vision of civil defense. Especially after the 9/11 terror attack when the United States reorganized their civil defense sector and established the Department of Homeland Security, all countries have put a "generic or all-hazard or risk reduction" in their focus. In general there is no great strategic and conceptual difference between risk reductions from external attacks than that from natural disasters.

As an outcome of the UN Decade on Disaster Risk Reduction (IDNDR 1990–1999) the importance of policy and its resulting implementation has now become generally accepted. By law, the envisaged disaster risk reduction is defined at first concerning the national "level of protection" and secondly under which legal framework the objectives should be achieved. The definition of the level of protection is a matter that requires a nationwide consensus of all stakeholders (national governmental authorities, local governments, nongovernmental organizations, research institutions, business, and the representatives of the social groups at risk) in an embracing dialogue. To find such a consensus is not a matter of weeks or months but should be implemented as a permanent review process. Next to the law, risk reduction requires a set of regulations defining the operational environment that enables the authorities to bring risk reduction into being. Guidelines are to follow that explicitly describe who is doing what, where, and who is benefiting, but also who may not be benefiting from the countermeasures. Thus the policy sets the frame by defining regulations, authorizations, prohibition, provisions, sanctions, declarations, or restrictions but does not outline activities actually to be carried out.

Mitigation, as demonstrated in the "Emergency Management Cycle" (see Sect. 5.3) is the step where policies and strategies are institutionalized before a disaster strikes. Thus mitigation has a very close causal relationship with the step of "preparedness and prevention" where the mitigation policies are actually implemented. Although there is no generally accepted definition of mitigation, most disaster risk managers nevertheless stick to this definition rather than to define it as an overall term describing all the measures from disaster assessment to real implementation of risk reduction countermeasures. Focusing on mitigation as a central part of national development strategies has been driven by several factors. First, due to the ever-increasing economic costs of disasters and disaster relief, societies are increasingly expecting and demanding that their governments protect them from disasters before a disaster strikes rather than just reacting to its impact. Second, the increasing understanding of the cause–effect relationship between hazard exposure and poverty calls for state intervention long before a crisis. In developing countries moreover risk reduction is traditionally seen as the state's task. Altogether the arguments raised give a distinct rationale to link disaster planning to development planning, a linkage often reiterated by the World Bank and other international donor agencies (Kreimer and Munasinghe 1991).

Natural hazard mitigation comprises every step taken to contain or reduce the effects of an anticipated or already occurred disastrous event, regardless of whether these steps are taken by an individual, a social group, the public, or states' official emergency management agencies. The aim of hazard mitigation is sustainably to reduce or better eliminate the long-term risk to life and property from hazards. Mitigation is taking action (in general) before a disaster strikes to reduce human and financial consequences later. But often actions that are taken right after a disaster occurred are called mitigation. Nevertheless mitigation is used exclusively here in context with prevention and preparedness as has already been described above. Mitigation when successful increases the resilience of formerly vulnerable societies. Effective mitigation requires that all stakeholders understand local risks, and address the root causes as well as the consequences and define the political will to invest in long-term community well-being. This definition distinguishes actions that have a long-term impact from those that are more closely associated with immediate preparedness, response, and recovery activities. Hazard mitigation is the only phase of emergency management specifically dedicated to breaking the cycle of damage, reconstruction, and repeated damage. The primary purpose of mitigation is to identify community policies, actions, and tools for implementation that will result in a communitywide reduction in risk of future losses.

Figure 8.2 gives a generalized overview on how risk mitigation is addressed today. In the past emergency management agencies very often started disaster management as an ad hoc relief and recovery scheme. Experiences from the last decades, however, revealed that the long-term effects of severe disasters require a shift in paradigm: to assist hazard-prone communities in a holistic way, before,

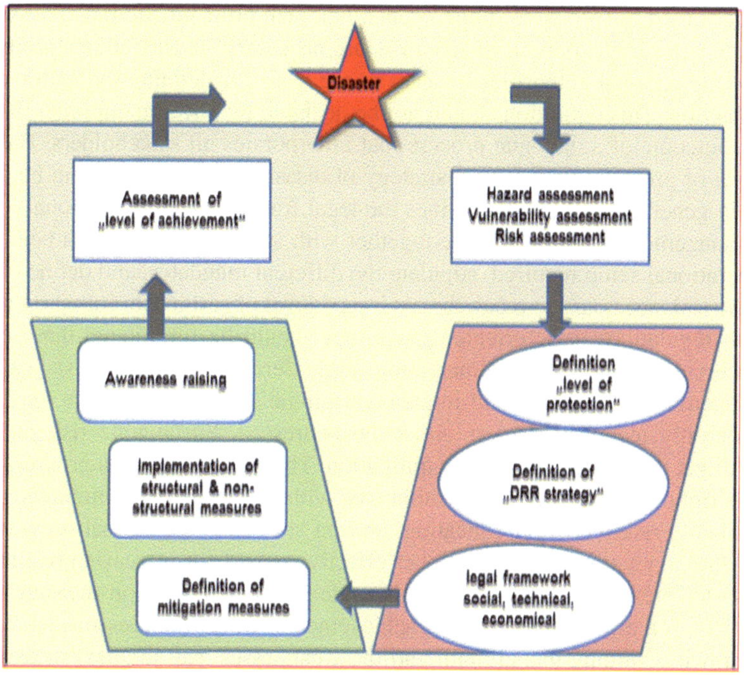

Fig. 8.2 Disaster risk mitigation planning structure (*light brown* scientific sector; *Red-brown* state levels; *light green* local implementation) (Own graph)

during, and after a disaster. Moreover there was a strong drive in the management focus towards cost-effectiveness. The long-time prevailing evidence-based disaster response and crisis reaction thus got a strong movement towards a culture of prevention. All these demanded a new disaster risk management strategy framing the legal and operational base. The new approach is that today the factor of prevention became an indispensable part of any strategy.

The strategy shows disaster risk management as a cycle. As with every cycle it has neither a beginning nor an end. Nevertheless let's start with the onset of a disaster, the impact of which, root causes, people at risk, and their affected livelihoods have to be assessed in a scientifically sound, careful, and neutral manner. In these tasks geoscientists, technicians, sociologists, and development planners are called upon to combine their expertise to come up with a reliable assessment. A legal frame for a systematic and rigorous disaster risk assessment on the national level must be established, that must furthermore comprise a stipulation on data collection systems and research and analysis by scientific institutions to ensure a knowledge-based disaster risk assessment. Only such a multisector knowledge-based assessment reveals the best opportunity for an in-depth assessment of the disaster that occurred and thus lays the basis for sustained disaster mitigation. But such an assessment alone will not give a realistic assessment of future events. Therefore based on the deterministically assessed single event, it is necessary to draw scenarios for probable future events. For such probabilistic risk assessments, well-educated and experienced scientists are necessary. Their findings are handed over to the national authorities to serve as the basis for defining the national level of protection. This definition should not be declared by government order. It has to be the outcome of a dialogue process that incorporates all stakeholders. Based on the level of protection, a national strategy of disaster risk reduction can be formulated: in general by a law that defines the legal framework of the national disaster risk management. The legal frame together with adjoining regulations benchmark the operational setup required, stipulate the different mandates, and define (in general) the task necessary to reach the envisaged level of resilience. Based on the law and the regulations topic-oriented guidelines should be formulated that describe the different tasks at the implementation level. Here again geoscientists' expertise is indispensable to develop and implement mitigation measures at the local level. But not only technical matters are to be addressed. Sustainable risk reduction needs to get the people involved in mitigation. Therefore mitigation programs have to start right from the beginning initiatives with the focus on community awareness raising, risk-reduction education, and on strengthening the self-help capacities of the populations at risk. Another effective means for increasing resilience is by disaster preventive land-use planning. Next, every mitigation measure has to be subject to a critical review of its achievements. Only when lessons learned are been drawn regarding the strength and the weaknesses, can the next measures be designed more efficiently.

To institutionalize implementation at the local level the provision of financial and technical support for cost-effective natural disaster mitigation measures as well as an effective expenditure monitoring is required. Without sufficient adequate financial support and the "binding forces of law" any attempt to increase resilience by planning and organization will be not successful. Therefore the state and its local authorities must endorse cost-sharing principles that nevertheless include a focus on the responsibilities of individuals, businesses, and insurers. Only when all decision-making levels are collaborating and are coordinated according to national strategy, can the government take up its original responsibility in guiding natural disaster risk management but not being actively involved in local mitigation.

The preceding statements clearly point out that risk-based planning is a prerequisite to shift the paradigm from a culture of reaction to a culture of prevention. Such a shift allows planning for future risk reduction based on a coherent risk assessment, defining safety levels that are applicable to the respective risk level, and moreover facilitating the setting of priorities and defining the level of public and private interactions. Risk-based planning is the main step in achieving this. In order to develop a long-term strategy to protect people from the adverse impact of future natural disasters, risk-based planning creates the general framework on the national as well as on the local level. Therefore mitigation planning has to involve the science sectors, the legally mandated authorities, and the implementation sectors, as well as the populations at risk. The envisaged benefits are that exposed people understand the nature of risk, the problems are addressed properly, and that reduction measures lead to an increase in societal resilience. The most important aspect of risk-based planning is that understanding risk is deeper and more broadly anchored in the society.

In Tables 8.1 and 8.2 an arbitrary selection of the major activities related to risk-based planning are cross-plotted with the major actors. It reveals that first of all, the legal framework for a risk-reduction strategy has to be established. Based on the framework the scientific sector sets in, assessing hazard, vulnerability, and the level of risk of the areas under investigation. Based on these findings the planners and local administrators are able to define the specific reduction measures and can initiate and pursue the implementation. National, international, private, or public financing schemes are then required to allocate the necessary funds. Risk reduction will not be successful and sustainable if the risk-exposed population is not involved adequately. The link between the official development decision-making process and implementation at the local level is part of the communication process carried out by the media and the official publication channels. The tables clearly demonstrate that all these interactions function as what can be called a "risk planning path." However, it should be noted that the table highly

Table 8.1 Administrative and societal sectors affecting activities disaster risk management

	jurisdiction	geoscientists	environmental scientists	Social scientists	economists	civil engineers	local experts	local administration	landuse planner	people at risk	private banking	risk insurance	international donors	officil financing schemes	official risk communication	print media, TV, radio
Legal Framework	■	■	■	■	■	■		■	■							
Law enforcement	■					■		■	■		■	■	■		■	
Definition of security level	■	■	■	■	■	■	■	■	■	■						
Definition of hazards		■				■	■	■	■	■						
Hazard zone cataster/data list		■	■			■		■	■							
Hazard assessment		■	■			■		■	■	■						
Definition of vulnerability	■	■	■	■	■	■	■	■	■	■						
Technical vulnerability cataster		■	■	■	■	■	■	■	■	■						
Human vulnerability cataster				■	■	■	■	■	■	■						
Vulnerability assessment		■	■	■	■	■	■	■	■	■						
Risk assessment		■	■	■	■	■	■	■	■	■						
Worst case scenarios		■	■	■	■	■	■	■	■	■						
Risk simulation		■	■	■	■	■		■	■							
Risk reduction planning		■	■	■	■	■	■	■	■	■						
Identif. of prevention areas							■	■	■	■						
Identif. of reduction measures						■	■	■	■	■						
Implement. reduction measures								■								
Assessmnt. of level of achieve.		■				■	■	■	■	■						
Cost - benefit assessment					■	■			■							
Risk education		■	■	■	■		■	■	■	■						
Risk acceptance							■	■		■						
Risk splitting	■				■						■	■	■			
Financing reduction measures											■	■	■	■		
Risk communication															■	■
Emergency drill		■		■		■	■	■	■	■					■	■

Table 8.2 Administrative and societal sector's relation to disaster risk management procedures

	Jurisdiction	Natural/social sciences	Local administration	Population at risk	Risk financing agency	Media/Communicator
Legal Framework	■					
Law enforcement	■					
Definition of security level	■					
Definition of hazards		■				
Hazard zone cataster/data list		■				
Hazard assessment		■				
Definition of vulnerability		■				
Technical vulnerability cataster		■				
Human vulnerability cataster		■				
Vulnerability assessment		■				
Risk assessment		■	■			
Worst case scenarios		■	■			
Risk simulation		■	■			
Risk reduction planning		■	■			
Identif. of prevention areas		■	■			
Identif. of reduction measures		■	■			
Implement. reduction measures			■			
Assessmnt. of level of achieve			■			
Cost - benefit assessment			■			
Risk education				■		
Risk acceptance				■		
Risk splitting					■	
Financing reduction measures					■	
Risk communication				■		■
Emergency drill				■		

generalizes the different functions, responsibilities, and interests the actors have in the field. And that there exists in the actual planning process a number of very specialized and sophisticated interactions between the different actors and different organizations.

Risk-based spatial planning lay the base for a long-term management of a disaster risk reduction strategy. The planning comprises elements from hazard identification to the definition of risk reduction measures and their respective implementation. Furthermore it allows the monitoring of the effectiveness of the countermeasures long after the actual measures have been completed. Spatial risk reduction planning should also entail conceptual means to identify alternative or additional uses for the area at risk under the assumption that these newly introduced land-use patterns will not lead to further endanger the area. To achieve completeness in planning, all possible alternative reduction measures have to be considered and have to be checked against their technical realization as well as against the envisaged economic benefits. These many requirements define the integral nature of any risk planning concept. As already mentioned, risk planning is in general based on a snapshot of the actual risk exposure situation, but it has to

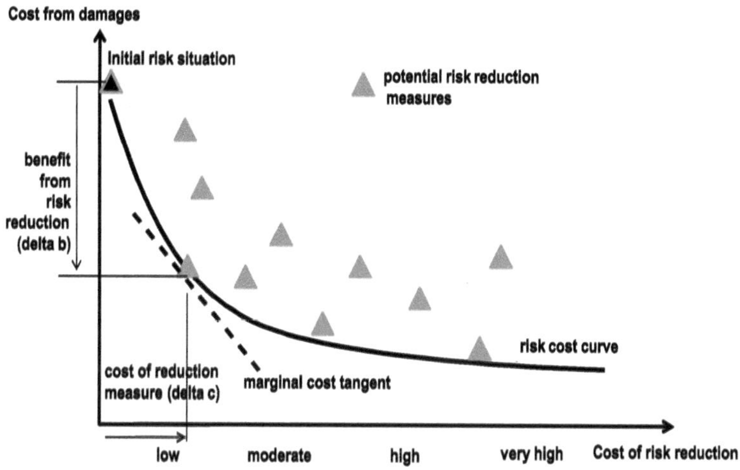

Fig. 8.3 Benefit–cost relationship of risk reduction measures (*Courtesy* Bründl 2009)

be considered in the planning that the implementation measures may be in effect for some time. A time span of up to several years may have to be considered, for instance, for stabilizing a slope against avalanche debris falls or even decades in the case of reforestation measures.

The risk-planning strategy has to comprise a methodological approach for time-differentiated quantitative risk assessment, based on the technical implement ability and the benefit–cost relation as given in Fig. 8.3. Generally the envisaged level of resilience should be oriented at the state of technology. Such a demand holds true especially for strong economies, whereas in many developing countries such a claim is often not achievable. In these countries the implementation in fact is mainly based on the low cost principle, although the measures should at least serve a minimum level of resilience and should be locally adoptable and socially accepted. In this context, the issue of translocation has to be mentioned. This describes the strategy to locate risk-affected people in other no risk areas, an instrument that is often practiced in developing countries. Translocation is often seen by the authorities as the sole measure as it is easily implemented, based on hierarchical decision-making structures, and is generally taken by technocrats. But such decisions mostly disregard the emotional "soft" aspects of regional planning. It turned out that as long as such measures are not rooted in the society at risk, the effectiveness proved to be very questionable.

The operationalization of the risk reduction measures should be laid down in the planning documents defining the main steps of the implementation process:

- Define the local prevention level (individual, public).
- Carry out risk assessment of the area under investigation.
- Outline the risk reduction measures.
- Define the cost regime with respect to the envisaged level of resilience (benefit).

- Enhance acceptance of the reduction measures by early involvement of the people at risk.
- Assess compliance of the risk reduction measures with other claims on the area, for example, industrial activities, agricultural production, environmental protection, social and cultural heritage, and ethical values.

The complex relationship between the risk-generating factors and the individual or societal group's perception of such kind of threats, brings risk governance to the crossroads of scientific, technical, and probabilistic assessment. Fact-oriented risk assessment collides with psychological, cultural, social, and ethical perspectives. Responsible and meaningful risk governance therefore has to take these different experiences into account. The natural scientists mostly see this context as a technical challenge, but as the risk definition originated from the social sector, the societal perspective based on cultural heritage, social status, ethnic and gender relations, and the participation perspective prevail. Risk governance therefore is placed at the "crossroad between natural events and socio-economic and technological developments and policy driven actions" (IRGC 2010). The risk governance framework has to become an integrated framework that sets the rules for a comprehensive (technical and social) assessment and for the formulation of management strategies to cope with risks, at the local level as well as internationally. Based on the conviction that both the factual and the sociocultural dimension of risk must be considered during the risk governance process, the risk governance framework integrates scientific, economic, social, and cultural aspects and includes the effective engagement of stakeholders. Coping with risk "requires a holistic approach to hazard identification, risk assessment and risk management. This makes risk more than just multiplying the losses with the probability of occurrence; it rather deals with uncertainty about the occurrence and the consequences" (Aven and Renn 2009).

The IRGC (ibid) stated that

> [G]governance of global systemic risks needs to focus on a community of multiple institutions and disciplines, including natural scientists, social scientists, engineers, policy makers, practitioners and educators from around the world. It ought to be truly integrative, especially in the face of the ever more complex, inter-connected and cascading risks that society faces. The global risk community must develop and apply innovative tools and methods, including theoretical, mathematical and computational tools as well as enhanced management approaches.

An effective and sustainable risk management strategy able to comply with the multitude of different risk settings first needs a definition of the role of the major stakeholders.

The Role of Government

- To work out and make effective a national disaster management strategy that defines the national level of resilience that is to be achieved and that is conducive to the ecological sustainability of the country
- To develop and implement an effective risk-based land management and risk plan

- To develop and implement an effective disaster prevention and emergency management at all administrative levels
- To define the stakeholders and mandates of the different administration levels
- To allocate the funds and secure a comprehensive implementation of the strategy
- To inform people regularly, early, and in depth about how to assess risks and reduce their exposure and vulnerability to hazards
- To raise awareness of the disasters the nation is exposed to, and to make the people understand what options are available to respond effectively to a hazard
- To support individuals and communities to prepare for all kind of events
- To mandate, equip, and provide trained and skilled personnel for an effective and well-coordinated response of the emergency services
- To help communities to recover from devastation by technical assistance and the allocation of rehabilitation funds
- To help the society to learn and adapt in the aftermath of a disastrous event.

The Role of Individuals

- To take their share of responsibility for preventing, preparing for, responding to, and recovering from disasters according to their individual, financial, technical, and personal resources
- To work out guidelines to improve their household resilience
- To acknowledge that most of the immediate rescues are rendered by those affected
- To make a disaster resilience approach the basis of the mitigation and prevention activities at the household and community level
- To regularly inform regarding hazard exposures and disaster events to raise awareness of the individual and society
- To organize regular emergency drills to keep the community alert regarding the hazards
- To increase individual disaster resilience by active planning and preparation for protecting life and property
- To actively strengthen cooperation between the local community disaster and emergency management authorities
- To take part actively in working out a local disaster management strategy
- To organize rescue and rehabilitation in case a disaster strikes.

The Role of Business and Nongovernment Organizations and Volunteers

- Business can play a fundamental role in supporting a community's resilience to disasters as it can provide technical resources and expertise and many essential services on which the community depends. The provider of critical infrastructure (water, power, heavy duty machinery) can especially substantially contribute to rescue and rehabilitation by providing services during or soon after a disaster. Governments as well as local authorities should look for close cooperation with the private sector by involving it as early as possible in the general considerations to increase local disaster resilience,
- Nongovernment organizations and volunteers are often first at the forefront when a disaster strikes and often provide most urgently needed support. But NGOs can also help communities to cope with and recover from a disaster. Governments

as well as local authorities should look for an intensive and comprehensive involvement of the NGOs to make them partners in the process of awareness raising and prevention.

A good example for a statewide multihazard management planning is the State of California's Multi-Hazard Mitigation Plan of 2007 (OES 2007) with which the then-acting Governor Arnold Schwarzenegger transferred US national law into state law. He commissioned California's Office of Emergency Services (OES) to work out a plan that in detail explained how the Office of Emergency Services could integrate the many state hazard mitigation programs into one statewide emergency management plan. Much emphasis was given to the planning process and its legal, institutional, and policy framework. The plan defined the advisory function under the umbrella of California's Office of Emergency Services (OES). Hazard, vulnerability, and risk assessments were made compulsory for all counties, districts, and local administrations at least for the primary hazards of earthquake, flood, wildfire, landslide, and tsunami. Broad space was reserved on improving the managerial capabilities that are required for sustainable risk prevention and reduction measures. The plan foresaw a provision for a regular update.

8.3 The Risk Mitigation Cycle

There is a common understanding that there will be never a 100 % chance to avoid a catastrophe (Kaplan and Garrick 1981). But all experience in disaster management from all over the world revealed that there are realistic chances to avoid at least some of the major impacts when knowing more of the specific details of the disaster characteristics, defining its onset, trigger mechanism, and its location of impact (risk assessment). Hillary Clinton, in those days secretary of state of the United States, is right when saying: "Never waste a crisis." With this sentence she wanted to express that in every catastrophe, there lies a real chance to at least tackle some of the risks. There is only one language worldwide in which the word "crisis" is composed of two other words, "threat" and "opportunity": this is the Chinese language (see Fig. 8.4). The basic meaning of the Chinese character

Fig. 8.4 The Chinese character "crisis" (*Courtesy* UNISDR 2004)

"crisis" is that a crisis originates from a "threat" but opens the way to future preventive action: "opportunity".

The central and most important task in disaster risk management is to organize how a society reacts to a disaster. The moment a disaster strikes, a multifold set of activities starts that at first is oriented to rescue the victims. Then the societal functions have to be restored. But in order to increase the people's resiliency to withstand the next disaster, it is indispensable to improve the society's risk prevention capacity. The United States of America's Emergency Management Agency (FEMA) has set up a "Natural Disaster Risk Management Cycle" that comprehensively describes the different steps of disaster management from the disaster onset to the preparation for the next (FEMA 2003). As FEMA concentrates on the emergency aspects, the cycle starts with the disaster impact. The onset of a disaster, its remedial actions, restoration, and rehabilitation of the social system followed by preparedness for a possible future disaster form a sequential terminology that is given in the graph in Fig. 8.5. FEMA furthermore underlines the management aspects in an emergency by introducing the step "mitigation" to the cycle and deliberately placing it last. Thus the cycle demonstrates that mitigation is that one activity which finally prepares the society to face the new risks. This approach makes the FEMA cycle different from many other incident operation plans. The management cycle expresses the significant change risk management perception has undergone since the International Decade on Natural Disaster Reduction (IDNDR) addressed

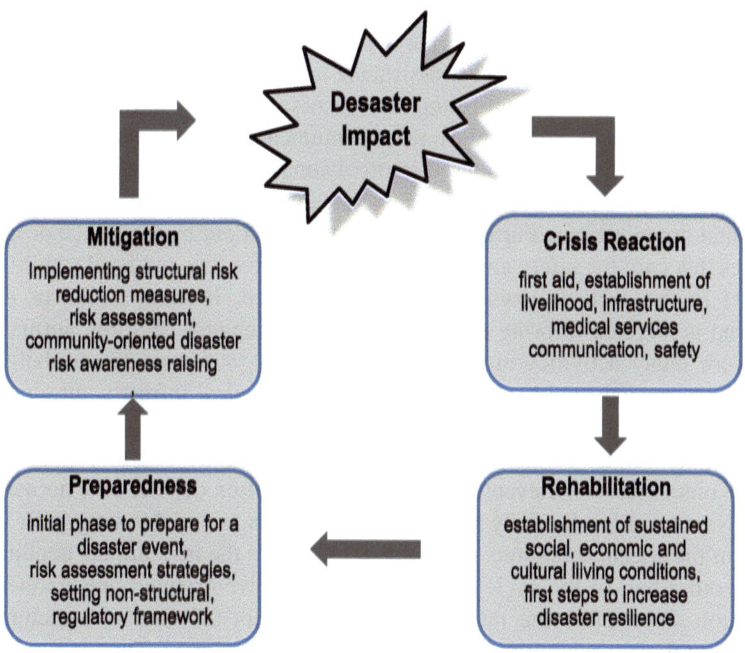

Fig. 8.5 Natural disaster risk management cycle (Own graph)

the issue in the 1990s. Since then disaster prevention became an integral part of emergency management. Rescue and restoration were seen as the main field of intervention of emergency management, mainly carried out by military rescue operations. By introducing prevention, the management is now called on to incorporate risk prevention in order to enable populations to become less risk exposed. Today in almost all industrialized nations and even in many further developed countries this approach is now part of the emergency management operational procedures.

Nevertheless, there are many developing countries still lacking this vision and ad hoc military interventions are seen as the main rationale. Even in Indonesia disaster prevention was not given a high priority until the tsunami happened in 2004. But since 2006 a new National Emergency Management Law strongly emphasizes the notion “to be prepared.”

The Natural Disaster Risk Management Cycle with the onset of a disaster and clockwise follows all the steps of disaster risk management. Regardless of where one starts, all steps form a cycle that encompasses the different mitigation steps either as a source of disaster or as the outcome (cause–effect relationship).

The cycle in Fig. 8.5 is divided in two sectors: on the right there are activities that describe the response to a disaster event and on the left activities are placed that describe the future-oriented disaster prevention. The steps on the right side, Crisis Reaction and Rehabilitation are both post-incident steps and start upon the disaster impact. The two steps of Preparedness and Mitigation form preventive actions to avoid the negative impact of future disasters.

Immediately after the disaster has happened, the management cycle starts with first-aid activities to save lives and to rescue people from the damaged environment (“crisis reaction”). It is that part of the management cycle where the well-known rescue teams and medical assistance are brought in that save the victims, build emergency shelters, and restore the necessary critical infrastructure such as electricity, water, and fuel supplies, telecommunication, and transportation. At this time most of the victims are housed in provisional camps, provided with food and water, and are given medical and social assistance by local authorities or nongovernmental organizations. Further activities are carried out by the police or military services that care for safety and security and to enforce law and legal stability.

Next in the cycle follows the rehabilitation step. It starts weeks or months later and is targeted to bring the affected society stepwise back to “normal” life. Therefore many activities are undertaken in order to restore life as before the event. This not only holds true for restoring the physical infrastructure including hospitals, schools, and bridges, but also for bringing the social system back to pre-disaster functioning. In the rehabilitation step the society is still exposed to the adverse outcomes of the event and the people still often depend on external assistance. Next to caring for daily living, rehabilitation is oriented to long-term stabilization of the social, economic, and cultural as well as the institutional conditions.

The other half of the cycle comprises the proactive oriented preparedness and mitigation steps. Here the management cycle enters the field of structural and nonstructural risk reduction. This sector is mainly oriented towards making the people more resilient to future disasters. The aim here is to carry out sustainable and effective actions that help to eliminate or at least to diminish negative long-term effects to the people, their property, or their living environment from future disasters.

The step preparedness mainly addresses operational procedures and brings the risk reduction plans into operation. Here risk assessments are carried out, lines for coordinating technical and political issues are established, and emergency operation plans are developed. These activities are either done by the national authorities or at a local level. Often the science sector is involved here. Based on experience from past disasters and findings of the risk assessments, local emergency managers will be enabled to review the existing spatial planning and will come up with improved plans. In this step also the communities themselves are addressed. Either the societies at risk or the individuals are called on to carry out their individual and local disaster reduction measures. Making the society more resilient to the impact of natural disasters will not be successful without the involvement of the population. It is the people who are vulnerable and with it the social system. Thus only by an early and comprehensive participation of the risk-exposed ones, can a sustainable resilience be achieved. There is a term in international disaster management saying, "Every mitigation is local," meaning that 90 % of the immediate action of a disaster strike is carried out by those who are affected. The development of emergency operation plans have therefore to be worked out together with the risk-exposed population. Experience has shown that such an operational approach can best define responsibilities and mandates and help to narrow the bridge of the administration and those affected. The better and the earlier the people are incorporated, the better they are informed and equipped, and the better they are integrated into the operational procedures, the better they will be able to save themselves. This holds true especially for an involvement of the people in emergency management. Experience shows that comprehensive participation is the best way to facilitate disaster response and help to improve the society's stability not only before a disaster strikes, but also to strengthen the self-help capacity in case of emergency.

Mitigation, as the last step in the cycle, mainly focuses on operational measures based on the experience gained from the last disaster and reflects the findings of the risk assessments carried out. Mitigation covers structural as well as operational activities such as improving the stability of roads and bridges, implementing the redefined building code, or reorganizing the operationality of the local health care system and social services. But not only technical matters are incorporated in mitigation, also many ecologically oriented aspects such as setting up recreational areas or planting trees and mangroves in near-shore areas. Rehabilitation not only concerns tangible matters to be brought back to functioning, but it moreover aims to make the society more resilient against future disasters. This can be either be achieved by technical matters such as improving the way houses are constructed to reach a higher degree of stability by introducing stronger pillars in the construction frame (seismic structuring), by just taking more cement, or giving the building an open floodable ground floor in flood-prone areas. Also allocating necessary funds for private rebuilding, or by counseling the victims on employment chances or for entering into private business will result in a higher degree of resilience. Nevertheless all actions taken on this side of the cycle are reactive in nature.

Although all the steps in emergency management aim at improving individual and societal resilience they nevertheless do not give a complete picture of the situation. Imagine all the prevention actions were successful, and then at the end of the cycle, the level of resilience would not be at the same level as in the beginning. So this book introduces instead of a cycle an "emergency management spiral," thus stating that at the end the population at risk will be better prepared and safer than before ("safety increment"). A flood dam that was rehabilitated and heightened by, say 3 m will hold off more water than before. Nevertheless a much stronger than normally experienced flood may destroy the dam. Even having the spiral end successfully reached, a certain risk still remains. The spiral concept is therefore seen as the task of the emergency managers to inform the people living behind the dam that their risk is now less than normal, but in case a much stronger than traditionally experienced flood level inundates an even larger area than before, it will be with a much lower probability. See Fig. 8.6.

But disaster risk management is not exclusively the domain of emergency management agencies, rather it is the result of the collective responsibilities of all sectors of society, including all levels of government, the nongovernment sector, and the science and technology sector, as well as the affected individuals to reach what is called a resilient society. A disaster-resilient community is one that works together to understand and manage the risks that it confronts. It is a notion of shared responsibility in strengthening the societal resilience to disasters. The concept of resilience helps to explain individual variations in response to risk in order to develop and make use of the capacity to recover quickly from difficulties.

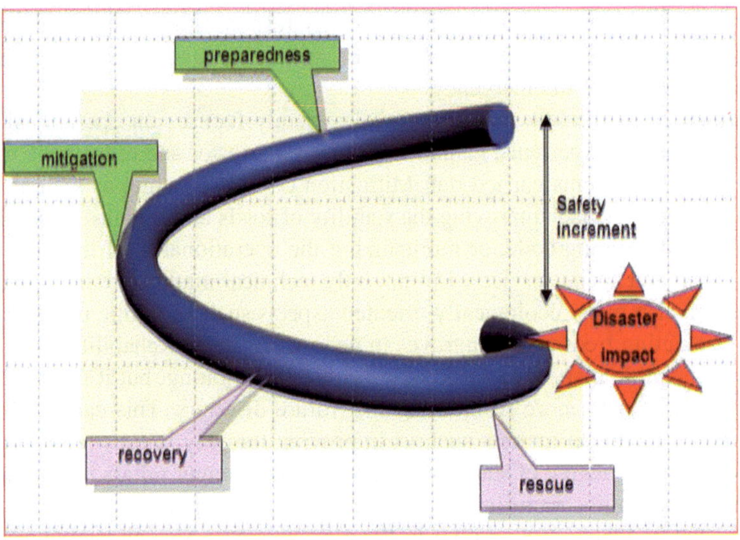

Fig. 8.6 Emergency management spiral (Own graph)

Broadly speaking, resilience is the capacity of a disaster-exposed society to adapt successfully to stressful situations and shocks. Defining resilience should always answer the question, "Of what and to what is a society resilient?" (DFID 2011). This answer identifies which level of resilience was envisaged as being reached and which implemented prevention measures successfully made the country, community, or household more resilient. But the material infrastructure is less of a concern for resilience than the social groups, socioeconomic or political system, environmental context, or institutions that benefit from prevention.

Resilience is believed to contribute to individuals' and communities' abilities to maintain their well-being in the face of a sudden impact from a catastrophe. Concerning the level of the individual, resilience is understood as a set of personal capacities to become an adaptable person, who is better able to frame stress-inducing situations in positive terms to return quickly to positive functioning, despite the challenging circumstances. Resilient communities are those able to integrate the individual's coping capacities and integrate these capacities into the organizational and legal framework of communal disaster mitigation activities. Although the term "resilience" is not very clearly defined and difficult to measure, it nevertheless helps to describe the situations disaster-affected people are confronted with and how they are able to overcome stressful situations in order to return to a normal and sustainable post-event functioning (Zahran et al. 2011). It turned out that certain groups are less resilient than others. At the individual level, factors including an age higher than 65 years, women, illnesses, prior mental and physical disorders, lower socioeconomic status, membership in a marginalized ethnic/racial minority, persons with low social support, abundant children, no or a low income make such persons especially vulnerable. Of all groups single females with many children are those who are most likely exposed to traumatic stress. Moreover there is quite an agreement among disaster managers that those who were exposed to higher risk and have already experienced a number of disasters developed a deeper understanding of the cause–effect relationship of hazard and risk, and thus gained automatically a higher level of resilience.

8.4 Definition of National Goals on Disaster Risk Management

With defining the national goals on disaster risk management, a state or a local government and even the international community declare what kind of material and ideal values the society wants to protect from damages, deterioration, or harmful impact. This process is called "subjects of protection." Defining such a target always originates from the impact that is envisaged to be avoided, but cannot be disconnected from the root causes that are responsible for the catastrophe. There is a clear relationship between the causes and the effects for any subject of protection. Defining a subject of protection is in the purest sense neither a matter of natural science nor of social sciences, rather than a jurisdiction (object of legal protection). But as in reality these discussions worldwide mainly concentrate on topics including pollution, climate change, and natural and man-made disasters, natural sciences developed to an integral part in these discussions. In general the subjects of protection are mostly related to the:

- Human being (health, security, personal integrity)
- Natural environment (fauna, flora, biodiversity, landscape)
- Natural resources (water, air, climate, soil)
- Cultural heritage
- Industrial and agricultural resources
- Ecological interconnectedness and relationship of natural and human resources

The general question to be answered is: "What kind of assets, subjects, or environmental objects are worth being protected" and if this is defined, the next question is: "To what extent should such a system be protected."

The discussion that follows regards how a system (e.g., water, soil, air, biological diversity) can be protected or empowered so that it will be able to fulfill its original functions in the future.

As the definition of a national goal of disaster risk management is at first a legal matter, all states worldwide meanwhile have enacted a variety of norms, rules, and regulations to guide the policies of human and natural protection strategies. In general the enforcement of these acts is handed over to state-owned authorities responsible for implementing the regulations. Sometimes, however, that authority needs to be further refined or explained, a task to which natural scientists are called on to contribute.

In the United States one such agency is the Environmental Protection Agency (EPA) that is mandated to write guidance, develop, and implement policies to encourage the national authorities as well as third parties to comply with environmental requirements. In Europe, the European Environment Agency (EEA, Copenhagen) is commissioned to provide sound and independent information on the environment for the countries in Europe. EEA is the EU major information source for developing, adopting, implementing, and evaluating environmental policy and also the general public. The EEA's mandate is:

- To help the European Community and EEA member countries make informed decisions about improving the environment, integrating environmental considerations into economic policies, and moving towards sustainability
- To coordinate the European environment information and observation network

The EEA is directed by a management board consisting of one representative of each of the member countries, two representatives of the commission, and two scientific personalities designated by the European parliament. A scientific committee advises the management board by providing advice and/or opinions on any scientific matter to be taken up and incorporated into the EEA work programs; among them specialist for life science, marine ecosystems, sustainable agriculture, biology, atmospheric processes, climate change mitigation and adaptation strategies, natural hazards and risk assessment, industrial risks, as well as spatial planning and management of natural resources and environmental law.

In Germany, the Federal Ministry for the Environment, Nature Conservation and Nuclear Safety, usually shortened to Federal Environment Ministry (BMUB) is the responsible authority for the environmental policy of the German government. The aim of environmental policy is to create or maintain a healthy living environment. The environmental policy of the German government encompasses many different areas for which the ministry sets goals, drafts legislation, and guides the implementation process. The ministry is primarily concerned with climate policy, conserving the diversity of fauna and flora, efficient use of resources and energy, as well as protecting people's health from environmental pressures. German environmental policy follows the principles of sustainability. This means being environmentally aware by simultaneously maintaining economic productivity as well as living and trading in a way that is socially fair. On the international agenda BMUB is active on all matters related to a sustainable living, especially the issue of global challenges to overcome climate change impact and the loss of biological diversity. BMUB has since then enacted a multitude of rules and regulations. The most important is the Law on Environmental Assessment (EIA) with which Germany has taken over an EU directive into national law. Other regulations concern, for instance, the Ordinance on Requirements for the Discharge of Waste Water, the Atomic Energy Act on the Peaceful Utilization of Atomic Energy, the protection against its hazards, and a regulation for the National Climate Protection Initiative. The World Bank report on disaster hotspots (Dilley et al. 2005) further stated that disaster preparedness has to become a standard element in development strategies, as the natural hazard cycles are repeating themselves faster every few years. The many floods in Germany since 2005 clearly underpin these findings. Management of disaster risk has therefore to become an integral part of development planning rather than only as a humanitarian issue. Moreover disaster prevention and preparedness in high-risk areas has to be strengthened by implementing multihazard risk management strategies from the very beginning. Only when effective efforts are undertaken to a proactive and preventive disaster management and by an early and comprehensive involvement of the people at risk, will it be possible to address disasters before they strike.

8.5 Society's Decision on Acceptable and Unacceptable Risk

The term risk assessment in general describes instruments to assess the risk from a hazard in respect to the benefits related to this hazard. To decide whether a risk is (according to Chap. 7) really posing a threat to a society, depends on the situation the society is living. A hazard that poses a threat to a certain society will not pose a threat to another one. Even within a society a threat, say from flooding, is not at all seen as a threat to people living higher up in the mountains or even in the lowland where many of them benefit from the nutritious material that was brought in. The rating that a hazard is risk-imminent is thus a result of a sum of different personal perceptions. In general the individual assessment of the level of risk is answered by the questions (Kaplan and Garrick ibid):

- Do I feel threatened by this hazard?
- Do I have experience with this kind of hazard?
- Do I feel experienced enough to cope with the risk?

Answering these questions automatically brings the discussion to the central aspect of risk assessment: the personal evaluation of risk exposure.

The next step in individual risk assessment is whether the individual accepts or tolerates that risk. Investigations of risk perception revealed that personal experience on, for instance, a flood standing one meter high in a private house or a risk from a faraway nuclear catastrophe, are generally assessed by a comparison of this particular risk event with events in the past: is this risk higher than last time, or less, and is the expected outcome rated acceptable or not. The division of acceptable (tolerable = a risk that we cannot avoid) or unacceptable (intolerable = a risk that we have to avoid) depends exclusively on the individual's risk perception, although to a large part the rating is influenced by risk assessments that are carried out by the official sector. Their judgment without any doubt influences the cognitive assessment processes and helps the society and the individual to determine personal risk rationally. But experience turned out that the normative assessment processes carried out by natural scientists, engineers, and sociologists only help the individual to get a personal view of the risk, but never replace the individual's final assessment (Fischhoff et al. 1981).

Different societies—even when facing the same type and severity of hazard—react completely differently to a disaster event. The acceptance of a risk is according to experience in the last decades fully dependent on the state of knowledge of such a disaster, from the historical background, as well as from the skills of prevention technique and of society's acceptance of finance countermeasures. Risk acceptance is furthermore in many places a matter of faith, belief, and sometimes fatalism. Every time a disaster strikes, the people ask one central question: do we have to accept such kind of tragedy or can we avoid it? The question generally addresses the authorities and seldom addresses the affected population itself. And there is another phenomenon to be recognized. Large deadly accidents such as a plane crash or a bursting dam see people reacting aggressively and with anger and immediately raising the question of who is responsible for the disaster, whereas on the death in a murder case of one young child, sadness and helplessness prevail.

Furthermore risk acceptance, or nonacceptance, is a social question: a manifestation of poverty. "Poor" societies very often have a different opinion than "well-off" societies on what is a catastrophe. Such a rating is in general based on high cases of recurrence of disasters and the realization that they are normally left alone with their problems, resulting in suffering and faith. For example, the Indonesian capital Jakarta's living quarter Kampung Malaju is every year subject to high flooding. The shelters are made up of high-rising bamboo poles that already reflect the increased water levels, although the residents yearly face a significant death toll, damage to their living shelters, and their adjacent (very small) vegetable gardens. Very often also the artisanal working places at home are destroyed. But when asked how they are adopting to the yearly floods, they answer, "Where is the problem; its only water."

The several floods in Western Europe in the last year mostly in England, Germany, and Poland in contrast were the starting point for an extensive upgrading of the Elbe dams and an increase in local and even transnational flood management. Thus different societies react differently to the risks they are exposed to, based on the state of technology, their financial capability, and traditional, cultural, and social values. In the last consequence it is society that declares what kind of hazard and with what frequency and severity it will accept threatening the levels of social welfare.

Compared to this, the task of the technicians and scientists is to determine how the societal demand for safety can be settled without compromising the demand for safety and welfare of other social groups. But is has to be mentioned here very clearly: risk acceptance, as seen by the public, normally does not consider the onset of a disaster. The risk from the standpoint of the population almost exclusively regards the damaging impact, the loss of lives, and damage of the property or living environment in general. Therefore when we talk about risk acceptance, we should be aware that the term "risk acceptance" is seen by the population at risk in most cases as a question of the acceptability of the disaster impact. Especially when natural/geologically generated disasters are considered, risk acceptance is restricted to the disaster outcomes. A totally different situation occurs when we discuss the risk acceptance of man-made and/or climatic-generated disasters. When human interference with nature causes a risk resulting in losses or damages, the public standpoint is totally different. In such situations the public sees the unexpected outcomes of, say the use of nuclear energy for power generation, and asks how we can organize society without such technology in order to be more on the safe side.

As risk governance has a responsibility for society in total or at least to certain exposed groups, risk governance has to find out what kind of risk and at what level of severity is accepted by the society. To give an example, there is a common understanding that even with the best available technology it will not be possible to prevent the Earth from being hit by meteoritic impact. Such impacts geologists estimate have hit the Earth several times in history, creating the Pacific Ocean and the moon and may have resulted in the massive extinction of fauna at the end of the Cretaceous period. A similar meteoritic impact is regarded as most probably leading to a complete extinction of life on Earth. Nevertheless there is no reasonable request to protect the Earth from such a catastrophe. Such a risk is weighted as "least probable" and thus is taken to be "acceptable," a risk that we cannot avoid. In contrast, climatic change originated a sea-level rise and is seen to be mostly man-made. In this case societies worldwide are asking for preventive measures that—even if they may not avoid local disasters—should at least minimize the impact.

The different risk exposure societies face every day make them nevertheless distinguish between risks that cannot be avoided, meaning risks they have to accept (tolerate) and those they are not willing to accept (intolerable). Consequently the question arises as to where to place the boundary between the two sectors, and moreover who is mandated to set this boundary. As elaborated in the previous chapters, there is no one boundary line that can be drawn here. Even in a small disaster-affected locality it may turn out to be difficult to define that boundary exactly. A flash flood that destroys the hutch of a Bangladesh rice peasant simultaneously provides fertile soils to his rice paddies, making one part of the disaster highly appreciated whereas on the other side it is not accepted. A scenario when transferred to an entire society or even to the world in general makes obvious how challenging it is to define such a boundary.

In practice the boundary constitutes a corridor rather an exact well-defined boundary line. In risk management this corridor has been named the ALARP corridor. ALARP is an acronym for "as low as reasonably practical" and thus points to the fact that the definition of the boundary line is fully dependent on the risk situation and will change accordingly from hazard to hazard. The term "reasonably" indicates that risk-reducing countermeasures should be governed by the idea of meaningfulness, giving the term a socioeconomic and cultural dimension. The term "practical" points to the issue that the envisaged mitigation measures should be technically achievable and follow scientific and technical standards and moreover that they be financially possible.

The situation of acceptable and unacceptable risks can best be explained by using the Risk Assessment Matrix already described in Chap. 7 (Fig. 8.7). The ALARP concept has been introduced into geology mainly by Fell and Hartford (1997) based on a series of structured risk assessment approaches. In his book, Morgan (1997) gave practical examples for the definition of acceptable and unacceptable risk in the field of landslide risk assessment in Canada.

The Swiss Ministry for Environment (BAFU) gave an example of how to define the boundaries between low and high risks (BUWAL 1996). In the manual BUWAL published a series of boundaries for the severity of a disaster impact as shown in the graph in Fig. 8.8. BUWAL points to the fact that posing the boundaries is an arbitrary and subjective decision that is mainly based on the risk experience of experts (Bayesian approach). Defining risk classes can be either given in a metric as well as in a logarithmic scale, depending on the variance of the data assessed.

The boundary line between acceptable and unacceptable is also called the line of acceptance. Moreover the line divides the sector of personal, individual risk responsibility (acceptable) and the risks that the society defines as unacceptable, where most people see the public sector to be responsible. But who defines that line? The definition of "acceptable" is based on the common understanding that the risk from a (major) hazard should not be significantly higher when compared with other risks to which a person is exposed in everyday life, a discussion that is mainly based on technical terms. Second, the debate comprises a social dimension. All groups of a society that are risk exposed, want to see their definition taken up by the society, thus defining the overall "level protection" for the entire society. There is only a legally authorized instruction (government) that has the mandate to settle the conflict and reconcile such different opinions. Third, who defines

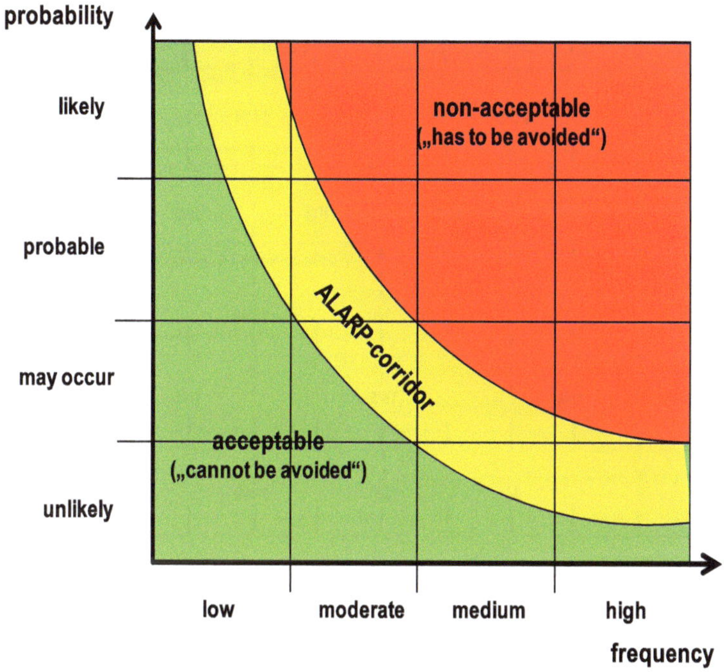

Fig. 8.7 Definition of acceptable and unacceptable risk: ALARP (Own graph)

what is "practical" and what is "reasonable"? Furthermore normative risk assessment should be based on the paradigm that once a society has accepted a certain level of risk as acceptable, than consequently all risks of lower severity have to be accepted also, although comparing risks emphasizes the difficulty in deciding what shall be compared: the disaster impact or the benefits from the use of a certain technology (WBGU 1999). Such a risk comparison moreover addresses certain emotional aspects of risk. Altogether the ALARP concept can help to reach a better understanding of risk. The debate finally resulted instead of drawing a line between accepted and unaccepted that it would be more practical to define the ALARP boundary by a corridor. But how wide the corridor has to be defined makes the assessment a scientific and political challenge. There is much agreement that it would be most worthwhile to narrow the corridor as far as possible by shifting the unacceptable sector down towards the sector rated unavoidable.

The ALARP corridor is the field where risk governance takes place. It is the sector where societies call on the natural scientists and engineers to provide sector-specific expertise. They acknowledge that scientists have the expertise to analyze and assess the state of risk and to offer implementable risk solutions. The provisions by the scientists are thus welcomed as a substantial and integral part of risk governance. But to gain such expertise, a society has to provide an

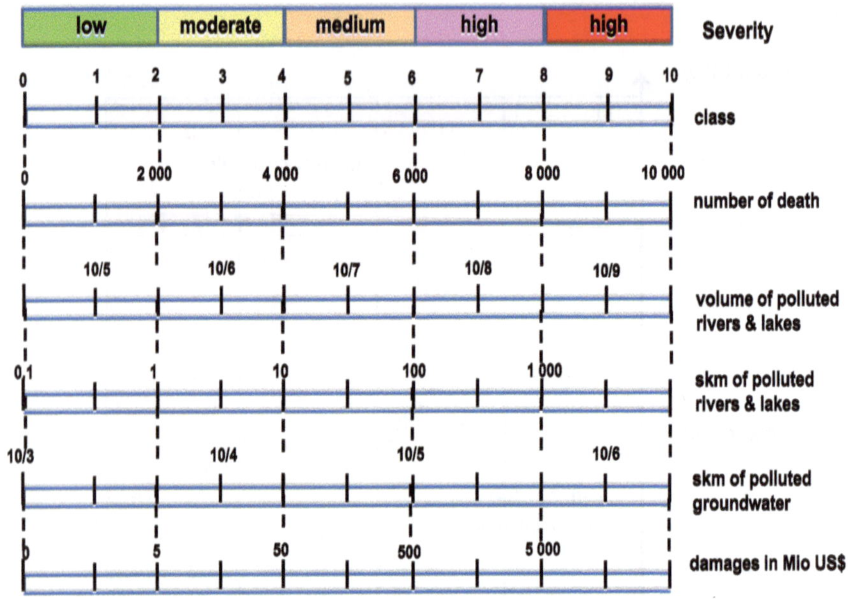

Fig. 8.8 Example of risk class definition (based on BUWAL 1996)

institutional framework conducive to scientific investigations by allocating the necessary financial funds and by providing manpower as well as the operational means. Without such an institutional framework scientists will not be able to work out reliable and robust risk assessments. The mitigation measures worked out by the scientists, engineers, and risk managers then have to be presented to the society and will lay the base of a societal dialogue. There is a common understanding that although such a dialogue should be based on scientific expertise, at the end the society has to make the development decision. The task of science and technology in risk governance is to provide technically feasible solutions but not to make the decisions. Such an assumption moreover reveals that without scientific input, sustainable disaster risk mitigation and prevention will not be successful.

Making use of risk definitions based on acceptable (tolerable) and unacceptable (intolerable) risk meanwhile has gained widespread acceptance. Today it is a commonly used instrument in disaster risk management and has proven its applicability in a series of different issues. One prominent example is the risk assessment on nuclear power plants of Switzerland (Fig. 8.9; v.Piechowski 2000) and Germany (BMU 2004) (Fig. 8.10). Also the German Commission for Environmental Change Assessment (WBGU) adopted the approach in its Year 2000 report; although it made some minor adjustments. Nevertheless these risk assessment concepts are also based on the fundamental understanding that a society has to decide what kind of disaster it is willing to accept. In this regard it is recommended to extend the term unacceptable by an additional explanation "what we shall avoid" and acceptable by the addition of "what cannot be avoided for the time being."

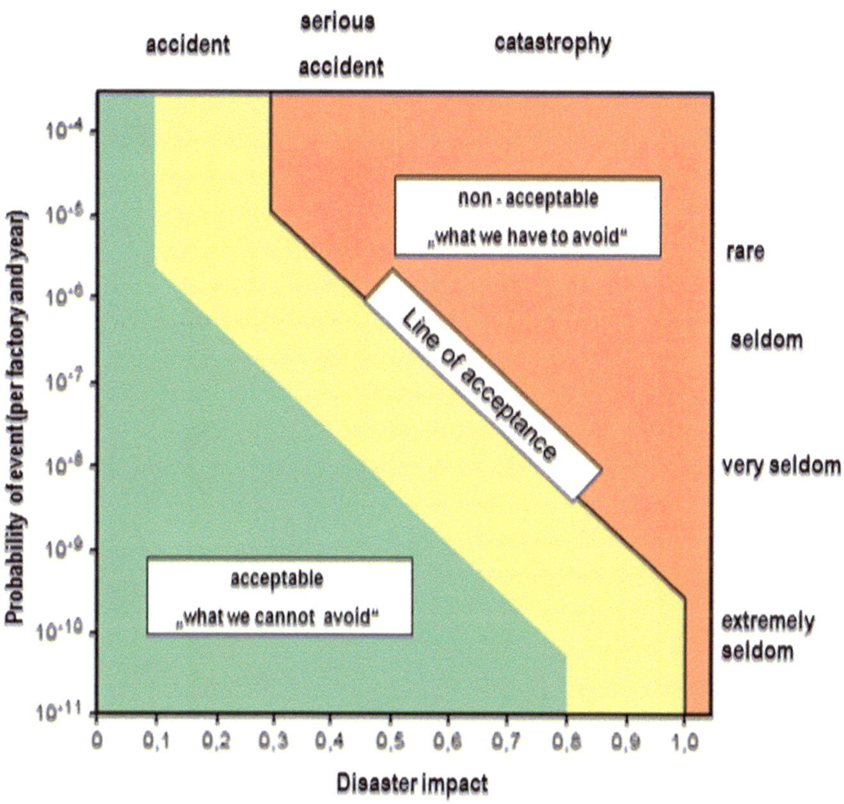

Fig. 8.9 Definition of risk acceptance on Swiss nuclear power generation plants (v. Piechowski, cited in WBGU, Report 1999; *Courtesy* Springer)

8.6 Risk Perception

Normally people tend to believe more in what they see and what they remember. Many research papers revealed this relationship is especially true in respect to individual risk perception from natural disasters and from climate change affected environments. Thus the availability of information that they have cognitively present and understanding the cause–effect relationship define what people recognize as a risk. Cognitive risk perception is the evidence-based perception of the likelihood and severity of a disaster whereas the emotional aspects of risk perception concern the impression of fear, helplessness, anger, distress, insecurity, hopelessness, sadness, and anxiety.

For long time it was felt (Combs and Slovic 1979) that the more salient the information is, the more dramatically it is displayed, and the more the risk at discussion is rated hazardous, the higher the individual sees it as a risk. This led to the assumption that the more often media—radio, television and movies—report

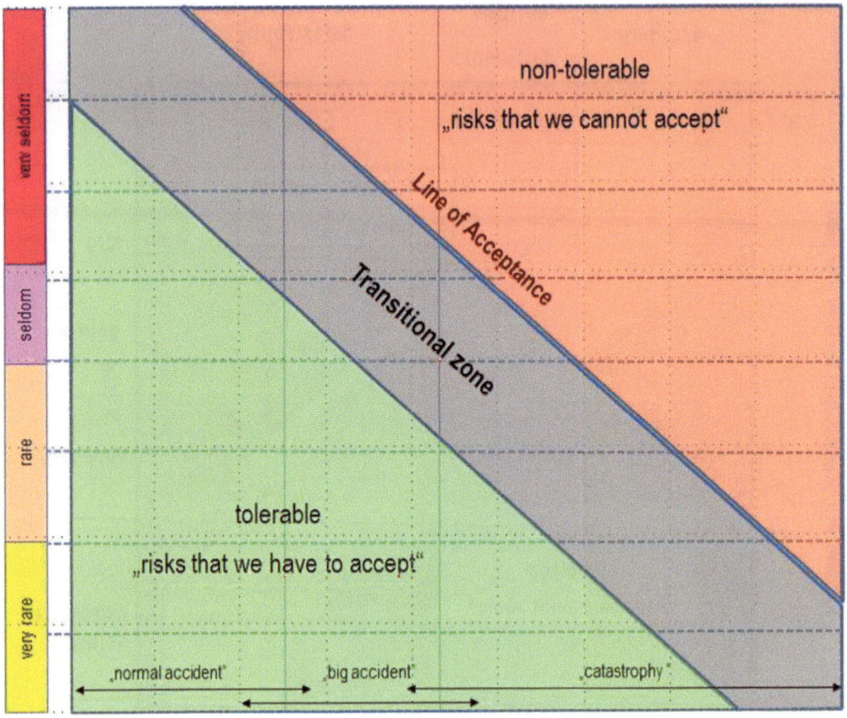

Fig. 8.10 Definition of tolerable and intolerable risk to the natural environment (*Courtesy* WBGU 1999)

on disasters the more likely media consumers reflect risk scenarios. But new studies point to the opposite direction. In countries where ample and comprehensive information on disasters, catastrophes, and risks is available, the less the people are influenced by this information. It turned out that even the nuclear accident of Chernobyl, the Indian Ocean tsunami 2004, or epidemics such as mad cow disease were not given higher relevance in the individual's perception when, for example, media coverage was intensified on occasion such as the tenth anniversary (Chernobyl). It rather turned out that individual risk perception is highly influenced (also) by the media when the information is new; it is felt to be trustworthy when it explains the generating mechanisms and transparently offers information on the anticipated effects (Sjöberg and Engelberg 2010).

Although risk is a "mental construct" (Slovic 1992), the natural and social sciences have developed specific concepts of risk. This construct also holds true for the part of the risk assessors as well as for the population at risk. Nevertheless, risk is closely connected with the impression people have of the risk they are exposed to, what they feel about or perceive: a situation that is called risk perception. Risk perception has much to do with how people behave in time of crisis. Such behavior is, according to social and psychological sciences, basically driven by personal experience, by the social, economic, and gender context, by expectations, fears, and hopes. Risk perception is thus a matter of emotions and not of facts (Rohrmann and Renn 2000), especially when it concerns technical or natural phenomena. According to Slovic (ibid) people's behavior facing a threat reacts along four cultural patterns: flight, fight, play dead, or experimenting trial and error. The individual perceives her specific risk deterministically: she recognizes the incidentally occurring event and attributes it according to a risk pattern based on experience, or the individual realizes her risk exposure as probabilistic, meaning unavoidable as "act of God" (Renn 1989). Many studies of individual and societal risk behavior revealed significant differences between the objective risk and the felt risk. The question arises as to the reasoning for such different perceptions. For the time being there are a multitude of descriptions and ideas presented to explain this. Kasperson et al. (1988) stated that even "After a decade of research on public experience of risk, no comprehensive theory (is available) that explains why an apparently minor risk sometimes produces massive public reactions." Some people feel threatened by a hazard for which another group of people does not see any risks at all. He questioned what makes one person perceptive to a risk and another not. As stated in Chap. 2, risks should be predominantly assessed in their social context rather than only from the natural science position. This book has already emphasized the position that natural/geoscience and civil engineering should play a substantial and integral part of such assessments. It is the combination of both, the sociological and science and technology sectors, that will lead to more reliable risk assessment methods than we have to date.

In this regard it is a well-known fact that individual risk perception lays the basis for how one copes with the risk he is exposed to and that such perception automatically defines the position on risk behavior. Based on a variety of sociological studies on "how people feel a risk" mainly carried out by Slovic, Kasperson, Renn, Beck, and many others, it can be stated that risk perception in general is:

- *Intuitively generated.* It is made up of psychometric risk indicators as estimating the risk by its severity, from the fears of death, or on personal integrity, of damage of property, loss of social and/or family bonds, loss of income, labor, and welfare, and cultural heritage. But risk perception is also a reaction of confidence or mistrust in the operational capacity of the disaster management structures.
- *Experience based.* This perception is based on knowledge of the risk characteristics, on the frequency of the event, an idea of the cause–effect relationship between onset and disaster impact, the disproportional memory of disaster events already made, as well as a self-assessment on how to be able to withstand the event. But it comprises a significant bias between "quantitative knowledge" versus the "quality of knowledge" as elaborated by Schütz (1971) and Malzahn and Plapp (2004).

According to the state of the art of risk perception analysis, two generally different approaches can be distinguished:

- The *psychometric approach* describes risks according to their perceived or anticipated characteristics.
- The *culture-oriented approach* describes risks according to how the individual is coping with the disaster.

In a study on the perception of frequency of lethal risk from accidents in the United States, Lichtenstein et al. (1978) came to the conclusion that in general even different kinds of people showed the same risk perception (Fig. 8.11). But this perception showed furthermore two different kinds of systematical biases:

- There is a clear tendency to (objectively) overestimate small frequencies of events and to underestimate large frequencies.
- To exaggerate the frequency of special causes.

The study could impressively demonstrate that external risk causes (natural disasters) are generally overestimated whereas personal disease-related risk causes (cancer, diabetes, etc.) are normally underestimated. So the risk from a tornado was estimated to claim a yearly death toll of about 8000–9000, whereas in fact it claims "only" 80. Similarly it holds true for the underestimation of death from heart disease: despite the anticipated toll of almost 100,000 the actual figure is more than a million.

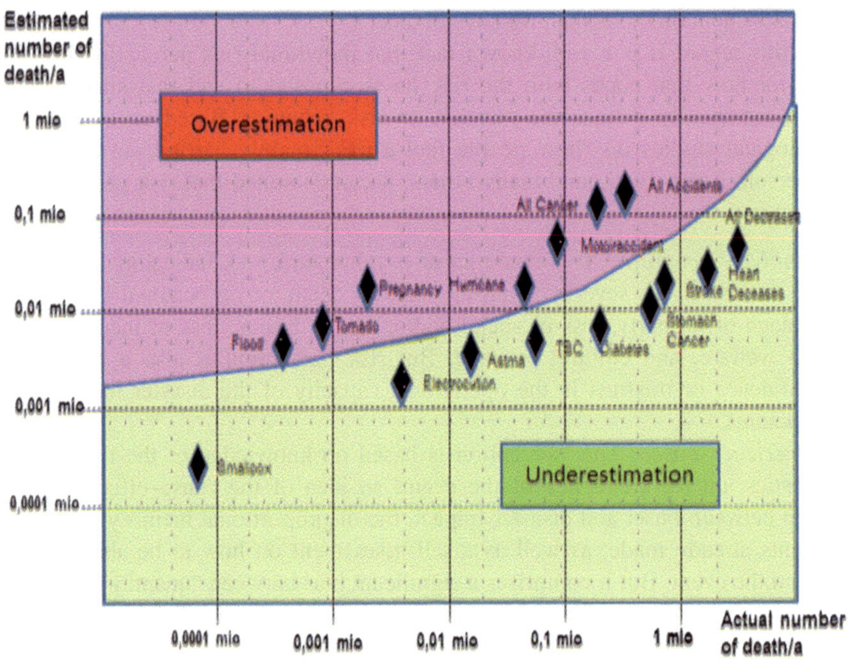

Fig. 8.11 Judged frequency of lethal events (Based on Lichtenstein et al. 1978)

In risk perception a great difference can be seen whether the population at risk is well informed on the risks with which they are confronted (Luhmann 2001). The distinction goes along with what he describes with terms "confidence" or "trust". People are confident when they believe in their experience and feel sure about their own expectations. They are confident in their "disaster coping capability" when they feel they will be able to withstand the impact. In contrast to that, "trust" is associated with risk, say from AIDS or BSE. Such risks bring the not-well-informed individual (Luhmann: "man on the street") to see a danger or a threat that is difficult to describe as the exposed has "not developed a well-founded opinion" on the outcome. The specific knowledge, the increased level of insight or experience, automatically enables the individual to consider the situation more technically. He develops an understanding to find the reasons for the risk causes and the probable effects that automatically make him increase his coping capacity. Luhmann further pointed out that trust will turn into confidence (downgrade) when the individual feels that he has no chance to alter the situation in his direction, yet on the other hand in the course of an increased coping capacity, confidence can also turn into trust. Plapp (2003) even goes a step further when summarizing that individual risk perception is in general deeply connected with self-assessed coping capacity, delineating the individual risk perception much more strongly than the experienced-based approach. She further pointed out that her analysis revealed that the social situation of the individual (income, social status, education) has only a minor influence on the psychometric risk perception. Only the age of the individuals turned out to alter risk perception significantly in a way that risks are anticipated to be more severe with increasing age.

A politically influential segment of society and therefore often referred to as a representative fraction of society queried on risk perception, are white males; a phenomenon that is called the "white male effect" (Olofsson and Rashid 2011). There are many studies on this subject among them by Rivers et al. (2010) and Slovic (1999) that revealed that in the United States white males generally have a different perception of any kind of risk. They tend systematically to perceive lower levels of risk (natural, technical, biological) when compared to other members of society (e.g., white women, nonwhite women, and nonwhite men). Their privileged demographic position in society makes them "a strong believer that at the end the things will turn out for the best" (Palmer 2003). The reason why such a group has a different risk perception is seen in sociopolitical factors such as power, status, alienation, and trust. White males are generally better educated, have a better career, better income, are living in well-developed quarters, are politically conservative, and show an extraordinary trust in technical and scientific expertise, thus making them strong supporters for administrative decisions. There is another striking notion: many policymakers fall into this group, and observations revealed a tendency that this phenomenon is the rationale behind this group becoming a preferential partner in risk perception assessment. These findings on the other hand revealed that women (white or nonwhite), ethnic minorities, and marginalized groups in the United States have a higher risk perception and that such groups are significantly less queried on their risk perception. Such a tendency, however, does not hold true, for example, for a state like Sweden, a state that is internationally known as strongly gender-equal. Here the findings revealed that Swedish men and women do not differ strongly in their risk perception. But also in Sweden ethnic minorities show significantly higher risk perception, a measure for sociologists of a still-prevailing inequality.

The results from Plapp's empirical assessment pledge fostering the vision that risk assessment inquiries should first make use of a qualitative risk assessment method as already proposed by Kaplan and Garrick (ibid), asking three central questions:

- What can happen?
- How likely is it that an event may occur?
- What are the consequences for the affected ones?

Such an approach is easy to perform but still a scientifically serious enough start for a risk assessment. This holds true especially in areas where the database is not broad enough to apply the many sociological and scientifically oriented risk assessments. It furthermore holds true for risk perception of a catastrophe that is threatening urban or rural areas. In rural areas risk perception is mainly characterized by a prevailing culture and practice-based knowledge from people's own experience or from the hearsay of the elderly. Because in urban areas the people are in general more technically experienced and define themselves often as "globally oriented," they immediately consider a risk situation from an operational and technical viewpoint. Risk behavior thus becomes impersonal and is seen more a matter of the authorities concerned than of the individual. In general an urban population has only very low risk experience resulting in a risk perception that is cognition based and much less intuitive than in rural areas leading to a lower risk perception. Social scientists have found that when the informed public is asked to order well-known hazards in terms of their potential number of deaths, they do so with reasonable success. However, when the same people are asked to rank those hazards in terms of perceived risks, they produce quite a different order (Slovic 1987).

The public normally is accustomed to rate the severity of a disaster event by the number of fatalities. Consequently that figure dominates the headlines of the news ("250 people killed in torrential flood"), making this event serious, dramatic, and important. But Fig. 2.3 clearly proves that in the course of the last century the death toll from natural disasters has significantly diminished, and at the same time the number of people exposed as well as the number of injured, made homeless, and socially affected has increased tenfold. Therefore there is a great need to change our risk perception from the mere number of fatalities to the overall number of victims, killed, missing, and affected. Especially those who are declared missing are a group of people very seldom treated as respectfully as necessary. In industrialized countries where many lives and much of the economic assets are insured, the life insurance payout for the loss, after the death of a person is clearly proven. Therefore many of those left behind have to fight sometimes for years to get the contracted subsidy. In developing countries the situation is even worse.

There, very often the husband is the sole legal representative of the family. When he is missing and as long he has not been declared dead, the widow has almost no legal protection to claim for the restitution of the house, plot, or the family heritage. Such a situation has taken place after the earthquake hit many villages along the North Anatolian Fault zone (USGS 1999) and especially in the aftermath of the 2004 tsunami in the Indonesian district capital Banda Aceh, where the local government documentation centers were also flooded and no legal documents were available anymore. As a symbol of acknowledgment of the rights of the surviving people, the local government initiated with the assistance of the German Max-Planck Institute for Anthropological Research (Arskal Salim 2010) a mediated reconciliation process that was able to solve most of the conflicts.

8.7 Risk Perception: "Gender Equality"

Women are considered the most vulnerable parts of society from natural disasters, together with children, aged persons, the ill and disabled, or other marginalized groups on the basis of ethnicity, culture, or class (Wisner et al. 2004). A key to understanding the particular vulnerability factors specific for women, is their social and cultural role in the family. Consequently their needs after a disaster differ from those of men. Their physical and psychological vulnerability arises from their inherent family instincts. Loss of shelter and family poses a tremendous pressure after any disaster. Despite their own loss, the women are traditionally expected to take the primary role of family care and to feed the children or other (most elderly) family members. Household duty and family responsibility make the women more vulnerable to physical, mental, and emotional stress that often leads to fears and trauma. After a major disaster, it is seen that women are more prone to depression and other emotional disturbances. A multitude of social, cultural, and economic consequences affect women either directly or indirectly. Women when they become widows often lose their social support and are often deprived of almost all social privileges. Often they become widows with young children. Having lost their spouses, women who have never worked outside their houses are suddenly confronted with administrative procedures, for instance, to obtain financial compensation for the death of their husbands, to provide food for their children, or in sending their children to school.

But women are not only most vulnerable from disasters; they also can contribute a lot to make a society more disaster resilient. Studies of the women's role in disaster risk management revealed that women contribute from the grass roots and household/family perspective. Their long-lasting traditional and cultural knowledge is recognized as valuable expertise that can improve the sustainability of risk management. The main contribution women can give to increase resilience mainly concerns their expertise in small-scale gardening and agriculture. Women are not "experts" on big catastrophes such as earthquakes or volcanic eruptions, rather on avoiding local soil degradation by wind erosion, flooding, and landslides and provide information for a positive diversification of livelihood on a variety of natural disasters that is widespread and responsible for serious impacts all over the world. As women are highly bonded to the villages and run effective social networks, those networks are seen as favorable starting points for local disaster reduction awareness-raising campaigns. Therefore natural disaster mitigation will not be successful if women are not made an integral part of the local disaster mitigation process. Such integration furthermore has the advantage that it fosters the changed role of women in a society in general. Furthermore as stated by Blakie et al. (ibid), gender inequality should not be addressed after a disaster, but rather should draw attention to the root causes for such type of vulnerability.

Although in industrialized nations the gender discussion predominantly refers to gender equality in the workplace (OECD 2005) and to bring the demographic structures into better balance, the gender discussion in the developing countries is concentrated on the issues of how to manage the survival of families. Natural disasters and even due to challenges following climate change, gender equality has been identified as a major approach to improve social resilience. This was the rationale to include gender issues in the strategies of natural disaster risk reduction and to consequently make them a substantial part of many concept papers. Thus the gender issue was incorporated in the Hyogo Framework (see Sect. 4.2) that has since been endorsed by more than 170 nations. The framework outlines the basic facts for a political strategy on gender-related disaster risk reduction and focused on the statement that the gender perspective has to become an indispensable part of all disaster risk management policies, from disaster assessment, the plans to increase resilience, as well as the follow-up processes of awareness raising, warning dissemination, and gender-neutral training. The framework points out that gender-equal disaster risk reduction is especially necessary at the household and community levels where women can contribute substantially in building a culture of disaster prevention. Gender equality in this respect refers to the equal rights, responsibilities, and opportunities of women, men, girls, and boys. Gender equality is achieved when the different behaviors, aspirations, and needs of women and men are equally valued and favored. The Hyogo Framework moreover stated that thus far the important capabilities of women are often neglected and therefore mostly left untapped. Women and men experience, perceive, and identify risks differently.

Although everyone can be exposed equally to a hazard, it is a well-known fact that women and girls experience different levels of vulnerability and that they are disproportionately affected by disasters. Worldwide more women die from disasters than men. Statistics from past disasters including the Indian Ocean tsunami and the 1991 Bangladesh cyclone have shown women are overrepresented in mortality rates. For example, in Indonesia and Sri Lanka, male survivors from the 2004 tsunami outnumber female survivors by three to one (UNISDR/UNDP/IUCN 2009). Almost one billion people in the world are illiterate, of whom two-thirds are women. Three in five of the more than 100 million children that do not go to school are girls. After a disaster or other stressful impact, many girls are forced to drop out of school to help with chores in the house, or to save money. Women are at greater risk of injury and death due to their societal restrictions and gender roles. The current gender relation between men and women in risk exposure originates from the roles and responsibilities women and men have at home and in society. They result in different identities, social responsibilities, attitudes, and expectations leading to an unequal socioeconomic development between men and women, including differences in vulnerabilities to disasters. For instance, swimming and climbing are not skills girls and women are encouraged to learn in some cultures. In some regions women's clothing limits their mobility. In some societies and cultures, women cannot respond to warnings by leaving the house without a male companion. Loss of crops and livestock due to a drought has a direct detriment to family food security and moreover such droughts make arable land barren with the result that many of the young males migrate to the cities to find jobs, leaving women alone in charge of the households. Women are given the responsibility of caring for children and the elderly, binding them to the house. No social acceptance of women as persons that are allowed to enter a contract makes them dependent on male decisions. Women's income of the lower social levels is more likely to be derived from the informal sector, which is often the worst hit by disasters and is least able to recover from the effects of disasters. And in the case of an emergency they are—when the men are not at home as on the occasion of the 2004 tsunami in Indonesia that happened on a Sunday morning when most of the men were gone to the markets—they have no chance to react properly.

Lower levels of education hamper women's access to information, and limit their ability to prepare and respond to disasters. In many cases women do not receive hazard warnings at all, as authorities disseminated warning information normally exclusively via the official channels (radio, newspaper, press conferences, meetings). Such information broadcasts exclude women from getting proper information thus making them again more vulnerable, and women have a generally low level of participation in decision-making bodies. Women's capacities are not applied, their needs and concerns are not voiced, and they are overlooked in policies and programs. Women are poorly represented in decision-making bodies as sociocultural norms and attitudes bar them from participation in decision making. Although women's vulnerability to disasters is often highlighted, the role of women in building a culture of resilience and their obvious contributions has not been adequately recognized. Benefiting from women's experiences and contributions to a gender-neutral disaster risk reduction starts with an incorporation of women's experiences at all levels of vulnerability analysis and by working out standard operation procedures that ensure:

- A comprehensive participation of men and women
- A gender-based identification of the inequalities between men and women
- A gender-differentiated vulnerability analysis (physical, social, economic, cultural, political, and environmental)
- Identification of the root causes for the many key determinants of vulnerability from historical analysis, disaggregated by sex
- Identification of women's needs and concerns

- Recognition of importance of the capacities and authority of women to conduct risk reduction
- Inclusion of women in training and awareness raising
- An active engagement of women's organizations in disaster reducing capacity building.

8.8 Risk Communication

8.8.1 Communication Processes

After a disaster strike, society is interested to know what happened and why and whether such a disaster may happen again. Moreover the news consumer is interested how it might put him personally at risk. Those affected in disaster events are the victims and the authorities on the other hand, with the media positioned in between. The media business is to sell news and in our daily life, either by disseminating information on what has happened, by presenting the human tragedy and a "moral assessment." The media help to outline the problem and describe the cause–effect relationship that led to the specific disaster, but normally do not see their responsibility in elaborating on problem-solving recommendations. The position of the media between the disaster and the external observers defines the media's position as favorable not only to influence the public agenda but also combine different bits of information into a broader context that can magnify, dramatize, or simplify the catastrophe, just as felt necessary (Vasterman et al. 2005).

To achieve this, the media operate in different modes. They often define their responsibility as a part of the socioeconomic and cultural system by taking up the task of information dissemination of official information to the public. But they can also follow quite another track, by strengthening people's perception of anxiety and fears, thus amplifying the perception of risk. Through these two modes, the media can have a huge impact on society. Whether the media act as agenda setters depends on their business policy. Nevertheless for the purpose of emergency management, the positive function of the media as an independent broadcasting instrument should be acknowledged, as it helps to disseminate official information to the public. The fact that moreover the media can also dramatize the news, leading to an amplification of risk perception cannot be excluded from the discussion. Emergency managers should be aware of the dichotomy the media can play. Risk management therefore has to develop a strategy that defines the role of the media to make it an ally not an enemy. It seems that when media and authorities collaborate to provide the public with reliable and neutrally formulated information, beneficial effects can be achieved and the well-being of the disaster community can be enhanced.

Risk communication is the strategy to disseminate information to people who require information. But communication should not be restricted to the channels by which the information is disseminated. Moreover the information disseminator is interested to know whether the news reached the consumer and whether it is understood in the way it was intended (Luhmann 1988). Therefore communication has a twofold direction: from the sender to the recipient and vice versa. According to the general rule of communication, everybody is communicating every day (Watzlawick 2009), even if someone intends not to communicate, she is still communicating. Watzlawick formulated this rule as: "There is never no communication."

Although there have been many advances in risk assessment in the last 20 years, until today the impact of a disaster has mostly been given in technical terms. Still today the number of death casualties is that one piece of information that has the widest spread in the media. Accompanying this information with impressive pictures and live interviews, based on fact, is what Kahneman et al. (1982) describe as the dissemination of "the paradoxical beauty of destruction" fascinating all those who are unaffected. The best examples were the pictures of rubble and debris of the dust-hidden latticework girders that gave the destroyed World Trade Center a worldwide symbolic image. Although 2880 people lost their lives in the attack, there were disasters in Pakistan and India with a magnitude higher death toll the very same year (Swiss Re 2002), which did not reveal a similar or even higher degree of attention. Or the tsunami of 2004 with its many pictures showing the flood-inundated shores and villages, resulting in a worldwide unprecedented wave of financial and technical help "showering Indonesia with US$7 billion aid money" (Indonesian Minister for Reconstruction and Rehabilitation, BRR; speech held 2010 at GTZ, Berlin).

Communication requires a specific regulatory framework with rules and laws that define who is mandated to assess and disseminate information to the public. An example of such a regulatory framework is the EU-EWFD (2000), which asks EU national authorities to assess flood risks regularly and to disseminate the information to the public. Moreover in order to be effective, communication should not be restricted to response to a crisis but to establish communication as a process even in times of no disaster. The mistrust of much of society is derived from the fact that the authorities only address the public in the case of an emergency, a situation that is often felt as disregarding the population's concerns. This gap has been filled by many "environmental groups" that specifically address the needs of certain societal groups. In order to strengthen the efficiency of the authority's fiduciary responsibility risk-related communication should be institutionalized as a continuous process.

The process of understanding information triggers on the side of the recipient an "information processing" that results in an individual reaction based on his individual socialization. The information can be understood as intended, but misunderstanding has to be taken into consideration every time. The main aspect in this regard is "acceptance." Whether a recipient accepts the information is generally based on personal experience and on trust in the honesty of the information disseminator. As a society is made up of a multitude of individuals it has a multitude of cultural socializations and risk experiences. Consequently the level of "acceptance" is multifold. But as political decision making depends on acceptance of the society in total, the question arises how far an individual reaction can be accepted by the "news sender" in order to serve the needs of the society or at least a fraction of the society. The question is described as the rule of "individual versus collective acceptance" (Ruhmann and Kohring 1996) and describes the different responsibilities and different demands of sender and recipient. Information broadcast by written or verbal communication is able to create interest, can exaggerate emotion, or revitalize emotions. In this regard the "circulation intensity" (Süselbeck 2013) of certain verbal expressions or the recurrent presentation of, for instance, emotional impressive pictures on certain events, makes the acceptance of such news as "reality" much easier. There is a multitude of studies that critically review the role of media in the aftermath of a disaster (Vasterman et al. ibid). In most studies, the media are portrayed writing sensation-seeking articles of enlarged anecdotal stories on topics such as who is to blame for the incident, who hinders the work of rescue workers, and so on. By repeating the same images (e.g., the airplanes hitting the World Trade Center's Twin Towers) over and over again such articles often create a "news syndrome" with the media consumer. The news receiver cannot in every case distinguish between news from a distant disaster event as a cognitive process making it quite understandable that the media are able to manipulate the news receiver.

The long-prevailing hierarchical communication policy, where the state has the overall responsibility to disseminate information has more and more been abolished by the idea of participation, giving both sides equal responsibilities in information generation and dissemination. Consequently the exchange of information is in addition to a cognitive, also a procedural, process that has to tailor a message to the concern of the recipient and the bouncing back of the acceptance or nonacceptance of the sender. As the exposure to stress during a crisis is much higher than normal, information has to be delivered quickly and with a high degree of accuracy. But very often just such reliable information is not available. This situation bears the risk of creating an information vacuum. The reaction will be rumors, stress, tension, and finally mistrust, very often exaggerated by the media. In the course of a crisis situation the way information content may develop in an adverse direction has been distilled by SAMSHA (2002) and outlined in Fig. 8.12. In order to avoid negative exaggeration of the disaster information content, risk communication has to be simply formulated and target-oriented messages, from which the news consumers are getting the impression that they are provided information that will enable them to better address the situation.

The communication channels used are either direct from sender to the recipient (e.g., a communication by the local authorities on flood evacuation) or via a transmitter, the media. The media works on one hand as a driving force for individual or collective perceptions and to approve concomitant opinions but also reflects and sometimes even exaggerates existing or inherited cultural and social discussions.

The media—print, voice, or visual—are an organized means of reaching a large number of people quickly, effectively, and efficiently. This includes newspapers, magazines, radio, television, and video, as well as movies. During recent times electronic media including Twitter, Facebook, and other Internet-based

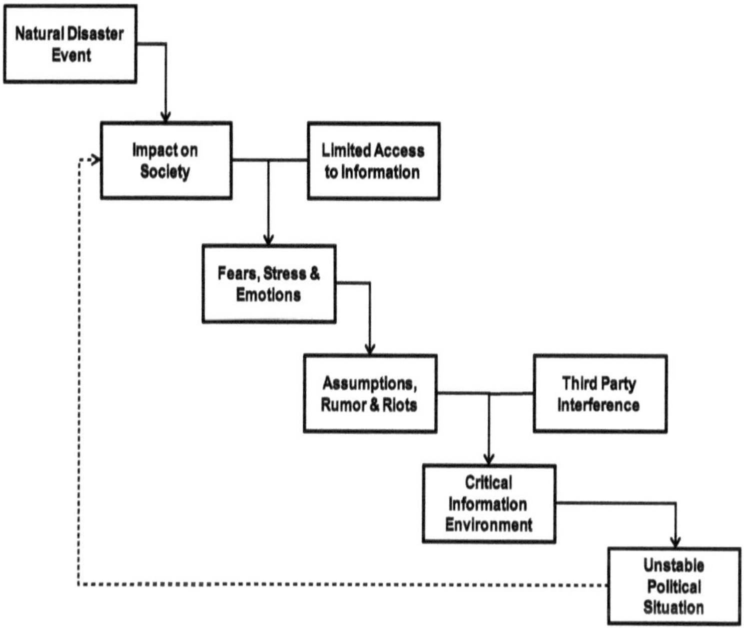

Fig. 8.12 Cascade of amplification of risk perception of socioeconomic marginalized populations due to limited information (based on SAMSHA 2002)

media networks have emerged as a major component. The informative and analytical role of the media has become a key component of information dissemination worldwide. The pressure of the media in a democratic setup is tremendous and it should be used in a responsible and constructive manner for the benefit of society. Media consumption is mainly based on cognitive effects such as education, experience, and belief as well as on affective effects, such as emotional responses. The "importance of these cognitive and affective effects rests on the degree to which they influence behavioral effects that define the social situation" of the individual (Beaudoin 2007).

Most impressive in media coverage are pictures and TV live reports on dog-rescue teams searching for buried victims, setting up tents, and installing water and electricity. Such kinds of information give the media consumer an impression of being part of the aid. Moreover very often, the high interest of the public in disaster information is often taken up by politicians. When at the location of the disaster politicians can demonstrate decision-making power, such as the day (shortly before a federal election took place) when the German chancellor visited flood-devastated villages in the Elbe/Oder River or when American presidents comforted earthquake victims in Turkey or consoled the Hurricane Sandy-affected New York area. But public interest in disaster reporting is short on time and generally concentrated on very symbolic events. And the momentum of public interest is mostly restricted to the rescue phase. After this, interest unfortunately diminishes very rapidly. An analysis by the University of Munich in 2001 revealed that more than 50 % of the people do not recall a disaster that happened 10 years back. The recall ratio of a disaster dated back 40 years is less than 10 %. Nevertheless in Germany there is still a large memory of the Elbe flood in Hamburg of 1962, as this was the first great disagreement after the Second World War and is still today a nationwide example of how a local politician brought in military services for rescue, although the operation was not backed by laws and regulations.

In time of disasters and catastrophes the mass media play an important role in reporting on the three primary characteristics of a disaster: short response time, surprise, and threat. The media become an indispensable part of disaster communication as impressively proven on the occasion of the Indian Ocean tsunami (2004) and the Haiti earthquake (2010). Media not only report on technical and scientific facts of a disaster, but also take over a responsibility to examine it in humanitarian, social, and economic terms. The print media have a major role to play in the pre-disaster prevention, mitigation, and preparedness activities, and the electronic media due to their direct and fast technical potential preferably report on the events themselves. In both cases the media need to be proactive in nature rather than reactive. In this regard the mass media, mainly the print media, television, and radio no longer take up the work of simple news distribution but meanwhile developed into a most potent means of educating society on disaster prevention, mitigation, and rehabilitation.

But what drives the media to take up one disaster while another is almost neglected? A study carried out by CARMA International (Franks 2006) analyzed the media coverage of Western countries of the Indian Ocean tsunami (2004), Hurricane Katrina in the United States (2005), earthquakes Bam in Iran (2003) and Kashmir (2005), Hurricane Stanley in Mexico (2005), and the racial conflict in Darfur, Africa (since 2003). The questions the study wanted to answer were, "What factors drive western media interest" and, "Whether these disasters are perceived equally and if not, why not?" The general outcome of the study was that "western self-interest is the main precondition for a significant coverage of a humanitarian crisis" (Franks ibid). Furthermore the "perceived economic impact of a disaster on western markets is clearly correlate[d] with the quantity of media coverage." It turned out that where the (western) economic interest is highest, there was the broadest coverage. For example, the expected economic losses from Hurricane Katrina were covered in 17 % of all analyzed media reports, whereas the impact from Hurricane Stanley or the earthquakes in Bam or Kashmir were perceived to be only small, resulting in a media coverage of only 1 %. Moreover there seems "to be no link between the humanitarian scale of a disaster and the media coverage." Of all the disasters, Katrina suffered the least death toll and had one of the lower population displacement rates, but generated more than 1000 articles, the highest figure in global media than any other disaster studied. On the other hand, the earthquake in Kashmir did not attract similar media interest as the quake of Bam although it suffered a 3.5 times higher death toll (90,000). The tsunami of 2004 resulted in nearly double the coverage of the Darfur crisis, although it had the same number of casualties (circa 180,000). Around 40 % of all reports that covered the tsunami in the western media were mainly directed to the 900 deaths and few displaced ones from Western countries. Although a direct association with the global economy was only low for the tsunami, the coverage nonetheless was dictated next to economic concern also from the strategic importance the region as a whole has to Western tourists.

Although the public wants to be informed quickly and sensationally, there is nevertheless a great demand for a serious, evidence-based information policy. The public generally do not accept it if their information demand is not settled properly and immediately question the authority's seriousness and sense of responsibility, a matter that once former US President George W. Bush experienced when visiting the Katrina-devastated city of New Orleans only after massive invitations from the local public and publications.

The media can be extremely effective in the following areas:

- *Educational*: Before a disaster, educating people about the hazard, prevention, and self-help. During the disaster, the media can be extremely helpful in providing accurate and unbiased information and post-disaster on individual impacts and needs.
- *Partnership*: Guiding people in preparing individual disaster prevention and mitigation, but also calling on the authorities to fulfill the resilience requirements of society.
- *Critical*: By critically evaluating the emergency planning and resilience targets that are envisaged by reviewing existing plans.
- *Suggestive*: The media can call for the strengthening of mitigation efforts. Thus the media have taken over the role as solicitors for society's needs and as an appeal to the people to come forward to render help. Generally the media have a responsible image in the public eye and this image can be utilized to claim resources to help disaster victims.
- *Watch-dog*: By playing the role of "giving the floor to the affected ones," especially after the disaster and when rehabilitation work is going on and for keeping the disaster machinery active.
- *Assistance*: Helping or even sometimes taking over or replacing the established lines of communication.

With such a multifold orientation the media not only transports fact-oriented information but ventures into the position to interfere in social matters by critically reviewing, for example, the disaster mitigation concept of the local risk managers (Rusch 2011). If necessary the media question specific rescue operations or the general concept of crisis management, or the concept of the many help agencies that automatically gather in the case of an emergency.

To successfully reach the populations at risk by radio, television, or the Internet requires that the information needs:

- To be reliable, trustworthy, and fact-based
- To be independent, impartial, and not driven by particular interest (open/hidden agenda)

- To be able to make the information channels transparent
- To have a fast, appropriate, and functional organizational setup
- To be mandated under a legal framework (rules, regulations, laws)
- To be financially and politically independent

According to Stallen (1991) risk communication either on man-made (technical/ industrial) risks or on natural disasters has different objectives to fulfill:

- *Practical*: Populations at risk require comprehensive and reliable information on their actual or potential risk exposure.
- *Moral*: A proof for a well-functioning society is when it is able to provide basic information to all social groups on risks to which they are exposed.
- *Psychological*: Information should be rendered in a way that enables the individual to assess the personal risk situation adequately, so that she is able to address the risk appropriately according to her attitude, belief, and capability.
- *Instrumental*: A society has to provide the basic functions by establishing effective means of risk assessment and risk communication (laws, rules, regulations) in a fact-oriented, freely accessible, independent, and effective manner.

The following seven (cardinal) rules for building trust and credibility have been proposed by Covello and Allen (1988) as a basic means to outline a risk communication strategy:

- Accept and involve the public as a partner. Understand the task as serving the needs of the public to be informed.
- Appreciate the public's specific concerns. Statistics and probabilities don't necessarily answer all questions. Be sensitive to people's fears and worries on a human level and respect their humanity.
- Be patient, honest, and open. Once trust and credibility are lost they are difficult to regain. Try to avoid misinformation, fears, and concern.
- Meet the needs of the media. Never refuse to work with the media. The media's role is to inform the public. Work with the media to ensure that the information they are providing is accurate and enlightening. Thus the media are not enemies.
- Cooperate with other credible sources and coordinate your information and communications with those of other organizations. Conflicts and disagreements among organizations create confusion and breed distrust.
- Arrange your messages according to a hierarchy of importance and deliver brief (20–30 s) statements that explain each one.
- Do not express personal opinions and never speculate.

In mass communication either on natural disasters, or on politics or sports events, the media are at first oriented towards the media consumer. In order to sell the news, their intention is to catch the interest of the media consumer. Thus eye-catching events that can be displayed through dramatic pictures, or that cover the fate and destiny of disaster-stricken families are much better to transmit than purely formulated descriptions of a catastrophe. Although media are by their understanding oriented to deliver information, they nevertheless see their position as an external and independent observer and when necessary as a critical analyst of the situation. In any case, the media wants first to settle the information demand of the public. Therefore they do not define themselves as having the official task of an information-disseminating agency. The fast nature of disaster information moreover results in a quick response attitude that requires the media not be too deeply involved in the event.

Therefore a catastrophic event brings the media into the bias situation, to give as much as possible of information about the facts, and on the other hand to have news that “sells”. Moreover the media have to understand the great expectations of the people during a disaster. People who are affected by a disaster expect widespread support from the media, by sound and reliable information that is not superimposed by the news agency’s political agenda. They expect the media to play a constructive role as being partners instead of critics by rendering proven, correct, and reliable assessments and not to exaggerate the news by subjective interpretation.

Another problem arises for the media that in general information interest on natural catastrophes is very short in time. A study was carried out in 2001 by the Forestry Institute of the Munich University, on how long a disaster event is preserved in the memory of the people. It turned out that only 10 years later only 50 % of the people can remember the event. A disaster of 1971 is only memorized by less than 20 %. This shows that disaster remembrance is extremely short and can even vanish within a couple of weeks. On the other hand Baudoin (ibid) reported on public behavior after the attack on the World Trade Center (9/11). The study revealed that the use of the CNN Website after the attack doubled every seven minutes between 9 AM and 10 AM and that within three hours more than 90 % of the American public was aware of the attack. More than 90 % of Americans got their information from television and radio that day. The “media dependency theory postulates that people’s dependency on the media and the effects of such dependency escalate during times of societal change and conflict” (Baudoin ibid). Thus far little is known about the effects of such news and media campaigns during the subsequent phases of catastrophe: restoration and recovery. But a media campaign with the goal of stimulating public safety following Hurricane Katrina raised safety concerns that resulted in an increased public and private flood prevention. But there is still quite a concern among emergency managers that thus far there is little evidence that such media campaigns can really change the “safety beliefs.” These “findings are counter to the idea that attitudes (beliefs) are easier to influence than behavior” (Beaudoin ibid).

A problem for the media to assess is that disasters are in many cases complex in nature. The media are thus reluctant to go into too much detail: what the triggering element of the disaster was; the geological characteristics that were disaster prone; how the people were affected; and what kind of prevention was undertaken, but failed, and what has to be done to prepare the people better for the next event. Another problem for media dissemination is the different interest of local and regional/super-regional media. Local media are much closer to the event and want to cover the information demand of the public that is more or less affected by the event. They show the victims, speak to the neighbors, the local officials, and visitors in a way that the readers/listeners become part of the situation, whereas regional media want to cover the general aspects of the event. Local media cover a limited area but the regional can distribute the news over a large area. The different news media—print, radio, TV—preferably cover those events that are dramatic, that display damages, loss of life, and uncover the victim's harm in the personal living sphere. It is thus the individual gets the feeling of direct engagement, in fact even with comprehensive coverage the news consumer will still remain a visitor, as he is not affected himself. On the other hand "concern" is the major emotional source of the many donations. Thus in the aftermath of the Indian Ocean tsunami about US$8 billion in aid money had been collected, for an estimated economic loss of about US$6 billion in Indonesia alone. The high amount of aid donations experienced in the course of the tsunami, made the German Alliance Development Works request the donors not to identify their donations "Indian Ocean tsunami," so as not to identify a specific aid operation for which the money should be spent. The consequence of the huge donations was that the group realized there was no money left for all the other emergency situations on Earth.

Next to post-disaster rescue and relief operations to provide water, power, food, shelter, and medicine the disaster victims also ask the authorities to protect them after the disasters and in general to restore their lives. Studies in the United States revealed that although the citizens often are well aware of the risk from natural disasters they themselves are nevertheless reluctant to take steps on their own in order to increase their personal risk exposure. It is assumed that only up to 30 % of Americans have an emergency preparedness kit at hand that suits the recommendations of FEMA. Many studies on risk perception wanted to enlighten the bias between risk understanding (attitude) and risk behavior. It turned out that the majority of people see their main interest in disaster preparedness primarily in not being caught by a disaster unprepared. The people in assessing their personal risk distinguish between the severities of a potential hazard (saliency), whether this hazard will really occur (immediacy), and how eventual this will happen (certainty). These factors are juxtaposed to the assessment of whether the individual feels susceptible to the potential outcomes and how well she estimates her disaster remedies (self-efficacy). These factors define what is called the vested interest in disaster preparedness (Miller et al. 2013) as well as the attitude towards risk perception. Thus risk preparedness concepts have to take into account that the risk-exposed individual will first see his personal interests covered and is thus not necessarily inclined to adopt the official technocrat-oriented resilience efforts immediately. As long as official risk prevention action is not addressing the individual vested interest most of the official risk mitigation efforts will be not successful.

In general the risk-concerning messages are based on official communication bulletins and press releases given by local administrations. But by such information paths only the official side of the catastrophe is covered and the side of the victims often is not addressed. Good risk communication thus comprises information gathered from experts and integrates the victims' experiences. Although the media claim to help bridge the gap between officials and the victims and that they help to increase the self-help potential of the population in the crisis, there still exists a great distance between those affected and the media consumer.

8.8.2 *Risk Communication as a Responsibility of Science and Technology*

Risk messages even when delivered by official sources may be given in a way that can lead consumers to misunderstanding and confusion and very often leads to distrust. Risk communication therefore especially when delivered by emergency management authorities must be understandable to everybody, accountable, trustworthy, and transparent and should cover the (frightening) risk perceptions of the exposed population. Still very often expert information does not reach the people as it uses a language and describes models the recipients don't understand. This holds true especially for information delivered from science and technology leading to a significant communication gap between scientists and society. There is at least one reason that causes this adverse public attitude: it is mainly due to pure misunderstanding of science and technological information content. This results from the normal inability of scientists to communicate ideas in plain language. The public often sees science and technology as something apart from their daily living experience and does not feel that such information can be helpful for them. The limited effectiveness of risk communication is mostly attributed to the lack of trust on the side of the news recipients (Slovic 2000). He pledges therefore that scientists should use a language that is commonly understood, for example, not to rate a probability of occurrence in a number like "0.000001 %" or a "100-year flood." For him, risk communication must be:

- Understandable to everybody
- Accountable, reliable, and fact-based
- Transparent and reproducible
- Should cover the information needs of the population

An effective warning to the public at risk of an imminent hazard should consist of four items (Wolgalter et al. 1987):

- A signal word ("danger"; "caution", "warning")
- A description of the hazard ("location", "severity", "frequency")
- A statement of the consequences that may happen ("death", "injury")
- An instruction on how to avoid the hazard ("contact local emergency help desk")

Good scientific communication via the mass media is especially important in those areas directly and strongly affecting people's lives, for example, before, during, and after natural disasters. But this also holds true for communicating on risks that have a very long onset, such as the global climate change or the depletion of natural resources. In communicating their ideas, scientists should make clear the limitations of their predictions and carefully explain the basis for their scientific conclusions or opinions. Scientists moreover should be aware that a publicly disseminated finding may contradict the public expectation, may raise fears and misunderstanding, and may place the scientist in juxtaposition to the news receiver. For coping with such a situation scientists should be prepared. They can address this problem by reiterating the point that their part in risk assessment is just to analyze and interpret the data and that drawing any further assumption lies beyond their responsibility. Nevertheless it is a clear fact that scientists as being part of society will not be able to "escape from the scene" and should keep in mind the adverse impacts from risk-prone situations.

As the UNISDR (2011) points out:

> One of the most important drivers of accountability is access to information, particularly to information on disaster risks. However, access to information is only effective when governments actively support the right to information and when citizens are aware of their legal right and are willing to assert it. A culture of social accountability directly improves the effectiveness of governance and service delivery.

UNISDR emphasizes this statement as risk communication has become one of the central tasks of government institutions in emergency situations. Consequently it is been laid down in numerous national and international laws and regulations on disaster risk management. The regulations emphasize that risk communication strategies should not be restricted to mere information dissemination but to include public participation. The prerequisite for this is an understanding of how the public perceives risk, how the media translate information, and how representatives of the public and private sector can integrate the information into the local rehabilitation, mitigation, and prevention measures. Effective risk communication must be based on state-of-the-art understanding of the scientific and technological cause–effect relationships of risk, a multihazard assessment comprising long-term data sequences as well as effective communication processes. "In democratic societies, decision-making processes have increasingly involved the public as legitimate partners" (Sinisi 2004). Successful risk communication is costly and needs to be integrated into overall risk management to be able to communicate policy decisions successfully (Covello and Allen ibid). Risk communication in the context of implementing policy options requires communication goals to be set before the communication strategy is put into practice.

The following basic considerations should define the risk communication strategy. It should:

- Be oriented towards an informed public, not to disseminate public concerns.
- Listen to the public's concerns. People care more about trust, credibility, competence, fairness, and empathy than about statistics and details.
- Accept and involve the public as an equal partner.
- Be honest, reliable, transparent, independent, and based on credible sources.
- Be aware that trust and credibility are difficult to obtain; once lost, they are almost impossible to regain.

- Be given in a commonly understandable, nontechnical language.
- Be given with compassion acknowledging the tragedy of illness, injury, or death.
- Accept that people even when understanding the risk information, may still not have planned carefully and later evaluate on the outcome of the communication efforts.
- Accept that different goals, stakeholders, and media channels require different actions.

Contributions from science in risk communication are highly welcomed to increase risk perception in the population at risk, although it became obvious that the people do not solely rely on scientific expertise. Studies on risk communication revealed that scientific advocacy is accepted as one pillar of the knowledge transfer, and the experience of laymen is often given a similar importance. Expert knowledge is highly accepted as far as it analyzes the root causes, the uncertainty, and consequences of natural disasters, whereas the laymen's vision is mostly related to the social impact. To increase efficiency in disaster risk reduction both types of expertise should be integrated into the decision-making process under the auspices of an institutionalized and legalized body. Such a body is then acting as a clearing house between natural sciences (risk assessment), the population at risk, and the government that is responsible for the well-being of society. Dombrowski (1992) describes this concept of a "society-oriented disaster management." This means it calls on accepting the people as an equal partner in risk reduction, thus increasing the legitimacy of the decision making. By a comprehensive and early integration of all three parties, the social acceptance of risk reduction measures, of which sometimes only a few groups are benefiting while society in total often has to shoulder the financial burden, will increase significantly.

8.8.3 Communication from the Administration

Information on risks is often released just to "reduce conflicts and smooth risk management." The German authors Ruhmann and Kohring (ibid) stressed that "the main objective of risk communication is not to disseminate information but to establish a reliable relationship between sender and receiver. Thus risk communication is a vital approach to build-up acceptance and accountability." Risk communication should therefore be established as a dialogue process that early and comprehensively incorporates those affected in the political decision-making process. The authors furthermore pointed out that it is not enough just to disseminate a qualified expertise and to take this as a sign for a successful communication rather than understand risk communication in the first order is an instrument to build up reliability and trust. But how to get the news accepted by the affected population? It turns out that acceptance is better, the better the political system is rooted in the society. Acceptance means that the information recipient is not only given the role of a mere news receiver, who is allowed to agree or disagree with the administrative directives. Normally many authorities at the national, provincial, as well as communal level still take the stand that their insight and directives are indisputable and should be followed uncritically for the sake of the people. Such a standpoint does not give the population at risk the status of an equal partner. The way the people take up the contents of the risk messages can differ very much. It turns out that the more the people are familiar with the kind of risk they are exposed to and the more they already experienced successful countermeasures and trust in the official risk mitigation capacity, the more they are inclined to follow the directives. Ruhmann and Kohring additionally stated that this lay the basis for a new definition of the term "acceptance," in a way that official risk communication is only to be achieved by engaging the population early, comprehensively, and at equal eye-level in the communication process. The basis for such a definition is that risk communication is open to social aspects. If a catastrophe is an event that endangers society then acceptance has to take this notion into account, making it not just a following of directives but attributing a political dimension to it.

"Without participation there will be no acceptance" (Dombrowski ibid). Communicating on risks is to establish a connection between authorities and the population at risk, by making risk communication an interactive process of information exchange among individuals, social groups, and institutions. As risk communication is a two-way-message instrument transporting messages from the experts to nonexperts and the other way round, the messages are difficult to formulate. Risk managers tend to formulate their messages purely in technical terms, making the messages autocratic and very often in a form of a directive. On the other hand, the information that the risk managers receive from the affected population is normally not very clear and in most cases very emotionally and not technically or operationally formulated. In order to make risk communication a successful tool, risk communication has to be established as a process. By a direct, event-related, and permanent exchange of information at the same eye-level, even when the information givers and receivers come from different social establishments, risk communication can raise the level of understanding on the risk exposure situations of the population. Having in mind that "[E]very mitigation is local," it is the population at risk that has to be incorporated in prevention and mitigation. This is best achieved when the people are provided with information that reflects the available knowledge and is accompanied with an offer for an information exchange. Thus risk communication will raise the level of understanding and satisfy the population's demand for security.

Renn and Levine (1991) emphasized the fact that many risk-exposed people have serious problems understanding technical terms such as probability, frequency, risk occurrence, and so on. Linked to this, "is the problem that success of risk management is difficult to measure." In the case where a mitigation countermeasure is successful, the population at risk often does not see a justification to discuss or implement risk mitigation any further: there is no risk; why do something. This situation brings risk management into a bias situation. Risk managers propagate further engagement in this matter; on the other hand they are proud of the higher level of resilience achieved (security increment; see Fig. 8.7). This book proposes for such a situation, the use of the term "prevention dilemma." The dilemma gets even stronger momentum when the amount of money for the mitigation measures becomes a matter of public debate and the managers have to justify their decision that the money on, for example, elevating the dam height was spent meaningfully and target-oriented, nevertheless having in mind that the probability of another flood event cannot be avoided forever. The nonappearance of an anticipated flood is the success; not the higher dam. In such a situation it is the task of the risk communication to convince the people in times of no risk that risk mitigation is worthwhile.

Risk communication in times shortly after a disaster is quite easy. Everyone is longing for information. The victims need information on how better to cope with the disaster impact. The news people need the information to cover their headlines and the politicians want to see themselves brought out. The time following a disaster is the time of the risk managers and risk communication. But the time before a disaster strikes or between two disasters when the memory has already vanished is the "real" time for risk communication. The problem risk managers are confronted with in that period is that no one is inclined to receive news of such kind anymore. It took only a couple of days for the terrific pictures on Hurricane Sandy in 2012 to vanish from TV screens. Risk managers have to keep this in mind when communicating disaster prevention initiatives.

8.8.4 Pre- and Post-disaster Communication

As pointed out already, the best risk communication is that taken by public officials in advance of an incident to better prepare communities at risk, risk managers, and the news media for a probable event. Below there is a series of questions, information, and messages given, that can serve as an initiative to create public interest, well before a disaster event might occur. If such messages are disseminated well in advance they can start an interactive process of information exchange and opinion among individuals, groups, and institutions. The American Mental Health Services (SAMSHA 2002) proposed a series of questions that should be posed to start a risk communication with the public:

- What are the messages to be delivered prior to an incident?
- What are the obstacles to effective communications and how can they be minimized?
- What are the opportunities for effective means of communication?
- What questions can we anticipate from the public in this nonrisk situation?
- What kind of information is supposed to be communicated by the media?

Post-event debriefing is a successful support mechanism for the staff to learn from the experiences garnered from the last emergency. In this sense "debriefing" can also be defined as a contribution to "pre-disaster" planning. Debriefing comprises a couple of structural elements, mainly on the:

- Technical elements of the emergency management operation
- Elements of the social and cultural situation to which the affected population was exposed
- Kind and the way assistance was rendered
- Lessons learned from the operation to improve the organization's performance in crisis reaction.

Debriefing furthermore improves the structural setup of the administrative procedures and the skill of the staff's organization. Post-event learning will help to improve crisis reaction capacity by an open critical reviewing not of those matters that went perfectly, but those where deficiencies occurred. Debriefing transports crisis-acquired knowledge to those who are not yet familiar with the experiences gained, making implicit knowledge explicit by bringing the information also to other groups in the organization. Thus the factor "debriefing" in the disaster management cycle has to be seen as a substantial element of the management capability. The general rule on debriefing is that it has to follow the principle of privacy and confidentiality, as most of the matters discussed are matters of individuals, often marginalized social groups, or governmental organizations that fail to provide the envisaged services. Therefore such debriefing has to be conducted openly, honestly, and with much respect for the rights of individuals. Furthermore all staff members involved in the emergency case has to take part in the debriefing as all aspects have to be elaborated.

8.9 Risk Financing

Although in industrialized countries the great majority of damages are covered by insurance, most of the developing and many advanced countries still lack appropriate means to cover financial losses from natural disasters. In such countries the traditional way of covering losses by the individual is to wait for the government to cover the losses, often by providing money for reconstruction of the damaged buildings. But as many of these countries (e.g., a country like Bangladesh) do not have appropriate means at their disposal to cover at least the minimum losses, most of the burden rests with the victims. Many of the countries at risk, however, have generally only a very limited financial resource base from their tax income Thus many of these highly risk-exposed countries depend on foreign aid either in the form of donations, like Indonesia that received about US$8 billion for rehabilitation of the tsunami damage, or in the form of long-term loans from the World Bank or from one of the regional development banks. But such loans are made at market conditions and one day have to be paid back. The deep dependency on external resources makes these countries even more vulnerable and was in many cases the reason that hindered a sustainable recovery.

The low income levels make such poor or deprived countries nontarget areas for the international insurance business, and if not, then it is restricted to the wealthier segment of the society or the highly productive industrial sector. The lack of many financial resources often hampers a fast recovery after a disaster as the necessary infrastructure rehabilitation measures are either not implemented or at best start with much delay and not to the necessary dimensions. To overcome this lack of money, risk-related financing schemes have to be implemented before a disaster strikes, especially in those societies at highest risk that have the lowest access to risk splitting or risk transferring mechanisms. This holds true for marginalized societies in industrialized countries (e.g., Hurricane Katrina) or entire societies such as Bangladesh. Both damages are, if covered, only to a limited extent.

The most appropriate way for societies at risk not to depend on external help from international donor agencies such as the Holdback, the International Regional Development Banks, or from private donations, is that the national governments declare their willingness to institutionalize a state, parastate, or privately organized risk transfer mechanism. Such a willingness opens good chances to uncover not only effective means for financing disaster losses but also can include risk prevention elements. Most important is that state authorities:

- Base their assumptions on realistic pictures of the local risk exposure.
- Understand sharing of risk to be a social challenge.
- Acknowledge that prevention pays off (reduces the costs by two to five times).
- Know about what kind of social, scientific, and technical elements disaster risk management should include.
- Be open to a dialogue with the private sector for identifying options for risk transfer mechanisms.
- Institutionalize a conducive legal framework that fosters cooperation of the government authorities with the private sector.

Sharing risk means that there is a mechanism that transfers or splits the risk from the people who are affected by a disaster to those who are not affected. All partners in the mechanism contribute a "low" amount of money (premium) to cover a high loss of an individual. But such kind of risk sharing is significantly different in countries with high and low income. There are many societies that cannot afford even the lowest premium, and those are often at the highest risk. In industrialized countries private as well as industrial buildings are generally comprehensively insured against damages. There are even countries where parts of the risk premium are taken over by the government or by tax exemption. In many developing countries, however, such risk- sharing mechanisms either do not exist or are only of a limited capacity and thus loss compensation is seen by the society as a national task, a situation that is well known also in many industrialized countries. In this context Kunreuther (1966) emphasized that instead of covering the losses by the states, the insurance companies should be "convinced" to assess regularly the building standards of private homes as a prerequisite for insurance. Another point in defining the premium should be not to base it exclusively on the individual's risk pattern but also take the livelihood conditions, the location, and the social environment as an equal determinant of vulnerability into account when defining the insurance premium.

Consequently risk mitigation should always be accompanied by an insurance industry. Insurance is by definition a form of disaster preparedness. It represents an important, if not decisive, prerequisite for many economic activities (Berz 1997). Without insurance coverage, for example, engineering projects in highly risk-exposed regions, such as power stations in earthquake zones, would expose such installations to uncontrollable risk of failure that is not acceptable to the investors as well as to regional development efforts. Insurance is a market-oriented instrument that allows for a more even distribution of the financial burden from disaster among the four parties concerned: the insured, the insurer, the reinsurer, and the government (MunichRe 2012). The insurer and the reinsurer bear most of the burden. But where the financial losses overstep the insurer's liability, in general governments step into help out as the last-resort insurer or provide financial incentives to the victims either by direct loans or by tax relief. The insurer moreover can contribute to increasing self-responsibility in disaster risk reduction behavior with the insured by introducing a rebate in the insurance policy: a substantial deductible as a clause in the insurance policy often initiates individual prevention measures. This cost-sharing principle is for both parties involved (insured and insurer) a vital means that rewards the insurer with a substantial premium rebate and on the other side reduces the risk of the insurer. The wealth of data make the insurance able not only in calculating premiums and in classifying hazard areas—known as rating zones—but also in tracing relationships between event intensity and loss intensity and estimating loss potentials from realistic disaster scenarios (MunichRe ibid). By this information they contribute a broad experience to worldwide cooperation with governmental, nongovernmental, and scientific institutions, industry, and the media. Also due to this information pool it was possible recently to reduce drastically the financial expenditure of the insurers, eligible for settling natural disaster losses by relatively modest deductibles as most of the natural disasters always entail a large, sometimes enormous, number of minor losses.

In general, people make choices on how much risk they are willing to bear and how much money they want to spend for the desired amount of prevention. But this choice is a matter of income. In this sense the people distinguish between a risk they are willing to tolerate and those they do not accept (see Sect. 8.5). It turned out that the individual often sees the sector for a personal involvement in the field of "tolerable" risks, and he attributes the intolerable risks generally as a state's responsibility. Reaching a higher level of resilience is thus not only a matter of personal experience but also of a substantial risk reduction commitment by society. In this regard risk mitigation is a prominent example for the "paradigm of social balance" that makes a society vital and sustainable.

Practically the populations at risk distinguish between self-insurance, when the person feels able to absorb a loss, and insurance coverage, which pays a specified sum when the event occurs. Prevention entails measures that have a cost, and insurance entails a financial premium, and a person chooses the level and combination that best moderates consumption fluctuations. The following tiers summarize effective measures of risk sharing at individual and household levels:

- Savings for loss prevention
- Personal hazard and risk assessment (individual disaster risk profiling)
- Investment to protect and maintain assets
- Insuring assets, property, and household goods
- Retrofitting stability of building structures
- Timely repairs
- Relocating to safer areas
- Awareness rising, training, evacuation drills
- Part of the early warning system
- Increasing participation in social networks

People who are insured for a risk often develop a different awareness of the risk to which they are still exposed. This makes them less sensitive to the hazards and makes them lose interest in taking their own loss prevention measures. Insurers thus see their role not only in just covering financial losses but also to counteract this mode of thought and behavior through awareness-raising campaigns. This scenario is for them the rationale to inform their clients comprehensively on the relationship between hazard, vulnerability, and risk. All major insurance companies therefore have established scientific and technical expertise in risk assessment. In so doing, the companies often find themselves at the forefront of scientific and technological development. They regularly publish information in the form of leaflets, brochures, and in the media in order to alert the public to risks and draw attention to the precautions they could take.

Three terms are fundamental to understand risk sharing:

- Deductible describes the amount or percentage of an insured loss that the policyholder must cover before any claims are paid by the insurer.
- Insurance pool is the collective pool of risk from multiple insurance companies. Pooling facilitates the development of insurance markets by spreading risk across insurers that would otherwise lack financial capacity to participate in the market. It enables insurers to provide affordable coverage for high-risk events.
- Reinsurance describes the mechanism to sell an insurance by an insuring company to another specialty insurance company (the reinsurer) for the purpose of spreading risk and reducing the insurer's own losses from large insurance claims.

8.9.1 Insurance (Self-Insurance)

Damages from natural disaster occur as we have seen worldwide, hitting industrialized nations as well as nations that are in a development status. But as in industrialized countries most of the values that are at risk are (normally) insured, so that losses did not extremely burden the economy of the individual as well the country in general, low-income households as we find in many developing countries are highly vulnerable to losses from natural disasters. An appropriate means to protect households from such kind of economic losses is through insurance. Insurance is an instrument to share the risk among a group of the society that is exposed to the same kind of risk. The principle of insuring losses from risks such as accidents, health, life term, equity price risk, or crop failure, as well as from natural disasters are based on assessing the possible financial losses of randomly occurring events based on statistically measurable and thus predictable distributions of disaster events. Such statistics allow the insurer to assess the risks and consequently to define the burden. "Ironically the widely accepted practice of insurances in the World's largest economies reflects a collective method of socializing losses" (World Bank 2001). As catastrophic events occur comparatively seldom, their potential loss is quite high, making the use of the traditional insurance practice of spreading risks over a large number of insured individuals difficult. The insurance risk management therefore developed a series of alternative risk financing concepts of which the most important are described below.

Turkey is one of the countries in the world that is most exposed to natural disasters, particularly earthquakes. Around 70 % of Turkey's population and 75 % of its industrial facilities are exposed to large-scale earthquakes. The 1999 Kocaeli-Izmit earthquake along the North Anatolian Fault Zone claimed a death toll over 17,000 and caused economic losses estimated at about US$5 billion; or around 2.5 % of gross domestic product. The nation's disaster hotspot is located at the city of Istanbul. There almost 15 million inhabitants—making the city the fourth biggest megacity on Earth—and a high industrial density, living on the highly active fault zone. According to recent assessments carried out by JICA (cited in Ilkesik 2002) the probability of a major earthquake affecting Istanbul in the next 30 years is higher than 60 %, resulting in a seismic risk exposure comparable to Los Angeles; but with damage potential that is much higher because of Istanbul's greater structural vulnerabilities. A seismic event of the same magnitude as that in 1999 would result in more than US$2.0 billion economic loss, up to 87,000 fatalities, 135,000 injuries, and heavy damage to 350,000 public and private buildings. Experts see this risk exposure as very dramatic, as it would burden the national economy to an extent that it would take Turkey years to recover again economically. But the Turkish exposure to natural disasters is not unique on Earth. There are many other places that also suffer economically from such disasters. Although the costliest disasters generally occur in developed countries, for instance, hurricanes since 2005 in the United States added up to losses of more than US$250 billion, fortunately mostly covered by risk insurance, in low- or median-income countries facing increasing economic losses only a few of them get insured. Thus disasters often significantly affect the national economies, leading to expenditures that normally were earmarked for social development projects and that then have to fund emergency and recovery needs.

Covering losses from natural disasters can either be managed before a disaster strikes, ex-ante, which means financial means are invested to prevent losses from occurring, or ex-post, which means losses from disaster events have to be covered.

Ex-ante disaster loss financing requires that the potential losses are known and have to be assessed prior to the event and a provision of money has to be made in the budgets for such an event. In fact, regarding loss assessment from many natural disasters of the 1990s (World Bank/GFDRR) it was possible to prevent economic losses of more than US$280 billion by investing US$40 billion in prevention, a ratio of 1:7. There are also other figures giving ratios of 1:5 or 1:8, but all of them bearing the same message: prevention pays. In this context another effect has to be considered, that in the case where prevention was successful, no loss occurred. Such a situation often brings the decision makers into a bias situation that can be described as the prevention dilemma, leaving the authority to explain why money has been invested, although nothing has happened.

Ex-post disaster coverage has the advantage that the costs can be quantified quite exactly but on the other hand this puts a huge financial burden on the national budget. In many countries with limited economic resilience, the financial means to be allocated were in general financed by new debts. Such financing has often heavily affected the country's debt service in the past and consequently could only be adjusted by raising taxes that once again strongly affected and discouraged new private investments. In addition, the lack of financial means in the aftermath of a disaster often has led to hamper recovery and forced governments to conduct an emergency budget reallocation often at the burden of other social development programs. Therefore many disaster-stricken low–middle income countries in the past had mostly to rely on foreign assistance. But such assistance is often rendered only for a short time (weeks or months) and is generally scheduled to finance relief and rehabilitation measures rather than to change the risk exposure. Moreover such donations do not cover the entire losses. An exemption on this was the overwhelming support by the international community on the 2004 tsunami in Indonesia, where more money was allocated by the international community than the actual loss of about US$5 billion. Furthermore in the future it is most probable that due to the rising frequency and intensity of losses from climate-related disasters, the traditional model of post-disaster financing and reliance in low- and middle-income countries on the donor community is no longer guaranteed (Cummins and Mahul 2008).

Despite the frequency and expenses of natural disasters in industrialized and developing countries, there is no internationally agreed-upon system in either the public or private sector for consistently compiling information about their economic impacts. Therefore for a long time, organizations involved in economic loss assessment pledge to establish an informed risk management policy that integrates any data on the direct as well as on the indirect losses from natural disasters. The committee proposed to base such calculations not on the reconstruction costs but on the losses from the disaster impact. This approach covers a much broader range of cost categories, as the term "costs" conventionally is understood to cover the losses that are reimbursed by insurance companies and governments. A calculation method should be developed that not only incorporates the local direct reconstruction costs but also the indirect costs such as price increases in far remote areas, death toll from medical service functioning failure, restricted energy generation due to disruption of the traffic connections (transport of coal), and many more. The algorithm furthermore should address the benefits of mitigation measures on the basis of a life-cycle investment, or at least for a decade in the future and not calculate the costs just by comparing the invested money and the cost of a normal or worst-case scenario.

The experience from the many risk events brought the developing countries and the international donor community to review the situation and was the rationale for many political approaches to economically secure disaster risk financing. The long-used practice just to rely on the international donor community to cover the losses is anticipated in regard to the many costly hydrometeorological disasters to be increasingly unrealistic. In addition, more and more countries realized that it would be less costly to invest in disaster prevention measures than to cover post-disaster losses. Therefore the UN Framework Convention on Climate Change and the subsequent Kyoto Protocol refer to the potential role of insurance in disaster mitigation and the Hyogo Framework for Action 2005–2015 (UNISDR 2005) identified the need to promote the development of financial risk-sharing mechanisms, particularly insurance and reinsurance against disasters, as a priority action for building the resilience of nations and communities to recover from disasters and recognized the need for innovative risk-financing mechanisms to be particularly relevant to the middle- and low-income countries.

Thus far funding for disaster mitigation and prevention in developing countries mostly was provided by industrialized country-financed international donor agencies such as the World Bank, the United Nations, or the many regional development banks (ABD, AfDB, etc.). In more advanced economies, losses from natural disasters are typically funded through a combination of private risk-financing arrangements and an efficient public revenue system. In middle- and low-income countries, with fiscal pressures, funding of post-disaster reconstruction strongly was based on ex-post borrowing and assistance from international donors.

Well-established forms of financing losses from natural disasters or of preventing future losses include different economic approaches. Risk compensation can be financed either through risk retention or risk transfer to an outside party.

8.9.2 Microfinancing

A special type of insurance that offers protection against the risks in life, especially for low-income people in the developing countries is called microinsurance. There have been in the past many successful attempts to provide financial means to such a group of people, who normally do not have access to the financial markets. The first to start such a credit scheme was Professor M. Yunus in Bangladesh, who invented the Grameen Bank model and was later honored with the Nobel Peace Prize. Since then the microfinancing market has strongly developed. Today many different credit schemes are offered by the international insurance business; all of them aim at risk financing for low-income groups and can be tailored according to the needs in developing countries. The world market for microinsurance is estimated to comprise more than two billion people, representing an economic potential of US$40 billion, according to information from Allianz, Germany (Allianz Insurance Company, Pro Vention Consortium).

The basic principles for microinsurance are that the losses must occur by chance, unexpected, and randomly. The losses must be pre-defined in terms of timing, region, type of risk, and severity. The rate of losses must be predictable, must embrace quite a large insured clientele, and should have quite a large areal distribution. The premium must be proportionate to the likelihood and costs of the risk involved but the mode of payment can be tailored according to the needs of the clientele. The insurance payments can be used to restore household and productive assets that were damaged during a disaster. The social and economic sectors that are subjects for microinsurance comprise the entire sector of insurance, from term life insurance, accidents, home insurance, to livestock insurance, as well as protection against natural perils including floods, rain, hail, or others.

Microinsurance can help split the risks especially of low-income households. Munich Re and ILO published in 2006 (Churchill 2006) a brochure that explained how microinsurance schemes can help to split risks from natural disasters. Traditionally such kind of informal risk splitting is known from many societies worldwide, but in fact such insurance schemes preferably covered agricultural losses, or economic burdens in the health sector or from accidental deaths. The outreach for such systems is limited and in reality only small. The poor are that fraction of a society who are more vulnerable to a crisis than all other societal groups and are those who are the least able to cope with disasters. The traditional coping strategies are normally restricted in the effects and low in return, thus providing only insufficient protection. According to ILO (2001) half of the world's population is excluded from any type of social security protection; in sub-Saharan Africa not even one in ten. Various experiences with microinsurance proved it a valuable instrument of risk sharing. Traditional microinsurance can mean that a peasant puts aside money from a good harvest for a time of emergency. Microinsurance in the sense ILO is propagating even goes a step further. It is rather designed to cover risks through a regular payment of premiums that is proportionate to likelihood, saliency, and the losses from natural disaster. Thus ILO pointed out microinsurance does not differ much from normal insurance but has a clear focus on low-income people, people who are in their majority defined by not having a regular income and not having access to formal social protection neither by employers nor from the government. The basic concept of microinsurance is "risk pooling" that means all insured participants pay the premium in a great pool from which a loss in a certain time period and of a defined nature will be covered. Thus all contribute a small amount of money but only a few benefit strongly. ILO pointed out that microinsurance schemes are best applicable when the risks are sudden, not predictable, and of a significant severity. Insuring such kind of risk is a means to give social protection for all those who are lacking respective government schemes. Microinsurance examples from Asia and Africa reveal that also the poor can increase their disaster resilience, thus making them interesting partners for the insurance business. The normal insurance business tries not to insure high-risk persons whereas the low-income oriented microinsurance policy runs the opposite way. It seeks to get many people into the portfolio thus sharing the risks to all and thus reducing the risk of a particular household. A reasonable way to share the risk is to identify groups of persons that share the same type of risk: agricultural cooperatives, small-scale entrepreneurs and religious groups or women associations. Such associations exist in all developing countries, making them a preferable target group for microinsurers.

Microinsurance schemes have already in many cases proven their general capability to provide security against natural perils. But "in practice there are only few successful experiences and it has proved extremely challenging to structure and implement affordable and high value micro-insurance products specifically for disasters" (Linnerooth-Bayer and Mechler 2009). They recognized that to implement a microinsurance system successfully, a couple of basic factors must be operational, including a powerful and diversified risk pool, low transaction costs and affordable premiums, together with a transparent and efficient mode of payouts. In addition microinsurance requires a highly specialized staff that operates under clearly defined procedures. Furthermore experience clearly shows that microinsurance deserves a backing by reinsurers, "as it is very difficult for most systems to provide insurance alone."

8.9.3 Risk Retention

Retaining a risk means that the individual, a company of (even) the government puts money aside from the annual budget in order to cover a loss when it occurs. This can be managed either according to plan or be done unplanned. If neither loss reserves nor disaster reconstruction funds have been established or designated, very often losses are just taken from available cash. From a risk-financing point of view, this technique is acceptable for losses that are small in nature and infrequent in occurrence. A more sustainable approach is to establish a loss reserve. Such an approach comprises a significant difference from the technique described above, as it recognizes a liability for loss and demands setting aside money or assets to fund that liability. Such a loss reserve is typically based on expected losses and is treated as a budget provision, requiring a pre-defined liability in the financial statements. The losses can be funded by cash, securities, or other liquid assets that are earmarked for the designated liabilities. Another element in risk financing is self-insurance which means that such an organization finances its losses through a planned strategy. The most typical forms of self-insurance are a self-insurance trust or a captive insurance company. A self-insurance trust is not insurance but a funding vehicle (e.g., a bank account with an independent third party/trustee) that is designated for the sole purpose of paying losses. The trustee administers the trust through a formalized agreement and a statement that outlines the type and limits of loss to be paid. The trusts were for long the most common vehicle for self-insurance, but they are gradually being replaced by captive insurance companies because these vehicles can more flexibly accommodate the various exposures and risk financing needs (Carroll 2001). A captive insurance company is an organizational structure established for instance by a large company or a private entity to cover their respective losses from natural disasters. The insurance business is primarily controlled by its owners who are also the principal beneficiaries. As before (self-insurance), captives are also insurance vehicles but with a greater flexibility to accommodate the many and different types of risk. As captives are obliged to tax and income statements, there is a great importance that they act in line with the company's risk-management program, which consequently will elevate risk management a part of the organization management.

8.9.4 Risk Transfer

Risk transfer by definition transmits an individual, party, or organization, and so on, risk to an insurance company which itself spreads it among many insurance holders. The most common method of risk transfer is a commercial insurance first-party insurance also called direct damage coverage of losses, providing financial reimbursement as the result of damage that also comprises all types of natural disasters. Insurance is a contractual relationship that exists when the insurer agrees, for a premium, to pay the insured a loss caused by a pre-defined event (peril). The risk premium is the amount of money the insured pays regularly to the insurer and depends on the agreed-upon level of returns. The premium thus describes the willingness of the insurance taker to accept a certain risk. From a practical view, insurance will nearly always involve some form of risk retention on a planned or unplanned basis and is generally subject to a deductible. A deductible is the percentage of an insured loss that the policyholder must cover by himself, before a claim will be paid by the insurer. The insurance companies themselves also seek to split their respective risks by passing their risk to a reinsurer or a group of reinsurers. In many countries the insurance companies are legally bound not to issue policies exceeding a maximum solvency margin of normally 10 % of their company net worth, unless those policies are reinsured. This significantly improves the insurer's capability to take on higher risks because some of that risk is transferred to collective risk pools with reinsurers like the Munich Re, the Swiss Re, or the Hannover Re. Over the years the reinsurers have developed sophisticated and reliable models to assess risk from natural disasters. One of the most famous is the Munich Re-Insurance Company that for a long time has established a powerful natural disaster risk assessment division and that has established the Munich Re Foundation to develop in-depth assessment of methods and strategies for risk assessment.

Today, risk financing is as described above viewed as a complex system involving economic aspects, contracts between insured and insurer, and a legal framework. The goal of risk financing is ultimately to protect assets and personal lives including some of the following:

- Identification of types of exposures and losses faced
- Anticipation of risks of the groups
- Financial provision to cover losses
- Pooling resources
- Spreading/transferring risks
- Risk prevention and retention
- Legally binding contracts
- Identification of ways to finance loss without jeopardizing the financial integrity of the contract partners

Transferring risks to the capital market often uses so-called insurance-linked securities (ILS) to reach a higher level of security by subsequently trading risks onto the secondary insurance market. This holds true especially for risks from natural hazards or to hedge against pandemic risks. This concept gave ILS a "foothold as an alternative asset category for investors and as an alternative form of reinsurance for insurers" (MunichRe 2012, 2013). And it is expected that this form of insurance will continue to gain in significance, because developments in supervisory law such as Solvency II are likely to give a further boost to their popularity.

Risk transfer also can be accomplished through the use of an indemnification provision. In natural disaster risk financing such a method can be a rationale if someone's interference with nature (e.g., the construction of a building) amplified the impact of a natural disaster leading to a claim for indemnification to be restored or reimbursed to make whole again. Nevertheless it should be noted that any insurance policy should never be viewed as a complete transfer of risk.

In developed countries private organizations and entities other than the government take over a large portion of the financial risk by insurance. Thus insurance is the primary tool for risk transfer in such countries. Risk transfer by insurance has several major advantages: it spreads risks between parties thus reducing the risk to the individual and it "allows the segregation of risk" (Freeman et al. 2003). In the higher-income countries about 30 % of the loss from natural hazards is insured, whereas in low- to middle-income countries insurance covers just 1 % of the losses.

8.9.5 Catastrophic Bonds (Cat Bonds, Cat Swaps, Risk Swaps)

Not only traditional financial insurance is an option for transferring risk. In the aftermath of the big disasters in the United States, for example, the Northridge earthquake and Hurricane Andrew, the insurance industry realized that the financial losses from such megadisasters can reach magnitudes that the insurance industry assumed not to be able to absorb in the future (Damnjanovic et al. 2010). The insurers therefore initiated a number of studies to estimate financial exposure based on the natural disaster experience. The anticipated financial losses from such disasters has led economists and geoscientists to develop alternative risk financing strategies, also known as alternative risk transfer techniques (ARTs). Among them are the so-called catastrophe bonds (cat bonds). Both instruments, insurance and cat bonds, are risk-management strategies potentially to embrace the impact of financial risks. Generally speaking, the difference of cat bonds and insurance is that cat bonds are paid by the insurance company when the economic losses from a natural disaster overstep the pre-disaster-defined level risk. Cat bonds are other than normal investment bonds not depending on the solvency of the creditor rather than on pre-defined type, location, and severity of a natural disaster.

The concept of cat bonds emerged from the intention of insurance companies to share the high to very high risks they would face if a major catastrophe occurred, and that could not be covered by the premiums. An insurance company therefore issues such bonds which are then sold to investors. If until the end of the contracted period no catastrophe occurs, the insurance company pays back the invested capital plus the premium and the interest to the investors. On the contrary, if a catastrophe occurs as defined in the bond's contract, then the incurred losses are paid by the insurance company to the claimholders, a situation that happened lately with the US$300 million cat bond of the Japanese Muteki Ltd catastrophe bond issued in 2008 by Munich Re after the damages of the March 11th Tohoku (Fukushima, Japan) earthquake cum tsunami were declared a total loss.

The advantage for the insurance company lies in comparatively high interest rates (when the losses are rated low), and those who want their risks from earthquake, flood, or hailstorm events to get covered, do not need to make respective budgetary provisions and can thus hand over their risks to the capital market. But not only private insurance companies are issuing cat bonds, governments also make more and more use of that financing scheme. Today the value of the worldwide assets reached about US$100 trillion, making that market highly interesting to the insurance business.

In response to increasing demand from risk-exposed countries, international donor organizations developed catastrophe bond issuance platforms that allow governments to use a standard framework to buy insurance. One of the best-known tools was launched by the World Bank, called the MultiCatProgram (World Bank 2014). The objective of the program is to facilitate access to insurance coverage for governments on terms that are better than normal market conditions, to help with disaster preparedness and to ensure governments' access to immediate liquidity to finance emergency relief and reconstruction work after a natural disaster. The first country to make use of the MultiCatProgram was Mexico that in 2009 had already sold US$290 million in catastrophe bonds that will cover up to US$140 million of earthquake damage, US$100 million against Pacific hurricanes and US$50 million against Atlantic hurricanes. The bond sale was managed under the lead of the World Bank by Goldman Sachs Group Inc. and Swiss Reinsurance Co. The World Bank's function in the program is to reduce the cost of issuing the bonds and make it easier to sell the bonds on emerging markets: "The bank will be playing a real catalytic role in getting some of these countries that have no access or are afraid to get this access to the markets."

Table 8.3 Selection of international catastrophic bonds executed by Munich Re Insurance (*Courtesy* Municn Re)

Client/company	Financial volume (million)	Perils covered/region
VenTerra Re Ltd.	USD 250	Tropical cyclones Australia/earthquake United States
Bosphorus 1 Re Ltd.	USD 400	Earthquake Turkey
Tar Heel Re Ltd.	USD 500	Named storms (tropical cyclones)
Lakeside Re III Ltd.	USD 270	Earthquake North America
Johnston Re Ltd.	USD 202	Hurricane United States
Queen City Re Ltd.	USD 75	US named storms
Queen Street VIII Re Ltd.	USD 75	Hurricane United States and Cyclone Australia
Queen Street VII Re Ltd.	USD 75	Hurricane United States and Windstorm Europe
Queen Street VI Re Ltd.	USD 100	Hurricane United States and Windstorm Europe
Queen Street V Re Ltd.	USD 75	Hurricane United States and Windstorm Europe
Queen Street IV Capital Ltd.	USD 100	Hurricane United States and Windstorm Europe
Queen Street III Capital Ltd.	USD 150	Windstorm Europe
Queen Street II Capital Ltd.	USD 100	Hurricane United States and Windstorm Europe
EOS Wind Ltd.	USD 80	Hurricane United States and Windstorm Europe

Table 8.3 was published by the Munich Re Insurance company in 2013 (MunichRe 2013) and gives some examples of reinsured catastrophe bonds the company is executing:

8.9.6 National Risk Sharing

- US National Flood Insurance Program (NFIP)

There are a multitude of examples on national insurance programs. One of the most popular programs worldwide is the US National Flood Insurance Program (NFIP) that was created in 1968 through the National Flood Insurance Act (FEMA 1968). Floods are the most destructive natural hazard in terms of economic loss to the United States of America. The program enables property owners to purchase insurance protection from the government against losses from flooding. The insurance is designed to provide a nonmarket-based insurance alternative to disaster assistance to meet the escalating costs of repairing damage to buildings caused by the yearly occurring floods mostly along the Mississippi/Missouri river path and in Florida. Since its inception in 1969, the National Flood Insurance has covered losses of more than US$40 billion in claims, of which more than 40 % of that money has gone to residents of Louisiana. The program moreover reiterates retrofitting the building standards that today save an estimated US$1 billion annually. In 2010, the program insured about 5.5 million homes in nearly 20,000 communities. Within the program, flood-prone areas are identified, specifically tailored flood insurance offered, and flood-prone communities are encouraged to implement flood-plain management activities. Originally, NFIP was meant to be self-supporting and intended that its operating expenses should be paid from the premiums collected for flood insurance policies. But it was found that there is a repetitive loss of about US$200 million annually that has to be covered by US taxpayers. Actually there is an initiative underway that aims at raising the premium in order to make the NIFP self-supporting.

Homeowners who want their property to be insured can participate in the NFIP if their local community has signed a legally binding agreement with the federal government that stipulates that if a community will adopt and enforce a "flood plain management ordinance" to reduce future flood risks, the federal government will make flood insurance available to the community. The agreement furthermore demands that flood risk maps be set up and regularly updated, for a sustained flood plain management and for the identifying local flood risk premium zones. The compensation of losses, the money spent on future flood reduction measures, as well as the implementation of the communal flood management plans is overseen by the FEMA. The compensation of losses provided by the program is oriented on the value of flood damage on houses and assets. The loss compensation is based on either the replacement cost value or the actual cash value. The replacement cost value is the cost to replace that part of a building that is damaged. To be eligible, certain conditions must be met: (1) the building must be a single-family dwelling; (2) it must be the principal residence, meaning the family lives there for at least 80 % of the year; and (3) the reconstruction costs are at least 80 % of the full replacement cost of the building. The actual cash value is the replacement cost value at the time of loss, less the value of its physical depreciation, and the replacement costs of personal property are always valued at actual cash value.

In order to encourage communities to do more and better in flood risk reduction a Community Rating System (CRS) was implemented in 1990 (FEMA 1990) as a voluntary program for recognizing and encouraging community flood plain management activities that is in full compliance with or even exceeds the minimum NFIP flood plain management requirements, and may apply to join the CRS. CRS-eligible communities can get their flood insurance premium rates discounted to reward community actions if the activities meet three goals:

- Reduce flood damage to insurable property.
- Strengthen and support the insurance aspects of the NFIP.
- Encourage a comprehensive approach to floodplain management.

Meanwhile more than 1200 communities participate in CRS, reaching nearly 3.8 million policyholders. Although CRS communities represent only 5 % of the over 20,000 communities participating in the NFIP, almost 70 % of all flood insurance policies are negotiated in CRS communities. Eligible for CRS support are communities that qualify a class rating system that is very similar to the fire insurance rating system of the United States. CRS classes are rated from 1 to 10. Each CRS class improvement produces a 5 % greater discount on flood insurance premiums. A community that does not apply at all for the CRS or that does not comply with the minimum requirements is considered a "class 10," and a "class 1" community thus receives the maximum 45 % premium reduction. But lowering the costs of flood damage is only one of the rewards a community receives from participating in the CRS. Other CRS benefits include:

- Citizens and property owners have increased opportunities to learn about risk, evaluate their individual vulnerabilities, and take action to protect themselves, as well as their homes and businesses.
- Flood plain management activities provide enhanced public safety and reduced damage to property and public infrastructure to avoid risk of lives, economic disruption, and loss.
- Communities can better evaluate the effectiveness of their local flood programs against a nationally recognized benchmark.
- Provision for technical assistance in designing and implementing flood reduction activities.
- Communities have incentives to maintain and improve their flood programs over time.

Although the NFIP is widely accepted as significantly increasing flood loss reduction in the United States, the program itself is under great criticism. Most of the criticism refers to the financial situation of the program. The cash-based budgeting is seen to obscure the program's actual costs and does not provide transparent information on emerging financial problems. A system that allows for an accrual-based budgeting is anticipated to better address the revenues and expenses situation. Moreover it has been estimated that less than 50 % of eligible property owners in flood plains participate in the program. In addition, even when the purchase of insurance is mandatory, the extent of noncompliance with the mandatory purchase requirement is unknown and remains a concern. In the past organization-introduced reduction of the subsidies often caused policyholders to cancel their policies or reduce their program participation, thus leaving them vulnerable to financial loss from floods. Furthermore, placement of the program within the Department of Homeland Security and no longer with NFIP itself, bears the risk of decreasing the attention, visibility, and public support the program receives. Moreover, homeowners who have built their houses before the flood zone was defined are also eligible for reduced premiums of up to 40 % lower than the normal risk premium. The incorporation of properties with two or more losses in a 10-year period has also added to program losses. This group of persons represents 38 % of claims losses, but accounts only for 2 % of insured properties.

On October 1st, 2013, the *New York Times* reported that the National Flood Insurance Program had changed its insurance policy. The report is summarized here. From that day on, the insurance premium will start going up steadily by 25 % per year for regions that are severely or repeatedly flooded, until the rates balance the actual risk expenditure. That means property owners in flood-prone areas who might have once been paying around US$500 a year (rates that were well below what the market would charge) will go up by thousands of dollars over the next decade. This took many homeowners affected by Hurricane Sandy to the streets to call for a "Stop FEMA" rally. Congressional representatives from states such as Louisiana and Florida that are likely to be hit by the NFIP changes have called on FEMA to delay the implementation of the new rule, although the law got overwhelming support from all political parties. FEMA says its hands are tied, as the Biggert–Waters Act obligates the program to adjust flood premium rates accordingly. By November 2012 the NFIP was more than US$20 billion in debt, a number that would take the NFIP 100 years to recoup its losses. The changes were aimed at those 1.1 million policyholders who were paying far less than what the market value for flood insurance would have been. Thus quite a number of policy owners have essentially been subsidized with public money for years, even decades. Therefore property owners are confronted with unexpected outcomes of the subsidized flood insurance policy, although the NFIP was once created to support these people. Before NFIP, the private insurance industry was unwilling to provide flood insurance simply as it wasn't profitable for them. The premiums did not cover the payouts following the many big floods. Thus the government stepped in, offering subsidized flood insurance to property owners, often at below market rates.

But shifting the burden from the private market to the government didn't really lower the costs of major floods, especially as more and more Americans moved to coastal areas. From 1970 to 2010, the population of shoreline counties increased by almost 40 %, to 120 million and is projected to increase by an additional 10 million people by 2020. Some critics mention that just the subsidized flood insurance, by shifting the risk from the individual to the public, had perversely incentivized building in flood-prone areas. And it is anticipated that things will get even worse if the consequences of sea-level rise continue as in the last decades. A recent study found that if no actions are taken to reduce flooding risk, losses could approach US$1000 billion by mid-century, assuming a sea-level rise of just 40 cm. The sea level around New York City has risen by about a foot and a half over the past century, which added to the devastating flood damage during Sandy. Investing in mitigation such as raising homes and protecting coastal communities with sand dunes and seawalls is therefore seen as the only alternative.

8.9.7 Turkish Catastrophe Insurance Pool (TCIP)

Turkey is one of the most risk-exposed countries in the world to earthquakes. Around 70 % of Turkey's population and 75 % of its industrial facilities are exposed to large-scale earthquakes. Most of the earthquakes occur along the North Anatolian Fracture zone (bordering the Marmara Sea) and along the East Anatolian Fracture zone. Since 1984 more than 120 earthquakes occurred with a magnitude higher than M5 that resulted in direct property and infrastructure losses frequently exceeding US$5 billion per event. The last major earthquake in the Marmara region in 1999 resulted in the loss of 15,000 lives and placed a financial burden of about US$6 billion on the economy and the government, also due to the fact that only less than US$1 billion in losses were covered by risk insurance.

Earthquake insurance coverage was relatively low at the end of the 1990s in Turkey. Only around 3 % of residential buildings were insured, as households traditionally relied on the government to finance the reconstruction of private property after major natural disasters. In the aftermath of the Marmara earthquake, the government decided to develop a property catastrophe risk insurance mechanism to reduce its fiscal exposure to natural disasters arising from the traditional government-funded reconstruction of private property. In 2000, the Turkish government created by Law No. 587 an earthquake insurance system compulsory for all residential buildings on registered land in urban areas. The World Bank (Gurenko et al. 2006) provided financial and technical assistance by the Global Facility for Disaster Reduction and Recovery (GFDRR). The TCIP has become the first national catastrophe insurance pool in World Bank client countries that provides standalone earthquake insurance coverage to homeowners and small and medium enterprises. The catastrophe risk-financing strategy of the TCIP relies on both covering the losses by their own financial resources and by transferring the risk to the reinsurance market. About US$80 million of losses will be covered through TCIP reserves; this part of the expenditures is initially complemented by a US$100 million World Bank contingent loan facility. The overage will be transferred to the international reinsurance markets. Moreover the Turkish government covers losses that would exceed the overall claims, which is currently sufficient to withstand a 1-in-350-year earthquake.

The main objectives of the TCIP are to:

- Ensure that all property dwellings have affordable earthquake insurance coverage.
- Create a culture of prevention and resilience.
- Reduce citizens' dependence on government to fund the reconstruction of private property.
- Reduce government's fiscal exposure to earthquake and fire damages.
- Transfer catastrophe risk to the international insurance markets.
- Encourage physical risk mitigation and safer construction practices.

In August 2000 the TCIP became a legal public entity targeted to lower government expenditure for catastrophes. Moreover the government intends with the TCIP to improve the risk prevention culture and insurance consciousness in the public by incorporating the three stakeholder groups into a public–private partnership: the risk-exposed individual, the national mandated authorities, and the insurance cum reinsurance market for a socially affordable risk sharing. The program is not subsidized and the premium rates are oriented at levels for people with an average income with a deductible of 2 % with a contract duration of 30 years (around US$62 per homeowner; the maximum coverage lies at approximately US$92,000 per policy). This financial scheme will lay the base for long-term fund accumulation and aims at sharing the financial burden between the individual and the international insurance market. The program offers a variety of insurance possibilities according to building type and property location. The risk coverage includes earthquakes and fire damage to residential structures but no household contents. Since the year 2000 the TCIP public–private partnership has stimulated the growth of the catastrophe insurance market in Turkey significantly. The number of earthquake policies sold increased sixfold from 600,000 in 1999 to more than 3.5 million in the year 2010. Nevertheless the TCIP still needs more time to achieve deeper market penetration. Today, the insurance coverage is at about 23 % of dwellings countrywide and about 40 % in particularly disaster-prone areas. Still the expectation prevails with homeowners that the government will pay for damages regardless of the insurance program. It became obvious that a program such as TCIP relies on a strong communication strategy (Gurenko et al. ibid) to ensure that residents are aware of earthquake risk, mandatory insurance laws, and the program's excellent claim-paying record. The World Bank has drawn furthermore the assumption from the TCIP, that catastrophe insurance requires high state-of-the-art catastrophe risk modeling techniques to price premiums that accurately reflect the underlying risk.

8.9.8 National Agricultural Insurance, India

The National Agricultural Insurance Scheme in India (NAIS 1999) is another impressive example of how insurance-based market conditions can help to reduce damages from natural disasters. With two thirds of the Indian population depending on agriculture for a livelihood, crop insurance has long been an important element of agricultural risk management. The government of India has historically defined crop insurance as a national responsibility to mitigate the risks of natural perils on farm production. In 1999 therefore the government established the insurance scheme that offers insurance for food crops, oilseeds, and selected commercial crops through the state-owned Agriculture Insurance Company (AICI). With about 25 million farmers insured, and a premium volume of US$650 million in 2011–2012 NAIS is the largest crop insurance program in the world.

The risk coverage by NAIS is based on the crop yield of a defined area known as the "indexed approach," where the actual yield of the insured crop, measured by crop-cutting experiments in a so-called "insurance unit" (IU) is compared to historical yields. If the former is lower than the latter, all insured farmers in the IU are eligible for the same rate of insurance compensation. This strategy proved to be technically much more operational as compared to individual crop insurance. The large number of very small landholdings in India made it virtually impossible to base premium and risk coverage on such a type of yield-based assessment. As NAIS is funded exclusively by the government this commitment leads to a highly variable fiscal exposure for the national budget often exceeding 100 % of the allocated budget. At the end of the crop season, aggregate claims exceeding the farmers' premium are funded 50–50 % by the state and central governments. The long-used post-disaster funding arrangement was defined by the lack of an actuarially sound premium rating methodology, which means that estimating payouts was not really feasible (Raju and Chand 2008). A revised funding and compensation scheme was therefore seen necessary. A new NAIS that would operate in a way that the government's financial liability predominantly would be rendered in the form of premium subsidies given to AICI and by funding ex-ante a disaster occurred, was found long overdue. Such a new mechanism would reduce the yearly budget management problems of this sector, as well as help to reduce the delays in claims settlement. The actual system of a comprehensive involvement of the national government was found not to be optimal. To address the identified challenges, the government requested the World Bank in 2005 to provide technical assistance to second a modification of the NAIS program. With this initiative the Indian government moved from a social crop insurance scheme to a market-based crop insurance program with actuarially sound premium rates and with a deeper participation of private insurers. The World Bank provided technical assistance to support modification of NAIS based on international best practice and in-country experience.

The New NAIS combines traditional and modern methods for crop yield assessment: a Weather-Based Crop Insurance Scheme (WBCIS) has been piloted for 11 growing seasons since 2007, with 11.6 million farmers and US$370 million covered in the most recent season and the modified NAIS has been conducted for four growing seasons since 2010, with greater than 1.1 million farmers and US$67 million covered in the most recent season.

The New NAIS launched for the 2010–2011 growing season now comprises among others:

- A best practice, standardized actuarially sound pricing system
- An experience-based approach for area-yield insurance
- Intensive institutional capacity development that serves the needs of the highly specialized agricultural insurance business
- Introduction of commercial weather-based crop insurance products
- The introduction of mobile technologies for improving crop-cutting data quality and timeliness
- The institutional capacity conducive to transfer NAIS to a market-based program
- Prototype of actuarial software that allows the pricing of 200 crop insurance products

After modification, NAIS was introduced in 12 districts, covering more than 300,000 farmers, with an expected claims ratio within 50 %. The program was targeted at 400,000 farmers for the growing season 2011–2012. Furthermore the policy dialogue with various line ministries about the fiscal impact of the modified NAIS was institutionalized as well as a dialogue on the welfare implications of the modified scheme. The program today underlies the insurance products by using a suite of technological and statistical innovations including GPS with video recording capabilities and mobile phone software. The use of weather and remote sensing data will lead to better crop-cutting experiments. As a result, the modified NAIS has increased reliability of insurance products for farmers by the introduction of checks and balances into yield indices to enable private insurers and reinsurers to take on the risks that were previously retained by government. New NAIS significantly reduced the total cost for the government through remote sensing and weather data and moreover increased the speed of claim settlement for farmers.

8.9.9 General Conclusion

In many developing countries, however, such an institutional and regulatory framework is still missing or is in its infancy. The lack of such structures hinders an insurance market from developing. And the traditional way the banking system is organized in many developing countries makes necessary institutional reforms that allow risk insurance to operate properly very difficult, although the strategies, structures, and operational setup for implementing insurance markets are already well understood. In addition to the regulatory framework, there are concerns related to the fundamental structure of the market for insurance, for example, many countries are ranked too small for a national insurance market to survive. Experience from Central America revealed that the region's small middle classes and medium-sized businesses represent the most frequent purchasers of insurance in developed countries. One option to initiate a risk insurance market is to make insurance mandatory by law and to demonstrate the benefits of insurance at the government level, for example, by insuring government-owned buildings and infrastructure construction. Another option is the creation of regional or transnational insurance markets. If a market comprises a large number of license holders, the market becomes attractive for the insurance industry as this would lower the cost of insurance. The larger the market is, the more the international insurance industry will enter the business. Regional proposals, such as the World Bank's initiative for a Central American insurance market, are based on overcoming barriers to the supply of insurance.

The advantages and limitations of commercial risk transfer are generalized based on Freeman et al. (ibid):

Advantages

- Predictable compensation of losses
- Higher reliability than foreign disaster relief assistance
- Strengthening of the private sector
- Payment of premiums reflects the actual risk
- Equitable distribution of costs and benefits
- Encouraging the adoption of measures to minimize damages

Limitations

- Property owners pay premiums that do not reflect their actual risk.
- Insufficient hazard database to predict average annual losses.
- Premium rates do not bear the full cost.
- Private insurance unobtainable in high-risk areas.
- Significant number of policyholders are underinsured.
- Low voluntary uptake.
- Unlikely fully reimbursement of a loss.

8.10 Critical Infrastructure

Critical infrastructure comprises assets, systems, and networks, whether physical or virtual that is vital for a proper functioning of society and thus defines a key responsibility of a state. An undisturbed functioning of a society is greatly dependent on the provision of essential goods and services. The threat can be caused by natural disasters, technical failure, man-made disasters, international terrorism, criminal acts, or civil war. The challenge to arrive at a higher level of resilience of the critical infrastructure has developed as a critical target in the last decades especially as many infrastructure systems are of a highly interconnected nature. Moreover most infrastructure assets and systems seen as critical for our society are privately operated. A system's interruption can thus result in unanticipated and cascading impacts from events across infrastructure sectors and geographical areas. Generally speaking, critical infrastructure consists of systems and assets that:

- Supply and distribute (energy, oil, gas, electricity, water, sewage).
- Allow communication (telephone, radio/TV, IT technology).
- Provide transportation (roads, railways, air transport).
- Maintain law and order (police).
- Secure personal integrity (fire brigade).
- Provide medical health services and other livelihood-related services (rescue).
- Secure the functioning of public institutions (schools, etc.).

National authorities are in general responsible for setting rules and regulations for the protection of infrastructure facilities in their territories, involving measures oriented to prevent disruptions, mitigate damage, and restore supply under the best conditions. Critical infrastructure comprises facilities that are moreover characterized by high capital intensiveness and high public investment at all levels of government. Moreover they are directly critical to the nation's welfare as well as to the nation's economy. The United States' "National Strategy for Homeland Security" listed the following critical infrastructure sectors:

- Agriculture
- Food production
- Water supply and sewage
- Public health

- Emergency services
- Government (national, federal, local)
- Industry (hazardous chemical industry, biological substances)
- Information and telecommunications (public and private media)
- Energy generation and distribution
- Transportation (road, rail, air)
- Banking and finance

In addition to these critical sectors, the strategy also introduces the concept of "key assets" as a subset of nationally important key resources. The strategy defines "key assets" as

> "[I]ndividual" targets whose destruction would not only endanger vital systems, but could create local disaster or profoundly damage a nation's morale or confidence. Key assets include symbols or historical attractions, such as prominent national, state, or local monuments and national symbols. Key assets also include individual or localized facilities that deserve special protection because of their destructive potential or their value to the local community.

The number of sectors included under the definition "critical infrastructure" has expanded in the last decades from the most basic public works to a much broader set of institutional, social, economic, ecological, and even psychological issues. Today due to the emerging threat from international terrorism the issue "critical infrastructure" risk assessment of virtual assets and cyberspace systems has become a dominant issue, especially due to its technological and operational interconnectedness. Forty percent of international value creation today is based on the computer and communication technology, a situation that is envisaged to increase even further, securing the IT-sector indispensable for sustainable economic and social development. Moreover in many countries, the energy supply is based on nuclear power generation that poses a serious threat to technical and man-made failure as well as to terrorist attacks. As due to globalization many such systems are often no longer "at the hands" of one nation or proprietor—such as the transnational energy networks that result in extensive cross-border infrastructure—an interruption immediately affects the functioning of several states.

Incapacitation or destruction therefore are of great concern to nations as well as private companies as this might have a debilitating effect on national security, the economy, public health, or any combination thereof. The fact that a vast majority of critical infrastructure is owned and operated by the private sector, any risk reduction strategy has to be based on building partnerships to ensure security and resilience of the critical infrastructure assets and systems. In order to increase resilience of the infrastructure, for instance, the United States of America has set up a specific legal framework. As in most countries of the world, the central responsibility was handed over to the ministry of the interior mandating it to work out the legal base and to assure its respective implementation. The central objective of this was to guide the everyday work in protecting the system to enhance resilience. A prerequisite to achieve comprehensive infrastructure protection is seen in the identification and validation of the critical assets. Moreover the strategy stipulates the local authorities, private companies, and federal agencies have a shared responsibility for the critical assets and demands they jointly contribute to the "National Infrastructure Assurance Plan" by assessing the vulnerabilities of the sector to natural disasters, or physical or cyber-based attacks. The assurance plan targets developing and maintaining lists of critical assets with at least a minimum importance to become an essential part of the respective protection strategies. A main pillar of any risk assessment strategy is to establish and operationalize a "Risk and Crisis Management Plan" (BBK 2008) to evaluate the respective company's elements at risk and put a value on each of the key assets. The impact of loss is estimated at least by using a rating system based on user-defined criteria. The risk and crisis assessment should cover all phases of the production cycle from planning to operation. The task should comprise a check whether the particular risk profiles can be managed and whether preventive measures taken are capable of increasing resilience. Such an evaluation should be carried out regularly, preferably once a year. Additional evaluations are necessary after countermeasures have been implemented, but are indispensable if the threat situation changes. When risk and crisis management are taken seriously it can provide long-term added value and thus are a management tool to lay the basis for a continuous optimization of the organization's level of security.

References

Arskal Salim (2010): Konflikt nach der Katastrophe. – in: Umwelt & Klima – Rechtswissenschaften, Max Planck Forschung, Hefte, Nr. 1, Berlin

Aven, T. & Renn, O. (2009): The Role of Quantitative Risk Assessments for Characterizing Risk and Uncertainty and Delineating Appropriate Risk Management Options, with special emphasis on Terrorism Risk. - Risk Analysis, Vol. 29.4, p.587, Wiley Online Library

Aven, T. & Renn, O. (2010): Risk Management and Governance: Concepts, Guidelines and Applications.- Risk, Governance and Society, Springer Dordrecht, London, New York

BBK (2008): Protecting Critical Infrastructures – Risk and Crisis Management. – German Federal Ministry for the Interior (BMI), Berlin

Beaudoin, C.E. (2007): Media effects on public safety following a Natural Disaster: Testing Lagged dependent variable models. - Journalism & Mass Communication Quarterly Vol. 5, p.695 –714, Sage Journals (online: http://intl-jmq.sagepub.com)

Berz, G. (1997): Insurance: halting an ominous trend - Natural disasters be prepared. - The UNESCO Courier, United Nations Educational, Scientific and Cultural Organization (UNESCO), p.50, Geneva

BMU (2004): Risikomanagement im Rahmen der Störfallverordnung - Bericht des Arbeitskreises Technische Systeme, Risiko und Verständigungs-prozesse der Störfall-Kommission; SFK-GS-41.- Bundesministerium für Umwelt, Naturschutz und Reaktorsicherheit (BMU), Berlin

Böschen, S. Dressel, K., Schneider, M. & Viehöver.W. (2002): Pro und Kontra der Trennung von Risikobewertung und Risikomanagement –Diskussionsstand in Deutschland und Europa. - Gutachten im Rahmen des TAB-Projektes „Strukturen der Organisation und Kommunikation im Bereich der Erforschung übertragbarer spongiformer Enzephalopathien (TSE)".- TAB Diskussionspapier Nr. 10, Berlin

Bründl, M. (2009): Risikokonzept für Naturgefahren – Leitfaden (eds). - Nationale Plattform für Naturgefahren (PLANAT), Bern

BUWAL (1996): Richtlinien – Beurteilungskriterien zur Störfallverordnung StFV. - Bundesamt für Umwelt, Wald und Landschaft (BUWAL), Bern (online: www-umwelt.schweiz.ch.publikationen. Code VU-3818.D)

Carroll, R. (ed)(2001): Risk management handbook for health care organizations. - American Society for Healthcare Risk Management, 3rd edition, Jossey-Bass Inc., Publishers, John Wiley & Sons, Hoboken NJ

Churchill, C. (2006) Protecting the poor – A microinsurance compendium. - International Labour Organization (ILO) in association with Munich Re Foundation, ILO-Geneva (online: www.ilo.org)

Combs, B.S. & Slovic, P. (1979): Newspaper coverage of cause of death. – Journalism Quarterly, Vol. 56, p.837-843, Chicago IL

Covello, V.T. & Allen, F. (1988): Seven Cardinal Rules of Risk Communication.- Unites States Environmental Protection Agency (USEPA), Office of Policy Analysis, Washington DC

Cummins, J.D. & Mahul, O. (2008): Catastrophe Risk Financing in Developing Countries, Principles for Public Intervention – Overview.- The International Bank for Reconstruction and Development / The World Bank, Washington DC

Damnjanovic,I., Aslan, Z. & Mander, J. (2010): Market-Implied Spread for Earthquake CAT Bonds - Financial Implications of Engineering Decisions.- Risk Analysis Vol. 30, No. 12, p.1753-1770, Wiley Online Library (online: DOI: 10.1111/j.1539-6924.2010.01491.x)

DFID (2011): Defining Disaster Resilience - A DFID Approach Paper. - Department for International Development, London

Dilley, M., Chen, R.S., Deichmann, U., Lerner-Lam, A.L., Arnold, M. with Agwe, J., Buys, P., Kjekstad, O., Lyon, B. & Yetman, G. (2005): Natural disaster hotspots - A Global Risk Analysis - Synthesis Report.- The International Bank for Reconstruction and Development / The World Bank and Columbia University, Washington, DC

Dombrowski, W.R. (1992): Bürgerkonzeptionierter Zivil- und Katastrophenschutz. Das Konzept einer Planungszelle Zivil- und Katastrophenschutz.- *Bundesamt für Bevölkerungsschutz und Katastrophenhilfe (*BBK), Bonn

EU-EWFD (2000): Directive 2000/60/EC of the European Parliament and of the Council establishing a framework for the Community action in the field of water policy" (EU Water Framework Directive-EWFD).- European Commission, Brussels

Fell, R. & Hartford, D. (1997): Landslide risk management. - in: Cruden, D. & Fell, R. (eds): Landslide Risk Assessment, p.71, Balkema, Rotterdam

FEMA (1968): The National Flood Insurance Program (NFIP).- Federal Emergency Management Agency (FEMA) (online: www.fema.gov)

FEMA (1990): National Flood Insurance Program Community Rating System (CRS). – Federal Insurance and Mitigation Administration, Federal Emergency Management Agency (FEMA), Fact Sheet March 2014, Washington DC (online:www.fema.gov/national-flood-insurance-program-2/community-rating-system)

FEMA (2003): Principles of emergency management. - Federal Emergency Management Agency (FEMA), Emergency Management Institute, Independent Study IS230, Washington DC

Fischhoff, B., Lichtenstein, S., Slovic, P., Derby, S.L. & Keeney, R.L. (1981): Acceptable risk.- Cambridge University Press, Cambridge

Franks, S. (2006): The CARMA Report: Western Media Coverage of Humanitarian Disasters. - The Political Quarterly, Vol. 77, Issue 2, p.281–284, Wiley Online Library

Freeman, P.K., Martin, L.A., Linnerooth-Bayer, J., Mechler, R., Pflug, G. & Warner, K. (2003): Disaster risk management - National Systems for the Comprehensive Management of Disaster Risk and Financial Strategies for Natural Disaster Reconstruction. - Inter-American Development Bank, Sustainable Development Department, Environment Division Integration and Regional Programs Department, Regional Policy Dialogue, Washington DC

Gurenko, E., Lester, R., Mahul, O. & Gonulal, S.O. (2006): Earthquake Insurance in Turkey-History of the Turkish Catastrophe Insurance Pool. – The World Bank, Washington DC

Ilkesik, M. (2002): Istanbul study - disaster prevention/mitigation basic plan in Istanbul Including seismic microzonation in the Republic of Turkey. Metropolitan Municipality Istanbul

Earthquake Master Plan. – prepared for the Istanbul Metropolitan Municipality; Japan International Cooperation Agency (JICA) and the Middle East Technical University, Istanbul, Technical University, Bosporus University and Yıldız Technical University, Istanbul
ILO (2001): Guidelines on occupational safety and health management systems.- International Labour Office (ILO), ILO-OSH 2001 (2nd edition), Geneva
IRGC (2009): Risk Governance: Coping with Uncertainty in a Complex World. – prepared for the International Risk Governance Council (IGRC), Natural Hazards, Vol. 48, No. 2, p.313-314, Springer-online
IRGC (2010): What is Risk Governance. - International Risk Governance Council (IGRC), Lausanne
Kahneman, D., Slovic,P. & Tversky,A. (eds.) (1982): Judgment Under Uncertainty: Heuristics and Biase.- Cambridge University Press, Cambridge MD
Kaplan, S. & Garrick, B.J. (1981): Risk Analysis, Vol. 1, No 1, Wiley Online Library
Kasperson, R.E., Renn, O., Slovic, P., Brown, H.S., Emel, J., Goble, R., Kasperson, J.X. & Ratick S. (1988): The social amplification of Risk: A Conceptual Framework. – Risk Analysis, Vol. 8, No. 2, Wiley Online Library
Kreimer, A. & Munasinghe, M. (1991): Managing the Environment and Natural Disasters.- The International Bank for Reconstruction and Development/The World Bank, Washington DC
Kunreuther, H. (1966): Economic theory and natural disaster behaviour . - Economic and Political Studies Division, Institute for Defence Analyses, Arlington VA
Lichtenstein, S., Slovic, P., Fischhoff, B., Layman, M. & Combs, B. (1978): Judged frequency of lethal events. – Journal of Experimental Psychology, Human Learning and Memory, Vol. 465, p.551-578, Washington DC
Linnerooth-Bayer, J. & Mechler, R. (2009): Insurance against Losses from Natural Disasters in Developing Countries. - United Nations, Department of Economic and Social Affairs (DESA), Working Paper No. 85, New York NY
Luhmann, N. (1988): Soziologie des Risikos. – Springer, Berlin/New York
Luhmann, N. (2001): Sozial Systeme-Grundriss einer allgemeinen Theorie. – Suhrkamp Taschenbuch Wissenschaft, neue Auflage 2001, Frankfurt
Malzahn, D. & Plapp T. (Hg.) (2004): Disaster and Society – From Hazard Assessment to Risk Reduction, Logos, Berlin
Miller, C.H., Adame, B.J. & Moore, S.D. (2013): Vested interest theory and disaster preparedness. – Disasters, Vol. 37, No. 1, Blackwell, Wiley Online Library
Morgan, C. (1997): A regulatory perspective on slope hazards and associated risks to life. – in: Cruden, D. and Fell, R. (1997): Landslide Risk Assessment.- P. 285-295, Balkema, Rotterdam
MunichRe (2012): Topics Geo online – 2012. - Munich Reinsurance Company, Munich
MunichRe (2013): Topics Geo online - 2013. - Munich Reinsurance Company, Munich
NAIS (1999): National Agricultural Insurance Scheme in India (NAIS).- Dept. of Financial Services Ministry of Finance, Government of India, Delhi (online: financialservices.gov.in/insurance/gssois/nais.asp)
OECD (2005): Does gender equality spur growth?- OECD Observer, No. 250, Paris
OES (2007): State of California – Multi-Hazard Mitigation Plan. - The Governor's Office of Emergency Services, Mather CA (online: www.oes.ca.gov)
Olofsson, A. & Rashid, S. (2011): The white (males) effect and risk perception: Can equality make a difference?. - Risk Analysis, Vol. 31, No.6, p.1016-1032, Wiley Online Library
Palmer, C. (2003): Risk perception: Another look at the 'white male' effect. - Health, Risk & Society, Vol. 5, No. 1 (online: DOI: 10.1080/1369857031000066014)
Plapp, S.T. (2003): Wahrnehmung von Risiken aus Naturkatastrophen - Eine empirische Untersuchung in sechs gefährdeten Gebieten Süd– und Westdeutschlands.- Dissertation, University of Karlsruhe
Quarantelli, E.L. (1995): Disaster planning, emergency management and civil protection: The historical development of organized efforts to plan for and to respond to disasters. – Disaster Research Center, University of Delaware, Delaware NW (online: eduelqdrc@udel.edu)

Raju S.S. & Chand, R. (2008): Agricultural Insurance in India - Problems and Prospects. – National Centre for Agricultural Economics and Policy Research, Indian Council of Agricultural Research, NCAP Working Paper No. 8, Delhi

Renn, O. (1989): Risikowahrnehmung und Risikobewertung in der Gesellschaft – in: Hosemann, G. (ed): Risiko in der Industriegesellschaft, Analyse, Vorsorge und Akzeptanz. - Erlanger Forschungen, Vol. 19, p. 176-192, Erlangen

Renn, O. & Levine, D. (1991): Communicating risks to the public - International perspectives . – Kluver Academic Press, Dordrecht

Rivers, L., Arvai, J. & Slovic, P. (2010): Beyond a simple case of Black and White: Searching for the white male effect in the African-American Community.- Risk Analysis, Vol. 30, No.1, p.65-77, Wiley Online Library

Rohrmann, B. & Renn, O. (eds.)(2000): Cross-Cultural Risk Perception - A Survey of Empirical Studies. - Risk, Governance and Society, Vol. 13, p.241, Springer Heidelberg

Ruhmann, G. & Kohring, M. (1996): Staatliche Risikokommunikation bei Katastrophen – Informationspolitik und Akzeptanz. - Zivilschutz-Forschung, Neue Folge Band 27, *Bundesamt für Bevölkerungsschutz und Katastrophenhilfe* (BBK), Bonn

Rusch, G. (2011): Medien in der Katastrophe – Katastrophen in den Medien. - Center for Interdisciplinary Research (ZiF), University of Bielefeld, Bielefeld (online: communicating_disaster@uni-bielefeld.de)

SAMSHA (2002): Communicating in a Crisis: Risk Communication Guidelines for Public Officials.-U.S. Department of Health and Human Services Substance Abuse and Mental Health Services Administration, Rockville MD

Schütz, A. (1971): Gesammelte Aufsätze 1. Das Problem der sozialen Wirklichkeit .- cited in: Geenen, E.M. (2004): Social Structures, Trust and Public Debate on Risk; Disasters and Society: from hazard assessment to risk reduction International Conference: Disasters and Society- from Hazard Assessment to Risk Reduction, p.251-256, Logos Verlag, Berlin

Sjöberg, L. & Engelberg, E. (2010): Risk perception and movies: A study of availability as a factor in risk perception. - Risk Analysis, Vol. 30, No. 1, p. 95-106, Wiley Online Library

Sinisi, L. (2004): Public concerns and risk communication. - National Environmental Protection Agency, Roma (online: sinisi@anapa.it)

Slovic, P. (1992): Perception of risk: Reflections on the psychometric paradigm. – in: Krimsky, S. & Golding, D. (eds.): Social theories of risk, p.117-152, Preager, New York NY

Slovic, P. (1987): The perception of risk. – Science, Vol. 236, p.280-285, Washington DC

Slovic, P. (1999): Trust, emotion, sex, politics and science- Surveying the risk assessment battlefield. – Risk Analysis, Vol. 19, No. 4, Wiley Online Library

Slovic, P. (2000): The Perception of Risk.- Earthscan Publications, p.473, London

Stallen, P.J.M. (1991): Developing communications about risks of major industrial accidents in the Netherlands. – in: Kasperson, R.E. & Stallen, P.J.N.M. (eds.): Communicating risks to the public, p.55-66, Kluver Academic Press, Dordrecht

Süselbeck, J. (2013): War sells, but who's buying? Zur Emotionalisierung durch Kriegsdarstellungen in den Medien. - Aus Politik und Zeitgeschichte, Jahrgang 2013, No. 32-33, Das Parlament, Bundeszentrale für Politische Bildung, Bonn

Swiss Re (2002): Natur und Man-made Katastrophen 2001. - Sigma, Nr. 1, Swiss Reinsurance Company, Zurich

Twigg, J. (2002): Lessons from disaster preparedness. – International Conference on Climate Change and Disaster Preparedness, (26-28 June 2002), The Hague

UNISDR (2004): Living with Risk - A Global Review of Disaster Reduction Initiatives". – United Nation International Strategy of Disaster Reduction (UNISDR), Geneva

UNISDR (2005): Hyogo Framework for Action 2005-2015: Building the Resilience of Nations and Communities to Disasters. - Extract from the Final Report of the World Conference on Disaster Reduction (A/CONF.206/6), p. 25, Geneva

UNISDR (2011): Global Assessment Report on Disaster Risk Reduction: Revealing risk, redefining development. – United Nation International Strategy of Disaster Reduction (UNISDR), United Nations, Geneva

UNISDR/UNDP/IUCN (2009): Making Disaster Risk Reduction Gender-Sensitive. – United Nations International Strategy for Disaster Risk Reduction (ISDR), Policy and Practical Guidelines, United Nations, Geneva

USGS (1999): Implications for Earthquake Risk Reduction in the United States from The Kocaeli, Turkey, Earthquake of August 17, 1999.- Unites States Geological Survey (USGS), Circular 1193, Reston VA (online: http://pubs.usgs.gov/circ/2000/c1193/)

Vasterman, P., Yzermans, C.R. & Dirkzwager, A.J.E. (2005): The Role of the Media and Media Hypes in the Aftermath of Disasters. - Epidemiological Reviews, Vol. 27, p.107-114, Oxford University Press, Oxford

v.Piechowski (2000): Störfallverordnung der Schweiz – Handbuch I, Anhang G; cited in WBGU: Welt im Wandel–Strategien zur Bewältigung globaler Umweltrisiken.– Jahresgutachten 1998, Springer, Heidelberg

Watzlawick, P. (2009): Anleitung zum Unglücklichsein. -15. Auflage, Pier, München

WBGU (1999): Welt im Wandel – Strategien zur Bewältigung globaler Umweltrisiken. – Wissenschaftlicher Beirat der Bundesregierung Globale Umweltveränderungen, Hauptgutachten 1998, Springer, Heidelberg

Wisner, B., Blaikie, P., Cannon, T. & Davis, I. (2004): At Risk - Natural hazards, people's vulnerability and disasters. - Routlegde, London

Wolgalter. M. S. et al. (1987): Effectiveness of Warnings. – Human Factors, Vol. 29, p.599-622

World Bank (2001): World Development Report 2001 – Attacking poverty. - The International Bank for Reconstruction and Development / The World Bank, Washington DC

World Bank (2014) MultiCat Program. - The International Bank for Reconstruction and Development / The World Bank, Treasury, Washington DC (online: www.worldbank.org)

Zahran, S., Peek, L., Snodgrass, J.G., Weiler, S. & Hempel, L, (2011): Economic of disaster risk, social vulnerability and mental health resilience. - Risk Analysis, Vol. 31, No.7, p.1107-1119

Chapter 9
Community Awareness and Participation

9.1 Community-Based Disaster Risk Management (CBDRM)

A "community" in general comprises both rural villages and urban neighborhoods. But what exactly defines a community is often difficult to describe (UNISDR 2006). This definition also holds true for disaster risk management, although often only a segment of the community can be exposed to a hazard whereas the other part can even benefit from it. But a community cannot be defined just from geographical location alone, but by the area where people share the same living conditions, have a common social structure and cultural interest, and speak the same language. The complex nature of indicators makes it difficult for an outsider to identify where a community starts and ends; however, generally speaking, the geographical area often comprises people that can be grouped under the above-given indicators.

A community-based organization is thus defined as an organization that is either publicly or privately organized, such as a political party, a sports club, a religious group, and/or being a member of a social network (i.e., elderly, orphans, disabled, children, etc.) or that is representative of the community as a whole, or of a significant segment of it. Community-based disaster risk reduction is a process that initiates activities and actions to increase resilience to reach a better and more sustained life, either within a community or of a community in total. Such actions and activities can therefore vary greatly from one community to another. Rifkin and colleagues (Rifkin et al. 1988) defined community participation as a "social process whereby specific groups with shared needs living in a defined geographic area actively pursue identification of their needs, take decisions and establish mechanisms to meet those needs." Thus community participation is a social process whereby specific groups are not restricted to being observers but can be actively involved in decision making and implementation. Since Arnstein (1969) introduced the "ladder" as an indicator for the level of community involvement, it has been possible to compare participation within and between different societal groups. Like the steps of a ladder, the degree of participation can be identified

© Springer International Publishing Switzerland 2016

U. Ranke, *Natural Disaster Risk Management*,
DOI 10.1007/978-3-319-20675-2_9

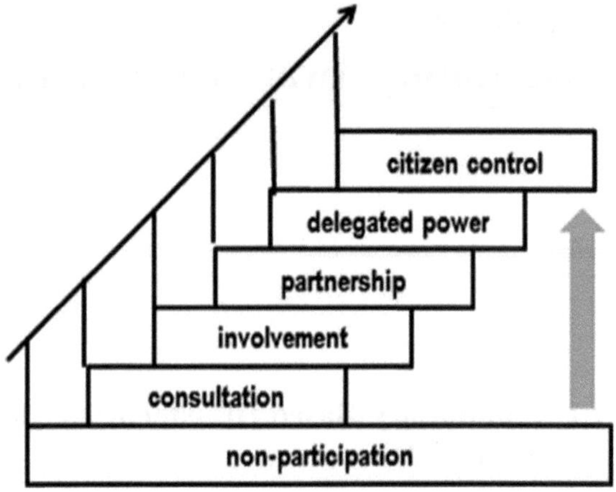

Fig. 9.1 Ladder of participation (Based on Arnstein 1969)

from "nonparticipation" (lowest step) ultimately to reach the top of the ladder, "citizen control" (Fig. 9.1). The ladder example, although it impressively describes the anticipated increase in participation, cannot neglect that it is not compulsory that step No. 2, for instance, leads to reaching step No. 3. The interdependencies in communal reality are much more complex and therefore it should be acknowledged that improvements at one step will result in changes at other steps.

There are a number of terms that describe society's involvement: "participation," "empowerment," "ownership," "capacity," or "competence" and some others. All these terms, even when their exact definitions may differ, all have in common that they are oriented to strengthen the ability of risk-prone societies to withstand possible adverse future impact. However, there is a certain logical interrelationship suggesting that community "participation" enhances community "ownership". "Ownership" in turn leads to increased "capacity" or "competence" and vice versa (Robertson and Minkler 1994). An emergency management that invites interest group representatives to participate in a census-based collaboration process by involving stakeholders changes risk-management behavior from a traditional public-meeting format to methods that involve relatively small groups of people (Beierle 2002). By such an intensification of societal involvement public participation is extended and deepened and will increase "capacity" on the individual and community levels as well. In this regard, the literature of disaster reduction is full of information on the importance that women play in disaster risk reduction.

All too often women are not seriously incorporated in development decisions, particularly in the preparatory process of defining the rules and regulations of the decision-making process. They are often invited to sanction the already taken decisions. Although it has been published for quite a long time, the *Gender Terrain of Disaster* (Enarson and Morrow 1998) clearly indicates that one of first steps towards an effective organizational and policy change has to reflect the realities of women's lives, an approach that no longer excludes 50 % of a society from developing planning decisions.

But shifting the focus from the individual to the community to reach a widespread behavioral change is best achieved by changes in the norms of acceptable behavior at the community level. It turned out that the moment society is involved as whole, changes are more likely to occur rather than if only individuals are addressed. The envisaged increase in resilience is best achieved by enabling individuals and communities to identify the cause–effect relationship of hazards and disasters as well as by participating in defining their specific vulnerability exposure and by an integrated working out of solutions to those problems.

The level of competence achieved can be valued by assessing the links of personal elements (engagement, trust, social cohesiveness) in socioeconomic and cultural contexts in strengthening:

- Social networks, social learning
- Capacity building
- Conflict resolution
- Community participation and stakeholder involvement
- Collaborative decision making
- Leadership and capability to assess the root causes
- Sustainability of the organizational structure
- Technical, operational, and financial resources allocation
- Identification of the social dimension of the root causes
- Accepting the role of seconded experts as mediating agents

Experience of disaster risk reduction programs all over the world indicates that it is not reasonably possible to quantify the success ratio of community participation in disaster resilience. Such an assessment is considered not meaningful as the dynamics and changes associated with "participation" are hardly to be quantified. Therefore resort should be made to the above-given set of indicators. Thus a more realistic picture can be drawn from the situation and the outcomes. Community participation and empowerment have to be seen as a dynamic process rather than a measurable static outcome.

Moreover programs on disaster risk reduction should not be designed far from the places of risk, as this implies a theoretical—although systematically, technically, and implementation driven—attempt. Such attempts often lack the inclusion of the "real" needs of the exposed population. Even when a society is at high risk, does this mean that it is helpless and lacks long-standing mitigation experience? Many communities have gained much experience from the grass roots-level crisis reaction. This approach is often referred to as a "bottom-up" process, because many a solution comes from the community itself and not as a directive from higher authorities. Disaster-stricken communities, although when they obviously were disenfranchised by a disaster, are not unable and unwilling to provide locally adoptable mitigation concepts. Therefore planners and managers that make full use of citizen expertise and energy will be more effective improving the safety and survival chances of their communities. Based on experience from the Northridge earthquake, where it was observed that a combination of local activism coupled with financial and organizational support from government and nongovernment agencies can be highly effective, FEMA integrated public and private partnerships as compulsory in the current US "Earthquake Hazard Mitigation Handbook" (FEMA 2002).

Community-based disaster risk management is an organizational and educational strategy to increase resilience through a set of community-oriented measures. It is people and development oriented and focuses on prevention and mitigation, preparedness, emergency response, and recovery from natural disaster impact. The aim of CBDRM is to create resistant societies by reducing the probability of disasters through target-oriented risk reduction measures, to reduce the social and economic consequences (death, casualty, injuries, direct, indirect losses) from damage, while simultaneously increasing the coping capacities to strengthen the recovery (ADPC 2006a).

The starting point to reach a sustainable level of disaster coping capacity is to define what makes a society vulnerable and what indicators define the capacity to withstand adverse effects from disasters. It is a well-known fact that people all over the world have learned to adapt permanently to crisis, and with every crisis they continuously come up with creative solutions (ADPC 2006b). Thus resilience is something like a moving target, but realistically it may not be possible for communities to achieve absolute resilience against hazards or other risk factors neither in industrialized nations nor in developing countries ("zero risk is out of reach in the contemporary world"). However CBDRM is a vital means to help communities achieve a certain level of sustainability. The level envisaged should be defined on indicators the stakeholders have agreed upon. Indicators that describe a disaster-resilient society are:

- A function community organization
- A disaster risk reduction strategy with an all-hazard risk assessment
- An early warning system
- Knowledgeable people on the cause–effect relationship of natural disasters
- Trained manpower for risk assessment, rescue, and relief
- Defined escape routes and evacuation centers
- Medical services in disaster-safe places
- A disaster-safe main transport and communication system
- Sustainable water and power supply
- Functioning relationship between disaster-exposed societal groups and local authorities
- Knowledge of risks and risk reduction actions

CBDRM moreover should have as a target to create a conducive although enabling environment for locally adoptable risk reduction. This should include the establishment of organizational schemes that allow independent forms of monitoring of the success ratio, generating objective judgment (not only on the side of the experts), and measures to overcome the often inherent imbalance of power between those who are recipients and those who are donors. Thus CBDRM should be the central institution to initiate an active search for a best fit with the particular local circumstances of the risk-affected areas by providing tools and topics for a better understanding of the cause–effect relationship of hazard and disaster, for defining ways and means to a sustainable risk reduction, to define supporters and objectors, and to outline scientific and technical expertise required. Experience from other risk-affected regions may also be included in the communal decision making as well as transferring expertise gained to other risk-affected regions. Moreover the decision should be targeted to achieve an equilibrium between the "technically necessary" segments of disaster reduction and the "maximum honoring" of the people's concerns. Such an equilibrium can best be achieved by broad involvement of the public representatives and an normative equalization of participants' power (Renn et al. 1995). Regardless of the different experiences with community-based disaster reduction, they all confirm that stakeholder involvement proved to better address the needs of the populations at risk. The technical expertise should be generally based on scientific evidence, elaborated independently, neutral and transparent, and the personal, emotional, and social aspects of the problem-solving mechanism should be brought in from the representatives of the different social groups in a process guided by an external mediator. To solve the dichotomy between science and politics, Beierle (ibid) proposes iteratively to bridge both spheres to accommodate the social aspects as well as technology and science. He points out that "technocrats argue that mitigation can best be achieved by science-based tools like risk assessment, building codes, wh while the advocates of the stakeholder involvement argue identifying the needs of the people at risk is a prerequisite for a higher people motivation and reflects a much broader the interests of the risk exposed."

However, stakeholder involvement has also raised some concern that mostly originates from the fact that for instance spatial planning and definition of rules and regulations for natural disaster reduction generally refer to the technical and operational sector. Therefore the critics emphasized that a too broad involvement of too many stakeholders from all sectors of society may "water down" the evidence to a "mixture" of aspects that are technically easy and socially difficult to define. The claim is that many "stakeholder[s] make inadequate use of scientific information and analysis and are too ready to sacrifice technical quality for political rationality" (Beierle, ibid). Beierle could, however, prove from a number of environmental protection planning processes in the United States, that this concern does not exist in reality and that an early and comprehensive stakeholder involvement in general led to technically adequate decisions supplemented with social aspects. There are no examples where community involvement resulted in a low-quality decision.

As it turned out to be an effective measure of disaster reduction, community-based and community-driven development projects have become an important form of development assistance worldwide; the World Bank's portfolio alone is approximately US$7 billion. A review by the World Bank (Manzuri and Rao 2004) of the conceptual foundations and effectiveness of such community-based programs showed that projects based on community participation often have not been that effective in reaching the needs of the poor as planned. There is some evidence that such projects, although they often created an effective community infrastructure, did not provide clear evidence of having reached the poor precarious sector. The reason is seen in the fact that most such projects are carried out by external specialists that automatically cooperate with local elites, resulting in getting the intimal project orientation out of focus. Several qualitative studies indicate that the sustainability of community-based initiatives depends crucially on an enabling institutional environment, which requires government commitment, and on accountability of leaders to their community to avoid "supply-driven demand-driven" development. External agents strongly influence project success, but facilitators are often poorly trained, particularly in rapidly scaled-up programs. The naive application of complex contextual concepts such as participation, social capital, and empowerment is endemic among project implementers and contributes to poor design and implementation. The evidence suggests that community-based and -driven development projects are best undertaken in a context-specific manner, with a long time horizon and with careful and well-designed monitoring and evaluation systems.

Although there is quite some concern about the effectiveness of community-based disaster risk reduction, the concept nevertheless proved a vital instrument of disaster risk management. Community-based means the community and its people are the central stakeholders and thus are to be comprehensively and actively engaged in the identification, analysis, treatment, monitoring, and evaluation of their specific risks. This means risk mitigation is based on the people's experience on those making decisions and implementating disaster risk management activities.

A handbook published by the Asian Disaster Preparedness Center (Abarquez and Murshed 2004) identified seven sequential stages that can be executed by disaster risk managers to reduce future risks before the occurrence of a disaster or after one has happened. Each stage grows out of the preceding stage and leads to further action. Together, the sequence builds up a powerful and proved planning and implementation tool:

- *Selecting the Community*: This is the process of choosing the most vulnerable communities for assistance on risk reduction to be rendered should be based on a predefined set of criteria.
- *Understanding the Community*: This step basically builds up a mutual, trustful, and transparent relationship with the local people. By such relationship, the general position of the social, economic, political, and economic aspects of the community is understood.
- *Participatory Disaster Risk Assessment*: This is a diagnostic process to identify the risks the community faces and to define the potential of the people to overcome those risks. The process involves hazard, vulnerability, and capacity assessments. In doing the assessments, people's perception of risk is considered.
- *Participatory Disaster Risk Management Planning*: In this step the people themselves identify risk reduction measures that will reduce vulnerabilities and enhance capacities. These risk reduction measures are then translated into a community disaster risk management plan.

- *Building a Community Disaster Risk Management Organization (CDRMO)*: For an effective disaster risk management a community organization that serves the specific local requirements is indispensable. Therefore it is imperative to build a community organization that includes representatives of all stakeholders Next to the organizational setup training should also be rendered to the stakeholders and the members of the organization.
- *Community-Managed Implementation*: The CDRMO should lead in working out and implementing a community risk reduction plan. This should include means to motivate all members of the community for their support.
- *Participatory Monitoring and Evaluation*: An assessment and evaluation strategy has to be institutionalized to secure a steady and unrestricted exchange of information among all people involved in the project, as well as to lay the base for a critical review of the achievements made. This project evaluation serves as the factual base for designing future activities.

As the basic intention of CBDRM is to get all stakeholders involved in the mitigation process, automatically quite a large number of individuals and social groups and political decision makers become part of the process. The actors in this multiple stakeholder scenario can be divided into two broad categories:

Insiders

Those who are either affected directly by a disaster or those who have the legal mandate to initiate and carry out the identified mitigation activities. This comprises individuals, families, representatives of the civil societies, as well as public services and the private sector. Very often these two groups are assisted by seconded external advisors. Indispensable to harmonize the different actor's interests, expectations, and fears, is to have an organizational setup conducive to the local situation: a "community disaster risk management organization." This organization is to act as the focal point of the management of disaster risk. It moderates between the different stakeholders and those responsible for the development decision making. Later it organizes and evaluates the implementation of the countermeasures. In order to establish effective working relations, the management main objective is to facilitate a broad consensus on targets, strategies, and methodologies among the multiple stakeholders in the community.

Outsiders

The term "outsiders" refer to those individuals, organizations, and stakeholders located within the community, but not directly risk exposed. Outsiders can include the government departments and agencies, but also the private sector and other agencies necessary to reach a consensus. Although they themselves are only indirectly related, they should not be excluded from the process. Such groups often have a great interest in reducing the community's vulnerability, for instance, labor unions or the business sector that employs many people in the area. Their role can be to support the local efforts in reducing vulnerabilities and enhancing coping capacities for the longer term. They can do this through providing their own technical, material, financial, and political support that may be initiated as part of their agenda. The management organization should include them from the very beginning; but such incorporation should be based on an all-stakeholder's consensus. Often outsiders dispose a variety of financial and technical resources and managerial expertise and even more important, often have great political influence that may help to push forward the community priorities.

In countries where emergency management is well organized and technically equipped to a high standard, the people mainly rely on help and support from mandated authorities, however, the risk-exposed in developing countries very often stand on their own in the case of an emergency. It is a well-known fact 90 % of all immediate rescue operations are done by the people who have been affected by a disaster themselves. Such help is often rendered by neighbors and volunteers without any technical resources and plan. Thus emergency management created the saying, "every mitigation is local," which already points to the fact that mitigation without the involvement of the population at risk will hardly be successful. In disaster risk management it became apparent that top-down approaches, when standing alone, fail to address the specific local needs of vulnerable communities. This assumption holds true especially for those countries where local emergency management is poorly established and when implemented, the mitigation and prevention measures often ignore local capacities and experience.

The key aspect of community involvement is its orientation on the sustainability of a community. It aims at empowering people to address the root causes of their vulnerability in order to change adverse socioeconomic and political structures into a sustainable paradigm of livelihood. There can be many reasons behind the often recognized lack of sustainability of disaster reduction measures. Some of them are the result of a lack of ownership, low participation, and a missing empowerment of communities at risk. Unless the disaster risk management efforts are not incorporating individual and community level, it would hardly be possible to reduce the vulnerability and losses. There is therefore no alternative to an early and comprehensive involvement of the people right from the risk assessment up to the following decision-making strategies. Different experiences in disaster is the rationale that the risk-exposed population has to be involved in any mitigation and makes it indispensable to address the group's particular demands. The only way to achieve this can be done by an early and comprehensive involvement of all stakeholders in the policy decision-making process. The reason why so many mitigation actions fail is that too often those who are designing and implementing although technically meaningful countermeasures are administrators that—often not coming from the affected regions and often not speaking the local vernacular—rely on recommendations from scientists and technicians but neglect the social dimension of disasters. Moreover mitigation measures are often implemented that proved perfect in one location, but are not suitable for another, even when the region faces the same type of hazard. For example, people regularly exposed to storm surges at the coast have a different perception of this kind of hazard than people living in narrow valleys of the Alps often threatened by avalanches in winter. So if a coastal riparian is taking a winter holiday in the Alps, he may either be over-worried about avalanche risk or not even feel worried at all, due to lack of experience; the same holds true for a Bavarian taking a holiday at the North Sea coast during storm times.

As personal experience, education, and skills significantly influence vulnerability, effective hazard mitigation and emergency response begins with an understanding of the complex ways in which social, economic, and political structures define the level of vulnerability. This understanding (Morrow 1999) can "only be achieved, when planning starts at the local level and when it engages even the most marginalized stakeholders in the process. Cultural norms regarding family size and household composition vary throughout the world, influenced by economic conditions, demographic trends and housing availability." In most rural societies technical and economic reasons still call for large extended families, whereas industrialization and urbanization result in small-sized families. Thus cultural differences can cause misunderstandings and mistrust between the disaster victims and the emergency rescue workers. The situation is even more complex in countries where people have to fight to make their daily living and do not have enough resources left to consider the possibility that a disaster might threaten their lives, even when they were complaining of the loss of lives from the last event. Furthermore many such societies are strongly traditionally organized and thus often rely on recommendations and advice from the elderly. Furthermore the elderly often base their recommendations on ethnic values or traditional beliefs rather than on a natural scientist's or technician's advocacy. For example when Mt. Merapi (Java) was about to erupt in the summer of 2006, the local administration ordered the evacuation of the settlements on the flanks of the volcano. But the local population rejected leaving the area. Leaving the house poses for many agriculturally oriented societies serious problems. Mostly these people just own a small hutch, made from local construction material, having a cow and a couple of goats and some poultry. The moment they leave the village, they fear their houses can be destroyed or no longer be protected against looting and their animals will not get food and water. So within a couple of days, the family food basis and shelter can vanish. Therefore at Mount Merapi the locals were asking for the Sultan of Yogyakarta, as their spiritual and political leader to come and tell them what to do ("We only go, if the Sultan tells us to go"). Soon after the Sultan was at the location, the people evacuated the area without any delay. But such conflicting situations not only occur in developing countries. In Germany on the occasion of different Elbe floods (2002, 2005, 2013) experience showed, that although the water level was steadily rising, a large number of residents rejected strongly evacuating their houses, with the consequence that they had to be evacuated later under much greater technical effort. This situation triggered a discussion whether the German public administration should be allowed to have the additional costs for the evacuation reimbursed by the residents when emergency management administrators must get their orders enforced in due course, therefore asking police or military forces to execute the order. The outcome is in many cases a big riot or social turmoil. This stress situation often develops although both groups have the same objective: to save the people from the potential disastrous event. So the basic problem is not the evacuation process itself and its reasoning for doing so, the main problem is that both sides speak different "languages" and do not understand each other: automatically getting into a verbal juxtaposition.

Even in areas that are regularly at risk from natural disasters, the population is very often not alert enough to their personal risk. An Australian assessment (EMA 2010) on the status of public engagement in disaster prevention revealed that: "Very often the individual do[es] not see a personal risk, very often the advices given by the authorities, are not in line and understood with resident's values and a lack of understanding of hazard safety messages."

The study concluded that:

> [T]he content of communal engagement must be focused on clarifying to what kind of risk a household is exposed, to make the house-owners and their families understand the hazard behavior they are living in, to understand what limitations the national/local emergency management services have in protecting life and properties, to learn what everybody can do on their own to reduce their specific vulnerability without affecting their normal life style.

The study worked out six principles that responsible disaster mitigation should comprise:

- Localize programs and activities where possible.
- Develop a general plan for enhancing natural hazard preparedness in a locality or region that provides a template for detailed planning and implementation and a roadmap for evaluation.
- Develop a small suite of programs/activities that focus on intermediate steps (processes) along the pathway from risk awareness to preparedness.
- Wherever appropriate, consider an all-stakeholder integrating approach to planning, program development, and research.
- Conduct and report frequent evaluations of programs and activities continually to enhance the evidence base for what works in particular contexts in community safety approaches.
- Seek to optimize the balance between central policy positions, agency-operational requirements, and specialist expertise, and community participation in planning, decision making, preparation, and response activities.

The study furthermore came to the conclusion that:

Public awareness is arguably the least and most poorly funded mitigation measure in Australia. With very few exceptions, it is undertaken as a limited auxiliary activity to the disaster management initiatives, rather than as a sustained strategic measure to raise public consciousness and understanding of hazards, risks, impacts and minimization.

A responsible risk management has to be aware of the technical/administrative and social conflicts. In order to develop a higher level of resilience the following measures should be implemented for the sake of the population at risk.

- Define the mitigation efforts as a process.
- Identify and involve all stakeholders in the decision-making process.

- Involve well-experienced experts and the powerful political leaders.
- Institutionalize roundtables.

To achieve the desired community resilience the method of community-based disaster risk management has proven its general practicability. The Asian Disaster Preparedness Center (ADPC 2006b) emphasizes that any disaster risk reduction strategy should be based on the following two aspects:

- CBRDM should become an integral part in all sectors of national disaster risk reduction policies.
- Governments should make local participation compulsory in all planning and implementation of the national risk reduction programs

9.2 Risk Reduction Education

An earthquake on September 20th, 2009 in West Sumatra killed more than 1000 people and injured many more in the city of Padang, a city that is highly prone to earthquake hazards. The Indonesian National Tsunami Warning Centre (NTWC) in Jakarta immediately identified the magnitude (7.9) of the quake and that no tsunami developed. However, the strong tremor caused widespread panic and fear of a tsunami among the people of Padang. Field research in Padang (GITEWS 2010) was carried out to uncover public reaction to the event. The study focused on the first 30 min after the tremor and interviewed 200 individuals. The study revealed that the first information on the event reached the authorities in Padang within 5 min via the Internet. As the earthquake left several local cellular telephones unable to function, the Padang mayor could receive the information only via text message. It took another 25 min until the warning was disseminated by the mayor to the general public by radio. The study revealed that there was no communication between the mayor and the Emergency Control Centre in Padang. Both actors operated separately from each other. As official information was absent in the first 30 min, the vast majority of the people were without guidance. The information spread mainly based on rumors. Over time, the radio news gradually found its way and more and more people went back home. Meanwhile half of the people had evacuated the low-lying coastal areas within 15 min. The majority of the people left the city mainly due to the strength of the quake they felt. Locally the evacuation underwent was unorganized. Panic caused massive traffic congestion as most people tried to escape on motorbikes and cars. The designated evacuation routes turned out to be not large enough to channel the masses. Moreover the people did not consider vertical evacuation a meaningful option. Those who did not follow the evacuation advice claimed different reasons for doing so. An inquiry of 102 randomly selected people (GITEWS ibid) revealed two thirds of the (interviewed) persons were aware of a tsunami risk (Fig. 9.2). They acted according to visual inspection, official information, and an individual risk estimation. About 40 % rushed to the sea to see whether the water was retreating, which convinced them

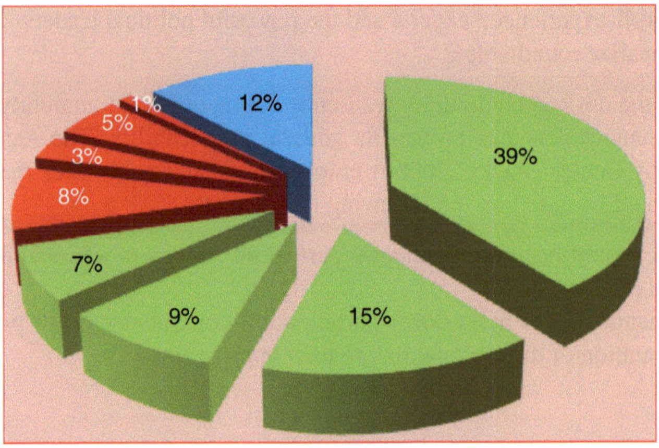

Fig. 9.2 Padang people's reasons not to evacuate (Based on: GITEWS 2010)

to refrain from evacuation. Another 20 % of the people hesitated to evacuate either to protect personal belongings, because of operational difficulties (traffic jam), or for fatalistic reasons. Only a minority of 1 % confessed not to having received any information, and 10 % claimed other reasons. The overwhelming majority of those interviewed acted on information and took their personal decisions on technical evidence. The Padang inquiry proved that information based on facts, rendered in time, and that is trustworthy is a key factor for an effective early warning. Nevertheless the situation in the city of Padang during evacuation did not fully correspond to the evacuation scheme envisaged.

A disaster such as the earthquake in Padang can affect people's lives and livelihoods in almost all parts of the world. And some societies are vulnerable and others not. Furthermore, within communities different households may be exposed differently and even within households the levels of vulnerability of individual household members may vary. There is also a great disparity in vulnerability between nations and communities, even when settling the same geographical locations, as the rescue and rehabilitation efforts for hurricanes proved that hit the island of Hispaniola (Dominican Republic and Haiti). After a disaster event, many of the societies and individuals at risk, especially those who live in far remote regions, such as the people on the island of Leyte (Philippines) when Taifoon Haiyan hit in late 2013, often do not have access to emergency service providers. Moreover before a disaster, different parts of a society are differently prepared even for regularly occurring hazards such as floods and tropical cyclones.

The findings from Padang, Haiti, and Leyte underscore the importance of providing specific and targeted education to enhance community capability to withstand a natural hazard with minimal harm and loss of property. From empirical analysis to the impact of natural disasters it could be proved that vulnerability has its causes at the macrolevel as well as at the microlevel as national-level factors including "good governance" and "educational setup" are the main determinants defining vulnerability. At the local level, "income" turned out to be the main predictor of vulnerability. Thus advocacy, education, and awareness are becoming the substantial cross-cutting components that aim at all possible actors. The core pillars for awareness raising are approaches based on natural and social science evidence that have to be condensed to standard messages, tools, and guidelines and that are in accordance with the disaster risk reduction efforts as shaped by the Hyogo Framework of Action (IFRC 2011a, b).

There are a large number of studies dealing with the impacts of education on social behavior in crisis situations. It especially turned out that there exists a strong relation between female education, for instance, on lowering fertility and population growth and the family exposure to external stress (Lutz and Samir 2011). Almost universally, women with higher levels of education have fewer children thus better education is associated with lower mortality, better health, and different migration patterns. However, this assumption that education (in general) will lead to a higher sensitivity to risk reduction related to natural disasters does not automatically hold true for the exposure to natural disasters. Nevertheless the basic relation that education leads to a higher income (Lutz et al. 2008) and that educated people are more aware or better informed of the risks as well as on ways and means of mitigation and adaptation has become a well-accepted fact. Although only few studies (Faupel et al. 1992; Striessnig et al. 2013; IFRC 2011a) have thus far considered education in the analysis of vulnerability in relation to natural disasters, it is nevertheless largely known that education is a prerequisite that strongly defines vulnerability in general. In the United States it was found that participation in disaster reduction training programs strongly defines individual and family disaster awareness. The authors found that pre-disaster education significantly improves a household's disaster preparedness, for example, in identifying safe escape routes, planning for what to do in the event of family separation, as well as on adaptive response measures such as stored food and water, battery-powered flashlights and radios, or securing items around the house, and a car with a full tank of gasoline. The authors conclude that disaster education is a powerful dimension to prepare a population for a disaster and that it makes much more sense to just empower people in the form of better general education than specifically to increase disaster resilience whereas education mostly probably will indirectly influence preparedness. A study by Samir (2013) even went a step further and tried to differentiate between the influence of education and of income and social status on the preventive risk behavior of village communities in Nepal. They studied the damage due to floods and landslides in terms of human lives lost, animals lost, and other registered damage to households. At all levels and under all statistical models, the results showed a significant effect of more education on lowering the number of human and animal deaths. With respect to the wealth indicator, the inquiry did not reveal such unambiguous results. Nevertheless the study gave a clear causal relationship that at the community level, the educational attainment of the young adult's statistically significant explanation for an improved coping capacity. Education was seen also to play a substantial part in bringing leadership to work effectively with the local people, and to communicate with higher authorities. Thus the educated ones were better prepared to address their specific concerns and to participate with their experience in community development, including issues related to flood and landslide risk reduction. In addition, educated people could have an opportunity to diversify their income by taking nontraditional jobs that were not affected by such events.

Inasmuch as emergency situations are atypical events for most people, it seems meaningful for individuals to gain appropriate skills that they could recall and practice in case of emergency. This finding requires external support until the reactions have become automatic. The recognition of the triggering moment of an emergency event can only be reached when often rehearsed. Numerous studies on people's reaction in crisis situations have shown that repeated practice of skills and behaviors and the recognition of the situations lead to an automatic performance. There is a strong, although theoretical, link between practicing sets of emergency skills and their respective performance in a crisis. Researchers have shown that emergency and disaster education as well as prior experience with the disaster hazard correlates with the chosen adaptive disaster responses. In repeatedly seeing the connection between two events, people learn to correlate a predictable relationship between them. Learning a new skill involves more than just the cognitive understanding of the relationship between two events. Skills require the rapid physical and mental perception and understanding of a set of specific circumstances, coupled with automatic performance of the skill itself. Skill learning therefore requires the ability to follow instructions plus to recall experience from former emergency situations. Emergency researchers have recognized repeated performance of a skill to be the most critical component of skill learning. In this regard it has to be acknowledged that the causal relationship between training and practicing disaster reduction behavior has only been dealt with in a few studies. Most of the findings given above are based on theoretical background analysis. A tracking back of actually taken actions to the respective training given is hard to prove. However, the large numbers of reports on educational programs, actions taken, and positive outcomes demonstrate that the successfulness of education to increase resilience can hardly be ignored (FEMA 1994).

Programs to increase public awareness and education for disaster reduction aim at applying existing knowledge of disaster reduction to local situations. Such actions are oriented to the populations at risk that normally comprise a variety of social groups of different age, gender, and ethnicity. Immigrants, homeless, disabled, unemployed, and marginalized are also common in such groups. Awareness raising has the task of bringing all these people together, to mobilize and encourage them to initiate their own activities. This can best be achieved by informing them about the hazards to which they are exposed. As the information is normally technical and knowledge based it has to be transferred into a vernacularly understandable language. The people at risk can usually quite well describe the problems they faced in the past, but they are seldom able to describe their needs and demands as they do not understand the complex cause–effect relationship of hazards and their potential impact on them. Several approaches have been successfully run to increase public awareness for disaster risk reduction. IFRC (2011b) listed training and education programs that have been successfully implemented, among them the following.

Campaigns

The campaigns are in general large-scale national training programs, such as campaigns to refrain from smoking, on child immunization, or the appeal to use seat belts. The main objective of such type of campaigns is to manage a change in social behavior on all levels of society. Campaigns are successful when they are built on consistent, easily formulated messages that are often repeated by trustworthy organizations and run by representatives of the public or the civic sector. Most memorable are standardized messages that are built around a single, unifying, and enduring slogan that may be implemented by:

- Publications, posters, newspapers or magazines, information cards, flyers, brochures
- Oral presentations, education modules, slide presentations
- Performing games and competitions
- Audio and video materials
- Web pages and activities
- Social media and telecommunication

The advantage of such an approach is that it reaches a large number of people in a short time. On the other hand it must be carefully planned and requires an excellent organization and much patience Moreover there is generally no direct success to be seen.

Participatory Learning

People can best be motivated by approaches in which they are the main actors. Together with experienced moderators, the participants can work out risk reduction countermeasures based on assessments of their respective vulnerabilities and in accordance with their coping capacities. The focus of participatory learning is to incorporate the risk-exposed populations in a dialogue and to call on them to bring in their specific experience. Participatory learning can be applied at all social levels from the level where the people are directly exposed to a disaster, and the level of the local communities that are responsible for implementing risk reduction strategies, up to the national level where such strategies are worked out and initiated. Elements for participatory learning can comprise:

- Action-oriented research hazard, vulnerability, and risk assessment, with a focus on the assessment of the coping capacity
- Disaster management planning
- Defining means and actors for risk reduction measures
- Joint elaboration of local emergency evacuation and simulation drills
- Monitoring drills and improving the evacuation plans

According to the *Red Cross Guidebook* the advantage of this approach lies in its incorporation of the local disaster-exposed people and calls to identify their respective fears and needs. It invites the people to bring in their expertise and develops their personal ownership. The disadvantage of the participatory approach is among others that it normally takes much time and effort. Moreover it often raises high expectations by the participants on the mitigation capabilities of the local authorities that in general require the approval of higher levels.

Informal Education

Informal education takes advantage of moments often in the aftermath of a disaster event to stimulate thinking and engagement of people in order to increase their safety and resilience. Informal education involves

> disseminating of standard messages but with the flexibility to accommodate the needs and concerns of specific local audiences.This is particularly effective because it offers the chance to get those directly involved who are highly at risk. According to IFRC (ibid) informal education in communities and schools turned out to be the "most flexible of all approaches" with respect to agenda setting, addressing a wider public, and the point of time after a disaster. The tools that can be used for informal education are more or less the same as those used for campaigning. Education modules that have proven its effectiveness are:

- Presentation by experts
- Brainstorming that encompasses the disaster experience of individuals or specific social groups
- Moderator-guided discussion in small groups
- Demonstration of case studies, supported by disaster impact simulations and role play

The method has the advantage that it helps organizing, sensitizing, and mobilizing communities. Thus it is building up relationships between different social groups and higher level authorities. Moderated and expert advocacy

> works equally well with adults, youth and children especially when it comes to a cross-generational dialog. Nevertheless the point of time often generates a highly euphoric momentum for the risk-changing behavior, a momentum that later can often not be fulfilled by the local authorities or by the participants themselves. Even when externally moderated, many in the workshop-defined requirements turn out to be unrealistic, creating much disappointment.

9.3 Examples of Community Disaster Risk Management

9.3.1 Indonesia: Banda Aceh an Example for a Peaceful Settling of Dispute

The tsunami of December 2004 that hit the northern part of Sumatran Aceh Province and destroyed almost its entire coastal area had its biggest impact in the Aceh capital, Banda Aceh, where about 80 % of the houses and of the physical infrastructure, including the only port facility in the region were completely destroyed. More than 100,000 people lost their lives in the city of Banda Aceh itself. The tsunami eroded the fertile topsoils of the area and covered the land with several centimeters of debris and rubble, leaving 23,000 ha of rice paddies and 120,000 ha of arable land no longer usable; altogether 300,000 land plots that hold confirmed land titles were destroyed. Furthermore all geographical orientation points, landmarks, and other markers were lost, along with the land title documentary of the Banda Aceh communal administration. Thus none of the landowners (in the case where he and his family had survived the catastrophe) was able to prove his claims. Immediately bitter conflicts arose among the landowners. The problem was how to establish an instrument that could settle these disputes in a meaningful, juridically sound, and socially acceptable way, knowing that the landowners had lost everything and therefore were definitively unable to present documents to prove their claims. And on the other hand, the settlement of disputes procedure should also guarantee that no illegal claims could be presented, especially as during the different political rule over the area and a more than 20-year-long military control over Aceh Province, many former land titles had been lost due to legal or illegal land reforms. The task of the Aceh administration was to establish such a process based on a dialogue that gave the plaintiffs a chance to sue for their claims objectively. Such a dialogue process has a long-lasting tradition in Indonesian history and is called "Adat," functioning as a kind of common law and has its legal base for administering such form of justice in the Islamic tradition. "Land is given by God and no one has the right to waste it," defining land as thus a common good. Based on that interpretation of the laws, the lawyers of the plaintiffs proposed a broad and comprehensive discussion forum. But after several long and very dramatic discussions no result was achieved. So the elders one day proposed to mandate one lawyer to settle the dispute. The lawyer invited representatives of both parties to a closed session, not before inviting all members, relatives, and friends of the filed case to a great party. And it took less than a day and the dispute was settled. The solution was that the party who was strongly benefiting from the settlement spent a considerable amount of money to restore a tsunami-demolished mosque, an offer "no good Muslim" can ever reject. In the long run, the money was taken to pave a parking lot of a new supermarket (Arskal Salim 2010).

The Banda Aceh example clearly demonstrates that it is not the administrative, technical, or scientific setup that defines the living environment of a region, rather the geographical, cultural, and traditional relations and socioeconomic conditions that make a region for the people a place to live. The society is a moreover defined by individuals or groups of people with personal and operational relations sharing the same like-minded interests, and having the same cultural heritage and political expectations and fears. Even when the society is made of different ethnic groups the patterns of relationships can enable its members to benefit in ways that would not otherwise be possible on an individual basis; both individual and socially common benefits can thus be distinguished, or in many cases found to overlap. Their overall activities and interests lead to develop a social, economic, and technical infrastructure that may be threatened by external impact such as a natural disaster.

The 2004 tsunami induced an economic crisis that threatened the existence of the entire Maldives nation. It was not the impact form the tsunami itself that created this crisis—the impact was comparably limited—it was the fact that in the aftermath the tourism sector collapsed totally. This forced the former government to create a "Sovereign Wealth Fund" with money earned from tourism to be used to purchase land elsewhere for the Maldives people. The money for land acquisition would be taken from the income produced from the more than 600,000 tourists every year that made up to about 30 % of the gross domestic product of 2008. The then-acting President Mohamed Nasheed's statement on the "Future of the Flood Threatened Maldives Islands" was the first political leader's address to the world that there are places on Earth where the climate-induced sea-level rise has shown the first serious signs. The water level rises steadily and starts threatening free existence on the islands, as in many other small island states: Tuvalu, Tonga, Fiji, Samoa, Vanuatu, Funafuti, and others. There is furthermore the well-known example of the island of South Talpati offshore Bangladesh, as the first islands that vanished from the Earth's map; the same will happen with the Tegua Atoll of Vanuatu. The same happened in Papua New Guinea where the government evacuated the 980 inhabitants of Carteret Island in 2005 (Jacobeit and Mettmann 2007). The more than 1000 islands forming the State of Maldives lie less than 1 meter above sea level, making the 380,000 inhabitants (and by the way the capital Male is with its 36,000 inhabitants the most densely populated area of the world) most vulnerable from flooding. But not only is the land size diminishing every day, also the ingress of sea water is destroying the island's fresh water reservoirs.

Thus the prime minister addressed his people that in the long term, "the Maldives will not survive as an island state." He urged his country that they have to face one day leaving the islands and seeking new homes somewhere. He therefore started to negotiate to buy land in the Indian Union, in Sri Lanka, and as well in Australia. The government of the Tuvalu Islands did the same. It negotiated with Australia and New Zealand to get shelter in those countries. But the Australian government will only accept up to 90 Tuvaluans every year, as it claims that there is no real risk from sea-level rise to Tuvalu and the Tuvaluans are coming as "economic refugees" and not for climate reasons. For Maldivians, India and Sri Lanka pose the first choice for evacuation, as they share the same language, culture, and ethnic heritage, especially with the Indian federal states of Tamil Nadu and Kerala. But the Maldivian population did not accept this "vision" and were afraid of being forced to leave the country. Therefore after heavy riots and political turmoil and following the opposition's charge that the Nasheed government was no longer able to govern the country in the way the people were demanding, Nasheed resigned in February 2012.

The Maledives example shows impressively what can happen when a reasonably founded and seriously thought over and meaningful oriented strategy failed, most probably due to not incorporating the population at risk in the decision-making process. Instead they were just presented a government's decision. Those who were deeply affected by the political decision, were not given the chance to express their views and especially in that sector overwhelmingly dealing with their everyday life. A broad and extensive discussion should have been institutionalized, giving everyone the feeling her fears, experiences, and "vision on livelihood" were properly taken up by the authorities. Next to the individual, representatives of all social groups (religious leaders, the political opposition, representatives from industry and science, etc.) should have been gathered and given ample time to express their views at a national roundtable. It is clear that such a discussion would have taken much effort and much time, but the time problem was then and is still today, not threatening the island's existence overnight. The government should have given such a socially comprehensive discussion enough space although respecting the fact that the rising sea level allows no way out other than a change in policy. Thus the fears of the population were neither focused on the technical matters of the sea-level rise nor on the financial aspects, but found its expression in the field of emotions and feelings; in the last consequence is was formulated on behalf of the Islamic belief and tradition.

But the problem of the "Maldive Vision" of resettling in other places also has an international dimension. A question that immediately arose was, what kind of political status will such persons be attributed after resettling. Are they still Maldivians, who are now living in India or will they be given Indian nationality or will they be treated as emigrants or ethnic minorities. The International Law on Refugees as it is laid down in the Charter of the United Nations distinguishes only between "refugees" and "internally displaced persons" (IDP), who are forced to leave their country or parts due to military, racial, or ethnic conflicts. Sometimes such refugees were forced to leave their country for decades (Afghanis settling in Pakistan in the 1980s) and for that time the people do not have any internationally accepted political representation. In such cases the United Nations High Commissioner for Refugees (UNHCR) takes over representing these peoples. A similar representation exists for IPDs. The basic legal definition on what forms a nation requires, according to international law, a nation's territory and a nation's population. But when territory is flooded due to rise in sea level and no territory exists anymore, the basic definition of what a nation is, is no longer valid. Although the United Nation Environmental Program (El Hinnawi 1985) introduced the term "climate refugee" into the public debate, there is no internationally accepted legal authorization for the UNHCR to take care of the climate refugees. The question for UNHCR is whether the refugees left their home deliberately or whether they were forced to do so. In the case of the Maldives the idea was to seek shelter in another country. The decision was supposed to be definitively taken voluntarily and not a subject of "forced migration." The United Nations since then has several times put the notion on the agenda of the United Nation Security Council but neither the five permanent members nor the industrialized nations, as well as many advanced countries, were inclined to take up the matter, although it was stated that already in 1990 the amount of "climate refugees" was estimated to be about 25 million, many more than those refugees of wars and conflicts (Myers 2001). The IPCC stated in 1990 that next to the climate-induced sea-level rise also desertification, soil erosion, and heat waves will make "climate refugees" a substantial problem of the future. In the Rio UNCED Agenda in Chap. 12 the notion of "climate refugees" was already made (see also Stern 2007). In order to object to the reluctance of the industrialized states the governments of the affected island states formed the Alliance of the Small Island States (AOSIS) to raise their voices in order to fight for a worldwide reduction of greenhouse gases. They argue that they release almost no CO2 into the atmosphere but are those who suffer first.

9.4 Community and Volunteers in Disaster Risk Management

The Hyogo Framework for Action 2005–2015 (UNISDR 2005) acknowledges that "civil society, including volunteers and community-based organizations are vital stakeholders in supporting the implementation of disaster risk reduction at all levels." The role of volunteerism in any disaster risk management strategy turned out to be of particular significance when oriented to strengthen community-related relief and recovery actions. UNISDR has addressed the role of volunteers (UNV 2007) and reiterates "all too often" disaster risk management neglects the many contributions volunteers can make, either in assessing the population's vulnerability or in using the imminent strengths to increase resilience. The central function of volunteers is not to act as the risk assessors, decision makers, or mitigation measure implementers, but to support the capacities of national and local actors in mitigating and coping with natural disasters. Both society at large and the individual benefit from volunteerism that thus can help to strengthen trust, solidarity, and reciprocity among citizens by purposefully creating opportunities for participation.

The UN General Assembly defines volunteerism as "undertaken of free will, for the general public good and where monetary reward is not the principal motivating factor" and recognized that "volunteerism proved an important component of any strategy aimed a poverty reduction, sustainable development or disaster prevention and risk management in particular to overcome social exclusion and discrimination." That will help to achieve the development goals and objectives set out in the UN Millennium Declaration.

Acknowledging that volunteerism is a fundamental source of community strength and resilience, the United Nations Volunteers Program (UNV) has been initialized to promote volunteerism to support peace and development. The program is to encourage local and national disaster risk managers to integrate volunteers into the local risk mitigation through a wide range of activities, including traditional forms of mutual aid and self-help, formal service delivery, campaigning and advocacy, as well as other forms of civic participation. The UNV strategy therefore is oriented to achieve a broader and deeper integration of volunteerism into disaster management strategies by identifying ways and means that volunteers can assist local disaster risk management. The United Nations Volunteers Program fosters building a "spirit of volunteerism" in community participation in disaster risk reduction strategies by an inclusion of volunteer resources in the local mitigation measures by establishing locally adopted specific mechanisms that enable engaging active stakeholder participation. This furthermore calls for the initiation of specific community-based training programs that indispensably consider the role of volunteers as appropriate to enhance local coping capacity.

At the national level volunteerism can among others assist to:

- Support the establishment of a common awareness and understanding of disaster cause–effect relationship among key stakeholders, including local authorities and local communities.
- Support governments in preparing, coordinating, and implementing crisis-sensitive development and recovery plans.
- Advocate for the recognition of imminent experiences the disaster-affected societies have in disaster mitigation.
- Facilitate the development of institutional capacities of national and/or local authorities.
- Strategically plan, negotiate, coordinate, and mobilize the support of local volunteers and civil society organizations.
- Provide coordination, operational, and technical support to district administrations, nongovernmental organizations, and volunteer-involving organizations.

At the community level volunteerism can among others assist to:

- Create a space to engage, empower, and mobilize community members at local disaster risk management efforts.
- Raise awareness, promote preparedness, and strengthen the local coping capacities.
- Assess hazard, risk, vulnerability, and capacity assessments.
- Develop local disaster plans that feed into district and national disaster plans.
- Mobilize community contributions (other volunteers, in-kind and other resources) for the implementation of local disaster mitigation countermeasures.
- Facilitate the inclusion and participation of affected communities, especially women and youth, in the planning and implementation of disaster risk management plans.
- Mobilize extended volunteer support to other districts/regions.

The program generally calls for developing national volunteer infrastructures, which could include the development of specific volunteer legislation and national volunteer schemes, the operational framework for mobilization, and coordinate the activities of volunteer-involving organizations, promote good volunteer management, and advocate the value of volunteers. Only when these policy forums are functioning accordingly, will disaster risk reduction be effective, ensure a good dissemination of information, and substantially increase public awareness.

In Germany a well-established volunteer service institution, the "Technisches Hilfswerk" (THW 2014) for decades has been committed to support people in need all over the world. The THW is administratively attached to the Federal German Ministry of the Interior (BMI). More than 80,000 volunteers are enrolled, but only 1 % of them are actually employed by the organization. Such an organizational setup is unique worldwide. And the status of volunteerism of the THW is its main objective and as such is highly appreciated and accepted in German society. The volunteers are stationed in almost 700 localities all over Germany. The specialists of THW cover all sectors of modern crisis intervention and relief and rescue operational demand. World famous are the sniffer dog groups that in many cases were able to rescue victims from under heaps of earthquake rubble. The technical equipment is state of the art and the people are well trained and many of them have years of experience. The THW specialist in her normal life is employed and in different private jobs and is released from work by law. THW has been involved in nearly all the bigger catastrophes of the world, like the tsunami in Banda Aceh, the earthquakes in China and Haiti, and with the many drought events in Northern Africa. But THW also helped in in Germany nearly all the catastrophes including the large floods of years 2002, 2003, and 2013. In the 2013 flood catastrophe the specialists executed 1.6 million hours of rescue work in strengthening dams with sandbags, erecting flood bridges, movable protection dams, and providing electricity, water, and medical services. But they were also involved in planning and administering the rescue operations.

A similar situation holds true for the volunteer programs that are rendered under the auspices of the European Union. Although there is a vast array of definitions and traditions concerning volunteering, the common denominator throughout Europe is that "wherever people engage in activities to help each other, support those in need, preserve our environment, campaign for human rights, or to initiate actions to help ensure that everyone enjoys a decent life—both society as a whole and the individual volunteers benefit and social cohesion is significantly strengthened" (EU-EAC 2010). The European Union therefore ranks voluntary action as an important component of the strategic objective of the European Union of becoming a competitive and dynamic, knowledge-based economy.

More than 100 million Europeans engage in voluntary activities, meaning that every third European claims to be active in a voluntary capacity and about 80 % feel that voluntary activities are an important part of democratic life in Europe.

But the European Union does not institutionalize volunteerism itself under its administration, but is engaged to orient and standardize volunteer initiatives of its member states strategically. Some member states already have well-organized volunteering landscapes whereas others lack broad support by the public. It turned out that such countries with a longstanding tradition of volunteerism such as the Scandinavian countries, the United Kingdom, or Austria are well established, but especially those countries that entered the Union late, are lacking a broad acknowledgment in society. There is no uniform regulatory framework defining volunteerism across the member states. Three distinctions can be made, states:

- Where a legal framework relating to volunteering is in place (Belgium, Cyprus, Czech Republic, Hungary, Italy, Latvia, Luxembourg, Malta, Poland, Portugal, Romania, and Spain)

- That do not have a legal framework, but where volunteering is regulated within other existing general laws (Austria, Denmark, Estonia, Finland, France, Germany, Greece, Ireland, Lithuania, Netherlands, Slovakia, Sweden, and the United Kingdom)
- Who are in the process of developing a legal framework for volunteering (Bulgaria and Slovenia)

The European Union identified funding to be the major obstacle to running a volunteer institution effectively. The main source of financing is from public funds. In some EU countries, however, this trend is starting to change. The state's capacity to fund the social sector has been declining and nongovernmental organizations began gradually taking over. Simultaneously, the proportion of financial resources coming from the private sector has been marked by a steady growth. However, the levels of financial resources will also in future present a significant challenge for the majority of voluntary organizations across the European Union. The economic value of volunteering varies greatly in the member states ranging from less than 0.1 % up to 5 % of the gross domestic product.

But more than the economic benefits, voluntary activities are focused on delivering social impacts to individuals, marginalized societal groups, or local communities that are deprived of free access to goods and services by a disaster. Many of these impacts contribute directly to a number of key objectives set out in EU policies such as promoting:

- Social inclusion and integration
- Social cohesion and social solidarity
- Self-satisfaction and dignity
- Involvement of citizens in local development
- Civil society development and democracy

providing:

- Education and training
- Support to the integration into the labor market
- Skills and competences
- Support to local sports clubs

Concerning an institutional development of the volunteer sector in the European Union the following challenges were identified. Although overall the level of volunteering has increased in most of the EU countries, the main difficulties seem to be related to the changes affecting the nature of voluntary engagement, as well as a mismatch between the needs of voluntary organizations and the aspirations of the new generations of volunteers. Factors include inadequate knowledge of the needs of the volunteer organizations, the general preference for short- rather than long-term voluntary commitments, and that most of the volunteers appear to be less willing to take on decision-making responsibilities. A dramatic increase in the number of voluntary organizations results in a spread of volunteers across a large number of organizations. Moreover the increasing professional nature of field applications means new challenges to the volunteers regarding the management of human resources and sometimes very specialized skills and expertise.

These increasingly demanding tasks create a tension between the required professionalization and the ability of volunteers to meet these demands, and their willingness to do so in an unpaid fashion. Moreover the European Union sees a certain risk of a political instrumentalization of the voluntary sector. In some countries the sector is increasingly seen as an instrument for tackling problems or providing services that the state cannot provide any more. These difficulties are expected to increase due to the economic crisis. A fundamental lack to recruit "the right volunteer" is furthermore hampered by a general lack of recognition by society. The European Union therefore emphasizes that voluntary activities become more broadly accepted by the public by a change in rewarding volunteers for their participation and thus becoming a matter of attracting new volunteers. One step towards this can be a higher validation of nonformal and informal learning from the volunteer's application home and abroad by society in general, by the private sector, as well as by state employers. This change in paradigm will also lead to abolishing stereotypes and negative connotations still existing in some civil societies. National strategies declaring "honoring appointment a social task" is seen by the European Union to be the central target to further develop volunteerism in its member states.

9.5 Early Warning Systems

Every year, on the first Wednesday in February at 1:30 PM all 7800 warning sirens are tested all across Switzerland (FOCP 2014). The aim is to ascertain the operational readiness of both the "General Alarm" and the "Water Alarm" sirens. The population is notified beforehand through announcements on the radio, on TV, and in the press. The sirens give a regular ascending and descending tone, which lasts for 1 min, and is repeated once after a 2-min interval. Where necessary, the sirens can continue to be tested until 2 PM. Then from 2:15 PM to 3 PM the "Water Alarm" signal is tested in those areas in the proximity of dams. It consists of 12 low continuous tones lasting 20 s and repeated at 10-s intervals.

In former times tolling a bell was a sign, understood by everybody that something extraordinary had happened. It was for centuries the usual way to warn people of approaching enemies, a fire, flood, and other disasters. The warning was originally based on information by eye-witnesses, for example, a fire brigade located in a forested area. As the messages are spread by mouth, the farther away people live from the location of the disaster, the longer it will take to reach them.

Today the warning of an impending disaster is in principle not very different, although the technology has advanced significantly (Ferruzi 1997). The warning systems we use today rely on measurement techniques, mostly from physics. Quantitative and mostly visual inspections of the changes in nature have been well established for more than 100 years, but quantitative assessments such as in meteorology date back to the 1940s, quantitative seismology to the 1960s, and quantitative volcanology is only about 30 years old. But the overwhelming breakthrough in early warning came with remote-sensing techniques that began when the first satellites started orbiting the Earth in 1970. Since then Earth observation science and technology developed extremely rapidly, so that together with the advances made in communication technology, world societies are provided every day with a multitude of information from any part of the globe. Whereas formerly a disaster could only be identified by visual evidence, today there are numerous networks installed all over the world that permanently record any changes in nature that may turn out to be hazardous. Together with the advances in disaster identification the science of risk prediction has also developed to provide the expression "early warning" is not only used to describe operational means to warn people of natural disasters but of all kinds of dangerous circumstances such as epidemiological disasters or even complex sociopolitical emergencies.

There are many emergency managers who define "early warning" as not to be restricted to the technical system to "toll the bell" but comprises also all actions in the forefront to identify the onset moment of the disaster as well as all actions that are to follow in the aftermath. Moreover the information provided should initiate sector-specific counteractions to be taken in advance to mitigate the anticipated risks. The UNISDR (2009) terminology (see Annex A), early warning is defined as "the provision of timely and effective information, through identified institutions, that allows individuals exposed to a hazard to take action to avoid or reduce their risk and prepare for effective response." To be effective and complete UNISDR recommended an early warning system should comprise four interacting elements (UNISDR-PPEW 2005):

- Risk knowledge
- Monitoring and warning service
- Dissemination and communication
- Response capability

Although this set of four elements appears to have a logical sequence, in fact each element has direct two-way linkages and interactions with each of the other elements. The second element, the monitoring and warning service, is the most well-recognized part of the early warning system, but experience has shown that technically high-quality predictions by themselves are insufficient to achieve the desired reduction in losses and impacts. Failures in early warning systems typically occur in the communication and preparedness elements. The human factor in early warning systems is very significant (Twigg 2002). This was true of Hurricane Katrina which affected New Orleans in late August 2005, although in this case there was the additional failure in respect to risk knowledge.

The primary goal for early warning systems is to protect lives and property. This goal can be achieved by giving a warning signal before the disaster strikes to give the people at risk ample time to bring themselves to safe places. Such forecasting comprises information on the place where the event is expected to strike,

Table 9.1 Assessment matrix to forecast probability of a disaster event according to time and location (Based on Feruzzi ibid)

Type of Hazard	Can occur at location	Lead time of occurrence		
		Short term (days/hours)	Medium (weeks/month)	Long term (years and more)
Earthquakes	Yes	No	No	Yes
Extreme weather	Yes	Yes	Almost	No
Flash floods	Yes	Yes	No	Yes
Forest fire	Yes	Yes	No	No
Landslides	Yes	Almost	Almost	Yes
Plain floods	Yes	Yes	Almost	No
Volcanic eruptions	Yes	Almost	Yes	Almost
Storm surges	Yes	Yes	Almost	No

the time of onset, and the severity that is to be expected. The main obstacle in disaster forecast is that even when the data collection and interpretation technologies are already highly advanced still only a few natural hazard types can be predicted with sufficient reliability. Table 9.1 gives a general impression of the probability to forecast a disaster event.

A reliable instrument in early warning is the daily weather forecast. Since a number of satellites have been geostationary placed or are polar orbiting the Earth (Meteosat, Eumetsat, and Geos) weather forecasts increasingly improved. It now possible to collect information on the changing weather on a very large scale in the form of images of cloud distribution over continents and oceans, while a multitude of ground stations continuously report on surface temperature, humidity, wind, and others. Thus weather forecasting has reached a sufficient level of reliability in regard to the next two to three days, although the medium- to long-term forecast is still problematic. The ability to generate a weather forecast effectively is mainly influenced by the factors of surface and air temperature, wind direction and speed, precipitation duration and amount, and by the local and regional sky conditions, as well as by the barometric pressure changes. Moreover weather observations are influenced by local terrain to varying degrees. This means a good knowledge of local effects on the weather regime is necessary to interpret and apply the observations correctly.

On the other hand, a scientifically meaningful and reliable forecasting of natural hazards is problematic; especially the prediction of earthquakes is still not possible. A study by Shearer (1999) stated that even from the surely best known region of earthquakes in the world (San Andreas Fault) where the most sophisticated instrument networks have been deployed, it is thus far not possible to predict the next big earthquake with a certain level of reliability. From earthquake records the next major earthquake was to be expected for 1990; but up to now it has not occurred. The chance to predict an earthquake even within a short term is poor, due to the fact that there are a multitude of the earthquake precursing seismicity, local changes in the electrical or magnetic fields, changes of groundwater levels, local emissions of radon, carbon dioxide, and other gases along fault lines. As such indicators also occur independently of earthquakes they do not pose a reliable measure that can be used to forecast such a hazard. In the case of the big earthquakes (Loma Prieta, Northridge, Wenchuan, etc.), "All such indicators have never all occurred together or have been observed by instrument networks." Still we do not know enough about the source of earthquakes, neither on the processes that occur along the plate boundaries nor about the exact meaning of a numerous earthquake precursors. So geoscientists are still very reluctant to claim "prediction of earthquakes is possible." There is "Every reason to believe that through technology, our disaster-warning capacity will improve and that soon it will be possible to organize a timely response to all natural risks" (Ferruzi ibid). From geological, geophysical, and geomorphological evidence and from historical data, it is thus far possible to identify areas at risk that have a higher probability for natural disasters (e.g., earthquakes) than others. For the purpose of emergency management such a prognosis is fundamental as it opens the chance to work out risk reduction strategies and to implement them accordingly, but such knowledge will not prevent people from being hurt in the case of a disaster.

Figure 9.3 is based on findings from "Day-Curve" (Day 1970), who "proposed that the tangible benefit of a flood warning system could be estimated as a function of warning time." The concept has been adopted for the purpose of the GITEWS Tsunami Early Warning Project and was presented by Joern Lauterjung on occasion of the annual GITEWS conference in June 2009 at GFZ, Potsdam. The concept is adopted here to outline a decision support system for the assessment of the natural

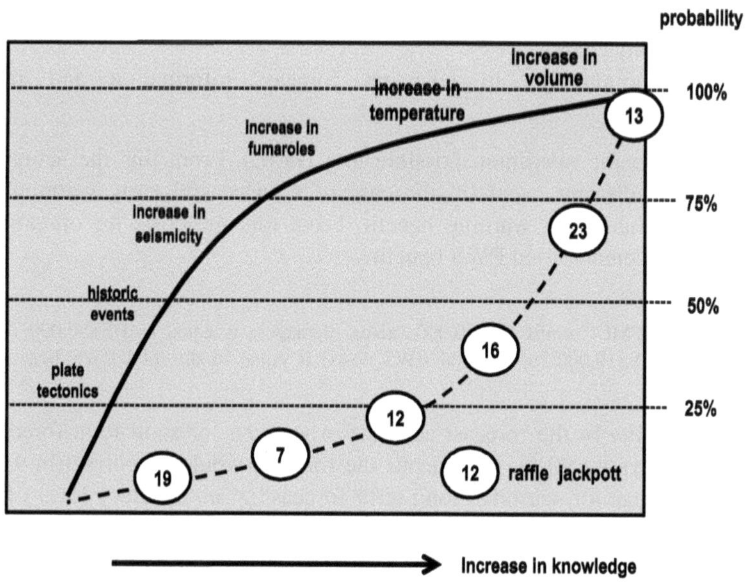

Fig. 9.3 Decision support system for a probability assessment of a "volcanic eruption" (Based on Day 1970)

disaster probability of an example of a volcanic eruption. The graph demonstrates very generally that the assessment of the probability for the onset of volcanic eruption fundamentally differs from say, the prediction of a raffle jackpot. Although natural disaster probability can be assessed from basic information of geological and historical evidence to say, 80 %, the ultimate success in a raffle is when the last ball has fallen. Thus as a general rule, the algorithm indicates that from internationally accessible information a prediction can be made that is a good approximation to reality. From evidence it is today already possible to define at least the area at risk and time horizon. With an increase of the international disaster monitoring systems, a real-time transmission of information and more sophisticated disaster probability simulation models, the probability assessment success ratio will definitively increase, although such a risk assessment is more directed towards formulating risk reduction strategies than serving the needs of a local early warning.

Following the tsunami of 2004 the Global Facility for Disaster Reduction and Recovery (GFDRR) of World Bank (World Bank 2006) and the United Nations International Strategy for Disaster Reduction (UNISDR 2005) analyzed the economy of Early Warning Systems that are in operation in Asia countries (Subbiah et al. 2008). Studies quantifying the effectiveness of early warning systems (EWS) have formerly been carried out mostly on flood warning systems in the United States.

The cost–benefit relation of EWS for selected Asian countries was calculated according to their input costs of:

- Scientific risk assessment and technical requirements to generate forecast information
- Institutionalizing the system and making it operational (training, capacity development running costs)
- Enabling communities to take up forecast information and respond appropriately

in relation to the maximum possible loss (PML). From this the actual damage costs of a disaster event (in the case of a successful early warning) were deducted = actual early warning benefit. From this the costs for operating the EWS were deducted = total EWS benefit.

Example:

US$100,000 (PML) minus US$50,000 (actual damages) is equal to US$50,000 (benefit); minus US$10,000 (operational EWS costs) is equal to the total EWS benefit of US$40,000

The probability of the forecast accuracy was taken for short-term forecasts of less than 10 days by 90 %; that means the forecast would be correct in 9 out of 10 cases, whereas for seasonal, long-term forecasting an accuracy level of 70 % was assumed. The economic benefits from successful warning were calculated by multiplying them by a factor of 0.8 (9–1/10), because there are 10 possible occurrences. In one of ten cases, however, the warning was assumed to fail. The possible losses from such "wrong" forecasting was deducted from the benefits due to a "correct" forecast. The calculation furthermore included the return rate by simple multiplication. The calculations were made based on the disaster impact. For this, worst-case scenarios were used when robust and reliable data were missing and such cases could be assumed to be representative also for the future. In cases where reliable data existed, such as on floods in Sri Lanka, the damage of a 50-year flood was calculated to be 5 % for an annual flood, and for a 10-year flood a damage reduction of 25 % was calculated. The next step was to assess the factor of damage reduction due to an early warning. The study assumed the damage reduction factors shown in Fig. 9.4 for household, agriculture, livestock, and official buildings.

The benefits were calculated according to the damage scenarios for selected Asia countries.

Bangladesh (tropical cyclone)	US$40.0 for every US$ spent
Sri Lanka (flood)	US$0.9 for every US$ spent
Thailand (flood)	US$1.8 for every US$ spent
India (drought)	US$2.8 for every US$ spent
Vietnam (tropical cyclone)	US$104.0 for every US$ spent
Regional tsunami forecast	US$200.0 for every US$ spent

Although the benefits calculated differ a great deal, they generally fit into figures that have been published before by the World Bank and Columbia University (Dilley et al. 2005) according to which successfully disaster prevention can result in damage reduction ranging between 1:5 to 1:7 US$. A similar figure was given by

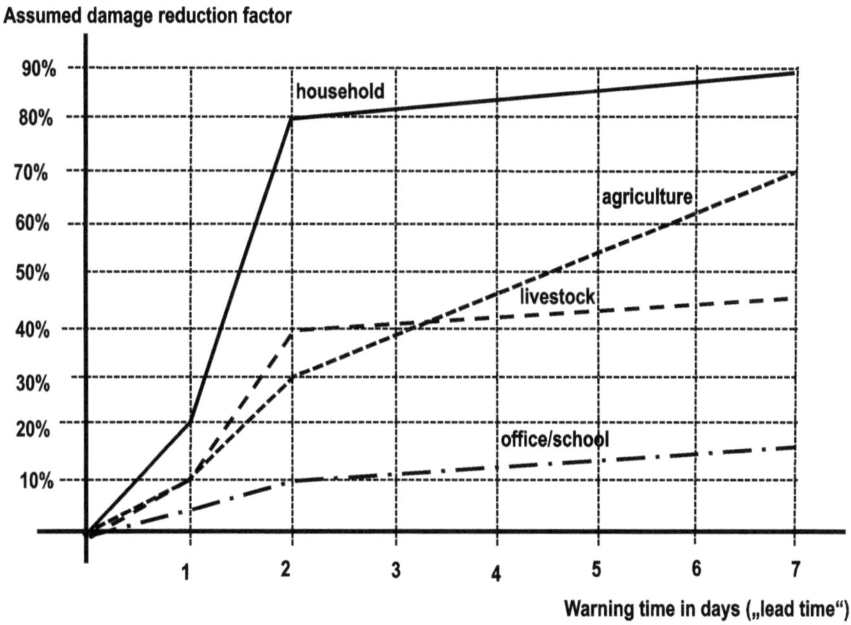

Fig. 9.4 Assumed damage reduction factors due to successful early warning for Asian countries (Based on Subbiah et al. 2008)

the Canadian Public Safety Agency on the "Manitoba Red River Flood Prevention Program" (APEGM 2014) which revealed a loss reduction ration of 1:8.

The great difference in figures for Asia is seen to have its origin in the highly different scenarios chosen that were based on very different datasets. Nevertheless the study came to the conclusion that in order to lower the costs for EWS systems a regional system would be able to optimize distribution of observation systems, reducing capital investment requirements and the implementation, as well as the annual operational costs Such transnational collective EWS systems should also integrate warning services for high-frequency and low-impact hazards including heavy rainfall local floods as well as low-frequency high-impact events such as earthquakes, tsunamis, and or volcano eruptions. The example of the "Regional Tsunami EWS" that is set up by Australia, India, Indonesia, and Malaysia can serve as a showcase on how early warning can be of benefit to risk-exposed societies.

9.5.1 The German-Indonesian Tsunami Early Warning System (GITEWS)

Tsunami early warning in the Indian Ocean became an issue after the December 26th, 2004 Indian Ocean tsunami, which killed more than 170,000 in the Indonesian province of Aceh Darussalam alone. The main scientific and technological challenge for the installation of such an early warning system in Indonesia was the tectonic setting of the so-called Sunda-Arc structure, an active continental margin that closely parallels the western coastline of Indonesia resulting in tsunami arrival times of less than 40 min after the occurrence of an earthquake (Rudloff et al. 2009). The challenge was to develop a technical and scientific concept that allows reducing early warning lead times down to 5–10 min escape time. Previously used systems, such as the Pacific Tsunami Warning System, are not optimal for Indonesia due to its geological situation along a subduction zone. If a tsunami is generated here, the waves will, in extreme cases, surge up the coast within 20 min, leaving only very little time for an early warning. This limiting factor was therefore the basis of the concept for the whole system.

For assuring a rapid recording of seismic parameters (Fig. 9.5) a dense network with monitoring stations, consisting of 160 seismic sensors all over Indonesian territory and sea floor seismic stations and buoy systems, tide gauge stations, were deployed as close as possible to the plate boundary, including a grid of real-time GPS crustal deformation monitoring devices. Introducing GPS buoys is a new invention that enables getting rapid information of a plate's deformation mechanism, seen as the trigger for tsunami generation. The dense seismic monitoring system is at the heart of the GITEWS project and is the cornerstone of the early warning system. Together with a set of GPS buoys it is possible to register any seismic activity that may lead to a tsunami. The interpretation of the data is managed through a series of newly invented, sophisticated modeling techniques and decision-supporting procedures to provide the exact epicenter, the focal depth, and

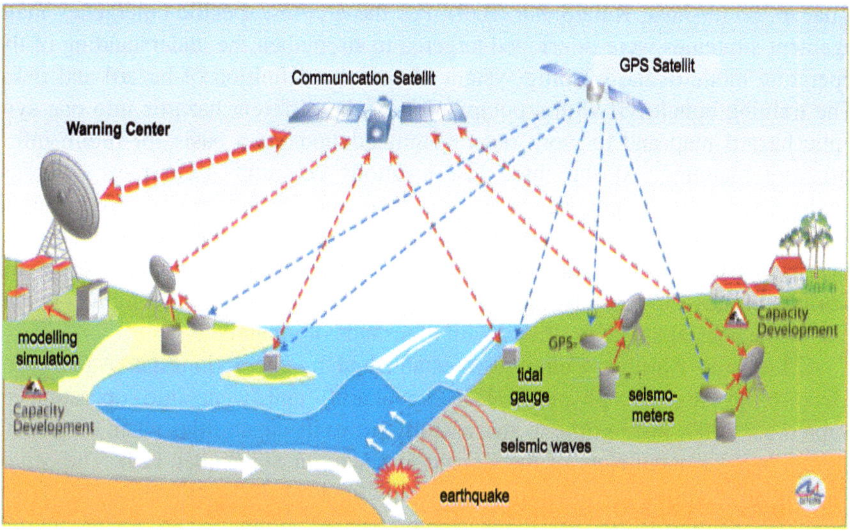

Fig. 9.5 Concept of the GITEWS tsunami early warning communication system (*Courtesy* Rudloff et al. 2009)

magnitude of the earthquake within a couple of minutes. After having identified an earthquake and tsunami the data are permanently completed with new incoming information from the sensor systems that allow a continuous upgrading of the probability assessment. A special evaluation software, called SeisComP 3.0 has been developed at the German Research Centre for Geosciences (GFZ) for data processing and tsunami simulations. The program has meanwhile been taken up by almost all Indian Ocean riparian nations. Several hundred pre-cast calculations on a probable tsunami generation were carried out in order to guarantee as fast as possible an assessment. In general such a calculation cannot be carried out within less than several hours. The interpretation programs now check whether the incoming data have already been calculated and then can immediately present the result.

In order to reach the people at risk ("last mile") as quickly as possible, the population along the shores of Northern Sumatra were trained how to react properly. For this GITEWS and the Indonesian Center for Earthquake Monitoring (BMKG) institutionalized the cooperation of the different national, local, and private stakeholders to organize the emergency management and to implement specific prevention measures. Thus the project has laid the foundation for the national disaster management strategy in Indonesia. Furthermore a series of training and education programs was developed to raise awareness of tsunami and earthquake response. These activities, termed "capacity development," have been an integral part of increasing the efficiency of the system. Scientists, disaster managers, national and local government, and local people underwent intensive training in what to do before, during, and after an earthquake and what special prevention has to be taken to cope with a tsunami. In order to increase tsunami resilience special training and education programs were carried out in three test regions (Padang, Sumatra;

Cilacap, South Java; Kuta/Sanur, Bali). For these cities, specific emergency management strategies were developed targeted to strengthen the understanding of the operating mode of the warning system and on the definition of hazard and risks. The training concluded with a compilation of the different hazards into one synoptic hazard map and to work out recommendations as a basis for future infrastructure planning. At the three test locations communication means (sirens, loudspeakers, police announcements, and radio and TV programs) were installed, respectively, initiated. Moreover in all locations local disaster management organizations (DMOs) were established and trained and given clear mandates in the warning process. Ultimately, the entire sequence from a warning up to the evacuation of the coastal area was drilled with thousands of inhabitants.

A Tsunami Warning Centre was established by the Indonesian Agency for Meteorology and Geophysics (BMKG, Jakarta) and went into operation in September 2007, just in time to successfully prove its capabilities when evaluating the earthquake parameters of the Bengkulu quake on September 12th. The center was able to identify a moment magnitude of 7.9 based on 25 stations within 4 min and 20 s, leading to the first tsunami alert ever disseminated by an Indonesian institution. In addition to the technical training and academic education which is of importance to operate such a system, GITEWS offers substantial capacity building programs. To ensure a permanent and independent operation of the warning system a public–private partnership model was established. Following this model BMKG was given the responsibility for the nationwide tsunami warning.

The GITEWS project was funded by the German government. The project was led by the GFZ and the BMKG although many German and international partners contributed substantially to the success of the warning system, mainly the Alfred Wegener Institute for Polar and the Marine Research (AWI), the German Aerospace Center (DLR), the GKSS Research Centre and the Federal Institute for Geosciences and Natural Resources (BGR), the German Agency for Technical Cooperation (GTZ), and the Leibniz Institute for Marine Sciences (IFMGEOMAR). The main international partner was the United Nations University's Institute for Environment and Human Security (UNU-EHS).

9.5.2 Katwarn "Use of Early Warning Systems"

The Katwarn ("catastrophe warning") is a Germany-wide concept to provide early warning for individuals, social groups, and companies on all kind of risk from natural, technical, epidemiological, and weather-related hazards or infrastructure failure (KATWARN 2014). The system furthermore provides recommendations on how to best address the identified risks. Katwarn is conceptualized as an additional dissemination tool for disaster occurrence information generated by local emergency management authorities (police, fire brigades, emergency services, etc.). The challenge of Katwarn was to have an operational system that reflected the state of the art of linking various information logistic technologies with market requirements. Thus it is now possible to help reduce the information chaos that often occurs during crises and disasters. The information is specifically tailored for the local risk situations based on the postal codes of Germany and is rendered via e-mail, fax, Internet, and smartphones and delivers advice on appropriate conduct to people wherever affected, in their home, office, or when out and about, and the so-called "last mile" in civil protection is overcome. Unlike previous approaches that mostly focused on how to identify a risk, Kartwarn aims at identifying information needs of the populations at risk and offers information logistics asking, "Who must know what and when in order to act properly." Katwarn was developed by the German Fraunhofer Institute on behalf of the public-sector insurance companies in Germany (Gesamtverband der Deutschen Versicherungswirtschaft) and has meanwhile been successfully implemented in quite a number of German cities and districts. The insurance companies are providing the system and the technical infrastructure to administrative districts and urban districts.

9.5.3 Hurricane Early Warning

The multitude of severe hurricanes in the North Atlantic alone cause billions of US$ of economic loss and kill hundreds of people every year in the Caribbean and along the southern and east coast of the United States of America (USNHC 2014). The 13 cyclones of year 2005 alone generated a loss of more than US$150 billion and killed 2800 people in the region.

The many concerns about hurricane risk led in the last century in the United States and in the Caribbean to institutionalize a hurricane assessment network. In the last decades there were many efforts to even further strengthen the monitoring and forecast capacity under the auspices of the World Meteorological Organization (WMO) and in cooperation with United States National Oceanographic and Atmospheric Agency (NOAA). For example, in the Atlantic Ocean and the Caribbean NOAA has established the National Hurricane Center and mandated the official body to issue tropical cyclone warnings for the Atlantic Ocean and the Gulf of Mexico, and the Central Pacific Hurricane Center is responsible for the Pacific Ocean from 140° west longitude to the International Dateline. The purpose of these organizations is preparing for and issuing forecasts and warnings of all tropical cyclones in the areas of interest. For this, specific extraterrestrial satellite observational platforms including land-based radar were installed to monitor sea surface temperatures, complemented by aircraft radar reconnaissance flights as well as upper-air observations from so-called "flying meteorological stations." NOAA furthermore operates Doppler radar to track tropical cyclone paths as well as monitor weather conditions both at the surface and in the upper atmosphere. Weather radars are able to locate exactly the areas of precipitation, calculate its motion, and estimate its type (rain, hail, etc.) and the amount, and to forecast future positions and intensity. Doppler radar is capable of detecting the motion of rain droplets in addition to intensity of the precipitation. Both data can be analyzed to determine the structure of approaching storms and hurricanes. Radar imagery monitoring during tropical cyclones is among the most important as it provides vital evidence from high-level (extraterrestrial) observation positions.

Worldwide the assessment is mostly based on the information provided by the Geostationary Operational Environmental Satellites (GOES 12, 13, 15) that follow geosynchronous orbits and are stationed over the west and east coasts of the United States and cover South America or from Meteosat-9 and 10. These operational spacecraft are located at 0° providing visible and infrared image data that are disseminated by Eumetsat or Eumetcast-America. The ocean surface topography is measured by the Jason-2 altimetry satellite that provides high-precision reference measurements of the ocean surface topography. Furthermore NOAA's mission is to conduct basic and applied research in oceanography, tropical meteorology, atmospheric and oceanic chemistry, and seawater acoustics. All over the world weather services such as the German Weather Service (DW) or the Royal Observatory in Hong Kong regularly monitor weather conditions and track storm movements by the visible and infrared spectra. The aim is to reach a better understanding of the physical characteristics and processes of the ocean and the atmosphere, both as a separate and as a coupled system.

The hurricane risk statements issued provide details of the storm's impact on the area such as the onset of winds, rainfall, and storm surge. They also call on the local emergency managers to start appropriate preparedness actions by providing specific information on evacuation preparations and on locations where emergency shelters are to be placed. In addition NOAA operates its own weather radio providing 24-h official weather information. The internationally agreed-upon Saffir–Simpson hurricane scale on windspeed is used to measure the hurricane's intensity (see Sect. 5.2). The scale is a five-class rating based on a hurricane's sustained windspeed and provides reliable indications that allow an estimation of potential property damage and flooding expected along the coast from a hurricane landfall. Hurricanes reaching category 3 and higher are considered major hurricanes, whereas category 1 and 2 storms are, although dangerous, require constant monitoring and often also preventive measures. Four different alert categories are distinguished:

- *Tropical Storm Watch* is an announcement that tropical storm conditions (sustained winds of 39–73 mph) are possible within the specified coastal area within 48 h.
- *Tropical Storm Warning* is an announcement that tropical storm conditions (sustained winds of 39–73 mph) are expected somewhere within the specified coastal area within 36 h.
- *Hurricane Watch* is an announcement that hurricane conditions (sustained winds of 74 mph or higher) are possible within the specified coastal area. Because hurricane preparedness activities become difficult once winds reach tropical storm force, the hurricane watch is issued 48 h in advance of the anticipated onset of tropical storm force winds.
- *Hurricane Warning* is an announcement that hurricane conditions (sustained winds of 74 mph or higher) are expected somewhere within the specified coastal area. Because hurricane preparedness activities become difficult once winds reach tropical storm force, the hurricane warning is issued 36 h in advance of the anticipated onset of tropical storm force winds.

In order to be able to track the many tropical cyclones worldwide and over time, each storm is identified by a name, beginning each year with A and ending with Z. The names given were female names until 1978. Then the system was abolished and alternating male and female names are in use. The names for year 2015 were already fixed in 2008 for the Atlantic Ocean starting with A (Andrea) to (W) for Wendy, and for the Pacific Ocean the names start with A (Alvin) and end with Z (Zelda). The names for the years until 2018 are also already fixed.

The central aim of the warning centers is to track adverse weather changes that may lead to hurricanes by an effective use of all available science and technologies to provide accurate and timely warnings and advisories on forecasts for hazardous weather conditions. The degree to which loss of life or harm can be reduced is dependent on how people receive the risk information and how they interpret the message to take the necessary decisions for protection. But it turned out after many hurricane warnings that the information provided was not used as intended. The problem starts when the forecasters at the warning centers, the local offices, or the local emergency managers create the message. Those who create the messages are in general technical people, whereas those the message concerns are lay people who only want their life and property safeguarded. Both groups are speaking different languages and are experiencing a broad range of social backgrounds. Although the forecasters are in general formulating their messages in technical terms, based on scientific evidence and operational facts, the information receivers are often unable to understand the message content (see Sect. 9.3). Risk information that is not understandable and that does not make transparent the cause–effect relationship of a probable hurricane track and its envisaged consequences creates mistrust and leads to a lack of trust in the hurricane risk assessment capacity in general. The next obstacle is via which information channels the message is to reach the populations at risk. The media have a responsibility to transfer information but also are highly interested in raising public interest often by overexaggerating news. A study carried out for the US National Weather Service (Demuth et al. 2012) therefore recommends that the partners in the warning system develop a culture of hurricane reduction responsibility by:

- Building up an understanding of each other's needs and constraints
- Ensuring formalized, yet flexible mechanisms exist for exchanging critical information
- Improving risk communication by integrating social science knowledge with design
- Testing and evaluating the message's reception with intended audiences.

9.5.4 Warning Chain

Table 9.2 gives an example of an effective alert communication system that is mainly based on the Australian Emergency Agency Manual No. 38. Extended by personal experience.

Table 9.2 Australian alert communication system (*Courtesy* Australian emergency agency manual no. 38)

Telecommunication	Advantages	Disadvantages
Telephone network (fixed wire) Network designed to handle normal traffic load; congestion can occur in case of emergency	Network already in place, versatile, direct communication, allows question and answers, reliable and stable network, low cost equipment, system can be extended to fax, uses lines and satellites	Only in fixed locations, vulnerable to physical network failures, disaster affected have to call, system can be overloaded in case of emergency, system needs intensive maintenance
Cell/mobile phones Should not be relied upon as the only communication system	High mobility, international roaming, GIS/emergency message add-ons, large distribution, very personal and direct information system, low cost equipment, uses lines and satellites	limited battery capacity, net congestion possible, coverage limited to network distribution, no broadcast capability
Facsimile	Low cost equipment, messages can be read, answers can be formulated in sentences, messages can be stored, maps and graphs can be transmitted, transmission to multiple addresses possible, easy to operate, widely used in homes, offices, schools, hospitals, and emergency centers	Paper necessary, system must be in operational mode, congestion possible, long messages take much time, system cannot receive a message while transmitting, congestion may occur when destination is busy
Satellite communication	Equipment small and light, ready to operate, low-cost equipment, reaching a large area, messages can be received by cellphone, individual messages and broadcast messages, versatile, voice and optical messages, message storage system, user friendly system widely distributed, user can be located with the system	Limited battery capacity, net congestion possible, coverage limited due to network distribution, satellite system can be affected by the disaster
Personal computer	Equipment small and light, ready to operate, low-cost equipment, reaching a large area, messages can be received, individual messages and broadcast messages, versatile, voice and optical messages, message storage system, user friendly, system widely distributed	Limited battery capacity, net congestion possible, coverage limited due to network distribution, transmitting can be affected by the disaster, hardware limitations, extensive hardware necessary, technology advances can make the system outdated, high cost of technology, system requires IT skills

(continued)

Table 9.2 (continued)

Telecommunication	Advantages	Disadvantages
Radio	Flexible, messages can reach houses, cars, schools, buses, no physical connection necessary, messages reach almost everyone, system is reliable and functional, system widely distributed, message can be repeated often, messages tailored for local demand, messages clear and understandable, low cost of equipment, reaching large areas	Radio system must be operational, messages must be understandable to everybody, message must be in time, radio messages must be formulated by emergency managers, messages often too general for the individual problem, complex technical system
Print media	Large distribution, medium is very cheap, medium is known to everybody, message can be read and stored, message can be distributed to others, medium comes regularly	Medium comes (normally) one time a day, message needs to be printed, message is normally one day old, messages need experienced personnel to be formulated, only one-way communication, reader cannot raise questions, messages give only generalized info, no direct communication possible, reaction of affected ones cannot be assessed
Sirens	Direct and event-related alert possible, fast and local, loud, technically reliable and easily established	People have to know what messages the system will transmit, system can generate panic, reaction of the people cannot be assessed, system depending on power supply, systems often stolen, too many failure alerts make people not listen

References

Abarquez, I. & Murshed, Z. (2004): Field Practitioners' Handbook.-Asian Disaster Preparedness Center (ADPC), Bangkok

ADPC (2006a): Participant's Workbook - Community-based Disaster Risk Management for Local Authorities. - Asian Disaster Preparedness Center (ADPC), Bangkok

ADPC (2006b): Critical guidelines - Community-based Disaster Risk Management. - Asian Disaster Preparedness Center (ADPC), Bangkok

APEGM (2014): Manitoba Flood Control And Protection. - Association of Professional Engineers and Geoscientists of the Province of Manitoba MA(online: www.apegm.mb.ca/Heritage/FloodControl.htm)

Arnstein, S.R. (1969):A Ladder of Citizen Participation. - Journal of the American Institute of Planners (JAIP), Vol. 35, No. 4, p.216–224(online: www.vcn.bc.ca/citizens.../arnsteinsladder.htm)

Arskal Salim (2010): Konflikt nach der Katastrophe. – in: Umwelt & Klima – Rechtswissenschaften, Max Planck Forschung, Hefte, Nr. 1, Berlin

Beierle, T. (2002): The Quality of Stakeholder-Based Decisions.- Risk Analysis, Vol. 22, Issue 4, p.739–749; Wiley Online Library (online:DOI: 10.1111/0272-4332.00065)

Day, H. J. (1970): Flood warning benefit evaluation-Susquehanna River Basin urban Residences.- ESSA Technical Memorandum WBTM Hydro-10,NOAA-National Weather Service, Silver Spring MD

Demuth, J.L., Lazo, K.J. & Morss, R.E. (2012): Assessing and improving the NWS point-and-click web page forecast information.- American Meteorological Society, NCAR Tech. Note NCAR/TN-4931STR (online at http://nldr.library.ucar.edu/)

Dilley, M., Chen, R.S., Deichmann, U., Lerner-Lam, A.L., Arnold, M. with Agwe, J., Buys, P., Kjekstad, O., Lyon, B. & Yetman, G. (2005): Natural disaster hotspots - A Global Risk Analysis - Synthesis Report.- The International Bank for Reconstruction and Development / The World Bank and Columbia University, Washington DC

El-Hinnawi, E. (1985): Environmental refugees. – United Nations Envronmental Programme, (UNEP), Nairobi

EMA (2010): Guidelines for the development of community education, awareness and management program. - Emergency Management Australia, Manual Series, No. 45, Canberra

Enarson, E, & Morrow, B.H. (1998): The Gendered Terrain of Disaster: Through Women's Eyes. –Preager, Westport CT

EU-EAC (2010):Volunteering in the European Union. – Final Report submitted by GHK, Educational, Audiovisual and Culture Executive Agency (EAC-EA), Dircetorate General Education and Culture (DG EAC), Brussels

Faupel, C.E., Kelley, S.P. and Petee, T. (1992): The impact of disaster education on household preparedness for Hurrican Hugo. – International Journal of MassEmergency and Disasters, Vol. 10, No. 1, p.5-24, Los Angeles CA

FEMA (1994): Basic training instructor guide. – Community Emergency Response Team (CERT), Emergency Management Institute (FEMA), National Fire Academy, Washington DC (online: www.fema.gov)

FEMA (2002): Earthquake Hazard Mitigation Handbook. – FEMA Hazard Mitigation Handbook Series, update June 2002, Federal Emergency Management Agency (FEMA), Washington DC

Feruzzi, F. (1997): Sounding the alarm - Natural disasters, be prepared. – The UNESCO Courier, United Nations Educational, Scientific and Cultural Organization, p.50, Geneva

FOCP (2014): Testing sirens. – Swiss Federal Office for Civil Protection (FOCP), National Emergency Operations Centre (NEOC), Bern (online: info@babs.admin.ch)

GITEWS (2010): 30 Minutes in the City of Padang: Lessons for Tsunami Preparedness and Early Warning from the Earthquake on September 30, 2009. - German-Indonesian Tsunami Early

Warning System (IS-GITEWS), German Technical Cooperation Agency (GTZ), Working Document No. 25, Case Study, Jakarta

IFRC (2011a): Public awareness and public education for disaster risk reduction: A guide. - International Federation of Red Cross and Red Crescent Societies, Geneva

IFRC (2011b):An evaluation of the Haiti Earthquake 2010 - Meeting Shelter Needs: Issues, Achievements and Constraints. – The International Federation of Red Cross and Red Crescent Societies (IFRC)(online: www.ifrc.org)

Jacobeit, C. & Methmann, C. (2007): Klimaflüchtlinge – Die verleugnete Katastrophe. – Greenpeace, Universität Hamburg

KATWARN (2014): Katastrophenwarnung. - Verband öffentlicher Versicherer / Fraunhofer Institut für Offene Kommunikationssysteme FOKUS, Berlin

Lutz, W., Goujon, A. &Wils, B. (2008): The population dynamics of human capital accumulation; in:Priskawetz, A., Bloom, D.E., & Lutz, W. (eds.): Population Aging, Human Capital Accumulation and Productivity Growth. Supplement to Vol. 34,Population and Development Review, the Population Council, p.149-188, New York NY

Lutz, W. & Samir, K.C. (2011): Global Human Capital: Integrating Education and Population .- American Association for the Advancement of Science (Science), Vol. 333, No. 6042 p.587-592, Washington DC

Mansuri, G. & Rao, V. (2004): Community-Based and -Driven Development: A Critical Review. - Development Economics Research Group, The International Bank for Reconstruction / The World Bank, The World Bank Observer, Vol. 19, No. 1, Washington DC

Morrow, B.H. (1999): Identifying and mapping community vulnerability. - Disasters, Vol.23 Issue 1, p.1-18, Wiley Online Library

Myers, N. (2001): Environmental refugees - A global phenomenon of the 21st Century. – Philosophical Transactions of the Royal Society: Biological Sciences, Vol. 357, p.167-182, London

Renn, O., Webler, T. & Wiedemann, P. (1995): Fairness and competence in citizen participation.- Evaluating models for environmental discourse.- Kluwer Academic Press, Dordrecht

Rifkin, S. B., Muller, F. & Bichmann, M. (1988): Primary health care: on measuring participation. Social Science and Medicine, Vol. 26, p.931–940, Elsevier B.V

Robertson,A. & Minkler,M. (1994): New health promotion movement.- Health Education Quarterly, Vol. 21, p.295–312, Bethesda MD

Rudloff, A., Lauterjung, J., Münch, U. &Tinti, S. (2009): Preface: The GITEWS Project (German-Indonesian Tsunami Early Warning System. - Natural Hazards Earth System Science, Vol. 9, p.1381–1382 (online: doi:10.5194/nhess-9-1381-2009)

Samir, K.C. (2013): Community vulnerability to floods and landslides in Nepal. – Ecology and Society, Vol. 18, No. 1, Acadia University, Wolfville, Nova Scotia (online: http://dx.doi.org/10.5751/ES-05095-180108)

Shearer, P.M. (1999): Introduction to seismology. – p.260, Cambridge University Press, Cambridge MD

Stern, N. (2007): The economics of climate change - The Stern Review.- p.692, Cambridge University Press, Cambridge MD

Striessnig, E., Lutz, W. & Patt, A.G. (2013): Effects of educational attainment on climate risk vulnerability.- Acadia University, Ecology and Society, Vol.18, No. 1,Wolfville, Nova Scotia

Subbiah, A.R., Bildan, L. & Narasimhan, R. (2008): Background paper on assessment of theeconomics of early warning systems for disaster risk reduction. Report submitted to the World Bank Group, Global Facility for Disaster Reduction and Recovery (GFDRR / Contract 7148513). - Regional Integrated Multi-Hazard Early Warning System, Asian Disaster Preparedness Center, Bangkok

THW (2014): Bundesanstalt Technisches Hilfswerk (THW), Bonn (online: thw.de)

Twigg, J. (2002): Lessons from disaster preparedness. – International Conference on Climate Change and Disaster Preparedness, (26-28 June 2002), The Hague

UNISDR (2005): Hyogo Framework for Action 2005-2015: Building the Resilience of Nations and Communities to Disasters. - Extract from the Final Report of the World Conference on Disaster Reduction (A/CONF.206/6), p. 25, Geneva

UNISDR (2006): A Guide to community-based disaster risk reduction in Central Asia. – International Strategy for Disaster Risk Reduction (ISDR), United Nations, Geneva

UNISDR (2009): 2009 UNISDR terminology on disaster risk reduction. - International Strategy for Disastr Reduction (ISDR), United Nations, Geneva

UNISDR –PPEW (2005): Project overview plan, Early Warning Strengthening Project. - UN Flash Appeal Project TSU-REG-05/CSS06-REGION: Evaluation and strengthening of early warning systems in countries affected by the 26 December 2004 Tsunami, update of 30 April 2005, International Strategy for Disaster Risk Reduction (SDR), ISDR Platform for the Promotion of Early Warning, Bonn (online.: http://www.unisdr.org/ppew/tsunami/ppew-tsunami)

UNV (2007): Contribution of Volunteerism to Disaster Risk Reduction: - UN Volunteers - Aspiration in Action, Practice Notes, United Nations International Strategy for Disaster Reduction (UNISDR), Geneva

USNHC (2014):US National Hurricane Center. - NOAA/ National Weather Service National Centers for Environmental Prediction, National Hurricane Center, National Weather Services (NOAA), Miami FL

World Bank (2006): Global Facility for Disaster Reduction and Recovery – Reducing Vulnerability to Natural Hazards. – The World Bank, Washington DC (online: www.gfdrr.org)

Chapter 10
Responsibility of Geoscience in Natural Disaster Risk Management

10.1 Science and Technology

10.1.1 The Role of Science and Technology

Progress in the science and technology of natural hazards and of related coping mechanisms have made it possible over the past years to introduce significant changes in our response to natural hazards. Science is a systematic approach of collecting knowledge about nature and condensing and synthesizing that knowledge into testable laws and theories, whereas technology is a term referring to a collection of techniques and the knowhow to make practical use of them. Science and technology significantly empower humans to control and adapt to their natural environments and moreover comprise how scientific research and technological innovation can affect social, political, and cultural values. In this context, it is the current state of humanity's knowledge of how to combine resources to produce desired products, to solve problems, and fulfill their needs; it includes technical methods, skills, processes, techniques, tools, and raw materials. When combined with the term "disaster risk management" science and technology, it refers to the state of the respective field's knowledge and tools regarding the assessment of root causes of natural disasters and their respective impact on the population at risk (WCS 1998).

Scientists in disaster risk management should:

- Assess the natural condition conducive for disaster. Such assessments should be carried out according to the state of the art and should comprise standardized quality management. Science should be independent, verifiable, and transparent and should be carried out under the assumption of a pre-set goal.
- Data collections, analyses carried out, and their interpretation should be undertaken carefully and prudently and be comprehensively documented and published accordingly. A problem with many scientific investigations is they are often third-party financed and commissioned under a certain topic. This often brings bias into the investigation of how far the financiers define the outcome and to what extent the findings can be made openly accessible.

© Springer International Publishing Switzerland 2016

U. Ranke, *Natural Disaster Risk Management*,
DOI 10.1007/978-3-319-20675-2_10

- Scientists should define their role to be honest "brokers" for the issue.
- The findings should be published in a way that opens space for alternative interpretations and carefully explain the implications and inherent limitations of the findings to the public.
- Multidisciplinary approaches should be promoted to encourage cooperation between the social and natural sciences, and draw lessons from local knowledge and promote the inclusion of experiences from international cooperative projects as well as from aboriginal wisdom.
- Encourage a holistic approach to incorporating the socioeconomic aspect.
- Whenever possible, the orientation of the investigations should be shifted from isolated local single-topic disaster assessment to a holistic assessment that reflects the dynamic of time and space scales.
- Scientists should be aware of the potential of communication. They should not restrict their responsibility to the scientific realm rather than take the chance to disseminate knowledge. This nevertheless requires awareness and a respective capability of transforming pure scientific wording into commonly understandable language. In many cases scientific outcome is given in very technical terms (risk probability of 0.0001 %) but not in a way that is understood by those to whom the information is addressed.
- The predictive power of science should be fully exploited to serve social needs with candid awareness of the limitations of scientific predictions.

Science is according to the German sociologist Horkheimer (1968) in general understood as the application of technical systems that uses theoretical propositions to analyze and interpret thematic fields, problems, and questions that are thus far not understood. Such scientific findings are then to be tested in practice, although the scientific approach is still of a strictly theoretical nature. Horkheimer emphasized that this traditional ideal of science is primarily applicable to the natural sciences. "Science in the traditional view seems to be neutral, independent, objective and a-historical and seems to be independent of person, place and time. Thus natural science itself is mainly viewed as a purely theoretical enterprise of which the practical possibilities of application should normally consider afterwards" (Klapwijk 1981). The restriction of responsibility only on the side of data creation and interpretation, leaving the social application to "someone else" is highly objected to by Horkheimer. He does not want scientists to define themselves as a specialist "who only thinks about his own area and thus favor the illusion of a neutral, unprejudiced science." For him consequently science is not neutral, and neither is scientific expertise. Scientists have to accept a fundamental responsibility towards society, because natural science functions within a society. All tasks, the field investigations, laboratory analysis, and data interpretation carried out by natural scientists are rooted in a pre-defined, specific, and target-oriented context in a way to "organize society in a scientific way, based on rationality and by reflecting the present state of technology" (Klapwijk ibid). "The idea of science must entail a critical consciousness of the possible positive and/or negative contributions to human life and culture. There is not one good reason for eliminating this social dimension from the idea of (natural) science, however much of this understanding has been ignored in university education and in practical research."

For natural scientists technical and naturally given facts generally govern the way of thinking. The motto is, "Knowledge is power," but Habermas (1968, 1994) also called on the natural scientist critically to reflect, "Why and to what end he is doing his research and how these interests relate to the social, economic, cultural and ecological situation of the society" (Klapwijk ibid), thus making traditional insights of natural "science an integral part of an all-encompassing social-critical view." Science must result in social responsibility (Horkheimer ibid) and emphasizes that theory always has to have a practical intent. He further points out that the "acceptance of social responsibility is more than having a perspective on the social relevance of social research."

But it should be acknowledged that making natural science political can only be achieved by a steady and iterative approach of both natural science and sociology. Finally there was a lecture given at the University of Goettingen by a sociologist with the title, "About the Naturalization of Sociology," indicating that it is not a one-way approach of natural science towards sociology, but that sociology also has to be open for "understanding nature" (see also: Beck 2011). The "responsibility of the natural sciences for society and politics cannot withdraw him from the political problems and moral issues he is integrated in" (Klapwijk ibid). The natural scientist must make political decision makers aware how and to what extent his or her findings can best be applied for the sake of society. On the other hand, the responsibility of political government for science is also only an indirect one. It is expressed, among others, in its care for the quality and financing of scientific activities. In any case policy should not direct or even not stimulate anticipated results to science, a sector that is for the time being highly controversial when we thinks of the discussion on nuclear waste disposal sites all over the world. Science has its own responsibility in the field of acquiring knowledge and in theory formation. The results of scientific activity could conflict with political planning, but this may not prevent the scientist from doing his work.

From the above given examples of natural catastrophes, the Indian Ocean tsunami, the earthquake at L'Aquila, and the another example of a landslide in the small German town of Nachterstedt clearly document the mandate and responsibility geoscientists are involved in risk identification and in increasing societal resilience.

Thus in the aftermath of the Nachterstedt event, the German Society of Geologists gave the following statement in a press release on the role and function of geoscientists in disaster situations. DGG made clear that the "Geologists see their responsibility not only in hazard assessment and working out expertise on the geologic factors leading to a disaster but also in disseminating this knowledge to official regulators and to the affected individuals." The partially dramatic outcomes of natural disasters in Germany but also elsewhere require a much closer cooperation and networking of science and implementation as is practiced thus far. And that kind of cooperation should not be restricted to the aftermath of a disaster but should be established in times of no disaster. These are the times where trust and confidence can be generated with the population at risk, in order to keep up their vigilance. The geosciences "are ready to take up their fraction of the social, moral and economic responsibility" with their engagement in disaster mitigation. But it should also be noted that not every aspect of probable disasters or a catastrophe that had happened should be commented on by the science sector. Although acknowledging the fact that especially in the case of emergency, the public and in this regard very often strongly demanded by the media, geoscientists are urged to give their opinion. The geoscientists have to take this demand into consideration and make clear that their responsibility is to work on the facts. But on the other hand the geoscience sector should acknowledge that even with its scientific elaboration, geoscience automatically becomes involved in the public debate and that there is no escape possible. The geosciences have to be prepared for this and to be open to take over this responsibility. But it has to make clear that it will restrict itself to the natural facts, but this restriction does not mean that geosciences see themselves not a part of society.

A striking example for the potential that science and technology have in order to protect people against the impact from natural disasters is the December 2004 Indian Ocean tsunami. This event was the onset of recommendations made by the British government (UK-DTI 2005) to propose an increase in science and technology in disaster prevention. The study came to the conclusion that, although the international geoscientific monitoring networks provided ample evidence of a major tsunamogenic earthquake shortly after the event, this information had not been effectively communicated to decision makers. The reasons for this were manifold, but in general, it originated in the very short distance between the epicenter and coast as well as in the politically unstable conditions after 20 years of martial law over the Aceh province that hampered a rapid dissemination of information. The study emphasized that science and technology can help to understand the mechanisms of natural hazards and to analyze their potential risks. However, the state of understanding of differing natural hazards and their potential impact is not equal all over the world and significantly differs on the type of disaster. Still there are significant shortcomings in understanding and information relating to such potential catastrophic events. The scientific and technological disciplines that are involved include basic physical and engineering sciences, natural, social, and human sciences. They can provide a wealth of information to the policy environment (sociology, humanities, political sciences, and management science). Scientific knowledge of the intensity and distribution in time and space of natural hazards, and the technological means of mitigation and prevention have expanded greatly as well as in understanding of the root causes and parameters of natural phenomena, in modeling techniques for predicting their behavior and in technological means of resisting their forces have increased considerably.

But the general assumptions drawn from this event are that the international community has to establish better ways to communicate the scientific expert's evidence to the people at risk. The contacts among natural scientists, engineers, and the political levels are still not established in a way that supports a social and economic oriented decision making. The contacts are too weak, too long, and mostly not based on a mutual understanding. As long as scientists restrict their role in assessing, analyzing, and explaining the causes of the disaster and do not accept a social and economic dimension related to each disaster, the gap between science and politics will hardly be bridged. What is necessary is to improve the integration of science into disaster risk management by strengthening the mutual understanding of the cause–effect relationship of natural disasters as well as the establishment of robust communication lines. Moreover as many of the hazards and disasters have global or regional impacts a deep and strong internationalization of scientific cooperation is indispensable, especially concerning hydrometeorological events, heat waves, tropical cyclones, drought, but also for transboundary pollutants and other threats (nuclear, chemical, diseases).

The above-given examples clearly demonstrate how deeply natural science is interlinked with the social, political, and economic sectors. This also holds true for the geosciences in general and especially holds true for the geoscientists in the field of natural disaster risk management. Many such examples can be found: all of them prove that natural science in this field is not restricted to technical and basic geological investigations and therefore cannot be regarded as not being an "independent" factor. In this context it has to be stated that sociology as a science for a long time did not accept any influence by natural facts (Plapp 2003).

In the past, advances in science and technology often proved to be of social benefit (UNESCO 1998). But even when the output often has been admirable, research and technology have to acknowledge that it rather should define itself at first to serve the needs of society. Often the opinion among scientists prevails that society shall have to accept the fact that science has to be free and independent. This common understanding has sometimes isolated scientists from the mainstream of society, making it difficult for them to realize the true public needs. Science must accept that even "basic, neutral, and independent" science has a societal responsibility. On the other hand, there is much evidence that the policy sector is sometimes in need of very urgent advice on technical matters, and has often been unaware of the scientific expertise available due to mere lack of information. A greater, broader, and deeper dialogue among scientists, policy makers, and the public is needed in order to overcome the often-recognized opaque decision making. Such a pro-active involvement would bring scientists into policy making with much benefit for society. Another source of hindrance for scientists to develop social responsibility is the general perception of science. It is seen by most scientists to be an intellectual property, resulting in a retention of information, an attitude that hampers an open circulation of information as is required for political decision making (cited in: UNESCO ibid.). Moreover in the course of the globalization, the perception of science as a general "market good" worldwide increased, challenging the traditional perception of science as a "public good" with the consequence that already today there are signs that cutbacks in research expenditure in welfare, education, health, and natural disaster risk-related sectors have taken place, sectors that enjoyed broad acceptance when science was recognized as a public good (UNESCO ibid.). In order not to duplicate the mistakes of the past, UNESCO sees a serious risk that only the higher income classes will benefit from the technological innovations and the poorer not or only less. This holds true especially for those resources that in many countries have traditionally been supplied through the subsistence economies: an economic approach that today is referred to as the "commodification" of livelihood.

In 1998 WCS (1998) stated that the scientific influence on our daily life has already increased and will ever increase. In order to comprehend this responsibility scientists should be aware that many findings in modern technology may create fear in the people. Similar to the international debate on climate change there is also in the natural disaster management sector a clear tendency that scientists and technicians are more and more involved in political decision making. This involvement leads to a political dialogue that can be described as "scientific or evidence based policy" (Rayner 2006). Such a tendency undoubtedly leads to a "scientification of politics" and on the other hand to "politicization of science" (Beck 2011). The complex nature of many of the cause–effects of natural processes calls for a deeper involvement of scientists especially in the sectors that are often "obscure," barely visible and are characterized by complex spatial, time, and socioeconomic relationships with strong vertical and horizontal interdependencies and interrelations. In this dialogue scientists are especially called on to work out model cases and theories and to deliver multisectorial solutions based on scientific evidence. But scientists should be aware of their specific function, to give orientation to policy (WGBU 2014) but not determine it and that it should follow the empirical science tradition. Scientists moreover should "reflect on the social consequences of the technological applications of their work." They are called on to "explain the degree of scientific uncertainty or incompleteness in their findings and at the same time, they should not hesitate to fully exploit the predictive power of science, especially in cases of direct threats like natural disasters or water shortages." WCS (ibid) makes the following recommendations. Scientists and scientific institutions should among others:

- Promote multidisciplinary approaches to research.
- Encourage cooperation between the social and natural sciences.
- Draw lessons from local knowledge systems and aboriginal wisdom.
- Take a realistic range of socioeconomic conditions and effects into account.
- Carefully explain the implications and inherent limitations of their research findings to the public.
- Fully exploit the predictive power of science and be aware of the limitations of scientific predictions.

WCS (ibid) reiterates that scientists should be more "proactive in policy making." This could be done by a sort of "science–policy agreement" that would automatically recognize "the value of scientific advice, but also make clear that such advice is but one ingredient in decision-making." Such "contracts can also define the performance standards by which the inputs of scientists" can later be evaluated. Scientists can most effectively contribute to consensus finding when the involvement is formalized and long-lasting. Macfarlane (2012), newly appointed head of the US Nuclear Regulatory Commission, stated the difference in understanding between geoscientists and engineers is that so far it is not an established fact that "nuclear engineers can and do integrate knowledge of Earth processes adequately." Moreover more scientific communities should consider adopting an international code of ethical conduct for scientists, as it has been already been declared by many international organizations.

10.1.2 International Framework for Science and Technology

European Union

The European Union can serve as a model on the spectrum of how research and development (R&D) can be successfully organized (UNESCO 1998). Europe is the second most important scientific power in the world after the United States. Although the percentage of investment in R&D remained in the European Union at about 2 % of the GDP for decades, the absolute amount of money spent has, however, increased significantly. Nevertheless there are quite a number countries in the Union where the percentage of R&D expenditure is stagnant or has even fallen. It is a well-known fact that the younger generations in Europe are not that interested in natural sciences and in engineering that are necessary to keep up the high standards of European R&D in the future. Therefore many politicians pledge for an earlier, deeper, and better involvement of natural science in the secondary curricula in addition to cooperation between universities and industry to provide target-oriented training to students in laboratories and field studies.

Leading countries in the European Union in basic and applied research are France, Germany, and the United Kingdom. At the end of the 1990s the European Union had about 900,000 researchers of whom one third were working in Germany. The highest percentage of scientists and engineers per labor force is in Finland with almost 9 %, whereas the European average lies at about 5 %. Germany follows Finland with almost 8 %, and then France with more than 5 %. When, however, the amount of scientific publications is compared, then the United Kingdom leads with almost 10 % of the world total, and Finland contributes less than 1 %. According to the historical development of science in the European Union, the main focus lies in basic sciences, but in the United States and Japan the fields of applied sciences dominate. There has been a shift in orientation towards marketable products in the European Union on biomedicine, chemistry, and live sciences in the last decade, yet physics, engineering, electronics, and informatics are still underrepresented (UNESCO ibid).

European research is increasingly a matter of cross-border cooperation that can be organized in the form of bilateral cooperation between single researchers and universities, usually based on personal contacts, or in the form of multilateral cooperation. Such cooperation requires a much higher degree of formalization and structuring. A formal scheme for cooperation between research organizations known as the Associated European Laboratories (AEL) was launched a few years ago at the initiative of the CNRS. Another frame for R&D cooperation in Europe is the COST program (COST 2010). COST is an intergovernmental framework allowing the coordination of nationally funded research on a European level. COST aims at reducing the fragmentation in European research investments and helps to open the European research area to cooperation worldwide. Since 1971 COST has connected over 30,000 researchers in Europe and beyond through networks in all fields of science and technology including the earth science sector (ESSEM).

There are different models within the European Union to finance R&D. In Germany, due to its basic law, science, research and education are the main objectives of the federal states, although the national government has created a number of powerful R&D organizations including The German Science Foundation (DFG), the Max-Planck Gesellschaft (MPG), the Fraunhofer Gesellschaft (FhG), and many others, to delineate science and research activities generally. In the United Kingdom and Scandinavia R&D is financed via specialized organizations such as the British Research Council. An even stronger state involvement is practiced in France, Italy, and Spain where large and financially independent organizations are established to outline and carry out R&SD under their auspices. In France, for instance, about 50 % of the national R&D expenditure is undertaken via such organizations.

One great challenge in R&D policy in Europe in the last decade was the shift to more cost-effectiveness of science activities. Financial restrictions in all member states forced the governments to reorient research funding. Consequently the governments established so-called "centers of excellence" that are professionally organized and based on budget-oriented financing. This shift in paradigm, made many scientists and engineers complain of a "loss of freedom" yet on the other hand, the R&D activities are carried out with much more professionalism.

By far the biggest change in R&D in Europe, however, occurred in the field of connecting (pure) natural science with social and economic impact. Today societies want to see benefits from the tax money allocated to the science sector. Especially in the field of environmental protection, climate change, or energy generation, societies today ask for a higher degree of social responsibility even from geologists and geophysicists. Moreover, responsibility and ethics in R&D is emerging as a real topic. Scientists should be educated to an ethical code that denies misconduct, fraud, and the like. Laboratory experiments should not be carried out misusing animals, seeds for plants should be developed that are not (only) oriented towards high productivity but also that their use does not infringe on other agricultural uses. For example, the use of fossil fuels for power generation in cases where alternative, renewable sources of energy are available should not be followed without a deep discussion on the pros and cons. The use of data for scientific studies of other than the scientist's own sources should be made clear and transparent and be fully in line with the stipulations of the law of intellectual property rights claims.

Principles for the role of science and technology to achieve sustained development of the world's natural systems were set down in Agenda 21 and are summarized here as far as they concern the sectors of science and technology:

- States have the sovereign right to exploit their own resources pursuant to their own environmental and developmental policies.
- In order to achieve sustainable development, environmental protection should constitute an integral part of the development process.
- States should improve scientific understanding through exchanges of scientific and technological knowledge by enhancing the development, adaptation, diffusion, and transfer of technology.
- At the national level, each individual should have appropriate access to information concerning the environment and be given the opportunity to participate in decision-making processes.
- States should enact effective environmental legislation. Environmental standards, management objectives, and priorities should reflect the environmental and developmental context to which they apply.
- States should develop national law regarding liability and compensation for the victims of pollution and other environmental damage.
- In order to protect the environment the principle of "precaution" should be widely applied by states according to their scientific and technical capability.
- Lack of full scientific certainty should not be used as a reason for postponing cost-effective measures to prevent environmental degradation.
- National authorities should promote the internalization of environmental costs as laid down in the polluter pay principle.
- Environmental impact assessments should be undertaken as a national instrument for any development activity that is likely to have a significant adverse impact on the environment.
- States should immediately notify other states of any natural disasters or other emergencies that are likely to produce sudden harmful effects on the environment of those states.
- States should provide prior and timely notification and relevant information to those potentially affected.
- States on activities that may have a significant adverse transboundary environmental effect.
- Knowledge and traditional practices of indigenous people should be considered in environmental management and natural resources development.

10.1.3 Technological Innovation

According to Malthus (1798), the "increase of population is limited by the means of subsistence," meaning that a population can only survive when its access to natural resources is secured. And as population growth proved exponential, this means in our finite world, that the amount of goods to be used must also grow exponentially too (UNDP 2001). But how will it be possible to provide enough resources to all? In the last centuries this was first of all achieved by technological innovations. They are essential for human progress and therefore have been at the heart of human development since earliest times. Advances in science and technology and the dissemination of knowledge have become substantial components to societies. The "revolutionary" advances in digital electronics, in genetic-based food production, or in decoding the human DNA, have pushed forward the frontiers on how people can use technology to secure a proper livelihood, stimulated economic growth, and empowered people to participate in their community's welfare. Today technology has devised tools for improving health, raising productivity, and facilitating learning and communication. It is furthermore characterized by its interaction with other types of technologies that altogether form an international network of science and technology.

The challenge for technology is tremendous, as today—even more than in former times—technology forms the basis of human development. The potential but also the expectations on what technology can provide is overwhelming. But technology cannot take over its role as driver of human development, rather human development is a prerequisite for technological innovations. The increase in social and political freedom, participation, and access to natural and material resources creates conditions that encourage people's creativity. A higher level of education will lead to a higher level of scientific and technical ability. Research and innovations in the last decade created an ever-increasing demand for natural scientists, information technologists, and science information network managers. The store of indigenous knowledge and the communication systems invented make scientists and technologists today globally mobile. Not only in developed countries but also in developing countries, scientists and technologists now have fast, comprehensive, and cheap access to state-of-the-art science and technology, on information on finance and business, and are provided with a means for broad skill transfer for their home country.

However, the rising expectations of the societies regarding technological innovations can only be met if more scientists are engaged in resource management worldwide followed by better implementation.

Innovations in science and technology alone will not be able to secure social development as long as there are no simultaneous changes in socioeconomic and ecological policy to take place. The growing power of commercial interests, the declining role of governments, and a general retreat of governments from responsibility cannot replace real political solutions. In this sense UNDP (ibid) stated that the "market is not enough to channel technological development to human needs" rather than turning technology into a tool for human development. This automatically requires public risk money investment to create and disseminate the innovations that societies need. The big difference between industrialized and developing countries deserves creating technologies that provide especially poor people with affordable products. One of the steps towards achieving this target is the international agreement on intellectual property rights. This agreement is seen as a cornerstone that can bridge today's very uneven distribution of intellectual capacity in poor and rich countries. To achieve fair and efficient global access to basic scientific knowledge and technologies the establishment of a "world intellectual property clearinghouse" (UNDP ibid) is seen as a major step ahead. By identifying all relevant intellectual property for a given technology, indicating what is available for use and how, the clearinghouse could play an important role towards solving the collective problems in natural resources management. Another instrument to enhance dissemination of knowledge is the promotion of common standards either on technical innovations or on production technologies. Without them local markets will not find sustainable access to the world markets.

Advances in science and technology, such as the increase in medical sciences in the last 100 years, global communication, or satellite images of our planet have improved vision of the Earth in most of the world's societies. In the last decade science broadly initiated growing public awareness in the ecological aspects of life and on how human activities are presently threatening this environment. It became clear that if the current consumption of natural resources and increasing stress on the regional and local environments continues, this will lead to a breakdown of the natural systems and will diminish a further development of civilization. Science is now asked to take up its responsibility to design measures to help achieve a sustainable ecological and economic system.

It is foreseeable that the world societies will continue to rely heavily on easy and open access to natural resources, for example, on sufficient and hygienically clean water, affordable fossil fuels, and fertile soil. But such resources are, although available up to now in sufficient quantities are running low. There are countries that are flooded regularly whereas others are suffering from extended droughts. Energy is cheap in some countries whereas others can hardly afford it. The central conflict between the current economic forces and the vision of sustainable development can be reconciled also with the help of scientists and engineers. Science constitutes an important aspect of our cultural domain and historically played a progressive role in socioeconomic development. Today science and technology seem to be caught in a cross-fire; it is acknowledged to be a major instrument of the world economy, but on the other hand, science is increasingly being called on to produce knowledge and technology that promote environmentally sustainable, people-oriented development and long-term management of resources. However, to manage such a change often requires fundamental structural adjustment of policies by the states and also by their societies, for example, when a country has to reorient its agriculture on cash crops to generate foreign currency rather than on food crops for local consumption, due to significant budget problems.

Only when policy makers accept that development decisions must not be based on political wisdom alone, but also on forward-thinking research and development, and when scientists and engineers are open to developing a new vision of science, can the necessary sustainability of the natural system be achieved, bringing production and consumption into equilibrium. Only when the financing policies reorient the terms under which research grants are allocated, to be more neutral and flexible, and only when scientists are not continually pushed to find short-term solutions when long-term ones are needed, will the goal for a sustainable world get closer.

The time is already overdue for scientists and engineers to direct themselves to the topic of public good in more detail. This approach has already been intensively investigated by Elinor Ostrom, Nobel Prize Winner on Economy (Ostrom 1990) and Hardin (1968). But geoscientists are still reluctant to incorporate their findings into their work. Hardin points to the fact that many of the social, economic, and ecological problems have no technical solution. Scientists should not "try to win the game" over the other disciplines but rather help finding a solution. Scientists cannot always control the application of their findings. However, they have a responsibility to engage in public dialogue about the implications of scientific findings and to help distinguish between socially beneficial and socially harmful applications. Often it will not be possible to solve a problem without relinquishing a stakeholder from his traditional privileges as the example of a flood retention area cannot be identified and implemented without interfering with the terms of use of the proprietor.

As many of the problems in risk reduction are acknowledged to be intangible, many geosciences regarded themselves as unable to contribute to a solution. But in fact even intangible goods can and are already quantified (tangible) every day as described in Sect. 7.2, "Valuing Statistical Life." What is needed is an objective means of measurement. One of the most striking measures in this regard can be called "human security" (Scheffran et al. 2012). The way to solve this is by weighing the values or the variables (Hardin ibid): "The problem is to work out an acceptable theory of weighting. Synergistic effects, nonlinear variations, difficulties in discounting the futures, all this is an intellectual problem, but is not in-soluble." Following this statement, geoscientists are called on to provide their expertise and experience to help establish social arrangements that define the role and responsibility of the different stakeholders in risk reduction. This will definitively lead to some sort of individual coercion, but what is needed is a sense of societal agreement ("mutual coercion") (Hardin ibid).

10.1.4 Acceptance of Technology

Based on studies in sociology it seems that there has been a significant change in western societies on how the people address technology (Hennen 1994). It turned out that today it is getting more and more difficult for the political sector to wean westernized societies from the comforts of technology. The change in acceptance has led to the assumption that in modern societies former controversies on social and economic strategies are replaced by a debate on the achievements of technology, although it is acknowledged that both sectors are closely interlinked. After having achieved quite a sustainable level of provision of goods in these countries, the "nature issue" developed in the 1970s that still dominates the debate. But the controversy nevertheless reveals that there is a great deal of uncertainty

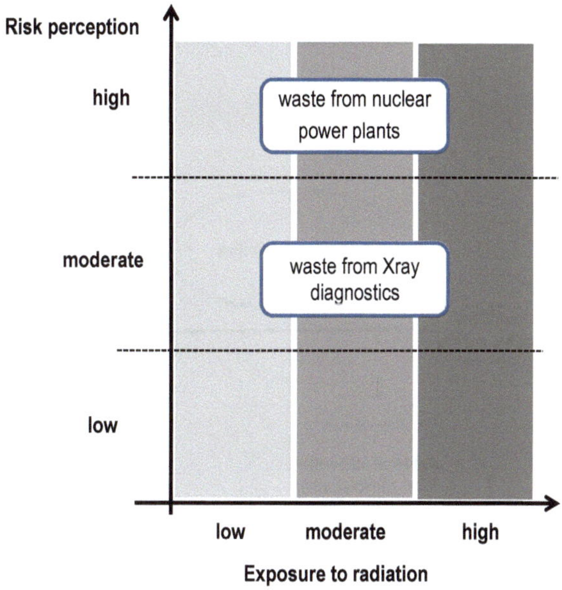

Fig. 10.1 General acceptance in Germany on radiation exposure dose from nuclear power plants and Xray diagnostics (Based on: Sigrist 2004)

in the societies that runs along the path of "how modern technology will affect the people." The study on acceptance of technology in Germany (Hennen ibid.) revealed that since the 1960s, the then quite positive acceptance of technology has diminished in general. The prevalent acceptance, for instance, of the use of nuclear energy for power generation has, although not refusing the technological advances, has been replaced by a tendency for critical distance. It turned out that technology is less accepted when it is seen in the context of globally interconnected companies and types of technology that are not transparent to society, whereas technologies that are applied in daily use are welcomed, especially when they serve the purposes of medical safety. Figure 10.1 illustrates this finding. Regarding the same radiation exposure dose (middle radiation) from nuclear power plants and from X-ray diagnostics, the acceptance of society is completely different. Although waste from medical treatment is cognitively related to health and thus enjoys broad acceptance, waste from nuclear power generation (although having the same radiation dose) is highly objected to.

A similar finding was made by the University of Bremen in a study on how the inhabitants of the cities of Hamburg and Bremen perceive different organizations' policies in regard to flood risk reduction measures (Fig. 10.2). It turned out the people place the hierarchically organized authorities and the political parties into the sectors of "traditional welfare" and "security," but they feel such organizations fail to address the risk and general policy critics-related sectors. This assessment makes clear that the people's obvious fears were only addressed by the alternative political groups, whereas the mandated organizations such as the technical services agencies are placed somewhere in between. The study moreover revealed that there is no difference in the people's perception on this in the cities.

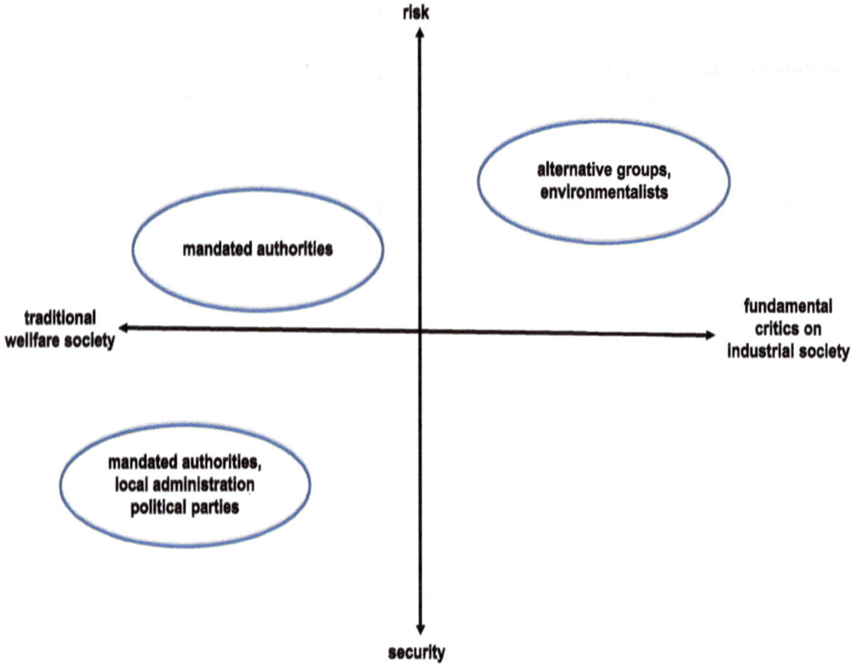

Fig. 10.2 Perception of inhabitants of the cities of Hamburg and Bremen on flood risk reduction measures (Based on: ARTEC 2007)

The surveys moreover revealed that people clearly distinguish between positive or negative effects of consequences from technology. When asked whether technology secures labor and income, more than 90 % answered the question with "Yes" (Hennen ibid). The German level of acceptance of technology is more or less the same as in other industrialized countries. For instance, in Japan technology is not seen a priori as good; although more than 50 % see it as a guarantee for labor opportunities. The study concluded that there are indications that the controversial debate on technology is unavoidable the more political decisions depend on the state of science and technology, which consequently raises the question of the legitimacy of technology. It sees furthermore that the heightened debate on technology acceptance is more an outcome of the increased expectations of the public from politics than an expression of technology refusal.

10.1.5 Code of Conduct

All research and investigations in science and technology and the pursuance of the implementation of the findings are agreed to be conducted upon an international obligation to three basic ethical principles. This ethic codex was originally established for the medical science sector (CIOMS 2002) in accordance to the Human Rights Charter, but in its true meaning, it also found application in the natural science sector in general. It can also be directly applied for geoscience engagement in natural disaster risk reduction management. The principles are:

- Respect for human beings
- Self-determination
- Social benefit
- Nonmaleficence
- Equality and justice

In 2007 the US Geological Survey (USGS 2007) enacted a policy for "scientific integrity" for ensuring such integrity in the conduct of scientific activities and procedures for reporting, investigating, and information dissemination. The policy transcribed the US Federal Policy on Research Misconduct as a binding regulation for all USGS employees. With this document the USGS employees commit themselves to respect unrestrictedly the integrity of science. A series of "rules" have been decreed that are to be agreed to by every employee and that are summarized below to:

- Contribute the best, highest quality scientific information.
- Honestly, objectively, thoroughly, and expeditiously process data.
- Fully disclose all research methods used and available data.
- Respect confidentiality and proprietary information.
- Not engage in falsification or plagiarism.
- Be constructive in criticism, participate in peer reviews.
- Be diligent in creating, using, preserving, documenting, and maintaining data.
- Establish quality assurance and quality control programs.
- Comply with federal law relating to security of confidential and proprietary data.
- Follow appropriate reporting standards.
- Respect the intellectual property rights of others.
- Differentiate among facts, opinions, hypotheses.
- Be responsible for the quality of data and for the integrity of conclusions.

USGS commits itself to pursue any allegation of scientific misconduct either caused by its employees or by others in a timely, open, and neutral manner. The prosecution of the alleged misconduct will follow a consensus-based process and will be based on three criteria:

- Significant departure from accepted practices of the relevant research community.
- The misconduct is committed intentionally or knowingly or recklessly.
- The allegation is proven by a preponderance of evidence.

The Max Planck Society (MPG) established in 2009 a set of "Rules of Good Scientific Practice" in order to define the ethical and moral bases for scientific activities carried out under their auspices (MPG 2009). The rules are fully in line with the recommendations given by the German Science Foundation (DFG 1998). They are binding on all persons active in research work at the Max Planck Society to observe scientific honesty and the principles of good scientific practice.

Good scientific practice is generally summarized as:

- Precise observance of scientific discipline
- Honesty and transparency in acquiring, selecting, and processing data
- Clear and comprehensible documentation of the methods employed and results achieved
- Storage of primary data for 10 years
- Systematic scientific skepticism
- Reproducibility of results
- Activity and analysis according to the standards of good scientific practice
- Systematic alertness to any possible misinterpretations
- Avoidance of overgeneralization
- No hindrance to the scientific work of others
- Openness to criticism
- Publication on principle of research results
- Fair evaluation and citation of any literature used
- Honesty in the recognition of the contributions of colleagues

Scientific misconduct is basically when false statements are made knowingly or as a result of gross negligence, when intellectual property of others is infringed upon, or when others' research work is impaired in some other way. The following list outlines the main aspects of misconduct:

- Fabrication of data and falsification of data
- Non- or selective reporting
- Rejection of unwanted results
- Manipulation of a representation or illustration
- Incorrect statements
- Unauthorized exploitation (plagiarism)
- Misappropriation (expert opinion, research methods, theft of ideas)
- Usurpation of scientific authorship or coauthorship
- Unauthorized publishing
- Sabotage of research work (including damaging, destroying, or manipulating experimental arrangements, equipment, documents, hardware, software)

Similar codes of conduct exist, for example, in Australia, the United Kingdom, and Canada. In Australia the Department of Mines and Petroleum (DMP) is committed to operating as a "competent, ethical, transparent and accountable organization which provides a satisfying, safe, and equitable work environment for its employees." DMP's code thus provides a clear framework for the standards of behavior and ethical decision making with which all DMP employees are bound to comply. Since 2006, DMP has provided all employees with specific training to create awareness of accountable and ethical decision making for all employees. The Canadian "Code of Ethics" reiterates the point that its "professionals must be qualified by trained by training and experience for the work they do and they must know the limits of their knowledge" (VanDine 2011).

In 2002 the Southern California Earthquake Center (SCEC) published recommendations based on the "California Seismic Hazard Mapping Act" (SHMA) as a companion to the California Building Code, that clearly define the role and function of engineering geologists and geotechnical engineers in investigating the country's seismic hazards and slope stability (SCEC 2002). The purpose of the Act is to protect public assets and the people from the effect of ground shaking, liquefaction, landslides, or other ground failures as well as from risks caused by earthquakes. The Act furthermore stipulates that the "site investigation report must be prepared by a certified engineering geologist or registered civil engineer." The experts must have proven expertise in the subject, must hold an university degree in the related disciplines, and must have several years' experience in the field of seismic hazard evaluation and mitigation. The expert has to have a certification as engineering geologist or be registered as a civil engineer by the State of California. Moreover the final report must be reviewed by a certified soil engineering geologist according to the SCEC (2008).

The Act further stipulates that each investigation has to be carried out by both geologists and civil engineers together in order to have both disciplines integrated to provide a broader based assurance in the finding and the recommendations. Furthermore the investigations have to be carried out under the "supervision of licensed professionals with many year[s] experience in the field." The involvement of both engineering geologists and geotechnical engineers is seen as a prerequisite to ensure that the hazards are properly identified, assessed, and mitigated. The recommendation even goes far beyond this general specification. It defines that an engineering geologist should investigate the subsurface structure of hillside areas and provide an appropriate input to the geotechnical engineer with respect to the potential impact of the subsurface geologic structure, stratigraphy, and hydrologic conditions on the stability of the slope. The assessment of the subsurface stratigraphy and hydrologic conditions of sites when underlain solely by alluvial materials should be performed by the geotechnical engineer, and the shear strength and other geotechnical earth material properties should be evaluated by the geotechnical engineer. The geotechnical engineer should perform the stability calculations. In order to identify the ground motion parameters, analysis may be provided by either the engineering geologist or geotechnical engineer, or a registered geophysicist competent in the field of seismic hazard evaluation.

Another type of code of conduct, for instance, is the "code of conduct of flood risk" of the Law Society of England and Wales (BLS 2014) that defines how solicitors should manage a lawsuit regarding a flood event. In this regulation it is reiterated that solicitors are not qualified to give advice on flood risk or interpret technical flood reports. In order to gain scientifically profound, reproducible, and neutral information on the risk from floods, the code proposes to commission specialists to provide the required information. In general the code points to contacting nationally authorized and mandated organizations such as the agencies for environment, water resources, and land use planning. Flood risk surveys have been developed by government institutions and by specialized private companies. The Law Society recommended using these standard templates for assessing the flood risk that in addition also inform about steps that can be taken to mitigate exposure to flood damage.

10.2 Geoscientists in Disaster Risk Reduction Management

When we accept that only by integrating natural sciences and social sciences, can the demand of the population at risk for a safer life be served optimally, we have to accept that the three groups have their specific claim in finding a solution. But it is exactly these different interests, viewpoints, expertise, and experience that aggregate to a comprehensive risk management and that make stakeholders become partners in change management and the recognition that natural disaster mitigation is a political process although it has a scientific and technical evidence-based basis.

For a natural scientist a hazard is just a natural phenomenon: a dormant volcano, an area susceptible to landslides, or a region often exposed to hailstorms. After investigating the area and having analyzed the cause–effect relationship of natural hazards and vulnerability factors, the scientists normally publish the findings in related publications. From this moment on she sees herself too often no longer part of the risk management process. The population at risk, however, perceives the natural hazard as a potential threat to their livelihood. According to previous experience (either personal or from hearsay), they know this hazard may turn out to become a disaster. Automatically those risk exposed try to assess the risk. Their central aim is to maintain their social and economic life not endangered by the threat. Moreover there is in general no functional relationship established between the risk exposed and risk assessors. The third stakeholder in the issue is the state. Other than natural scientists and the risk exposed, it is his intention to maintain the existing social and economic order of the entire society. It has, therefore, introduced different intervention levels: local level, provincial level, and on the national level to work out specific risk reduction strategies and their respective implementation. To achieve this the state has implemented a series of appropriate instruments. One of many concepts, for instance, is to set the legal framework to institutionalize hazard, vulnerability, and risk assessment whereas others are implementing awareness-raising campaigns or incorporating the population into the local disaster risk-management decision making.

As in the environmental sector, the task of geoscience is also cross-cutting and interdisciplinary in disaster risk management. And it is just this transboundary nature that makes this sector a real challenge. Either by identifying areas at risk or by working out risks preventing countermeasures on rockfall, hazards, or whether geoscience is involved in agenda setting in spatial and regional planning, its expertise is finally targeted at rendering services to make people less exposed to natural disaster threats. In this sense the geo-related evidence is the starting point for any development planning and is thus indispensable to reach a higher level of human security. As being a part of the cause–effect related decision-making process, geoscientists' decisions to refrain from taking "political" responsibility will not be accepted by the public. Otherwise risk-affected ones as well as political levels will lose confidence in their expertise. Geoscientists cannot deny this involvement by saying, "I am only providing the facts but drawing assumptions from them is for others," will be accepted by society. Nevertheless, geoscientists are not to make decisions, but are expected to take up their part in the decision-making process, by providing evidence and facts and proposing solutions, thus contributing to finding the right solutions

But geoscientific expertise is not only asked for in dealing with the "classical" geotectonically related natural disasters, but the broad range of climatological and meteorological disasters have a significant component that deserves geoscientific expertise. Entering into this sector gives the geosciences a different definition from what it was for centuries. For a long time, geosciences were defined as a "historical descriptive science," defining it to be a science investigating geological facts and deriving from them purely nature-related cause–effect relationships. With the oil-crisis in the 1970s and even more severely after the "water crisis" in the 1990s, societies were asking for information on how to settle the daily demand. This request brought geoscience into the situation to bridge historically acquired knowledge with social and economic problems of today. With climate change, solving this societal demand gained even stronger momentum. This also applies to disaster risk management; especially when we consider that the ongoing changes in climate will lead to more frequent and more extreme disasters that will be a threat to more people. This paradigm of a "multilevel assessment" approach, however, does not lead away from classical geoscience. It only pledges for a geoscientist to be involved at different levels of societal, ecological, and political action. Still we need—without any doubt—classical geoscientific investigations, analysis, and mapping. But today geology is asked to go a step further by transferring gained expertise to regions that are not investigated in such detail, in order to help to define technically implementable countermeasures. Societal demand even goes beyond this demand. Today, geologists are requested to enter into political decision-making processes and are asked to become "technical advisors" that assist the political decision makers by implementing political decisions to reach a higher degree of resilience of otherwise risk-exposed populations. This demands that the geologist on one hand to stay natural while being open to the societal and economic question. The geosciences, as described above, are and have to be taken as an integral part of the economic risk assessment our societies are facing daily. The geosciences are not the ones that create such hazards and risks, but it is their task to identify risk situations, work out effective countermeasures for risk mitigation, and also to be moderators between the population at risk and the technical prevention measures. Following the above discussion this book strongly pledges that also "natural science-oriented" disaster risk assessment has to comprise a distinct social component: a conceptual approach that is nowadays named "systematic risk" (Klinke and Renn 2006). The term risk "denotes the embeddedness of any risk to human health and the environment in a larger context of social, financial and economic risks and opportunities. Systematic risks are at the crossroad between natural events, economic, social and technological development and political actions." These new interrelated risk fields also require a new form of risk analysis, in which data from different risk sources are either geographically or functionally integrated into one analytical perspective. Systematic risk analysis requires a holistic approach to hazard identification, risk assessment, and risk management.

Figure 10.3 gives an impression of an example of disaster risk insurance on how the different stakeholder groups define their specific interest and how these interests are interconnected. The natural scientist sees her role in analyzing the hazard situation. Second, the insurance company sees its interest in insuring risk but can only realize this when the company is prospering, briefly stated, "not to make losses." The individual has to understand the hazard and risk to which he is exposed. When the scientific information covers his inflation demand, than he will be inclined to take the self-help approach. The states' interest is having a functioning social system: in order to secure the livelihood of the society he has to create a regulatory framework conducive to the private sector ("to make profit") but also to the individual for risk prevention at an affordable premium scale.

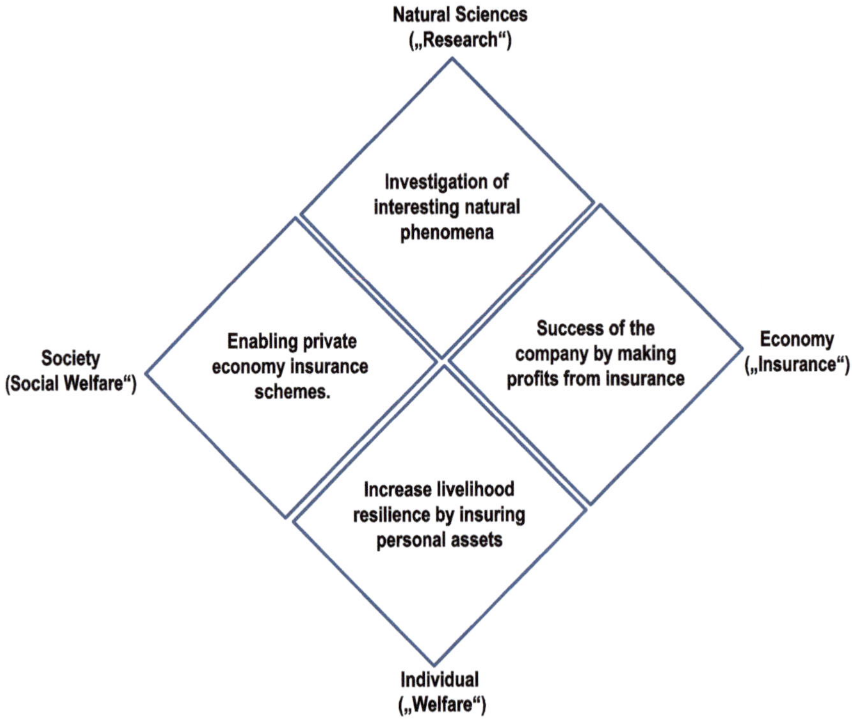

Fig. 10.3 Stakeholder interest in disaster risk management (Example: risk insurance; own graph)

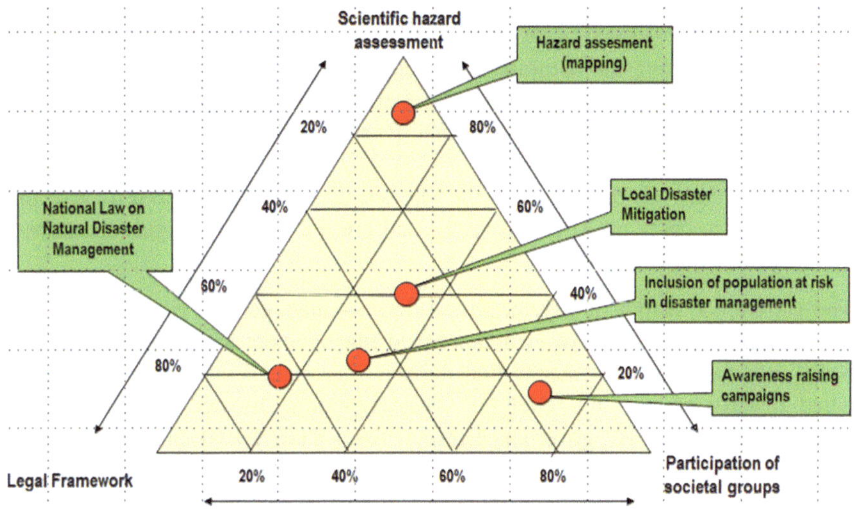

Fig. 10.4 Triangular diagram describing the position where different stakeholders in disaster risk management are positioned (Own graph)

To make this interrelation a bit more clear the different "stakeholders" can be entered into a triangular diagram (Fig. 10.4). The diagram identifies where they are positioned in the context of disaster risk management. The positions of the different actors are given for the examples (a) hazard assessment, (b) national disaster law, (c) local disaster mitigation (strategy) (d) the integration of the population at risk in the political decision, and (e) campaigns for awareness raising. It can be clearly seen from the diagram that even the extreme position of "scientific hazard assessment" far out at the triangle's tip is influenced by the legal aspects as well as by the population at risk. A hazard assessment that does not reflect such influence will not be able to provide a meaningful contribution to a sustainable risk management. Awareness campaigns are positioned near the "participation of social group's" tip. But every awareness raising needs reliable information to increase disaster resilience and can only be implemented within the existing legal risk reduction framework. A law of risk reduction, however, requires scientific input but also has to reflect the social, economic, and cultural situation of the society in general. A local (or national) strategy for disaster risk reduction, when successfully implemented, will find itself positioned right in the middle of the triangular diagram.

Wisner et al. (2004) created a model that describes the dynamic interdependence of the different stakeholders (Disaster Pressure and Release Model). The model explains how different trigger levels (root causes): the changing livelihood conditions (dynamic pressure) and the socioeconomic situation of the population at risk (unsafe condition) collide with the natural setting (hazards) to become disasters. The model is widely used in risk management as it can shed light on the interaction of the different trigger causes that make up a disaster.

In general the root causes can be distinguished into three different elements:

- *Nature*: describing the naturally given conditions that result from processes in the Earth's crust (plate tectonics) or from climate change (sea-level rise).
- *Legal order*: defining the organizational framework a society has given itself to govern societal life.
- *Population at risk*: describing those who are directly exposed to a hazard; they are in many cases not responsible for and are not able to mitigate the adverse impact without help.

The geoscience responsibility in natural disaster management is to foster a shift in paradigm from a reactive evidence-based management to a proactive science-based management. This means that disaster risk management is no longer oriented towards rescue and relief but towards putting disaster prevention in the focus. The IRGC's risk governance framework (IRGC 2010) therefore points to the fact that science-based disaster management has to be based on experience and expertise to assess the probable impacts of future events properly. Risk management should be concentrated no longer on an assessment of what type of hazard may occur, where, and to what severity, but to have the population at risk in focus. Geosciences thus have to realize that they are partners in this deal, although an indispensable one but the focus on securing livelihood is more than a matter of social scientists and economists IRGC (ibid). Managing risks requires both the geoscience sector and the social sector that must work together to take care of the different aspects of risk. The often prevailing separation in risk management in administering the impact and the scientific assessment of the natural root causes is not practical. The gap has to be bridged in order to bring disaster risk management out of the corner of pure actions. Risk management will only be successful when it integrates all different scientific sectors. Essential for a sustainable risk governance framework is to gain a thorough understanding of a risk and to develop options for dealing with it. Therefore IRGC's risk governance framework distinguishes between analyzing and understanding a risk and deciding what to do about it. This distinction reflects IRGC's strategy for the clear separation of the responsibilities for risk appraisal and risk management as a means of maximizing the objectivity and transparency of both activities. Those responsible for both should be jointly involved in the framework's other elements: pre-assessment, characterization, and evaluation and communication.

The task of the geoscientists in natural disaster management can be distinguished into a sector that is dominated by pure and traditional scientific orientation, and a sector where the scientific findings are taken up for technical and economically oriented activities, and finally into a sector where scientific findings are transformed into practical development decisions (advisory services). Figure 10.5 shows an example of how the different objectives are interwoven and that the advisory and applied sector especially depend substantially on geoscientific expertise. Although not be a part of the transform sector, traditional natural science in natural disaster management has to be aware that its objectives will lead to a series of forthcoming activities that finally are aimed at increasing societal

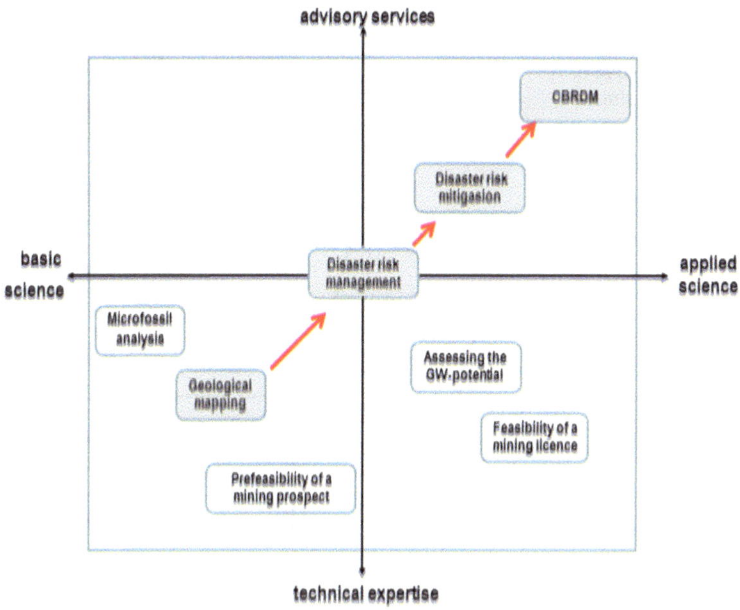

Fig. 10.5 Development path indicating how the scientific sector contributes to socioeconomic development decisions (Own graph)

resilience, thus defining a development path (red arrow). The path describes that even pure scientifically oriented activities might be of use for political development decisions. It is thus obvious that therefore the geoscientist has been aware that his activities in courses of development will leave the facts-dominated scientific realm and will more and more be taken up in the sector of advisory services. The political decision makers deserve that input in order to make their decisions. This nevertheless defines on the other hand the role of natural scientist to not become the decision maker himself, but that be aware that his findings are being used (see Sect. 2.2). This emphasizes that politicians, planning, and implementing authorities should be aware of the potentials and expertise the natural scientists are able to provide and that they should not try to make their decisions purely on political will and targets.

When we accept that scientists and engineers have a distinct role in the intention of society to increase its resilience from natural disasters, we have to define how the scientists should take up their responsibility in transferring scientific and technical information to their partners, or better, to those who depend on such information. Scientists acting in this sector should be aware that she has different fields to act in that are differently defined although their fields are highly interconnected and interrelated. Figure 10.6 describes the main fields of intervention a scientist has to venture in the intermediation of scientific knowledge. But moreover all the fields she is in should be acknowledged. Society will neither accept a scientist's restriction purely in the science sector, nor will it be possible just to choose

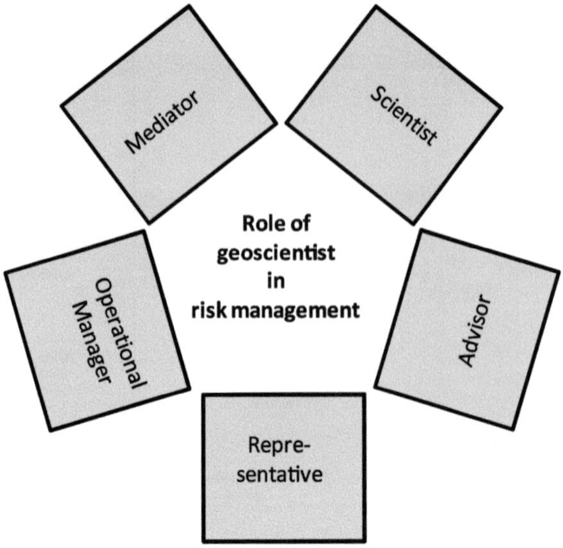

Fig. 10.6 The role of the scientist in intermediation of scientific knowledge in disaster risk management (Own graph)

one or two of them. Her function is to be scientist, advisor, and mediator. In general this function is further defined to be representative of a state, a public entity, or a private company. And in very many situations she is asked to provide the necessary financial, technical, and operational resources for the risk assessment and the implementation of the reduction measures with the people.

10.3 Final Remarks

Statistics have proved (see Sect. 3.4) that natural disaster events are increasing and are increasingly affecting more people. The main reasons for this are the general increase in population (especially in developing countries resulting in a poverty migration into regions that are unsuitable for settling) thus bringing many societies to the brink of social catastrophes. Moreover the population increase is accompanied with the accumulation of economic assets and technology in hazardous regions, even in countries that are not that advanced. A further driving force to more risks of disasters is the change in climate conditions. All these factors force the people to more and more go into areas that have formerly not been settled and that are according to their risk exposure not suitable for settlement. These circumstances lead to an increase in risk from natural hazards and pose the greatest challenge for societies. To be able to cope with the envisaged increasing risks, societies are demanding measures, instruments, and organizational frameworks to make their living safer. In this regard societies are especially addressing the natural scientists and civil engineers to analyze the root causes for the disasters and to find and work out appropriate mitigation strategies. Together with the authorities, scientists and engineers are asked to implement these measures accordingly. In this context WBGU (1999) stated that the option, "Refrain from benefits in order to minimize/avoid risk" is theoretically possible but will in practice not be achievable. Rather the opposite approach is favored: take up the challenge and derive from the disaster impacts technologies and mitigation strategies that make societies capable of coping with the harmful threats. In this respect the geosciences have an obligation to deliver implementable mitigation measures. Therefore in order to better address societal demand, the geosciences should take the position of the people at risk as the starting point for their investigations and not bow to geological analysis under the disaster paradigm. This does not mean that "sciencc" should be underrated in its ability to master the challenge, but it will rather help to focus on the "real" problems. With rational, reasonable, and reliable analysis and recommendations the geoscience sector will take its responsibility in disaster prevention. But on the other hand this needs a framework that is inducive to science. The official sector has the obligation to allocate necessary financial means and an organizational and time framework allowing scientific investigations at the required intensity. But scientists are in a certain dilemma in this aspect. Natural disasters occur every day and thus need answers at short notice. But in-depth investigations often require time spans of at least many years. Thus the scientists and engineers must be courageous enough in some situations just to assess a disaster based on their although limited but ad hoc available knowledge and not evade the responsibility by transferring the problem to future findings. The excuse of lack of appropriate knowledge is often perceived by the decision makers and the risk affected as science backing out of its responsibility. The geosciences should furthermore be made an equal partner in disaster risk management. And the complexity of natural phenomena and social implication will not be solved by one discipline alone. There must be a shift in paradigm: natural scientists must be open to issues beyond their original disciplines and on the side of the decision-making sector to acknowledge that nature issues are a substantial and integral part of every disaster resilience strategy.

It is a fact that all over the world the population at risk is more and more demanding of the political decision-making levels to set up and operate mitigation measures that are sustainable in reducing society's disaster exposure. Regarding the threats society is exposed to, risk assessments are required that enable the society to develop effective prevention measures in order to increase societal resilience. But asking authorities to do the "necessary" is one side of the coin. The other side is that the populations at risk themselves are taking over their responsibility for a safe future. Thus there is a great demand in societies to share the risks between the social groups equally and fairly. This requires that all social groups develop the same vision on who shall be protected, to what extent, and how to distribute the burden of risk prevention within the society. It is a common understanding that in case of an emergency, the individual assess his personal risks in general based on former risk experience. Therefore those heavily affected have a different perception of the risk than those far off. Nevertheless they have to share the burden. There is only one solution to this dilemma, all groups of society have to develop the same vision of what is seen as risk and what is not (see Sect. 2.2). Such a consensus is what Beck postulated as a "society of risk" or what has also been described as a "culture of risk." A culture of risk requests all sectors of society to do their best in order to increase the societal risk-coping capacities. It requests from all nations a stronger involvement of disaster risk management in political decision making than experienced today. Best practice examples proved that prevention by technical countermeasures is seen one of the best means to safeguard people and economic values from the impact of disasters. But not only technical countermeasures are substantial, rather it proved that nonstructural, organizational, and managerial aspects of risk management such as risk-oriented special planning or awareness-raising campaigns with the population at risk are identified to be of high effectiveness. Moreover a culture of risk also demands from the industrialized countries to take over their responsibilities, at first to reduce their CO_2-emissions, in order to reach a worldwide level of 2 % increase in greenhouse gas emissions, as agreed upon in the many United Nations climate conferences. But risk prevention is a matter of costs. Many studies and especially that of Dilley et al. (2005) proved that every dollar spent for increasing disaster resilience prevents damages of at least US$5. Therefore in the long-term risk financing saves a nation's value. But such an understanding has not yet fully developed in every nation.

Although a multitude of risk reduction strategies are at hand and proved their prevention potential, it is nevertheless a known fact that it is not possible to make a society risk-proof against each and every disaster. On the other hand waiting for a disaster to occur is rejected as an unethical approach. The fundamental dilemma is that human activity is generally accompanied with unintended implications and simultaneously it will not be possible to settle these demands without interfering with the natural system (WBGU 1999). And human development has a history of learning from failures (disasters, catastrophes). The acceptance of a baseline of risk and the knowledge of how to prevent unacceptable risk is the challenge of today.

The only way to make the public and the political decision makers aware of risks is by indicating the loss reduction potential of prevention and by simultaneous valuing the increase in welfare. Risk managers therefore have to monetize the investment spent for prevention and to compare this with the savings from the level of resilience that is envisaged to be achieved. Furthermore any risk reduction effort will not be successful when there is no similar perception on what a risk is by the individual or the societal groups as well as by the authorities. Only when all groups share the same risk perception, will it be possible to define the common levels of resilience to be envisaged. Next is to reach a consensus on the willingness to pay for the increased coping capacities. This can only be achieved by a dialogue process in which the population at risk has to be involved from the very beginning and that cannot be left to risk managers and politicians only. Finding consent needs reliable, sound, and technically and scientifically based data and information on the different risk types and an evaluation instrument that enables the participants to compare the different disaster impacts. As a result, a ranking of disaster types defines what kind of disaster, at which location, and to what extent, should be mitigated in order to secure an optimal investment of the normally small budgets. And it furthermore enables the authorities to comply with the German national law's request to assure the same living conditions, to secure life and social welfare all over Germany. But it is also an instrument that helps to ask for the individual's personal engagement.

In the course of the climate change debate and the increasing world population, a discussion is coming up on how losses and victims will develop from natural disasters. It seems obvious that the population increase itself will not be the factor that will matter, but rather that the trend of poverty migration into the big megacities will be getting stronger. In search of labor and improved living conditions these migrants just deteriorate the already difficult living conditions. Being the last in the chain in the search for living quarters they are forced to settle areas that are, from their geological and geomorphological pattern, not suitable for living, a behavior that just increases the hazard exposure. A similar outcome is envisaged from the changing climate. Holzer and Savage (2013) in a comprehensive study came to the conclusion on the future risk from earthquakes for people and their living environment that "More people will die from earthquakes even when the statistical occurrence of earthquake remained more or less constant over the centuries." The study analyzed earthquakes with death tolls of more than 50,000 from the time span since 1500 AD. Comparing those events to estimates of world population, they found that the number of catastrophic earthquakes has increased as population has grown. After statistically correlating the number of catastrophic earthquakes in each century with world population, they predict that total deaths in the century to come could more than double to approximately 3.5 million people if world population grows to 10 billion by 2100 from 6 billion in the year 2000. The study underscores the need to build residential and commercial structures that will not collapse and kill people during earthquake shaking.

Although benefits from science and technology to humanity are undoubtedly unparalleled in the history of the human species, the influence of science on people's lives is growing. But as on the other hand many of the scientific achievements turned out to be harmful and revealed negative long-term effects, a considerable measure of public mistrust in science and technology developed. Especially in marginalized societies there is a growing belief that many of them will suffer from such (direct or indirect) consequences of technical innovations, that from their viewpoint are beneficial only to a privileged minority. Scientists should reflect on the social consequences of the technological applications or dissemination inasmuch as the power of science to bring about change places a duty on scientists to proceed with great caution both in what they do and what they say. Even when scientists discuss a "comprehensive" insight to a problem, their state of knowledge will in any case be partial. Thus they have to explain to the public and policy makers about the degree of scientific uncertainty or incompleteness in their findings (L'Aquila, Italy; see: Chap. 1). At the same time, though, they should not hesitate to fully exploit the predictive power of science. The impact of technological interventions on individual people, communities, and the environment must be carefully considered. To do this, science needs to become more multidisciplinary and its practitioners should continue to integrate social and natural sciences (see Sect. 2.2).

The World Conference on Science (WCS 1998) summarizes this call to the scientists as follows:

> [A] major challenge for global science is to find institutional arrangements conducive to address the multi-faceted problems. The large number of international networks and programs, reflects either the great demand in scientific communication but also reveals a certain narrowness of the established scientific institutions and the lack of strategic, integrated support by national governments in areas like global change or international aid. What is needed is the formation of true international partnerships that allow scientists in different disciplines and countries to fully support each other's aims and share resources and management duties to mutual advantage.

Human beings tend just to react to the various challenges of nature rather than to look into the future, based on experience. There is a nice story of a fire brigade that dug a water well after the house has caught fire, a typical sign of mere helplessness, an action that replaces the lack of preparedness and prevention. The relationship between the perception of a natural hazard and appropriate prevention as well as providing relief in the aftermath of disaster both addresses the social and official risk behavior. The responsibility of natural scientists, technicians, and emergency managers therefore has to be at the cross-point between purely technical matters and social implementation and has to be oriented to develop a culture of prevention rather than a culture of relief.

The multitude of reports and statistics available on natural disasters proved that natural disasters although generated by natural processes are not the sole origin of damage and loss of lives, but a combination of natural, social, and economic causes. Generally speaking, to become a victim from a natural disaster is strongly linked with poverty. Poverty is a result of social, economic, and cultural vulnerability. Those who cannot afford living in safe places are forced to settle, for instance, in poorly constructed shelters on steep slopes around the megacities, regions not suitable for sustained living, whereas those who can afford an alternative living have significantly lower exposure to threats from natural disasters. Natural scientists and engineers, when dealing with disaster management therefore are called on to address this dichotomy, even if many of them feel this sector lies beyond their traditional responsibility. Figure 10.7 gives an impression of how strongly poverty and disaster exposure are correlated. The figure correlates the annual death toll for all disasters against the death toll in relation to the population and compared this with the Human Development Index (UNDP 2004). Although the data show a large variation, the shaded envelope, however, indicates a clear relationship of poverty and mortality risk from disasters.

The task for scientists in natural disaster risk management is to analyze the complex system of nature and to work out recommendations that define the root causes of hazards and disasters, thus providing reliable facts for political decision making to develop strategies for the way out. But scientists are fulfilling this demand in a somewhat restricted manner. Their traditional focus on information dissemination is

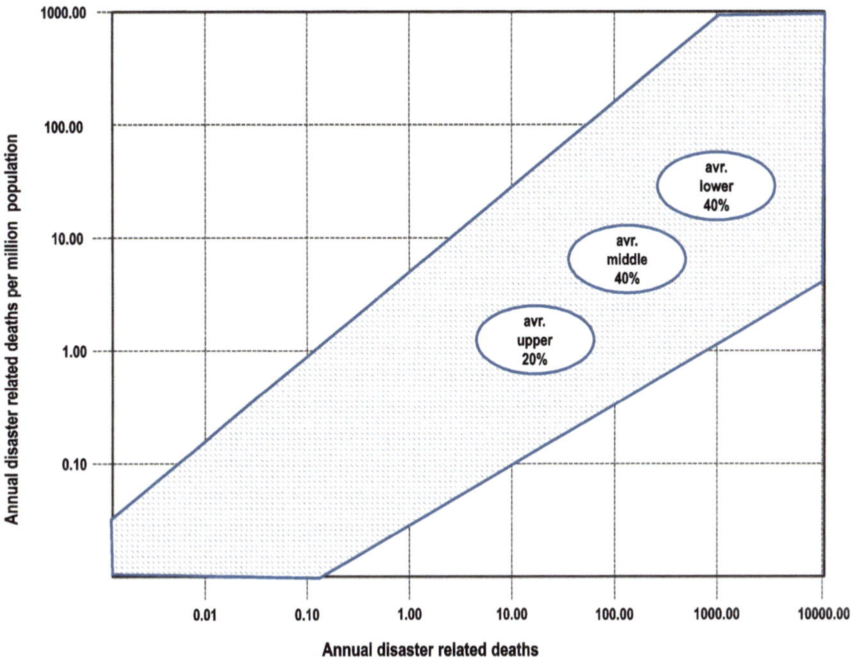

Fig. 10.7 Disaster-related mortality risk and human development (Based on Mutter 2005)

still too much supply-driven. But science-related information policy has to become more society-driven. Science in regard to assessing natural disasters is moreover based on measuring the frequency and severity of a particular event. Thus science sets in when the event has taken place. But natural disaster risk management is a process that requires information and recommendations on the probability of future events. This calls on scientists to develop a dynamic risk modeling instrument with roots in exact monitoring but allowing for a prediction of the future. The changing climate conditions, the increase in world population, the increase in poverty migration, as well as the change in Asian food habits from a cereal base to a meat base are dynamic elements that have multifold repercussions on natural disaster risk exposure. More and longer time series for data interpretation on disaster events are necessary, for instance, on earthquakes. Many of them are too short for drawing reliable future scenarios. Moreover often today data acquisition is based on highly different instruments and measuring standards, and often local structures are not in place for a centralized collection of data at a national level. Thus not all over the world is the time of an earthquake onset measured in universal time, but in local time. The density of the data acquisition networks is locally that scarce that it cannot claim to be international, forcing scientists to resort to similar data, but not from the area. Even in strong economies such as the United States, for instance, the instrumental network for monitoring volcanoes does not comply with the quality levels the country has set for itself. Only a few of the volcanoes rated most dangerous are monitored to the standard USGS deems necessary. At the international level many of the seismic monitoring instruments in use are 50–100 years old and no longer fulfill the technical standards required for an international compatibility, such as in Indonesia, where seismic instruments are located at easily accessible places and often around university cities. Such a network leaves large parts of the world not covered by modern, reliable, and comparable monitoring. Furthermore the interpretation of the measurements is lacking internationally agreed-upon standardization and harmonization. An international initiative is necessary to define for each of the natural disasters a binding interpretation procedure, "enhancing global indexing of and vulnerability enabling more and better intercountry and interregional comparison" (UNDP 2004). Only that will allow us to correlate and compare the findings for the sake of all of us. What is lacking is an international task force for a rapid assessment of damages that has to begin its duties right after the disaster has taken place. The days after bring much more information on the impact. When the bulldozers have cleared the rubble it is often too late to identify clearly the level of amplification the peak ground acceleration underwent by the (poor) structural conditions. Such a task force would be best institutionalized under the auspices of the United Nations. But also an initiative by the European Union would be of much help to increase resilience.

Sachs (2005) emphasized that not enough has been done to combine science expertise and ethical commitment. Natural science, economics, and social sciences are still "running more or less parallel" to each other and are not interlinked to the degree necessary. He pledged to combine "scientists, engineers, public health specialists, sociologists and economists with an ethical commitment to attend the poverty problems of the world." This requires the expertise of "scholars and scientists and committed practitioners from across the range of human knowledge." Synergies between natural sciences and social and economic development are currently underutilized.

The area where both natural scientists and socioeconomists meet is a transition zone where questions are raised on the natural root causes for disasters and their respective input on the population. In this zone questions have to be answered how a society can be made disaster resilient (Fig. 10.8). Moreover, in the course of implementation new questions arise leading to a cycle of interaction between these two sectors.

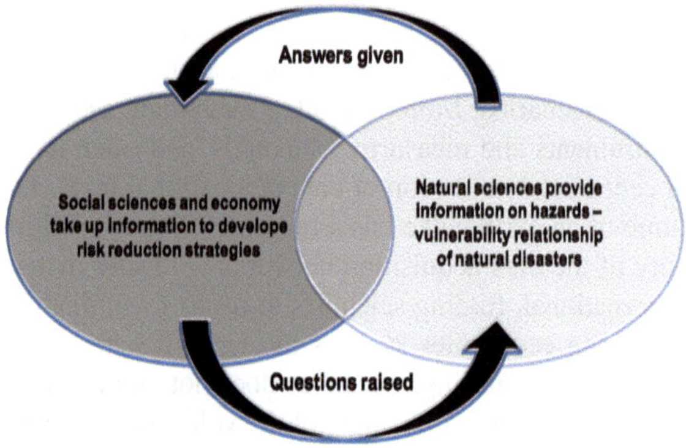

Fig. 10.8 Natural science and development decision-making cycle (Own graph)

Moreover the natural scientists that are working in disaster risk management are today already strongly interconnected and embedded into multifold networks so they no longer work alone. The international monitoring networks require and provide a permanent exchange of ideas, knowledge, and expertise. And this level of interrelation will deepen further. The increasing amount of natural disasters to occur worldwide and especially in the course of climate change, will affect the developing countries much more than other nations and will result in strongly increasing casualties. This perspective calls for strengthening the international exchange of experience and expertise as shown in Fig. 10.8. The state of the art of disaster mitigation today allows answering most of the questions. The main obstacle seen is how the authorities are implementing the solutions worked out. But science and politicians very often do not regard in their solution finding, the experience of local specialists and that of the disaster-affected people. What is required is a much closer and interdisciplinary cooperation between these stakeholder groups.

Experiences gained by the disaster managers from a certain region will simultaneously increase the overall state of knowledge that can be applied elsewhere. For example, on the occasion of the volcano eruption of Mt. Pinatubo (Philippines), PHIVOLCS and USGS started monitoring the volcano right after the first signs of unrest, months before the actual eruption. Both organizations sent out a group of geophysicists, volcanologists, and geologists to monitor all signs of the volcanic activity until months after Mt. Pinatubo calmed down. This early start and the long duration analysis make Mt. Pinatubo one of the best analyzed volcano eruptions in history (Newhall and Punongbayan 1996). Another example is that of the Koceali earthquake in northern Turkey (1999) where geophysicists from USGS and their colleagues from Turkey analyzed the earthquake event in great detail. A great amount of geoscientific data especially on liquefaction was collected that allowed a detailed and convincing analysis of the quake origin and its impact. The assumptions drawn moreover were applied to analyzing the San Andreas Fault Zone that is tectonically very similar to the North Anatolian Fracture Zone. These two examples show how in the future cooperation between scientists from different countries can be joined to both sides' benefit.

Natural scientists are called upon to be open to address the needs of risk-exposed people by joining the Munich Re Insurance Company sponsored "Hohenkammer-Charta" that describes how risk reduction can be better addressed in future. The Charta was declared in 2005 when the Munich Re Insurance company gathered celebrity scientists, politicians, economists, and also representatives of risk-exposed societies to outline the main challenges for sustainable risk mitigation (Munich Re 2005). Among the issues listed, all stakeholders were called to "enter into a dialog between the risk affected and the political decision makers in order to derive at the same understanding of the problems and solutions" in disaster risk management. Such a "dialog can only be successful when all stakeholders: politicians, economists, scientists and the populations at risk combine their expertise, experience and form public private partnership. Governments are called on to provide issue-related legal, economic or social incentives to foster local risk reduction."

The challenges from natural disasters are being addressed by many different governments and donor agencies. However, in order to achieve sustainable development these challenges are increasingly recognized as being interconnected (wicked problems). And there is a common understanding that socioeconomic improvements should not be achieved at the cost of environmental degradation or by leaving specific groups of a society without proper protection. Societies that have risks from natural disasters significantly reduced result in improved human well-being and social equity. Achim Steiner from UNEP emphasized on the occasion of launching UN REDD + Programme (UNREDD 2014) that what is needed is "to create the enabling conditions to succeed from good governance and sustainable financial policies to (arive at an) equitable distribution of benefits." This request also holds true for natural disaster risk management. When all stakeholders are aware of the potential of synergies that can be made from intertwining social and natural sciences, the necessary transition will be accelerated. To be successful the transition needs an enabling political environment that includes elements such as good governance, law enforcement, sustainable supporting financial mechanisms, and an equitable distribution of benefits. Moreover the integration of all stakeholders is required to provide innovative ideas for supporting a resilient development and can help to develop and institutionalize certified standards on risk reduction to safeguard the society, can lead to institutional reforms, regulations, and norms, and include public education campaigns. Quantifying the benefits can help to specify the opportunity costs of prevention measures and thus can underline the benefits of disaster prevention for social and economic development. The more groups that participate in the decision process the better the tradeoffs of the money spent for prevention will be understood. One important measure to increase acceptance of the required changes is to initiate research on the quantification of the costs of inaction. The "Zero-Option" model, an increased coordination of cross-sectoral planning and resource management will create a solid foundation for the general goals of natural disaster risk management.

The described and listed instruments, initiatives, and experiences have all proved their capability to change the behavioral attitude of a society in order to come from a "culture of risk" to a "culture of prevention."

References

Artec (2007): Governancestrukturen und Handlungswissen im Kontext: Hochwassergefahr, Küstenschutz und Klimawandel. - Artec, Forschungszentrum Nachhaltigkeit, Bericht 2003-2007, Universität Bremen.

Beck, S. (2011): Zwischen Entpolitisierung von Politik und Politisierung von Wissenschaft: die wissenschaftliche Stellvertreterdebatte um Klimapolitik; In: Schüttemeyer, S.S., (Hrsg.): Politik im Klimawandel: keine Macht für gerechte Lösungen? Nomos, Baden-Baden, p. 239 – 258.

BLS (2014): Bristol Law Society (online: www.bristollawsociety.com/.../legal.../meet-the-lawyer).

CIOMS (2002): International Ethical Guidelines for Biomedical Research Involving Human Subjects. - Council for International Organizations of Medical Sciences (CIOMS) in collaboration with the World Health Organization (WHO), Geneva.

COST (2010): Annual Report 2010.- European Cooperation in Science and Technology (COST) (online: http://www.cost.eu).

DFG (1998): Recommendations of the Commission on Professional Self Regulation in Science Proposals for Safeguarding Good Scientific Practice. – Executive Board of the Deutsche Forschungsgemeinschaft (DFG), Berlin.

Dilley, M., Chen, R.S., Deichmann, U., Lerner-Lam, A.L., Arnold, M. with Agwe, J., Buys, P., Kjekstad, O., Lyon, B. & Yetman, G. (2005): Natural disaster hotspots - A Global Risk Analysis - Synthesis Report.- The International Bank for Reconstruction and Development / The World Bank and Columbia University, Washington, DC.

Habermas, J. (1968): Technik und Wissenschaft als Ideologie.- Suhrkamp, Frankfurt am Main.

Habermas, J. (1994): Erkenntnis und Interesse (mit einem neuen Nachwort, 1994). - Suhrkamp, Frankfurt am Main.

Hardin, G. (1968): The Tragedy of the Commons. - Science, New Series, Vol. 162, No.3859, p. 1243-1248, Washington DC.

Hennen, L. (1994): Technikakzeptanz und Kontroversen über Technik - Ist die (deutsche) Öffentlichkeit 'technikfeindlich'?. - Büro für Technikfolgen-Abschätzung Beim Deutschen Bundestag (TAB), *TAB-Arbeitsbericht Nr. 024, p.57ff, Berlin.*

Holzer, T. & Savage, J. (2013): Global earthquake fatalities and population.- Earthquake Engineering Research Institute, Earthquake Spectra, Vol. 29 Issue S1,Oakland CA.

IRGC (2010):What is Risk Governance. - International Risk Governance Council (IGRC), Lausanne.

Klapwijk, J. (1981): Science and social responsibility in Neo-Marxist and Christian perspective. - in: Blockhuis,P. et al. (eds): Weteenschap, wijsheid, filosoferen: Opstellen aangeboden aan Hendrik van Riessen; Van Gorcum, Assen, Chapter 7, p.75-98, Rotterdam.

Klinke, A. & Renn, O. (2006): Systemic risk as challenge for policy making in risk governance.- Forum Qualitative Sozialforschung, Berlin

Macfarlane, A.(2012): Fukushima lessons: The disconnect between geology and nuclear engineering.- Elements, International Magazine of Mineralogy, Geochemistry and Petrology, Vol. 8, No. 3, p.165.

Malthus, T.R. (1798): An Essay on the Principle of Population - An Essay on the Principle of Population, as it Affects the Future Improvement of Society with Remarks on the Speculations of Mr. Godwin, M. Condorcet, and Other Writers.- Printed for J. Johnson, in St. Paul's Church-Yard, London.

Horkheimer, M. (1968): Kritische Theorie: Eine Dokumentation.- Fischer Verlag, Frankfurt a.M.

MPG (2009): Rules of Good Scientific Practice.- The Max-Planck Society, Munich (online: https://www.mpg.de/…/rulesScientificPractice.pd).

MunichRe (2005): Topics Geo Edition Wissen - Jahresrückblick Naturkatastrophen 2005. - Munich Reinsurance Company, Munich.

Mutter, J.C. (2005): The Earth Sciences, Human Well-Being, and the Reduction of Global. Poverty.- Eos Transactions of the Royal Society, Vol. 86, No. 16,19, p.157,164-167, Wiley Online Library.

Newhall, C. G. & Punongbayan, R.S. (1996): Fire and mud: eruptions and lahars of Mount Pinatubo, Philippines. – Philippines Institutes of Volcanology. (PHIVOLCS) and United States Geological Survey (USGS), University of Washington Press, Washington (online: http://pubs.usgs.gov/pinatubo).

Ostrom, E. (1990): Governing the Commons - The evolution of institutions for collective action. - Cambridge University Press, Cambridge.

Plapp, S.T. (2003):Wahrnehmung von Risiken aus Naturkatastrophen - Eine empirische Untersuchung in sechs gefährdeten Gebieten Süd– und Westdeutschlands.- Dissertation, University of Karlsruhe.

Rayner (2006): What drives environmental policy ? - Global Environment Change, Vol.16, Elsevier (online: www.sciencedirect.com).

Sachs, J.D. (2005): *The End of Poverty: Economic Possibilities for Our Time.- The Earth Institute, Columbia University (online:* www.earthinstitute.columbia.edu/endofpoverty).

SCEC (2002): Recommended procedures for implementing of DMG Special Publication 117 Guidelines for analyzing and mitigating landslide. hazards in California.- Los Angeles Section Geotechnical Group (ASCE), Document, Southern California Earthquake Center, Los Angeles CA.

SCEC (2008): Guidelines for evaluating and migrating seismic hazards in California. – Special publication No. 117 A, Revised and Re-adopted September 11, 2008, p. 108, California Geological Surveys Public Information Offices, Los Angeles CA.

Scheffran, J., Brzoska, M., AFES-PRESS, H.G., Link, P.M., Schilling, J. (eds.) (2012): Climate Change,Human Security and Violent Conflict - Challenges for Societal Stability. – Hexagon Series on Human and Environmental Security and Peace, Vol. 8, Springer Verlag Berlin, Heidelberg (online: DOI 10.1007/978-3-64228626-1_1).

Sigrist, M. (2004): Die Bedeutung von Vertrauen bei der Wahrnehmung und Bewertung von Risiken.- Arbeitsbericht der Akademie für Technikfolgenabschätzung in Baden-Württemberg, Universität Stuttgart, Band 197 (ISBN 3-934629-50-4), Stuttgart.

UK-DTI (2005): The role of science in physical natural hazard assessment, report to the UK Government by the Natural Hazard Working Group (June 2005), UK Department of Trade and Industry, London.

UNDP (2001): Human Development Report 2001 - Making new technologies work for human development.- United Nations Development Program (UNDP), Oxford University Press, New York NY.

UNDP (2004):International patterns of risk . – in: Reducing disaster risk - A challenge for development, Chapter 2; United Nations Development Program (UNDP), Bureau for Crisis Prevention and Recovery, United Nations, Geneva.

UNESCO (1998): World Science Report 1998. - United Nations Educational, Scientific and Cultural Organization (UNESCO), Elsevier, Paris.

UNREDD (2014): UN REDD Programme. - The United Nations Collaborative Programme on Reducing Emissions from Deforestation and Forest Degradation in Developing Countries. – United Nations Development Programme / United Nations Environment Programme, Geneva, Nairobi.

USGS (2007): Manual 500.25 - Scientific Integrity 01/05/07. -United States Geological Survey (USGS), Office of Human Resources Instruction, Reston VA.

VanDine, D. (2011): Professional practice and insurance – Canadian technical guidelines and best practices related to landslides: A national initiative for loss reduction. – Geological Survey of Canada, Open File 6981, Ottawa ON.

WBGU (1999): Welt im Wandel – Strategien zur Bewältigung globaler Umweltrisiken. – Wissenschaftlicher Beirat der Bundesregierung Globale Umweltveränderungen, Hauptgutachten 1998, Springer, Heidelberg.

WBGU (2014): Homepage „Wissenschaftlicher Beirat der Bundesregierung Globale Umweltveränderungen" (WBGU).- Minister für Bildung und Forschung (BMBF) / Minister für Umwelt, Naturschutz und Reaktorsicherheit (BMU), Berlin.

WCS (1998): Toward a New Contract between Science and Society. - World Conference on Science, Report of the North American Meeting, Kananaskis Village, Alberta.

Wisner, B., Blaikie, P., Cannon, T. & Davis, I. (2004): At Risk - Natural hazards, people's vulnerability and disasters. - Routledge, London.

Annex A: Glossary of Definitions of Hazard, Vulnerability, and Risk

Since the UNISDR publication *Living with Risk* (2004) the terms that are in use in disaster risk management have been standardized and harmonized to a certain degree. In the following a list of terms is given that have been collected from well-known world leading organizations in disaster risk management, mainly from UNISDR (2009) and from publications made by UNDRO-UNOCHA, UN-ISDR, UNEP, UN-EHS, Australian Emergency Management Agency (EMA), Federal Emergency Management Agency of America (FEMA), and the Springer *Encyclopedia of Natural Hazards*.

Although we have now reached a certain level of harmonization there are many glossaries published that indicate for many of the terms a larger band of interpretation. The list is a selected choice of terms and outlines the sometimes large definition variations and also indicates that many of the terms have changed their definition in the course of time. The UNISDR (2004) publication gives each of the terms special comments. The list below moreover presents a number of different definitions that will be helpful to better understand the true meanings of the terms. It is, however, recommended for the sake of a common international understanding, to make use of the definitions published under UN-ISDR whenever possible.

Adaptation

Adjustment in natural and/or human systems to a new or changing environment that exploits beneficial opportunities and moderates negative impacts.

The adjustment in natural or human systems is in response to actual or expected climatic stimuli or their effects, which moderates harm or exploits beneficial opportunities. This definition mainly addresses the climate change topic but also holds true as an application to nonclimatic factors such as soil erosion or surface subsidence. Adaptation can occur in autonomous fashion, for example, through market changes, or as a result of intentional adaptation policies and plans. Many disaster risk reduction measures can directly contribute to better adaptation.

© Springer International Publishing Switzerland 2016

U. Ranke, *Natural Disaster Risk Management*,
DOI 10.1007/978-3-319-20675-2

Awareness

The extent of common knowledge about disaster risks, on the factors that lead to disasters and the actions that can be taken individually and collectively to reduce exposure and vulnerability to hazards. Public awareness is a key factor in effective disaster risk reduction. Its development is pursued, for example, through the development and dissemination of information through media and educational channels, the establishment of information centers, networks, and community or participation actions, and advocacy by senior public officials and community leaders.

Catastrophe

A serious disruption of functioning of a community, causing widespread human, material, economic, or environmental losses that exceed the ability of the community to cope with using its own resources. A great catastrophe is defined if the event results in a death toll of >2000, the number of injured and homeless >200,000, the total economic loss exceeds 5 % of the GDP or the country appealed for external help.

Coping capacity

Positive factors that increase the ability of people and the society they live in to cope effectively with hazards that increase their capability and capacity to withstand or otherwise reduce their susceptibility from natural hazards.

The ability of people, organizations, and systems, using available skills and resources, to face and manage adverse conditions, emergencies, or disasters. The capacity to cope requires continuing awareness, resources, and good management, both in normal times as well as during crises or adverse conditions. Coping capacities contribute to the reduction of disaster risks.

Climate change

A statistically significant variation in either the mean state of the climate or in its variability, persisting for an extended period (typically decades or longer). Climate change may be due to natural internal processes or to external forcing, including changes in solar radiation and volcanic eruptions, or to persistent human-induced changes in atmospheric composition or in land use.

Critical infrastructure

The primary physical structures, technical facilities, and systems that are socially, economically, or operationally essential to the functioning of a society or community, both in routine circumstances and in the extreme circumstances of an emergency. Critical facilities include such things as transport systems, air and sea ports, electricity, water, and communications systems, hospitals and health clinics, and centers for fire, police, and public administration services.

Cumule

A disaster event can be the result of many different effects, that all are generated by one single root cause. The cumule is the addition of all the different effects.

Disaster

Severe alteration in the normal functioning of a community or a society due to hazardous physical events interacting with vulnerable social conditions, leading to widespread adverse human, material, economic, or environmental effects that require immediate emergency response to satisfy critical human needs and that may require external support for recovery.

A serious disruption to community life that threatens or causes death or injury in that community and/or damage to property which is beyond the day-to-day capacity of the prescribed statutory authorities and which requires special mobilization and organization of resources other than those normally available to those authorities.

A serious disruption of the functioning of a community or a society causing widespread human, material, economic, or environmental losses that exceed the ability of the affected community/society to cope using its own resources.

Early warning

Provision of timely and effective information, through identified institutions, that allows individuals exposed to a hazard to take action to avoid or reduce their risk and prepare for effective response.

Elements at risk

Inventory of people, houses, roads or other infrastructure that are exposed to the hazard.

Emergency

A present or imminent event that requires prompt coordination of actions concerning persons or property to protect the health, safety, or welfare of people, or to limit damage to property or the environment.

Emergency management

A range of measures to manage risks to communities and the environment; the organization and management of resources for dealing with all aspects of emergencies. Emergency management involves the plans, structures, and arrangements established to bring together the normal endeavors of government, voluntary, and private agencies in a comprehensive and coordinated way to deal with the whole spectrum of emergency needs including prevention, response, and recovery.

The management of emergencies concerning all hazards, including all activities and risk management measures related to prevention and mitigation, preparedness, response, and recovery.

Emergency service

An agency responsible for the protection and preservation of life and property from harm resulting from incidents and emergencies. Syn. "emergency services authority" and "emergency service organization".

Exposure

In the context of vulnerability to climate change, refers to the climate-related stressors that influence particular systems, and can include stressors such as droughts (e.g., in the context of water resources, agriculture, forestry) or sea-level rise (e.g., coastal flooding, habitat loss).

People, property, systems, or other elements present in hazard zones that are thereby exposed to potential losses. Measures of exposure can include the number of people or types of assets in an area. These can be combined with the specific vulnerability of the exposed elements to any particular hazard to estimate the quantitative risks associated with that hazard in the area of interest.

Extreme weather event

An event that is rare at a particular place and time of year. Definitions of "rare" vary, but an extreme weather event would normally be as rare as or rarer than the 10th or 90th percentile of the observed probability density of weather events.

Frequency

A long-term average time-rate of an event in a given time duration.

The number of occurrences of an event in a defined period of time.

Geoengineering

Deliberate large-scale manipulation of the planetary environment as a strategy to counteract anthropogenic climate change.

Greenhouse effect

Trapping and buildup of infrared radiation (heat) in the atmosphere (troposphere) near the Earth's surface. Some of the heat flowing back towards space from Earth's surface is absorbed by water vapor, carbon dioxide, ozone, and several other gases in the atmosphere and then reradiated back toward Earth's surface. If the atmospheric concentrations of these greenhouse gases rise, the average temperature of the lower atmosphere will gradually increase.

Greenhouse gas

Any gas that absorbs infrared radiation (heat) in the atmosphere. Greenhouse gases include, but are not limited to, water vapor, carbon dioxide, methane, nitrous oxide, chlorofluorocarbons, hydro chlorofluorocarbons, ozone, hydro fluorocarbons, perfluorocarbons, and sulfur hexafluoride.

Hazard

A source of potential harm or a situation with a potential to cause loss; a potential or existing condition that may cause harm to people or damage to property or the environment.

A potentially damaging physical event, phenomenon, or human activity that may cause the loss of life or injury, property damage, social and economic disruption, or environmental degradation. Hazards can include latent conditions that may represent future threats and can have different origins: natural (geological, hydrometeorological, and biological) and/or induced by human processes (environmental degradation and technical hazards). Hazards can be single, sequential, or combined in their origin and effects. Each hazard is characterized by its location, intensity, and probability.

A potential threat to humans and their welfare that can vary in terms of magnitude as well as in frequency, duration, area extent, speed of onset, spatial dispersion, and temporal spacing.

A potentially damaging physical event, phenomenon, or human activity that may cause the loss of life or injury, property damage, social and economic disruption, or environmental degradation.

Hazard analysis

Identification, studies, and monitoring of any hazard to determine its potential origin.

Hazard (geological)

Geological hazards include internal Earth processes of tectonic origin, such as earthquakes, geological fault activity, tsunamis, volcanic activity and emissions, as well as external processes such as mass movements (landslides, rockslides, rockfalls, avalanches, surface collapses, and debris and mudflows).

Hazard occurrence probability

Probability of occurrence of a specified natural hazard at a specified severity level in a probability-specified future time period.

The probability of occurrence of a specified natural hazard at a specified severity level in a specified future time period, elements at risk, an inventory of those people or artifacts that are exposed to the hazard and vulnerability, or the degree of loss to each element should a hazard of a given severity occur.

Hazardous physical event

Events that may be of natural, socionatural (originating in human degradation or transformation of the physical environment), or purely anthropogenic origins and that have the potential to deteriorate living conditions and pose a threat to the people (see: Tobin and Montz 1997; Wisner et al. 2004).

Hazard physical exposure

The total number of technical, material, and ecological elements at risk, or an inventory of those people or even (old) artifacts exposed to the hazard.

Human system

Any system in which human organizations play a major role. Often, but not always, the term is synonymous with "society" or "social system," for example, agricultural system, political system, technological system, or economic system.

Human–natural system

Integrated systems in which human and natural components interact, such as the interaction between socioeconomic and biophysical processes in urban ecosystems.

Hydrologic cycle

The flow of water through the Earth system via the processes of evaporation, vertical and horizontal transport of vapor, condensation, precipitation, and the flow of water from continents to the ocean.

Hydrologic systems

The systems involved in the movement, distribution, and quality of water throughout Earth, including both the hydrologic cycle and water resources.

Land use

The total of arrangements, activities, and inputs undertaken in a certain land cover type. The term land use is also used in the sense of the social and economic purposes for which land is managed (e.g., grazing, timber extraction, and conservation).

Likelihood

The chance of an event or an incident happening, whether defined, measured, or determined objectively or subjectively.

Mitigation

Actions taken to reduce the impact of disasters in order to protect lives, property, and the environment, and to reduce economic disruption.

Measures taken in advance of a disaster aimed at decreasing or eliminating its impact on society and the environment.

Structural and nonstructural measures undertaken to limit the adverse impact of natural hazards, environmental degradation, and technical hazards.

An intervention to reduce the sources or enhance the sinks of greenhouse gases and other climate warming agents. This intervention could include approaches devised to reduce emissions of greenhouse gases to the atmosphere; to enhance their removal from the atmosphere through storage in geological formations, soils, biomass, or the ocean.

Monitoring

A scientifically designed system of continuing standardized measurements and observations and their evaluation. Monitoring is specifically intended to continue over long time periods.

Mortality

The number of fatalities divided by the number of people exposed.

Natural Hazards

A source of potential harm originating from a meteorological, environmental, geological, or biological event.

Preparedness

Measures to ensure that, should an emergency occur, communities, resources. and services are capable of coping with the effects; the state of being prepared.

Activities and measures taken in advance to ensure effective response to the impact of disasters, including the issuance of timely and effective early warnings and the temporary removal of people from a threatened location.

The knowledge and capacities developed by governments, professional response and recovery organizations, communities, and individuals effectively to anticipate, respond to, and recover from, the impacts of likely, imminent, or current hazard events or conditions. Preparedness action is carried out within the context of disaster risk management and aims to build the capacities needed to manage efficiently all types of emergencies and achieve orderly transitions from response through to sustained recovery. This must be supported by formal institutional, legal and budgetary capacities.

Prevention

Measures to eliminate or reduce the incidence or severity of emergencies.

Activities to provide outright avoidance of the adverse impact of hazards and related environmental, technological, and biological disasters.

Probability

Frequency with which an event occurs given a long sequence of identical and independent trials in playing dice.

In statistics, a measure of the chance of an event or an incident happening.

Recovery

The coordinated process of supporting emergency-affected communities in reconstruction of the physical infrastructure and restoration of emotional, social, economic, and physical wellbeing.

The restoration and improvement wherever appropriate of facilities, livelihoods, and living conditions of disaster-affected communities, including efforts to reduce disaster risk factors. The recovery task of rehabilitation and reconstruction begins soon after the emergency phase has ended, and should be based on pre-existing strategies and policies that facilitate clear institutional responsibilities for recovery action and enable public participation. Recovery programs, coupled with heightened public awareness and engagement after a disaster, are a valuable opportunity to develop and implement disaster risk reduction measures.

Remote sensing

The technique of obtaining information about objects through the analysis of data collected by instruments that are not in physical contact with the object of investigation. In a climate context, remote sensing is commonly performed from satellites or aircraft.

Resilience

The capacity of a system, a community, or a society to resist or absorb the negative impact of a disaster in order to regain an acceptable level in functioning and structure of the society after it has been exposed to external stress. In addition resilience determines the capability of a social system to organize itself or to increase its capacity for adaptation to future hazardous situations and includes the capacity to create foresight to defend external pressure.

The ability of a system to recover its capacity to function after a disturbance.

The capacity of a system, community, or society to adapt to disruptions resulting from hazards by persevering, recuperating, or changing to reach and maintain an acceptable level of functioning.

The ability of a system, community, or society exposed to hazards to resist, absorb,

accommodate to, and recover from the effects of a hazard in a timely and efficient manner.

The ability of a social or ecological system to absorb disturbances while retaining the same basic structure and ways of functioning, the capacity for self-organization, and the capacity to adapt to stress and change.

The capacity of a system to absorb disturbance and reorganize while undergoing change.

The ability of countries, communities, and households to manage change by maintaining or transforming living standards in the face of shocks or stresses, such as earthquakes, drought, or violent conflict, without compromising their long-term prospects.

The starting point for reducing disaster risk and promoting a culture of disaster resilience that lies in the knowledge of the hazards and the physical, social, economic, and environmental vulnerabilities to disasters that most societies face.

Response

Actions taken in anticipation of, during, and immediately after an emergency to ensure that its effects are minimized, and that people affected are given immediate relief and support

Risk

The likelihood of harmful consequences arising from the interaction of hazards, communities, and the environment; the chance of something happening that will have an impact upon objectives. It is measured in terms of consequences and likelihood; a measure of harm, taking into account the consequences of an event and its likelihood.

The probability of harmful consequences, or expected losses (deaths, injuries, property, livelihoods, economic activity disrupted, or environment damaged) resulting from interactions between natural or human-induced hazards and vulnerable conditions. Conventionally risk is expressed by the notation:

- Risk = Hazard × Vulnerability: Coping Capacity

The term risk refers to the expected losses from a particular hazard to a specified element at risk in a particular future time period. Loss may be estimated in terms of human lives, buildings destroyed, or in financial terms.

The combination of the likelihood and the consequence of a specified hazard being realized; refers to the vulnerability, proximity, or exposure to hazards, which affects the likelihood of adverse impact.

An abstract concept closely related to uncertainty with different definitions in different disciplines. In disaster risk reduction, risk is considered a function of hazard, exposure, vulnerability, and values of elements at risk.

Risks are mental "constructions;" they are not real phenomena but originate in the human mind.

Risks are abstract and cannot be managed; only the outcome can be managed.

Risks may have impact far beyond the initial incident location due to the interwoven network of the nation's critical infrastructure.

Risks have the potential of being exacerbated by changes in environmental conditions and failing infrastructure.

Risk acceptance

The level of potential losses that a society or community considers acceptable under the prevailing social, economic, political, cultural, technical, and environmental conditions. In engineering terms, acceptable risk is also used to assess and define the structural and nonstructural measures and systems to a chosen tolerated level.

Risk assessment

A process to determine the nature and extent of risk by analyzing potential hazards and evaluating existing conditions of vulnerability that could pose a potential threat or harm to people, property, livelihoods, and the environment on which they depend. The process of conducting a risk assessment is based on a review of both: the technical features of hazards such as their location, intensity, frequency, and probability; and an analysis of the physical, social, economic, and environmental dimensions of vulnerability. The risk assessment does so while taking into particular account the coping capabilities pertinent to the risk scenarios.

A risk assessment method that assigns statistical values to risks.

Risk Communication

The imparting, exchanging, and/or receiving of clear, credible, and timely information about the existence, nature, form, likelihood, severity, acceptability, treatment, or other aspects of risk to improve decision making in risk management.

Risk communication is carried out among public authorities, risk assessors, risk managers, the public, and all other interested parties. It is intended to achieve a better understanding of risks and risk management.

Risk Identification

The process of finding, recognizing, and recording risks.

Risk Management

The use of policies, practices, and resources to analyze, assess, and control risks to health, safety, environment, and the economy.

Consists of identifying and deterring threats, reducing vulnerabilities, and minimizing consequences.

Process of identifying, analyzing, assessing, and communicating risk and managing it considering associated costs and benefits of any actions taken.

Risk Perception

A stakeholder's view of a risk. Risk perception reflects the stakeholder's needs, issues, knowledge, beliefs, and values.

Risk Profile

A description of an entity's existing management practices, common vulnerabilities, tolerance, and key interdependencies concerning its particular risks, as well as an assessment of their relative likelihood, consequences, and priority.

Risk Tolerance

The willingness of an organization to accept or reject a given level of residual risk. Risk tolerance may differ across an organization, but must be clearly understood by those making risk-related decisions.

Scenario

A coherent description of a potential future situation that serves as input to more detailed analyses or modeling. Scenarios are tools that explore, "if…, then…." statements, and are not predictions of or prescriptions for the future.

Sensitivity

The degree to which a system is affected, either adversely or beneficially, by climate-related stimuli. The effect may be direct or indirect.

Severity

A measure of the disaster impact = loss value.

Stakeholders

Individuals or groups whose interests (financial, cultural, value-based, or other) are affected by climate variability, climate change, or options for adapting to or mitigating these phenomena. Stakeholders are important partners with the research community for development of decision support resources.

Storm surge

The temporary increase, at a particular locality, in the height of the sea due to extreme meteorological conditions (low atmospheric pressure and/or strong winds).

Structural/nonstructural measures

Structural measures are physical, technical, and structural measures to increase the physical capacity to withstand a disaster impact.

Nonstructural measures concern the regulatory and organizational setup for implementing disaster mitigation activities.

Susceptibility

Capacity for taking, receiving, being affected by, or undergoing of an entity by external impression, impact, or influence.

How the characteristic of an entity is influenced by or interacts with external factors that may change them in a way that the result may not be neutral or beneficial, but may result in adverse events (injuries, degradation, deterioration, etc.).

Sustainability

Balancing the needs of present and future generations while substantially reducing poverty and conserving the planet's life support systems.

Sustainable development

Development that meets the needs of current generations without compromising the ability of future generations to meet their own needs.

Synoptic

Pertaining to motions of whole weather systems, on spatial scales of hundreds to thousands of kilometers and timescales on the order of a few days.

System

Integration of interrelated, interacting, or interdependent components into a complex whole.

Threat

The presence of a hazard and an exposure pathway.

Threshold

A point in a system after which any change that is described as abrupt is one where the change in the response is much larger than the change in the forcing. The changes at the threshold are therefore abrupt relative to the changes that occur before or after the threshold and can lead to a transition to a new state.

Trust/confidence

Trust = sharing the same values; confidence = having "trust" in the performance competence of an organization.

Uncertainty

An expression of the degree to which a value (e.g., the future state of the climate system) is unknown. Uncertainty in future climate arises from imperfect scientific understanding of the behavior of physical systems, and from an inability to predict human behavior.

Vulnerability

A set of conditions and processes resulting from physical, social, economic, and environmental factors, which increase the susceptibility of a community to the impact of hazards.

Reflects the range of potentially damaging events and their statistical variability at a particular location.

The degree to which a system is susceptible to, or unable to cope with, adverse effects of climate and global change, including climate variability and extremes, as well as climate change in conjunction with other stressors.

From 1991 until today the focus on vulnerability has shifted from:

Potential damages: Vulnerability is expressed as the degree of expected damage as a function of hazard intensity (UNDRO 1991)
to
People's capacity to recover: Represents the interface between exposure to the physical threats to human well-being and the capacity of people and communities to cope with the those threats (UNEP 2002)
to
Factors that make a society vulnerable: The conditions determined by physical, social, economic, and environmental factors or processes, which increase the susceptibility of a community to the impact of hazards (UNISDR 2004)
to
The system's capability to absorb risks: The extent to which the risk-absorbing system reacts to stress (The International Risk Governance Council 2005)

Vulnerability Assessment

The process of identifying and evaluating vulnerabilities, describing all protective measures in place to reduce them and estimating the likelihood of consequences.

Weather

The specific condition of the atmosphere at a particular place and time. It is measured in terms of parameters such as wind, temperature, humidity, atmospheric pressure, cloudiness, and precipitation.

Annex B: International Disaster and Emergency Management Agencies

Australia

Australian Emergency Management Institute

In order to prevent Australian communities from losses caused by disasters, the Council of Australian Governments (COAG 2004) adopted a whole-of-nation resilience-based approach to disaster management. Subsequently in 2011, the Australia-New Zealand Emergency Management Committee (ANZEMC 2014) developed the National Strategy for Disaster Resilience (NSDR). The strategy recognizes that a national, coordinated, and cooperative effort is needed to enhance Australia's capacity to withstand and recover from emergencies and disasters. It will be executed by a newly established National Emergency Management Committee that includes experts from commonwealth, state, and territory and local governments. Disaster resilience is defined as the collective responsibility of all sectors of society, including all levels of government, business, the nongovernment sector, and individuals. The strategy furthermore stipulates the collective incorporation of the principles of disaster resilience into all sectors of public life.

The key messages of the National Strategy for Disaster Resilience are:

- Disasters will happen.
- Disaster resilience is a government's as well as an individual's business.
- All sectors of society have a role to play and should be prepared.
- Connected communities are resilient communities.
- Every Australian should know how to prepare for any natural disaster.
- Individuals should get ready to reduce the effects of future disasters.
- Individuals as well as official risk assessors should learn from experience.

The strategy stated that if public and private sectors work together with a united focus and a collective but shared sense of responsibility, they will be far more effective than the individual efforts of any one sector. The NSDR is the first step in a long-term, evolving process to deliver sustained behavioral change and enduring

© Springer International Publishing Switzerland 2016
U. Ranke, *Natural Disaster Risk Management*,
DOI 10.1007/978-3-319-20675-2

partnerships that are essential to building disaster-resilient communities. It is expected that state, territory, and local governments will use the NSDR to inform local action and business and community leaders, as well as the not-for-profit sector are also encouraged to embrace this approach. Strengthening Australia's disaster resilience is not a standalone activity that can be achieved in a set timeframe, nor can it be achieved without a joint commitment and concerted effort by all sectors of society. But it is an effort that is worth making, because building a more disaster-resilient nation is an investment in our future. Exercising in a multiagency environment is an integral part of preparedness. It allows organizations to practice working together in a simulated environment. A needs-based, objective-driven approach delivers exercise outcomes that can be effectively used to assess policies, procedures, and systems as part of a continual improvement and/or regulatory process. To design multiagency emergency management exercises, it is important to understand the structured exercise management process, which includes concept development, planning, conducting, and evaluating.

The role of government is to strengthening the nation's resilience to disasters by:

- Developing and implementing effective, risk-based land management and spatial planning
- Developing and carrying out a nationwide hazard assessment
- Developing and implementing disaster prevention and mitigation based on the risk assessment
- Effective information of the population at risk about how to assess their particular risks and how to reduce exposure and vulnerability to hazard
- Ensuring the most effective disaster response from emergency services and volunteers
- Supporting individuals and communities to prepare for extreme events
- Supporting communities to recover quickly from devastation and to learn, innovate, and adapt in the aftermath of disastrous events

The role of individuals follows the vision that disaster resilience is in general based on individuals. They should take their share of responsibility for preventing, preparing for, responding to, and recovering from disasters. In order to strengthen their coping capacities the strategy emphasizes that they should be provided with specific and locally adoptable disaster risk reduction guidance, with technical and scientific resources, and with organizational means to administer disaster reduction activities at the community level. The disaster resilience of people and households can significantly increase by active planning and preparation for protecting life and property, based on awareness relevant to their locality. It is also increased by knowing and being involved in local community disaster or emergency management arrangements, and for many being involved as a volunteer.

The role of nongovernment organizations and volunteers is defined in the strategy as to be at the forefront of strengthening disaster resilience in Australia. The dedicated work of these agencies and organizations is acknowledged as critical to helping communities to cope with and recover from a disaster. It is to them, that

Australians often call for support or advice. The strategy appreciates the role non-governmental organizations and volunteers have as partners to spread the disaster resilience message and to find practical ways to strengthen disaster resilience in the communities they serve.

In order to strengthen Australia's risk reduction capacity the Australian Emergency Management Institute (AEMI) was established and mandated as the Centre of Excellence for knowledge and capability development in national emergency management. As a part of the Attorney-General's Department, Emergency Management Australia division, AEMI provides a range of education, training, professional development, information, research, and community awareness services to the nation and our region. In executing the national strategy AEMI has been given a significant role in building the capacity and professionalism of the emergency management sector in Australia. The institute supports broadening the national security capability by a number of educational and awareness-raising activities: the most prominent is the Advanced Diploma of Public Safety "Emergency Management," a nationally accredited and professional development program. Moreover AEMI hosts a range of workshops based on the national research and innovation agenda agreed by the Australia-New Zealand Emergency Management Committee.

Canada

Public Safety Canada (PS)

Public Safety is the national Canadian agency that was created in 2003 with the responsibility for protecting Canadians and helping to maintain a peaceful and safe society. Public Safety Canada is mandated to coordinate all disaster and emergency management in order to ensure national security and safety, from natural disasters to crime and terrorism. Formerly known as Public Safety and Emergency Preparedness Canada (PSEPC 2014), PS is legally incorporated in the federal Department of Public Safety and Emergency Preparedness. All together PS agencies have an annual budget of $6 billion and more than 52,000 employees working in every part of the country.

The department is in many ways similar to the US Department of Homeland Security and covers in addition to natural disaster and emergency management, the responsibilities for the Canada Border Services Agency, the Royal Canadian Mounted Police, the Canadian Security Intelligence Service, the National Parole Board, and the Office of Critical Infrastructure Protection and Emergency Preparedness (OCIPEP). PS is the national hub of the national emergency management system and serves as the center for monitoring and coordinating the federal response to an emergency. The network of 11 regional offices provides day-to-day regional operational support to the Government Operations Centre. The risk reduction efforts in Canada are implemented according to the shared responsibilities of the federal governance structure (national–provincial–territorial) that assigned each of the administrative levels specific mandates to ensure the development of sustainable and resilient communities.

Geographically Canada borders on three oceans, stretches across six time zones, encompasses mountains, plains, forests, and tundra. With all its landforms and weather types, the possibilities of severe weather and geological events are a constant reality. Canada hosts weather patterns that range from Arctic cold to heat waves, with precipitation that is seemingly endless, to droughts. The geological situation favors the occurrence of all kinds and types of natural hazards that regionally can lead to risks for people and their livelihood. As in all over the world, also in Canada the frequency of natural disasters is increasing, leading to significant personal, material, and economic strain on individuals, communities, and the fiscal capacity of all levels of government. The historic disaster record of Canada reveals the regular occurrence of blizzards, earthquakes, floods, hailstorms, icebergs, sea ice, landslides, and mass movements as well as snow avalanches, tornados, tsunamis, storm surges, and volcanic eruptions.

The PS emergency management structure comprises four interdependent risk-reducing functions: prevention/mitigation, preparedness, response, and recovery, and has therefore laid down its disaster management principles in "Canada's National Disaster Mitigation Strategy." This strategy was developed collaboratively by the federal, provincial, and territorial governments, sets out a comprehensive, multidimensional approach that anticipates joint contributions, community-based partnerships, and national-level initiatives. It is based on the agreed-upon vision of all national and provincial stakeholders, that mitigation should play the central part of the emergency management framework. The strategy emphasizes the common vision for disaster mitigation activities in Canada, it integrates disaster mitigation into Canada's emergency management framework, and it identifies primary actions that will be undertaken by federal, provincial, and territorial partners to support implementation of the national strategy. Moreover it recognizes that to reach full implementation, the national strategy will require a long-term effort. Through the implementation of the National Disaster Mitigation Strategy, disaster risk reduction of individuals, communities, and infrastructure can be achieved.

The prevention and mitigation strategies as defined by the National Strategy will reduce or prevent disasters, losses, and emergency response and recovery costs that would otherwise be incurred. Experience with risk reduction in Canada and other countries clearly demonstrates that mitigation actions provide significant return on investment. For example, benefit–cost ratios for flood prevention measures in Australia, the United States, and the United Kingdom are 3:1, 4:1, and 5:1, respectively. Already in 1960 Canada invested CN$63.2 million in the Manitoba Red River Floodway and has thus saved an estimated CN$8 billion in potential damage and recovery costs. Especially by specified mitigation measures, it is envisaged that Canada's built environment (e.g., public utilities, transportation systems, telecommunications, housing, hospitals, and schools) can be improved to withstand the adverse impacts of natural forces.

The goal of the National Disaster Mitigation Strategy is:

To protect lives and maintain resilient, sustainable communities by fostering disaster risk reduction as a way of life.

The principles reflect the essence of what the National Disaster Mitigation Strategy aims to achieve and how it should be developed. The principles are:

- Preserve Life—protect lives through prevention
- Safeguard Communities—enhance economic and social viability by reducing disaster impacts
- Fairness—consider equity and consistency in implementation.
- Sustainable—balance long-term economic, social, and environmental considerations
- Flexible—be responsive to regional, local, national, and international perspectives
- Shared—ensure shared ownership and accountability through partnership and collaboration

The PS set up the Canadian Disaster Database (CDD) that today contains detailed disaster information on more than 900 natural, technological, and conflict events that have happened since 1900. The CDD tracks "significant disaster events" which conform to the Emergency Management Framework for Canada definition of a "disaster" and meet one or more of the following criteria:

- 10 or more people killed
- 100 or more people affected/injured/infected/evacuated or homeless
- An appeal for national/international assistance
- Historical significance
- Significant damage/interruption of normal processes such that the community affected cannot recover on its own

The database describes where and when a disaster occurred, the number of injuries, evacuations, and fatalities, as well as a rough estimate of the costs. As much as possible, the CDD contains primary data that are valid, current, and supported by reliable and traceable sources, including federal institutions, provincial/territorial governments, nongovernmental organizations, and media sources. Data are updated and reviewed on a semi-annual basis.

Germany (Federal Office of Civil Protection and Disaster Assistance)

The protection of the population at risk in Germany lies according to the Constitution (§70) in the sovereignty of the 16 German federal states and comprises disaster mitigation as well as prevention measures. In order to strengthen the efficiency of the mitigation and prevention efforts both, the federal states and the central government established the central command in risk management in times of natural and man-made catastrophies: the Federal Agency of Civil Protection and Disaster Assistance (Bundesamt für Bevölkerungsschutz und Katastrophenvorsorge (BBK 2008). The agency is commissioned with the classical set of civil protection duties including hazard assessment and awareness raising of the risk-exposed population, including warning and information. The agency organizes relief, rescue, and mitigation activities that are to be implemented together with the authorities at local levels. In order to raise synergies at the national and local level, the agency has built a Mutual Crisis Management Center (GMLZ) and the central German Crisis Information System (deNIS), where information on disastrous or crisis-affected regional situations is gathered and evaluated at the local as well as national and international level. With the GMLZ and deNIS, the agency coordinates and operates the crisis response activities of the federal states and renders technical assistance on demand. The agency furthermore is highly involved in prevention planning, disaster emergency management education and training, as well as in disaster research mainly in disaster medical assistance.

In Germany the implementation of rescue and relief operations is organized in general by volunteer organizations such as the Technical Help Services (THW), local fire brigades, and several other organizations rendering medical and humanitarian assistance including the Red Cross, Malteser Hilfsdienst, and others. The federal government covers the expenditures for the equipment and running costs for risk mitigation activities according to the Federal Law on Civil Protection and Catastrophe Assistance (§11,1). In Germany there is a long-standing tradition of volunteer assistance in crisis response. More than 80,000 mostly young people committed themselves under the auspices of the agency alone in the THW (2014) to help in times of crisis. Due to standardized training and nationwide standardized equipment, the volunteers of THW can be employed at very short notice. THW also provides technical–humanitarian aid outside Germany. At almost all major disasters worldwide THW was present to support the United Nations High Commissioner for Refugees (UNHCR) and the United Nations Children's Fund (UNICEF) with development.

In the aftermath of World War II sirens were installed as the state's only operated means of information in case of a crisis. The flood events along the Elbe and Oder rivers revealed that system was well functioning. Together with the radio and TV stations, information was disseminated to the public early and comprehensively. Nevertheless it was found that a new form of information dissemination should be introduced in order to secure very fast communication all over the country. A satellite-based information system (SatWas) was established that enables the agency to communicate information in case of a crisis very quickly. Since its introduction in the year 2000 the BBK Crisis Center in Bonn and the 16 federal state crisis centers, the main TV stations, and all major radio stations are operationally harmonized for immediate and comprehensive information transmission.

The many disaster events of the last years with their enormous economic losses and the comparably high death toll have demonstrated how vulnerable even modern societies such as Germany are still today. In response to that the Federal Ministry for the Interior invented in 2002 together with the German federal states a new strategy to reduce the risk from disasters in Germany. The strategy stipulates emergency management as a joint responsibility of national as well as federal and local governments. One of the many pillars of the strategy is to carry out regular joint emergency management exercises (LÜKEX) in order to optimize the protection of the population based on various threat scenarios. LÜKEX exercises are understood as "strategic exercises" that primarily aim at the strategic decision-making level of the national and federal administrations, in particular the inter-ministerial crisis staffs and the political-administrative staffs. The main objective is to improve the common response capacity and to promote the development of a coordination and decision-making culture in crisis management organizations. The LÜKEX exercises deal with varying combinations of federal authorities, providers of critical infrastructure, relief organizations, and other NGOs. Five strategic LÜKEX exercises have been carried since 2004, based on the following scenarios:

- LÜKEX 04: Extreme winter weather conditions with extensive electric power failure
- LÜKEX 05: Terrorist attacks in connection with the 2006 FIFA World Cup
- LÜKEX 07: Global influenza pandemic
- LÜKEX 09/10: Terrorist threat involving conventional explosives, chemical and radioactive weapons (dirty bomb)
- LÜKEX 11: Security of information technology threatened by massive cyberattack

The wide range of participants and the meanwhile long-standing tradition resulted in an increased trust of the participants and have helped to improve crisis management in real national crisis situations.

The threats exposing Germany daily to the different hazards are steadily monitored and interpreted at the Federal Emergency Assessment Center (GMLZ). The center's main target is to collect, assess, and forecast identified risk scenarios. The center than disseminates the information to the local authorities. For an appropriate and immediate response to the population on risk events the Federal Office of Civil Protection and Disaster Assistance has established the information system "deNIS" that collects all relevant data on hazards, disasters, and ermergency cases and renders information direct and online to the authorities and population.

On the occasion of the flood in 2013 in the Upper Elbe of the Danube region private initiatives started using the networks to disseminate information and requests for help. In the cities of Passau and Leipzig (Germany) the initiatives were able to reach more than 100,000 supporters and private helpers within two days. Technical developments such as the Internet today make a new kind of support possible that was not even thinkable years ago: the use of social networks such as Twitter or Facebook in case of disasters. The social networks detailed the locations where helping hands were needed, where to fill sandbags, or where to clear flooded houses of rubbish and to help transport equipment or personnel to certain locations. Even the local authorities acknowledged the support and integrated the initiatives into their rescue operations and were thus able to reach and mobilize more private helpers than ever before in similar events.

European Union (European Commission's Humanitarian Aid and Civil Protection Directorate General (ECHO)

The European Commission's Humanitarian Aid and Civil Protection Directorate General (EU-ECHO 2014) provides rapid and effective support to victims of disasters beyond the European Union's borders. ECHO has therefore in 1996 launched a specific program "DIPECHO" (Disaster preparedness dedicated to disaster preparedness). The importance of disaster preparedness is clearly recognized in ECHO's mandate and in the European Consensus on Humanitarian Aid adopted in 2007. Disaster preparedness has been placed high up on the agenda for Good Humanitarian Donorship agreed-upon in 2003 in Stockholm. ECHO strongly supports all international efforts, including those coordinated by the United Nations, to increase disaster risk reduction worldwide. ECHO has actively participated in the development of the "EU Strategy Supporting Disaster Risk Reduction in Developing States" adopted in February 2009. This strategy commits the European Union to integrate disaster risk reduction effectively into every EU development and humanitarian policy. It obligates advocating the European Commission, the European Union member states, national governments, international financial institutions, and other development partners to ensure that disaster risk reduction becomes an integral part of sustainable development policy. ECHO especially contributed to formulating the Hyogo Framework for Action 2005–2015 and participated strongly in establishing the International System for Disaster Reduction (ISDR). During the United Nations sessions of the Global Platform for Disaster Risk Reduction, ECHO actively supported its strategy in Community Based Preparedness and Disaster Risk Reduction (CBDRM).

The DIPECHO program focuses on highly vulnerable communities living in some of the most disaster-prone regions of the world and is targeted on saving lives, providing relief, and rendering assistance to natural disaster vulnerable groups by people-oriented preparedness measures at the national or regional level. Moreover it comprises specific measures to complement strategies in disaster-affected countries, that enable local communities and institutions to better prepare for, mitigate, and respond adequately to natural disasters. The programs are targeted to increase local resilience and reduces their vulnerability by enhancing the coping capacities. DIPECHO projects are implemented through a wide range of partners, including local organizations that provide access to the most marginalized and vulnerable people. They typically emphasize training, capacity-building, awareness-raising, establishment or improvement of local early-warning systems, and contingency planning. There are numerous examples where community-based preparedness measures enable the communities to be better prepared to save their own lives and their livelihoods when disasters strike. As any other relief provided by ECHO, DIPECHO projects are carried out by European-based aid agencies in close cooperation with UN organizations and local NGOs and national authorities. The best results are only achieved when there is effective cooperation between citizens, civil society groups, and local, regional, and national authorities.

Since the launch of the DIPECHO program in 1996, ECHO has invested more than €255 million in disaster preparedness. The program had been expanded over the years and now covers eight disaster-prone regions: the Caribbean, Central America, South America, Central and South Asia, Southeast Asia, Southern Africa, and Southwest Indian Ocean and Pacific Region. The projects funded by the program include simple preparatory measures, often implemented by the communities themselves.

ECHO's contribution to disaster preparedness goes well beyond the DIPECHO program as many of ECHO's major humanitarian programs include disaster preparedness or mitigation of disaster impacts. ECHO acknowledges that disaster risk reduction is a long-term challenge and is therefore encouraging other stakeholders to integrate disaster risk reduction systematically in their strategies. Especially post-disaster emergency responses are seen to comprise significant risk reduction elements.

United Nations

Centre de Recherches Epidemiologiques et des Disastres (CRED-EMDAT), The Centre for Research on the Epidemiology of Disasters (CRED 2014), was established in 1973 with the aim of providing standardized data compilation, validation, and analysis on natural and man-made disasters. The center is located within the School of Public Health of the Univerity of Catholique de Louvain in Brussels (Belgium). CRED activities focus on humanitarian and emergency situations with major impacts on human health. This includes all types of natural disasters such as earthquakes, floods, windstorms, famines, and droughts; and man-made disasters creating mass displacement of people from civil strife and conflicts as well as the burden of disease arising from disasters and complex emergencies. CRED also undertakes research in the broader aspects of humanitarian crises, such as human rights and humanitarian law, socioeconomic and environmental issues, early warning systems, the special needs of women and children, and mental health care.

According to its charter as a nonprofit institution CRED provides free and open access to its data through its website. The center has become a World Health Organisation Collaborating Centre in 1980. Since then, it has increased its international network substantially and became a collaborative partner with the UN Department of Humanitarian Affairs (UN-DHA) and mainly with all leading disaster relief and emergency management organizations of the world, such as the Office of Foreign Disaster Assistance (OFDA-USAID), the European Union Humanitarian Office (ECHO), or the International Federation of the Red Cross and Red Crescent, as well as with nearly all nongovernmental agencies. In addition, the center has established a dense network with a great majority of international academic institutions. Systematic collection and analysis of disaster-related data provide invaluable information to governments and agencies in charge of relief and recovery activities. It also aids the integration of health components into development and poverty alleviation programs. However, there is a lack of international consensus regarding best practices for collecting these data. Together with the complexity of collecting reliable information, there remains huge variability in definitions, methodologies, tools, and sourcing. EMDAT provides an objective basis for vulnerability assessment and rational decision making in disaster situations. For example, it helps policymakers identify disaster types that are most common in a given country and have had significant historical impacts on specific human populations. In addition to providing information on the human impact of disasters, such as the number of people killed, injured, or affected, EMDAT provides disaster-related economic damage estimates and disaster-specific international aid contributions.

CRED's activities cover applied research, the development of management tools, logistics, and training on all disaster risk management-related topics. CRED enjoys political and financial autonomy and flexibility which allows it to respond rapidly to situations and to maintain a variety of activities in agreement with the developing needs of its field. Its location in Brussels moreover allows active links with different European Commission programs. The center organizes courses and workshops all over the world on different disciplines such as medical and public health fields (epidemiology, planning, environment, biostatistics), disaster database management, medical anthropology, nutritional sciences, and documentation as well as mass communication. The working languages of the staff are French and English.

One of CRED's core products is the Emergency Events Disaster Database (EMDAT). EMDAT is the only public domain natural disaster database, next to the two other global private sources of Sigma from Swiss Reinsurance Company and NatCat from Munich Reinsurance Company. The database main objective is to serve the purposes of humanitarian action at the national and international levels in all sectors of disaster management to rationalize decision making for disaster preparedness, as well as provide an objective base for vulnerability assessment and priority setting. For example, it allows one to decide whether floods in a given country are more significant in terms of its human impact than earthquakes or whether a country is more vulnerable than another for computing resources.

Today CRED has cooperation agreements with 129 countries and specific institutions worldwide and has become in the course of time the official provider for statistical data on natural disasters worldwide. EMDAT contains essential core data on the occurrence and effects of over 16,000 mass disasters in the world from 1900 to present. The database today comprises more than millions of data million pieces of information on disasters. Data are only entered when information has been transmitted to them by the collaborating governments and agencies and when they fulfill at least one of the following criteria:

- 10 or more people reported killed
- 100 people reported affected
- A call for international assistance
- Declaration of a state of emergency has been given

The EMDAT database contains disaster profiles on all countries of the world for the 10 main natural disasters, including a summary of events from 1900 to 2014. Separate profiles for natural and technological disasters exist. The "Disaster Profile" provides data specified according to the main disaster types from droughts to wildfire. The "Disaster List" gives data that allows the generation of a list of disaster events for a particular country and region and a certain period of time.

In addition to providing information on the human impact of disasters CRED also runs a number of different databases and research activities, including:

- CE-DAT (Human Impact of Complex Emergencies). The database provides access to a series of health indicators, as an essential source of nutritional, health, and mortality data.
- emBRACE (Building Resilience Amongst Communities in Europe) that aims to elaborate a European-wide conceptual framework of disaster reduction. It develops a conceptual and methodological approach to clarify how the resilience capacity of a society confronted with natural hazards and disasters can be characterized, defined, and measured by specific indicators.
- EM-BIB (Emergency Management Bibliography Database) was established in 1988 and contains more than 15,000 documents, articles, and books on disaster and conflict-related topics.
- MICRODIS (Integrated Health Social and Economic Impacts of Extreme Events—Evidence, Methods and Tools) was envisioned to improve the understanding of health vulnerabilities and risk factors for disaster impacts.
- MICROCON (Micro Level Analysis of Violent Conflict), analyzes violent conflicts of the world by an in-depth, micro-level analysis of the conflict cycle.

In addition to the EMDAT database, CRED publishes a series of different journals and press releases. The most prominent are the newsletter CRED Crunch published biannually about the general disaster situations of the world. Meanwhile more than 30 such newsletters have been published, or special country risk profiles given in the CE-Dat Spotlights or the CE-Dat Scene.

United States of America

Federal Emergency Management Agency (FEMA)

Based on the Homeland Security Act of 2002, the Department of Homeland Security was established in 2002 to integrate all organizations related to national security into a single agency (FEMA 2014). Establishing the DHS was the largest US government reorganization in 50 years. The department has three primary objectives:

To prevent terrorist attacks within the United States
To reduce America's vulnerability to terrorism
To minimize the damage and recover from attacks that do occur

DHS officially began operation early in 2003 and established a multicolor-coded "Terrorism Risk Advisory Scale" to provide a comprehensive and effective means to disseminate information regarding the risk of terrorist acts to federal, state, and local authorities and to the American people. The Department of Homeland

Security is headed by the Secretary of Homeland Security who is appointed by the president of the United States with the consent of the United States Senate.

Within the department several component agencies and internal divisions are consolidated, among them the Federal Emergency Management Agency (FEMA). The agency was originally created by President Carter and dedicated to managing natural disasters. FEMA's primary purpose is to coordinate the response to natural, technical, or other man-made disasters that occur in the United States, including external territories such as Puerto Rico. The main rationale for FEMA to deliver its services is that the local or federal state authorities declare themselves not capable of coping with the disaster. As delineated in the National Response Plan, emergency response and planning is first and foremost a local government responsibility. When local government declares itself not able to cope with the diasters, it requests additional resources from the county level; if this capacity is exhausted, then from the state and then the federal government. For FEMA to take over, the governor of the state in which the disaster occurs must officially declare a State of Emergency and formally request FEMA to respond to the disaster. The only exception to the state's gubernatorial declaration requirement occurs when an emergency and/or disaster takes place on federal property or to a federal asset.

FEMA supported the nation in some of its former greatest crises. FEMA personnel have been engaged during the great Midwest floods of 1993, the Northridge earthquake in 1994, the attack on the World Trade Center and the Pentagon, as well as during the many hurricanes and flood disasters. FEMA has its headquarters in Washington, DC, and also runs 10 regional offices located throughout the country. These offices work closely with other federal agencies, strategic partners and tribal, state, and local officials in their regions. The annual budget of DHS is about US$50 billion of which about US$7 billion are allocated to FEMA, the biggest single budget provision. FEMA's core mission is to prepare the United States for every kind of disaster. It aims to reduce the loss of life and property and to improve the capability to prevent and mitigate all hazards. FEMA wants to be the first responder in an emergency case. The agency defines emergency management as "the governmental function that coordinates and integrates all activities necessary to build, sustain and improve the capability to prepare for, protect against, respond to, recover from, or mitigate against threatened or actual natural disasters, acts of terrorism or other man-made disasters." As of November 2007, FEMA has responded to more than 2700 presidentially declared disasters.

Recognizing the specific commonalities between natural hazard preparedness and civil defense activities (known as the "dual-use approach" to emergency preparedness planning and resources, FEMA developed the Integrated Emergency Management System, an all-hazards based on preparedness, response, recovery, and mitigation, and which provided direction, control, and warning systems common to the full range of emergencies from small isolated events to the ultimate emergency. FEMA enacted the National Disaster Recovery Framework (NDRF), a conceptual guide to ensure nationwide coordination and recovery planning at all administrataive levels before a disaster, and defines how to work following a disaster. Within the framework for the first time, all national agencies and administrations were coordinated, leadership, roles, and responsibilities defined and recovery planning before a disaster happens clearly assigned. It outlines how important state, local, and tribal leadership and participation of community members in decision making and coordinated engagement of organizations is critical for successful recovery. The framework emphasis is on core principles, such as individual and family empowerment and partnership and inclusiveness. To serve disaster victims and communities more quickly and effectively, FEMA builds on experience, applies lessons learned and best practices from field operations, gathers feedback from many sources, and constantly strives to improve upon its operational core competencies. FEMA established an Emergency Management Institute located in Emmitsburg, Maryland. Together with the National Fire Academy (NFA) both organizations meanwhile have trained more than 7000 residential students each year and thousands more in the field and through distance-learning courses. In pursuing the strengthening of the capacity to mitigate future disasters, FEMA is financing the acquisition of strategic buyouts of, for instance, high flood risk properties or is allocating funds and advocacy to encourage communities to adopt better building practices and codes.

But on the occasion of the most costly natural catastrophe of the United States, the landfall of Hurricane Katrina, FEMA did not perform according to its own vision and mission, creating an emergency situation that was later described a "failure of state." In the aftermath FEMA was heavily criticized for its slow response and inability to coordinate its efforts with other federal agencies' relief organizations. The criticism focused primarily on administrative mismanagement and the great lack of preparation in the relief efforts: for instance, many victims were left in New Orleans without water, food, or shelter, or the deaths of several citizens by thirst, exhaustion, and extensive violence by officials and as well as from bandits. Within days a general public debate arose about the local, state, and federal government's role in the response to the storm. This led the government to enact the Post-Katrina Emergency Reform Act signed by the president in October 2006. The Act substantially reorganized FEMA to become a "new" authority. The Post-Katrina Emergency Management Reform Act of 2006 created a "New FEMA." With an expanded mission more responsibility of homeland security preparedness was given to the regions. The new FEMA leads and supports the nation in a risk-based, comprehensive emergency management system of preparedness, protection, response, recovery, and mitigation. FEMA is forward-leaning, able, agile, and reliable. Businesslike in its approach, FEMA inspires public trust and workforce pride. Through timely information, resources, tools, and technical assistance, FEMA is helping families and communities overcome all hazards—natural and man-made—and helping America build an overall culture of preparedness.

Switzerland

Swiss Federal Office for Civil Protection (FOCP)

The Federal Office for Civil Protection (FOCP 2014) supports the cantons and municipalities as well as the partner organizations in their civil protection activities in the Swiss Confederation. With the creation of the FOCP in 2003, all areas of the Federal Department of Defence, Civil Protection and Sport, and others that are specialized in civil protection issues were grouped together. This restructuring reflects the growing importance of civil protection that integrates emergency management, protection, rescue, and relief systems. On January 1st, 2004, the Federal Council enacted the new "Federal Law on Civil Protection and Protection and Support" (BZG) that was agreed with 80 % of the votes by the Swiss citizens through a special referendum. The FOCP has around 300 staff and offices are spread over several locations. The FOCP headquarters with the National Divisions of Planning and Coordination, Training, Infrastructure, and Support is located in Berne. The partner organizations of FOCP are the cantonal police, the fire service, health service, and technical service of the different communities.

As Switzerland is periodically affected by natural and technical disasters that threaten the general population and its livelihood or at least impose significant constraints on everyday life, dealing with catastrophes and emergencies poses a great challenge to the national and cantonal administrations. FOCP therefore identifies, analyzes, and evaluates all potential hazards and risks based on probability of occurrence and the extent of damage following an integrated risk management strategy. The findings are to provide the basis for planning prevention measures and to delineate emergency provisions for coping with catastrophes and emergencies. The results are regularly made public in the form of emergency bulletins and are often laid down in maps that display the potentially hazardous regions. In civil protection, integrated risk management is understood as the systematic approach to cope with hazards and emergencies by using well- balanced measures of preparedness, response, and recovery.

The concept of integrated risk management carried out by FOCP encompasses prior to an disaster event:

- Assessment of hazards
- Evaluation of risks
- Planning of risk reduction measures
- Provision of advisory services and technical support

In case of an emergency:

- Planning and implementation of rescue and relief operations
- Planning and implementation to support the reconstruction and recovery phase

One of the partners of FOCP is the world-famous Swiss Institute for Snow and Avalanche Research (SLF) that for decades has been engaged in research on snow, natural hazards, avalanches, permafrost, and the mountain ecosystems as well as the analysis of climate and environmental changes. SLF is active in both basic and applied research targeted at developing practical instruments for national and cantonal authorities, industry, and the general public that can be used to manage the risk associated with natural hazards mostly derived from snow. In close connection with its research activities SLF also offers a range of services. These include consulting, expert opinions on avalanche accidents and avalanche protection, and the development of warning systems for natural hazards in the Alps. The best-known service is the avalanche bulletin or warning report for the Swiss Alps, which is published twice daily in wintertime. SLF employees furthermore teach at the University of Zurich (ETH) and various universities in Switzerland and abroad, and provide basic and further training for safety experts.

One of the main research fields of SLF is to improve knowledge of snow avalanches. As they constitute a type of very fast-moving mass movement, they often cause serious property damage and loss of life. Most snow avalanches are released from slopes steeper than about 30°, or are triggered by snow loading due to wind, by a temperature change, or are even triggered artificially, for example, by skiing. As it is so far not possible to predict the exact location, time, and extent of an avalanche event, SLF is carrying out extensive research on snow to better understand the underlying formation processes of avalanches in order to improve avalanche prediction. The research activities include the following topics:

- Snow failure and avalanche initiation
- Fracture mechanics of snow
- Snow slope stability and stability tests
- Critical snowpack layering
- Monitoring of instabilities on a slope
- Spatial variability of snowpack properties and its relevance for avalanche formation
- Precursor signals using seismic instruments
- Upward-looking radar technology
- Formation of wet-snow avalanches
- Modeling snowpack instability with the snow cover model SNOWPACK
- Stability evaluation and avalanche forecasting

SLF publishes regular bulletins and brochures on the snow and avalanche situation of Switzerland, especially the weekly Bulletin on Avalanche that was first published in 1954. The bulletin indicates on a scale from one to five a forecast on the snow avalanche hazard distribution. The scale is now standardized Europe-wide and is also adopted in the United States and Canada. The bulletin addresses public authorities (snow avalanche security services, public health services) but also the public that is exposed to snow avalanche threats as well as recreationists. The intention of the bulletin is to drop the fatalities of the present approximately 25 snow avalanche events every year, of which 90 % are recreationists. SLF further operates a Web-based, interactive avalanche prevention platform. With the bulletin SLF raises awareness on the avalanche danger that prevails outside marked and open pistes, offers a wealth of information on the subjects of avalanche science and avalanche prevention, and contains a new tour planning tool (White Risk).

Volcanic Ash Advisory Center (VAAC)

On June 24th, 1982 a British Airways B747 flight was on the route from London to New Zealand, crossing the Indonesian Archipelago when at about 10,000 m height passed volcanic ash from the eruption of the nearby Mt. Gulanggung (Java). First the cockpit crew noted a gleaming light outside the aircraft's windscreen. Then smoke began to accumulate in the passenger cabin, that had a strong odor of sulphur. Passengers noted that the engines were unusually bright, with light shining forward through the fan blades. One engine after other began to flame out. The reason for the failure was not immediately apparent to the crew or ground control. The aircraft was diverted to Jakarta and while gliding downwards finally succeeded in restart all four engines, allowing the aircraft to land safely. The gliding of the aircraft entered into the Guinness Book of Records as the longest glide of a commercial aircraft ever. Nine days later a Singapore Airlines flight passing the same ash cloud was also forced to shut down three of its four engines. Post-flight investigation revealed that the aircraft flew through a cloud of volcanic ash. The ash particles sandblasted the windscreen and clogged the engines. As the ash entered the engines, the mineral ash particles melted in the turbine (combustion temperature is 1400 °C and thus about 200 °C hotter than the solidus temperature of volcanic glass) and recrystallized on the turbine blades. A nearly identical incident occurred in December 1989 when a KLM flight from Amsterdam to Anchorage encountered the ash plume of the Mt. Redout volcano, Alaska.

Following these incidents a network of Volcanic Ash Advisory Centers (VAAC 2014) was set up in the 1990s by the International Civil Aviation Organization (ICAO), an agency of the United Nations to increase air traffic safety. Nine centers were installed all over the world, each one focusing on a particular geographical region, to assess the danger to commercial aviation from ash clouds and thus to alert pilots early to divert their flight around the cloud. In order to develop an integrated volcanic ash observing system the centers are established with national meteorological offices, for instance, in Britain (British Met Office), the United States (NOAA), or France (Meteo France) and others. There experts steadily monitor all available observations on volcanic activities such as from satellite, radar, lidar, and aircraft and continuously measure the volcanic ash concentration levels using aerosol radio sondes. The experts are responsible for coordinating and disseminating information and forecasting the ash cloud's dispersion. An accurate assessment of the height of the initial eruptive volcanic ash plume is critical for predicting the subsequent trajectory of a volcanic ash plume. Traditional weather radar technology is able to provide such an assessment, because volcanic ash in the eruptive plume is highly reflective. The experts further use the volcanic information bulletins rendered by the Smithsonian Institution and the USGS. The centers provide a real-time assessment of the horizontal and vertical extent of volcanic ash and on the possible associated volcanic ash concentration levels. Moreover the pilots are called on to report immediately to the centers on every sign of heightened volcanic activity, especially upon detection of a sulphur smell in the cabin as volcanoes are the only sources of large quantities of sulphur gases at cruise altitudes. Thus both SO2 and H2S are indicators of volcanic activity even if no other signs are detectable or reported. The issuance of an alert follows an internationally standardized procedure (see Sect. 2.1) based on a specific color code that gives the alert levels from green (normal) to red (eruption imminent). The alert communication moreover comprises information on the name of the volcano, the country, location, and crater elevation, the source of the information, such as satellite or pilot observation, details of the eruption including time of day in universal time and date of the eruption, details of the ash cloud including the flight level and size, details on the current movement of the ash cloud and the expected ash cloud trajectory.

Pacific Tsunami Warning Center (PTWC), Hawaii

During the early morning of April 1st, 1946, an earthquake of magnitude 7.4 occurred in an area of the Aleutian Trench located approximately 90 miles south of Unimak Island (Aleutian Island chain). During the earthquake, a large section of seafloor was tectonically uplifted along the subduction fault generating a large, Pacific-wide tsunami. Well-documented accounts of the tsunami come from Scotch Cap, located on Unimak Island, and from the Hawaiian Islands. The tsunami wave reached a height of about 40 m at the Uminak coast, destroyed a Coast Guard lighthouse, and killed the five lighthouse men. Five hours later the first tsunami waves reached the islands of Hawaii in the middle of the Pacific Ocean. The complete destruction of the lighthouse made any information dissemination on the event impossible. Thus the tsunami hit the islands without prior warning. The tsunami wave heights reached an estimated maximum of 15 m on the different islands. They inundated the coast region for some 100 m inland in some locations and produced extensive destruction along the shorelines of the Hawaiian Islands, especially at Hilo harbor, on the big island of Hawaii. There the entire city's waterfront was destroyed. In total 159 people were killed. Impressive photos and an amateur movie had documented the wave front and the intensive destruction at Hilo harbor. In addition to Hawaii other coasts of the United States were also strongly affected by the tsunami. So the community of Taholah, Washington State was struck by an approximate 1 m surge; Coos Bay, Oregon, reported a 3-m wave. The tsunami was also noticed in Santa Barbara and farther down to the Los Angeles area. The tsunami crossed the Pacific, producing waves up to 10 m high in some locations at the Marquesas Islands in French Polynesia, and even had the power to damage fishing boats in Chile.

Following the 1946 earthquake cum tsunami, the Pacific Tsunami Warning (PTWC 2014) was established in 1949 on Ewa Beach, on the island of Oahu (Hawaii). The Center is one of two Tsunami Warning Centers that are operated by United States National Oceanic and Atmospheric Administration (NOAA). PTWC is part of an International Tsunami Warning System (ITWS) and serves as the operational center for the Pacific Ocean. The other tsunami warning center is the National Tsunami Warning Center (NTWC) in Alaska, serving all coastal regions of Canada and the United States, the Caribbean Sea, and the Gulf of Mexico. The function of PTWC is carried out under the auspices of the United Nations Educational, Scientific and Cultural Organization (UNESCO) and the International Oceanographic Council (IOC) running the International Coordination Group for the Pacific Tsunami Warning System.

In the aftermath of the deadly tsunami of the Indian Ocean in 2004 PTWC has extended its area of investigation and today includes the entire Indian Ocean and the Caribbean into its warnings. In 2004 PTWC did not immediately gave a warning to the Indian Ocean riparian, although their instruments indicated a severe and tsunamogenic earthquake, as this area was not their responsibility. When a warning was released later it was too late. Nevertheless even if the warning had been disseminated in time, there would not have been enough time to evacuate the people at risk, as the tsunami traveled the distance from the epicenter to the shoreline of Sumatra within 15 min.

The PTWC uses the international grid of seismic data networks as its starting point. When an earthquake is generated with a magnitude higher than 5.0 and the plate tectonic situation of the area favors the generation of a tsunami, the regional network of tide gauges is used to pinpoint the occurrence. Then based on the regional oceanographic data (seabed morphology, ocean tides, etc.) the possible tsunami track is calculated. The interpreted forecast of the future of the tsunami is then issued to emergency managers and other officials, to news media, and the public all around the Pacific Ocean, the Indian Ocean, and the Caribbean. In addition to the international seismic grid, NOAA has deployed more than 30 of its own sea bottom tsunami detection buoys that are placed on the Pacific Ocean floor (DART buoys = Deep-ocean Assessment and Reporting of Tsunami). Each station consists of a sea-bed bottom pressure recorder that detects the passage of a tsunami and transmits the data via acoustic modem to the PTWC headquarter in Ewa Beach. The system has meanwhile demonstrated its usefulness and has considerably improved the forecasting and warning of tsunamis in the Pacific Ocean.

References

ANZEMC (2014): Australia-New Zealand Emergency Management Committee (ANZEMC). - Australian Government, Canberra (online: www.em.gov.au/.../NationalEmergencyManagement)

BBK (2008): Protecting Critical Infrastructures – Risk and Crisis Management. – German Federal Ministry for the Interior, Berlin

COAG (2004): Natural Disasters in Australia: Reforming mitigation, relief and recovery arrangements. - A Report to the Council of Australian Governments by a high level official's group, Australian Journal of Emergency Management, Mount Macedon, Victoria

CRED (2014): Homepage "Centre for Research on the Epidemiology of Disasters" (CRED) .- Universite catholique de Louvain (UCL), Brussels (online: cred.be/sites/default/files/MicrodisProjectReport.pd)

EU-ECHO (2014): European Commission for Humanitarian Aid and Civil Protection (EUECHO).- European Union (EU), Brussels (online: hhttp://ec.europa.eu/echo)
FEMA (2014): Executive Order 12127 on Federal Emergency Management Agency. Department of Homeland Security, Washington DC (online: www.fema.gov)
FOCP (2014): Swiss Federal Office for Civil Protection FOCP, Bern (online: www.civilprotection.admin.ch)
IRGC (2005): An introduction to the IRGC Risk Governance Framework.- International Risk Governance Council, White Paper No. 1, Geneva (www.irgc.org)
PSEPC (2014): Public Safety and Emergency Preparedness Canada's (PSEPC).- Public Safety Canada, Ottawa (online: www.publicsafety.gc.ca)
PTWC (2014): Homepage "Pacific Tsunami Warning Center" (PTWC) . – NOAA National Weather Service Pacific Tsunami Warning Center, Ewa Beach HI (online: ptwc.weather.gov)
THW (2014): Bundesanstalt Technisches Hilfswerk (THW) (online: thw.de)
Tobin, G.A. & Montz, B.E. (eds)(1997): Natural Hazards – Explanation and Integration. - TheGuilford Press, London/New York NY
UNDRO (1991): Mitigating natural disasters phenomena, effects and options. – A manual for policy makers and planners. – Office of the United Nations Disaster Relief Co-ordinator, United Nations, Geneva
UNEP (2002): Capacity building for sustainable development: An overview of UNEP environmental capacity development initiatives.- United Nations Environment Programme (UNEP), Nairobi
UNISDR (2004): Living with Risk - A Global Review of Disaster Reduction Initiatives". –United Nation International Strategy of Disaster Reduction (UNISDR),Geneva
UNISDR (2009): 2009 UNISDR terminology on disaster risk reduction. - InternationalStrategy for Disaster Reduction (ISDR), United Nations, Geneva
VAAC (2014): Volcanic Ash Advisory Centers. - International Civil Aviation Organization (ICAO), Geneva (online: http://www.ssd.noaa.gov/VAAC/vaac.html)
Wisner, B., Blaikie, P., Cannon, T. & Davis, I. (2004): At Risk - Natural hazards, people's vulnerability and disasters. - Routledge, London

Bibliography

CRS (2004): Critical Infrastructure and Key Assets: Definition and Identification. – Report for Congress, Congressional Research Service, The Library of Congress, Washington DC
Covello, V.T. & Allen, F. (1988): Seven Cardinal Rules of Risk Communication.- United States Environmental Protection Agency (EPA), Office of Policy Analysis, Washington, DC
Greenberg, M., Lahr, M. & Matell, N. (2007): Understanding the economic costs and benefits of catastrophes and their aftermath: A review and suggestions for the U.S. Federal Government. - Risk Analysis, Vol. 27, Issue 1, p. 83-96, Wiley Online Library (online: 13 MAR 2007 DOI: 10.1111/j.1539-6924.2006.00861.x)
Harmeling, S. (2012): Global Climate Risk Index 2012 – Who suffers most from extreme weather events? - Weather-related loss events in 2010 and 1991to 2010. – Germanwatch, Bonn (online: www.germanwatch.org/cri)
Lauterjung, J., Muench, U. & Rudloff, A. (2009): The challenge of installing a Tsunami early warning system in the vicinity of the Sunda Arc, Indonesia. – Natural Hazards Earth System Science, Vol. 10, p.641-646, Copernicus Publications, Goettingen
NBI (2014): Understanding the Nile Basin. - Nile Basin Initiative (NBI), Entebbe, (online: www.nilebasin.org)
NOAA (2014): Saffir-Simpson Hurricane Wind Scale. – NOAA, National Weather Service National Centers for Environmental Prediction, National Hurricane Center, Miami, FL (online: www.nhc.noaa.gov/aboutsshws.php)

NRC (1989): Improving risk communication. - Committee on Risk Perception and Communication; National Research Council, National Academy of Science, Washington, DC

Poller, J.D. (2001): Managing catastrophic disaster risks using alternative risk financing and pooled insurance structures. – International Bank for Reconstruction and Development / The World Bank, Technical Papers, No. 495, Washington DC

Rheinberger, C.M., Bründle, M. & Rhyner, J. (2009): Dealing with the white death: avalanche risk management for traffic roads. - Risk Analysis, Vol. 29. 1, p.76-94, Wiley Online Library

Sbarra, P., Tosi, P., DeRubeis, V. & Ferrari, C. (2009: Web based macroseismic survey of 2009 L'Aquila earthquakes sequence. - Istituto Nazionale di Geoifisica e Vulcanologia (INGV), Roma (online: www.researchgate.net; access: July, 2014)

SFK (2004): Risikomanagement im Rahmen der Störfall-Verordnung - Bericht des Arbeitskreises Technische Systeme. - Risiko und Verständigungs-prozesse der Störfall-Kommission beim Bundesministerium für Umwelt, Naturschutz und Reaktorsicherheit, Berlin

Spence, W., Sipkin, S.A. & Choy, G.L. (1989): Measuring the size of an earthquake - Earthquakes and Volcanoes, United States Geological Survey (USGS), Vol. 21, No. 1, p.58-63, Reston VA

Taira,T., Silver, P.G., Niu, F. & Nadeau, R.M. (2009): Remote triggering of fault-strength changes on the San Andreas fault at Parkfield. – Nature ,Vol. 461, p.636-639

Thywissen, K. (2006): Core terminology of disaster reduction: A comparative glossary; in: Birkmann, J.: Measuring vulnerability to natural hazards - Towards disaster resilient societies . - United Nations University Press, Bonn

TCIP (2013): Turkish Catastrophe Insurance Pool (online:www.tcip.gov.tr)

USEPA (2000): Value of a Statistical Life - Guidelines for Preparing Economic Analyses.- National Center for Environmental Economics, US Environmental Protection Agency (online: http://yosemite.epa.gov/ee/epa/eed.nsf/webpages/homepage)

USGS (1989a): Strong Ground Motion and Ground Failure. - The Loma Prieta Earthquake. - United States Geological Survey (USGS), Professional Papers 1551-A, B, C, E, and F, Reston VA

USGS (1989b): Performance of the Built Environment. - The Loma Prieta Earthquake . – Unites States Geological Survey (USGS), Professional Papers 1552-A through 1552-D, Reston VA

USGS (2007): USGS Volcanic Activity Alert-Notification System. - United States Geological Survey (USGS), Fact Sheet 2006-3139, Version 1.0, Reston VA

USGS (2012): Earthquake Facts and Statistics. - United States Geological Survey (USGS), Earthquake Hazard Program, Reston VA

World Bank (2007): India - National Agriculture Insurance Scheme: Market-based solutions for better risk sharing.- The World Bank, Global Facility for Disaster Reduction and Recovery, Washington, DC

Xia Fei Xie et al. (2011): The role of emotions in risk communication. - Risk Analysis, Vol. 31, No. 3

MIX
Papier aus verantwortungsvollen Quellen
Paper from responsible sources
FSC® C105338

If you have any concerns about our products,
you can contact us on
ProductSafety@springernature.com

In case Publisher is established outside the EU,
the EU authorized representative is:
Springer Nature Customer Service Center GmbH
Europaplatz 3, 69115 Heidelberg, Germany

Printed by Libri Plureos GmbH
in Hamburg, Germany